QUANTUM CONSCIOUSNESS RESEARCH LIMITED

QUANTUM GRAVITY and the ROLE of CONSCIOUSNESS in PHYSICS

Resolving the contradictions between quantum mechanics and relativity

Sky Darmos

2021/3/28

Explains/resolves accelerated expansion, quantum entanglement, information paradox, flatness problem, consciousness, non-Copernican structures, quasicrystals, dark matter, the nature of quantum spin, weakness of gravity, arrow of time, vacuum paradox, and much more.

Working titles: Physics of the Mind (2003 - 2006);
The Conscious Universe (2013 - 2014).

Presented theories (created and developed by Sky Darmos):
Similar-worlds interpretation (2003); Space particle dualism theory* (2005);
Conscious set theory (2004)
(Created by the author independently but fundamentally dependent upon each other).

*Inspired by Roger Penrose's twistor theory (1967).

Old names of these theories: Equivalence theory (2003);
Discontinuity theory (2005); Relationism (2004).

Contact information:
Facebook & Skype: Sky Darmos
WeChat: Sky_Darmos
Email: Skydarmos@qq.com or Skydarmos@protonmail.com

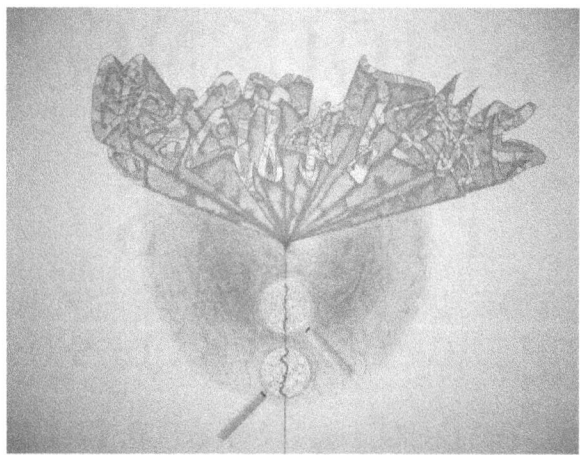

Cover design: Sky Darmos (2005)
Its meaning: The German word "Elementarräume", meaning elementary spaces, is written in graffiti style. The blue background looks like a 2-dimensional surface, but when we look at it through the two magnifying glasses at the bottom we see that it consists of 1-dimensional elementary "spaces" (circles). That is expressing the idea that our well known 3-dimensional space could in the same way be made of 2-dimensional sphere surfaces.
The play with dimensionality is further extended by letting all the letters approach from a distant background where they appear as a single point (or point particle) moving through the pattern formed by these overlapping elementary spaces.
The many circles originating from all angles in the letters represent the idea that every point (particle) gives rise to another elementary space (circle).
All this is to artistically visualise the space particle dualism theory developed by the author and described in this book.

First edition: 28.02.2014
Present edition: 28.03.2021
Time of research: 2003 – 2007; 2012 – 2021
Copyright © 2014 – 2021 by Sky Darmos

ISBN 978-1533546333

[Pre-2016-ISBN: 978-1496108586]

Financial support: Stavros Darmos, Guan Lăi-Kuăn, Pàng Săŭ-Yŭk, Zăŭ Zón-Yừn, Doros Kiriakoulis, Săn Gịn, Gino Yu (Yŭ Zik-Lẽi), Lin Yi-Song.
Main inspiration: Sir Roger Penrose.
Most likeminded person: John Archibald Wheeler.

This book is dedicated to *Sir Roger Penrose*, who has had the greatest influence on me through his works on consciousness [the Penrose-Lucas argument] and the nature of space (or space-time) [twistor theory].

It is also dedicated to all those giants of physics who have adhered to the orthodox consciousness centred interpretation of quantum mechanics, namely *Werner Heisenberg* [Copenhagen interpretation], *Wolfgang Pauli* [Pauli effect & synchronicity], *John Von Neumann* [consciousness collapses wavefunction], *Eugene Wigner* [Wigner's friend & Wigner's dilemma] and *John Archibald Wheeler* [delayed choice & participatory anthropic principle], but also to those who adhere to it till this day, such as *Paul Davis*, *Bernard Carr* and *Henry Stapp*.

"The word 'there is' is a term of human language and is referring to reality as it is reflected in the human soul; it is not possible to talk about another kind of reality."

"I avow that the term 'Copenhagen interpretation' is not happy since it could suggest that there are other interpretations, like Bohm assumes. We agree, of course, that the other interpretations are nonsense, and I believe that this is clear in my book, and in previous papers. Anyway, I cannot now, unfortunately, change the book since the printing began enough time ago."

"The atoms or elementary particles themselvs are not real; they form a world of potentialities or possibilities rather than one of things or facts."

"What we observe is not nature itself, but nature exposed to our method of questioning."

"The natural laws of quantum theory no longer deal with the elementary particles themselves, but with our knowledge of them."

Werner Heisenberg

"It takes so long to train a physicist to the place where he understands the nature of physical problems that he is already too old to solve them."

"Today's physics is a limit case, which is only valid for nonliving objects. It will have to be replaced by new laws, based on new concepts, if conscious organisms are to be described."

"It was not possible to formulate the laws of quantum mechanics in a fully consistent way without reference to consciousness."

Eugene Wigner

"The present quantum theory is unable to provide the description of a real state of physical facts, but only of an (incomplete) knowledge of such. Moreover, the very concept of a real factual state is debarred by the orthodox theoreticians. The situation arrived at corresponds almost exactly to that of the good old Bishop Berkeley."

Albert Einstein

"Actually, quantum phenomena are neither waves nor particles but are intrinsically undefined until the moment they are measured. In a sense, the British philosopher Bishop Berkeley was right when he asserted two centuries ago 'to be is to be perceived'."

"We are participators in bringing into being not only the near and here but the far away and long ago."

"No phenomenon is a real phenomenon until it is an observed phenomenon."

"It is my opinion that everything must be based on a simple idea. And it is my opinion that this idea, once we have finally discovered it, will be so compelling, so beautiful, that we will say to one another, yes, how could it have been any different."

"Nothing is more important about quantum physics than this: It has destroyed the concept of the world as 'sitting out there'. The universe afterwards will never be the same."

John Archibald Wheeler

"A scientific world-view which does not profoundly come to terms with the problem of conscious minds can have no serious pretensions of completeness. Consciousness is part of our universe, so any physical theory which makes no proper place for it falls fundamentally short of providing a genuine description of the world."

Roger Penrose

"Every great and deep difficulty bears in itself its own solution.
It forces us to change our thinking in order to find it."

"Everything we call real is made of things that cannot be regarded as real."

"Nothing exists until it is measured."

"When we measure something we are forcing an undetermined, undefined world to assume an experimental value. We are not measuring the world, we are creating it."

Niels Bohr

"A careful analysis of the process of observation in atomic physics has shown that the subatomic particles have no meaning as isolated entities, but can only be understood as interconnections between the preparation of an experiment and the subsequent measurement."

"I am now convinced that theoretical physics is actually philosophy."

Max Born

"I regard consciousness as fundamental. I regard matter as derivative from consciousness."

Max Planck

"Could it be that consciousness is an equally important part of the consistent picture of our world, despite the fact that so far one could safely ignore it in the description of the well-studied physical processes? Will it not turn out, with the further development of science, that the study of the universe and the study of consciousness are inseparably linked, and that ultimate progress in the one will be impossible without progress in the other?"
"Avoiding the concept of consciousness in *quantum cosmology* may lead to an artificial narrowing of our outlook."

Andrei Linde

The author with the German version of Penrose's "Shadows of the Mind"
[Greece; August 2005; The birth month of space particle dualism theory].

TABLE OF CONTENTS

5. CONSCIOUSNESS IN DETAIL

6. IMPLICATIONS FOR GRAVITY

APPENDIX

Personal background

Some readers might find it strange to encounter the word 'consciousness' in the title of a scientific work on theoretical physics. The strict materialists might even find this reason enough to label my book as esoteric or metaphysical. I want to remind these readers that most interpretations of quantum theory, even the very materialistic ones, don't work without consciousness playing an important role in them. In the very popular many-worlds theory for example, it is consciousness which separates the different worlds from each other.

The first approach of coming to an understanding of quantum theory, the Copenhagen interpretation, was also tied to the idea of a conscious observer. In their book "The Matter Myth" the two well-known scientists and authors Paul Davis and John Gribbin went even so far as to announce that materialism is proven wrong by modern physics. Their book and many other books on theoretical physics, discuss topics like the nature of time and its relation to consciousness, as well as the question of whether it is consciousness which ultimately creates reality.

One of the people who made the biggest effort in explaining both the puzzles of consciousness, as well as drawing a more honest picture of the status of present theories in physics to a large audience is Roger Penrose. Many may know him through his proof that within the framework of general relativity every black hole must contain a singularity within its event horizon. In 1968, this proof was extended by Stephen Hawking in order to prove that based on general relativity, the Big Bang must have started off from a singularity. As we will see in chapter 4.3 and 4.6 both of these proofs don't hold true in a full theory of quantum gravity.

Similar to Stephen Hawking, Roger Penrose became known to a larger audience through the publication of a popular book about his views on the universe, which was a bestseller; its name was "The Emperor's New Mind". It addressed the impossibility and incredibility of *Artificial Intelligence* (AI) and the relation of consciousness to certain issues in formal logic and theoretical physics. I was deeply inspired by Penrose's descriptions of the phase space, his space-time diagrams, his followings out of Gödel's incompleteness theorem and his twistor theory. Since early on I had a deep admiration and appreciation for Penrose, however there is one important point in which I deviate strongly from him, and this is the measurement problem, namely the problem where the bizarre overlapped superpositions of different seemingly mutually exclusive states in quantum mechanics turn into the classical realistic states we see around us. Penrose takes the view that this problem can only be answered with a theory of quantum gravity. He refuses both the many-worlds interpretation, which negates the choice of realities, as well as the Copenhagen or Wigner Von-Neumann interpretation, which makes consciousness responsible for the reduction of the wave function (also referred to as 'consciousness causes collapse'). In contrast to this he makes gravity responsible for this reduction of the wave function. In my opinion this is not going to explain consciousness, as it would make it very hard for consciousness to interact with the brain under these conditions (see section 2.1).

At this point I should say something about myself. How did I get into this field?

The first time I got interested in the universe was when my dad told me about the solar system. That was on a new year night on a beach near Skala (Greece). There were fireworks and jet skies, and he drew the solar system in the sand. Later he bought me a book on astronomy for teenagers.

It was about planets, red giants, neutron stars, black holes and quasars. I was so fascinated that I read it every day over and over. I wrote all the information I found in it onto a drawing of the solar system I made.

The reason my dad explained the solar system to me was probably because he noticed that I tend to often look up into the sky and also that the easiest way to make me stop crying had always been to show me the night sky. The names of people can sometimes have a huge impact on their character and interests.

My father Stavros Darmos (1952 – 2015) who supported my natural interest in the universe. [Greece; 04.08.2009 & 11.08.2009]

I remember being puzzled by some deep questions even before that. When I was a little boy I often asked myself why there is something instead of nothing. For instance, the question of how a creator which supposedly had existed since eternity, in emptiness, could possibly choose a particular moment for his creation without waiting an eternal amount of time for doing so disturbed me so much that it caused me sleepless nights. My dad, who was an atheist, didn't intervene in my mother's more or less Christian upbringing.

Sometimes when I was hanging upside down on my swing in front of our house in Skala, Greece, I saw left and right reversed and asked myself if there is any fundamental difference between left and right, a problem which I later discovered is also subject of some studies in particle physics.

I was born in Greece to a Greek father and a German mother in a remote area on an olive plantation. There were no other people living there, so I was often alone with my thoughts. At the age of seven my mother took me and my sister to Germany, but we still spent every summer on that olive plantation. I can remember lying awake in the same room with my sister, when she asked me what extra dimensions are, and if they exist. I remembered a newspaper article about Stephen Hawking talking about extradimensions, but at that time I could not give my sister a clear answer – I was just around ten years old, she was around eight.

Left: Me and my dad. Moments before he had said: "As you go on you will learn and discover more and more things." [18.02.2013; Greece]; Right: My dad at work [06.03.2013; Greece].

I had fun reading about my favourite topics, but had great difficulties dealing with the society around me, especially my classmates. It took me until I was 18 before I went to the bookstore to find out what extra dimensions actually are. In the following years I often made so called 'to-think lists'. I was reading on every bus, thinking on every road. In the railway station in Pforzheim (Germany) on the way to the train I was thinking about the phase space, when I had the idea for similar-worlds interpretation (2004). The idea for space-particle-dualism came to me when I was sitting in a car on the way to a beach near my family's home in Skala, Greece in 2005. On my first travel to China I combined these two ideas to form the present space particle dualism theory (Guangzhou, China; 2006). In the few years after 2007 not much happened for me in this field, because I started following my new interest in etymology. I had the idea to use etymology (word history) and Chinese characters to connect western and eastern languages. That led to the discovery of 3,800 similar morphemes between German and Chinese and took me six years to complete. After finishing compiling a book on these cognates, I slowly started working on my physics theory again in 2012 when I was in Macao. In 2013 I discovered a correspondence between local entropy increase and the expansion of the universe (Shenzhen; December, 2013). Then in the beginning of 2014 I summarised my theory in a first short version of this book.

Although the theory had existed since 2005, it was not until September 2016 that it became a really consistent theory. At that time I was back in Germany for a month when I found out the real size of elementary spaces, the most basic objects of the theory.

Sky Darmos
Guangzhou, 07.05.2015
Zhongshan, 21.07.2017

Overview

This book is about the nature of reality and the nature of our very self. Our consciousness. The new theory presented in this book has the potential to both unify physics and to explain consciousness. Although being a rather new theory it already makes a lot of correct predictions and is easy to verify. It also solves numerous problems which have been unsolved for about 80 years. Some of these problems do not have any solution being proposed over this period of time.

Problems solved by this theory include:

1. **Measurement problem.** (Schrödinger's Cat). Is the cat able to measure itself? Is the measurement apparatus able to measure itself? Does the measurement actually change physical reality?   [Chapter 1.2]

2. **What are superpositions?**   [Chapter 1.2]

3. **Einstein-Rose-Podolski Paradox**. How to interpret quantum entanglement? [Chapter 1.2]

4. **Spin Interpretation**. What does it mean for a particle to return to its original state after a half rotation, or after two rotations? Is there a way to interpret spin as rotation at all?   [Chapter 3.5]

5. **Accelerated Expansion**. Why is the universe's expansion accelerating?   [Chapter 1.2; 1.3; 4.1; 4.7]

6. **Expansion problem**. Why is it expanding at all?   [Chapter 4.1; 4.7]

7. **Dark matter problem**. What is dark matter?   [Chapter 3.3]

8. **Information paradox**. What happens with information falling into black holes? Is it lost?   [Chapter 4.3]

9. **Background dependence**. What is space or space-time?   [Chapter 1.3]

10. **Flatness problem**. Why is the universe so absolutely flat?   [Chapter 1.3]

11. **Horizon problem**. Why is the universe homogeneous and isotropic as far as we can see?   [Chapter 4.7]

12. **Chronology protection conjecture**. How can time travel paradoxes be prevented? [1.2; 6.1]

13. **Hierarchy problem**. Why is gravity so weak?   [Chapter 1.3; 3.7]

14. **Holographic principle**. Why does the maximal entropy of a region depend on the surface of the area enclosing it?   [Chapter 1.4; 3.4; 4.4]

15. **Mind-Body-Problem**. Why do we have consciousness? Where does Qualia come from? What is happiness, what is sadness? What means death?   [Chapter 5.1; 5.2]

16. **Causality problem**. How can it be that there is real randomness in the world? [Chapter 2.1]

17. **Generations of matter**. Why is it that nature is so wasteful that it repeats itself creating three generations of matter?   [Chapter 3.9; 3.10]

18. Why do elementary particles have the masses we observe?   [Chapter 3.11]

19. **M-sigma relation**. Why is it that the size of a galaxy and that of its central black hole are so closely related?   [Chapter 4.5]

20. **Entropy problem**. Why did the universe start off with such low entropy? - We will see that it actually didn't!   [Chapter 2.2; 4.1, 4.8, 4.15]

21. **Quasi-crystal problem**. How can quasi-crystals with a 5-fold-symmetry grow? [Chapter 2.1]

22. How can the human brain exceed mere calculations as required by Gödel's Theorem? [Chapter 1.2; 2.1; 2.3; 4.9]

23. **Fermi-Paradox**. Why do we not find any signs for alien life out in the universe? Grad 2 and 3 civilisations should be very easy to detect. Where are they?
[Chapter 4.2]

24. **Mind over Matter**. How much influence do we have on external states?   [Chapter 2.1]

25. **Matter-antimatter asymmetry**. Is our world depending on a slight matter excess at the beginning of time?   [Chapter 4.8]

26. What is the origin of the small irregularities in the cosmic microwave background (CMB)?   [End of chapter 4.4]

27. What is the origin of the various CMB anomalies and the periodic and concentric galaxy distribution?   [End of chapter 4.2]

1. BASIC CONCEPTS

1.1 One world, two views

Most people who study theoretical physics seek a picture of the world. They want to know what it is they are living in. They want to understand the true nature of reality. The understanding of this nature should tell them a lot about themselves, since they are part of this world. This seems to be a natural requirement for science. Yet all theories of the last century failed to do so. Today's physics makes contradictory statements about what it is we are living in. That's because physics is based upon two totally different mind sets. One is Einstein's theory of relativity and the other is quantum mechanics, developed by Bohr, Schrödinger, Heisenberg, Dirac and others. These two mind sets are so different that they would never fit together in their present form. According to relativity we live in a space-time with four dimensions; according to quantum theory, we live in a Hilbert space with an infinite number of dimensions.

In Einstein's theory of relativity, time is a dimension, so that everything which has happened and everything which is going to happen is already there, as an unchangeable entity. This is a completely computable and deterministic world view. In contradiction to that, quantum mechanics states that the future is open. It can only be calculated statistically on the basis of probabilities.

How can these two theories contradict each other so much and still both give correct and precise predictions? It is because they are used in different realms. Quantum theory is used describing the behaviour of particles in the microcosm, while general relativity is used for big things like planets, galaxies and black holes.

There have been numerous attempts to find a theory of everything, describing the big as well as the small. In such a theory relativity and quantum mechanics must appear as different aspects of a deeper underlying law. It should be commonly known that simply applying the laws of quantum mechanics to space-time will not get us anywhere. Yet all proposals for final theories of the universe are still working with these two different kinds of laws. In more than 80 years of collaborative efforts, no one had any idea on how to combine those laws into a unit so that they would appear as two different aspects of one thing, with worldviews not contradicting each other anymore.

Why is it so difficult? The main reason is that although physicists understand and embrace the principles of relativity, they have no clue what to think about quantum mechanics. There is no consensus about how to understand the underlying principles of it. Embracing relativity means believing that time is only a dimension; that the flow of time is an illusion and that simultaneity is relative. Yet quantum mechanics depends crucially on simultaneity and the flow of time as we will see later.

The essence of quantum theory is not hard to see: it gives the conscious observer a central role. But this role is so hard to accept for science that most simply gave up deriving any concrete world view from quantum mechanics. Therefore the first and most common interpretation of quantum theory, the Copenhagen interpretation, can stand for two kinds of views: one can be described as 'consciousness causes collapse', saying that conscious observation leads to a *choice between realities*, technically called *collapse of the wavefunction.* The other is just an accumulation of philosophical thoughts claiming that our normal logic and everyday language is not suitable for the quantum world; that we should not even seek out for a 'world view'; all there is are measurement results. This view is sometimes humorously referred to as 'shut up and calculate'.

Both versions of the Copenhagen interpretation have the central notion of the 'measurement'. It is hard to see how to define a measurement without conscious observers. If we do not give consciousness a central role here, we leave the Copenhagen interpretation without any criteria for what a measurement actually is. Niels Bohr seems not to have used consciousness as a criterion. He thought that it is somehow the measurement apparatus which causes the choice, simply because it is macroscopic. But what must count as macroscopic is also dependent upon the human. Bohr thought that fixed classical states could somehow be an emergent property arising from many quantum systems being entangled with each other. This is an idea which is usually referred to as *decoherence theory* and which, in opposition to the Copenhagen interpretation, doesn't involve a collapse of the wavefunction. Thus it seems the only way to get a clear-cut definition for the Copenhagen interpretation is to involve consciousness. Although around 50% of physicists would say that they stick to the Copenhagen interpretation, only 6% of them give consciousness a central role.[1] That means that most of them simply refrain from interpreting it at all. This is because most physicists think that there is no way to prove the existence of a superposition after interference is lost and that therefore there is no way to experimentally decide which interpretation is correct. As we will see later this viewpoint couldn't be further from the truth.

The 'shut up and calculate' approach to quantum mechanics reflects the view that all quantum philosophy is in vain and that we should not seek for any world view. Bohr embraced quantum mechanics, but Einstein could not accept it. He said to Bohr: "I just can't believe that the moon is not there if I don't look at it."

It is this mental barrier which caused the partial stagnation of theoretical physics. The origin of this barrier is a materialist mind frame and the belief that consciousness is only based on mere computation. Experts on formal logic know that this can't be true.[2] The reader should refrain from restricting him- or herself to this simplistic world view. Let's now see how to resolve the main conflict by considering some of the principles these two theories are based upon.

Note:
1. As an example for how self-evident this truth appears to people working in the field of formal logic see the last chapter of "Gödel's Proof", Ernest Nagel & James R. Newman; 1958. Nowadays the mainstream either turns a blind eye on these issues or tries to sneak around them in all kinds of seemingly sophisticated ways.
2. M. Schlosshauer; J. Koer; A. Zeilinger (2013)."A Snapshot of Foundational Attitudes Toward Quantum Mechanics".Studies in History and Philosophy of Science Part B: Studies in History and Philosophy of Modern Physics.44(3): 222–230.arXiv:1301.1069.doi:10.1016/j.shpsb.2013.04.004.

1.2 Similar-worlds interpretation [29.10.2003 (cloud in phase space); 22.07.2004 (solution of the Einstein-Rose-Podolski paradox); 08.05.2012 (planes of simultaneity); 19.05.2013 (entropic expansion)]

What is a superposition?

In quantum theory we can never both measure the position and the impulse of a particle exactly. If we are more interested in the position of a particle, we must use highly energetic photons to measure it, but then we disturb the particle so much that we cannot get any information about its impulse. The other way around, if we are more interested in the impulse, we must use low energy light with a long wave length, but since the wave length is equal to the position uncertainty, we will not know the position of the particle with any satisfying accuracy.

This is Heisenberg's uncertainty principle, which states that the product of the uncertainty in position Δx times the uncertainty in impulse Δp is greater or equals half of the reduced Planck constant:

$$\Delta x \times \Delta p \geq \hbar/2$$

One might think that this is just a restriction for our ability to obtain knowledge about nature: although we can't measure these properties exactly, they should still exist independently. Yet the disturbing thing is that they don't. Our borders for obtaining knowledge seem to be at the same time the borders of nature. Particles do not know their exact properties themselves.[1]

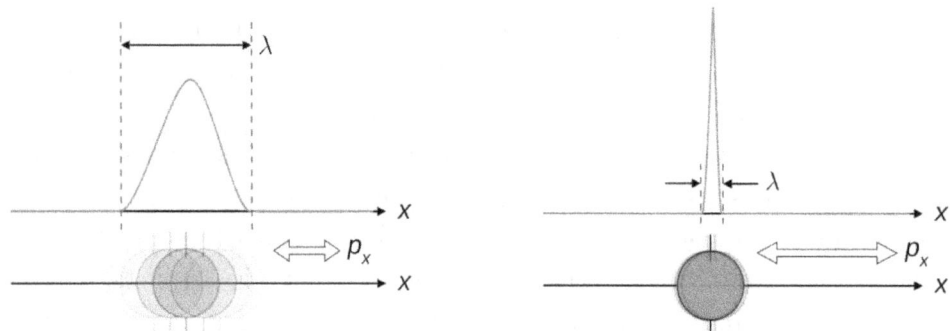

Fig. 1. First impression: there are borders of knowledge. Real truth behind: our borders/limits for gaining knowledge are at the same time the borders/limits of nature/reality.

Even more mysterious is the fact that the probability for a particle to be located at certain spots in a slice of a certain area in space has the form of a wave. The length of this wave depends on the impulse of the particle. These quantum waves can interfere with themselves if they are sent through slits in walls (see fig. 2, 3 and 4), or if they are split up and united by constellations of mirrors and beam splitters (see fig. 5). A single particle can follow many paths at once and interfere with itself. It does not choose its position before the very moment of measurement!

Not everybody agrees to look at these waves as mere probability functions. Some think that 'wave' and 'particle' are different aspects of a mysterious quantum entity that is principally incomprehensible to the human mind and that is not wave nor particle. Therefore this phenomenon is commonly referred to as *wave-particle dualism*.

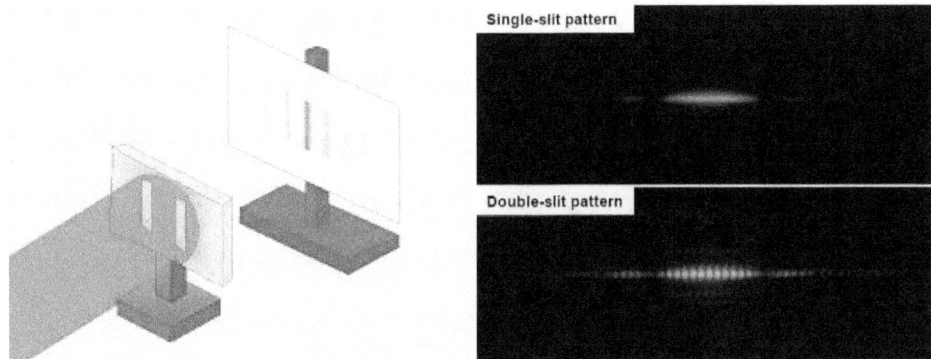

Fig. 2 & 3. The double slit experiment and a real interference pattern.

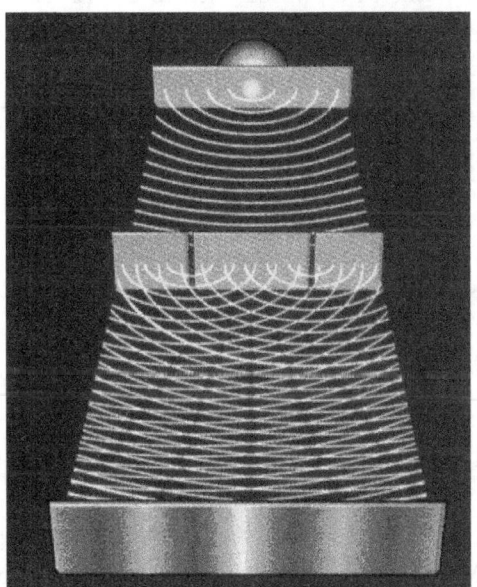

Fig. 4. The wave nature of light as exposed in the double slit experiment has been known for centuries. However, it was not until the beginning of the last century that it was found that the wave nature is not an emergent property of having many light particles (photons) accumulated together, but that it is an inherent property which manifests even when photons move alone. It is the wavefunction of a photon which moves through both slits, interferes with itself and tells the particle where to appear with the greatest likelihood. [Illustration source: http://abyss.uoregon.edu/~js/21st_century_science/lectures/lec13.html]

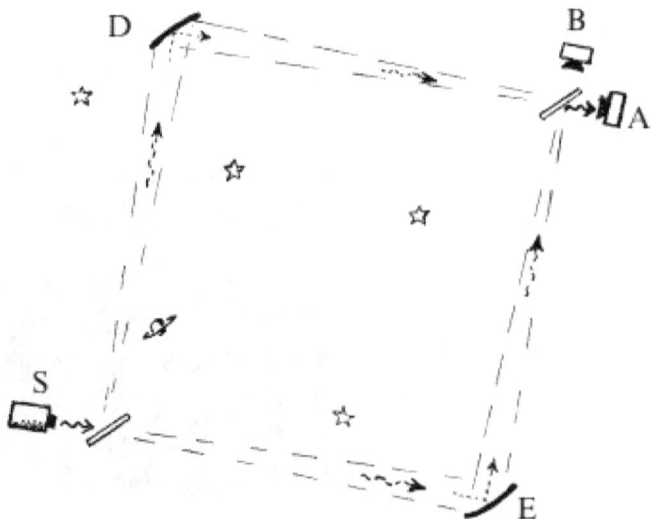

Fig. 5. Different possible paths of a photon interfere with each other. In this example laser beams are sent through a constellation of beam splitters and mirrors. On the path S-D-B the photon is reflected three times, which causes its wave to be set back to such an amount that it is cancelled out by the wave S-E-B. Therefore no photon arrives at detector B. The stars and planets illustrate that there are no distance limits for these effects. [Illustration taken from Penrose's "Road to Reality"]

But what is a measurement? If one particle hits another particle, it could know about the position and impulse of the other particle simply through the force and direction of the hit. But does a particle know something? Surely it doesn't. And if a second particle were enough to make the wave function collapse and to fix the position of the particle, then we would have no quantum world. So what does it take then?
A measurement apparatus is macroscopic, but it does not really know what it measures. It needs to be interpreted by humans who have some knowledge about its construction and the syntax it uses. Measuring has a lot to do with 'knowing', but that requires consciousness and yet only humans and other animals have consciousness. That means the measurement has to reach a conscious observer to count as such.

In reality, the measurement itself isn't part of quantum theory. It is only introduced to get verifiable data out of the theory. The measurement, and therefore the collapse of the wave function, does not follow out of the equations of the theory itself. The clue of most physicists who worked on the theory was that it is somehow the observation through the conscious observer which leads to the collapse of the wave function. This intuitive point of view is known as the *Copenhagen interpretation*. In this explicit form it is more precisely referred to as the Von Neumann-Wigner Interpretation. Some authors, such as Roger Penrose, use the term 'Copenhagen interpretation' synonymous to 'Von Neumann-Wigner interpretation'; however for more clarity I will distinguish the two for most parts of this book.

Quantum theory was established in 1926 and was almost exclusively interpreted using the Copenhagen approach, until 1947 when another quite contrary interpretation was introduced. Hugh Everett took the view that the superposition, the overlapping of different states, does not stop when reaching the human. He suggested that we humans could ourselves be part of such a superposition, without knowing it. The probability wave of an object has a length reciprocally proportional to its mass and that makes it pretty impossible for hypothetical wavefunctions of human being to lead to any measureable interference. That is why different paths we might choose would not interfere with each other even if they really co-exist in superposition. For instance, I can go straight through an open door even if another door beside it is open too. Thus all different versions of oneself would be quite isolated. This view is called the *many-worlds interpretation*. It is quite a popular interpretation, because it leaves consciousness, a phenomenon which cannot be explained within today's science, out of the formalism, or at least it gives it a more passive role: instead of consciousness leading to a choice between realities, it is just that every state of the observer has a separate consciousness, which is unable to see the other versions.

It somehow makes the theory deterministic again: if everything which can happen really happens, then the world is deterministic. We always know what will happen next: everything! And .. nothing. According to Everett, the probabilities in quantum mechanics are there only due to the restricted perspective of the observer. According to the Copenhagen view however, consciousness does something: it chooses between realities. Yet in the many-worlds interpretation consciousness does nothing and there is nothing to do, because nothing really happens. Now if consciousness does nothing, then why does it have to be restricted to one certain world and one certain time? If nothing happens, why do we feel a flow of time? Surely one can argue that every state in every moment and every world has a different consciousness, which is just there, existing and doing nothing. We could hold that view, but we would have to ignore all we know about consciousness.

So what do we know? The most important and most certain things that we know about consciousness we do not know from neuroscience but from theoretical mathematics and the field of formal logic. One of Einstein's closest friends, the mathematician Kurt Gödel, proved that the holy grail for mathematicians at that time, namely to find a formal system of logical statements which can be used to derive all other statements of mathematics, is unattainable. He did not only show that it is impossible to establish such a system, he furthermore showed that every such system would necessarily contain statements which could not be proven within the system; in particular sentences which refer to themselves. This has the implication that it is impossible to construct a robot mathematician which could substitute a human. In other words: logical reasoning cannot be automated. It requires access to meaning and truth and that is what we call consciousness. We can therefore define consciousness as an incomputable but deterministic element in human thinking. Why deterministic? Because logical reasoning, although not being simulable by a computer, always has the same outcome when given the same input.

Fig. 6. What about other interpretations? Many worlds? Many minds? Decoherence? No, Gödel's incompleteness theorem rules them out.

How could the human brain go beyond computation? According to the mainstream, it is not yet known when and if the wave function collapses. It is therefore legitimate to assume that it does not collapse before measurement information reaches the conscious part of the brain. The brain can therefore exist in superpositions of many different states. If consciousness could choose between these states, an effective mind-brain interaction could be established.

If instead the many-worlds interpretation were true, the choice between realities would be merely an illusion and therefore there would be no way for the mind to influence the body. If nothing really happens, there is nothing left for consciousness to do. To believe in the existence of something which does nothing, which is an *epiphenomenon*, is unscientific in my opinion. Consciousness is real and it does something; it is more real than everything else, because all that we see, we see as part of it. I furthermore believe that the world cannot exist without it.

Therefore we shall here adhere to the Copenhagen view and see where it leads us to.

What other conceptual problems do we face? What about *quantum nonlocality*, the phenomenon where a pair of particles created together remain mysteriously connected over unlimited distances and know about each other's state instantaneously? Such a pair of particles usually has no fixed spin. Their spin is a superposition of all possible spins, having them rotate in all possible directions simultaneously. However, once the spin of one of these particles is measured, its partner particle immediately 'chooses' a spin axis orthogonal to that.

Since nothing can travel faster than the speed of light, it seems impossible to find a classical realistic explanation for this behaviour. We shall see how this puzzle can be solved in just a moment, but first we shall take a closer look at the most fundamental concept of quantum theory; the *superposition*.

What does it mean for a particle to be located at many spots at the same time?

If we leave out the relativistic space-time from our picture of the world for just a moment, we can imagine the world to be a so-called *phase space*. It is a space with unlimited dimensions where each point stands for the universe in a certain state. Similar states are close to each other in this space. In a classical world one can describe our feeling of a flowing time as a moving point of collective consciousness within this space of states.

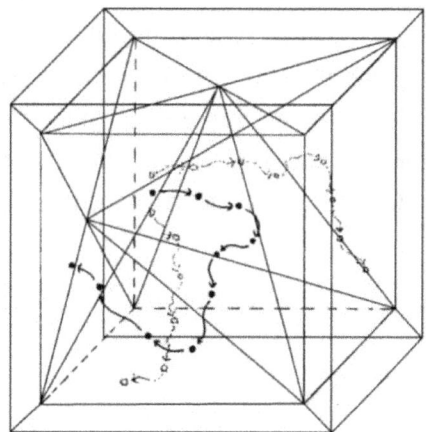

Fig. 7. Each point in phase space represents the universe in one moment. Similar states are close to each other. The passage of time can be represented by a moving point of consciousness in this phase space.

This is all good and nice, but isn't quantum theory telling us, that the universe is not in one certain state at a moment?

What if we try to describe the non-classical quantum world by simply <u>extending this point of consciousness, to a</u> *cloud of consciousness*? All states in a superposition would be part of this cloud. The cloud would grow in time, but every measurement would make it smaller again. Our Consciousness would be in all worlds which are indistinguishable at the present. I call this the 'similar-worlds approach', in contrast to Everett's 'many-worlds approach'.

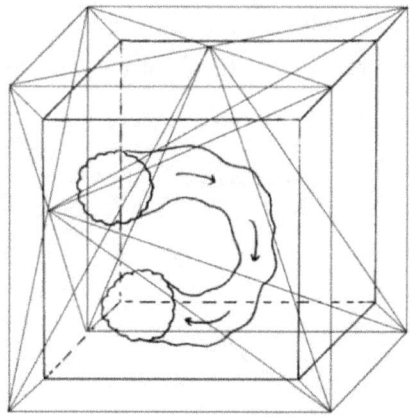

Fig. 8. Moving *cloud of consciousness* in phase space.

We can get superpositions this way, but it takes something more to get wavelike probabilities. This something we will come to later, in chapter 1.3.

Now this is very different from the general view. Usually it is the particles which thought of as to be fuzzy and unlocated. Now we turned everything around: it is now the observer, or better, the observer's consciousness which is in many similar and presently indistinguishable worlds at once. Can this help us to understand the mysterious connection between distant particles in quantum mechanics? Yes, it can. In this picture we can simply say that the spins of those entangled particle pairs are always orthogonal in each state of the universe. Since every state now is a state of the whole universe, it is easy to have the whole universe connected without any information flow between particles. We can call this the 'fuzzy observer principle'.[2]

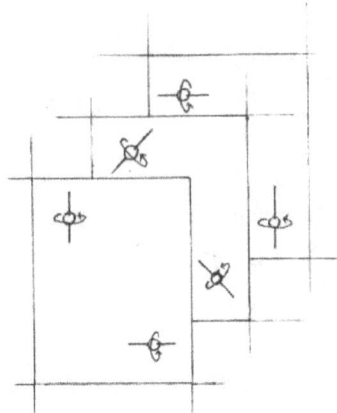

Fig. 9. Spins are orthogonal in each of the *similar worlds*.

Again we arrived at a nice new description of our world, but isn't relativity theory telling us, that the notion of a certain succession of moments for the whole universe doesn't make sense? Isn't the very notion of a flow of time disproven by the relativity of simultaneity and the subsequent introduction of the space-time paradigm?

In Einstein's relativity theory there is no concept of simultaneity. In other words, simultaneity is relative, not absolute. Two events might happen simultaneously for a stationary observer, but at different times for another fast-moving observer.

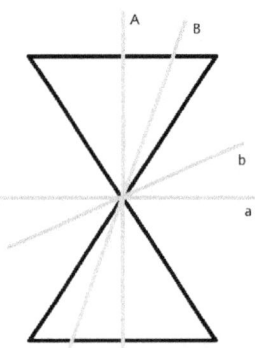

Fig. 10. The planes of simultaneity *a* and *b* of two a stationary observer *A* and a fast moving observer *B*.

Contrary to that, <u>quantum theory needs the concept of simultaneity</u>. The wave function collapses everywhere at the same time when something is measured (by a conscious observer).

Relativists tend to regard the past, the present and the future as equally real. If one takes relativistic planes of simultaneity (now-slices), an alien on our plane of simultaneity but at the other end of the universe (or just very very far away), would only have to move at a minimal speed into our direction to get Isaac Newton on his plane of simultaneity. Equally, he would just have to move in the opposite direction with an equally low velocity to bring our future onto his plane of simultaneity. But does that make our future real? 'Real' would imply that it is somehow fixed and that someone sees it now. Could that hypothetical alien in a remote part of our universe possibly see our future? No, he could not. He can only see what lies on his past light cone. That is the entirety of all light rays which reach him from the past (light always travels with a limited speed, so it is always reaching out from the past).

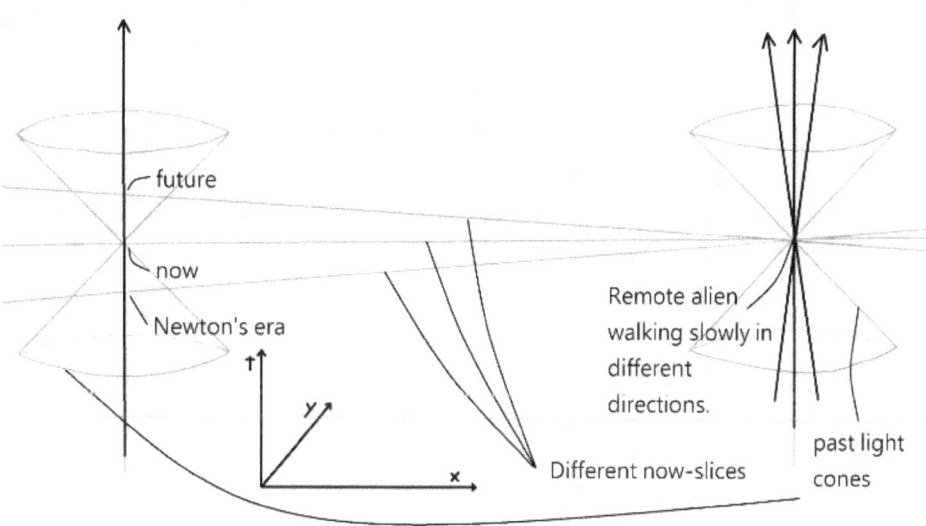

Fig. 11. For a remote alien on our now-slice slow walking would be enough to bring our past or our future onto his now-slice.

So the fact that observers in different *reference systems* have different planes of simultaneity doesn't imply that past, present and future are equally real. An observer on our plane of simultaneity can never see our future, even if it lies on his plane of simultaneity. Since no observer on our plane of simultaneity is able to measure our future, no one can bring our future into a certain quantum state. It is therefore open, as it has to be, according to the laws of quantum mechanics.

When considering interstellar distances we can identify our cloud of consciousness in phase space with an assembly of space-times or *past light cones* that are equivalent for consciousness. We can call this group of space-times or past light cones an 'equivalence group'.

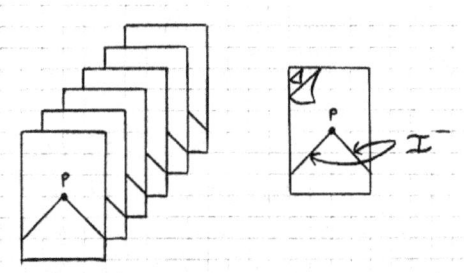

Fig. 12. An *equivalence group* of space-times in the similar worlds approach. The point *P* is the present moment of a conscious observer. All events on the past light cone *I⁻* of *P* are common to the equivalence group of space-times. [sketch from the 20th may 2005]

The introduction of light cones doesn't mean we have to give up the phase space picture. If we assign reality only to the past light cones of conscious observers, and <u>not to the whole space-time</u>, then each point in phase space can represent the state of the universe on these past light cones in one moment. We do not need to assume anything more to be real than these past light cones which are all originating at the so called *conscious now-slice* and are connected through entanglement (more about this in chapter 4.2).

When we talk about what is on the conscious now-slice, we talk about the state of rest-mass particles, while when we talk about what is on the past light cone of the conscious observer, we talk about the state of photons. For visualization it is helpful to imagine light cones being located in a space-time. However, <u>we have to refrain from assigning any reality to this space-time</u>. The cone shape is merely a visualization of the fact that the information we get from photons about the world is delayed in time. It doesn't mean we really live on a hyper-cone (a cone with a 3-dimensional surface). It also doesn't mean time is a dimension.

This gives us a very different perspective of the universe than relativity theory. In the similar-worlds interpretation only what lies on our past light cones is real. There is no *world line* of the observer like in relativity theory.

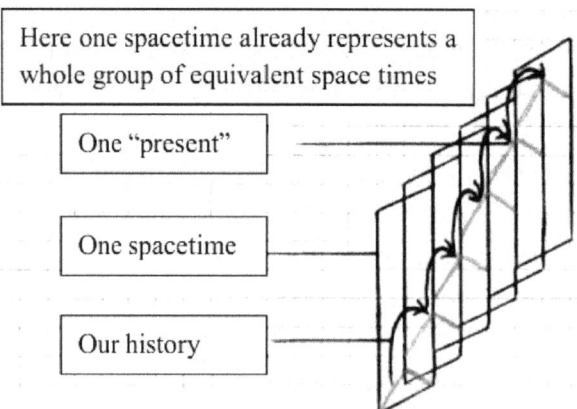

Fig. 13. The grey marked areas on these 'space-times', which represent whole groups of overlapped 'space-times', are the only areas where we exist at all, our past not being part of one physically existing 'space-time', but only of a series of ever-changing groups of 'space-times'. [sketch from the 19th May 2005]

That also <u>solves time traveling paradoxes</u>, because there is no past to visit even if one could stabilize a wormhole to perform time travel. However, as we will see later, a wormhole is not a possible geometry of 'space-time' (the second part of this theory, which will be introduced in chapter 1.3, does not allow wormholes).

However, assuming that only what is measured is real, leads to what I call *Wigner's conflict*, namely: if only the lightcone surface is real, then what about people around me? Are they unreal? The answer is: no, because entanglement connects different observers and allows their wavefunctions to change in a synchronized fashion!

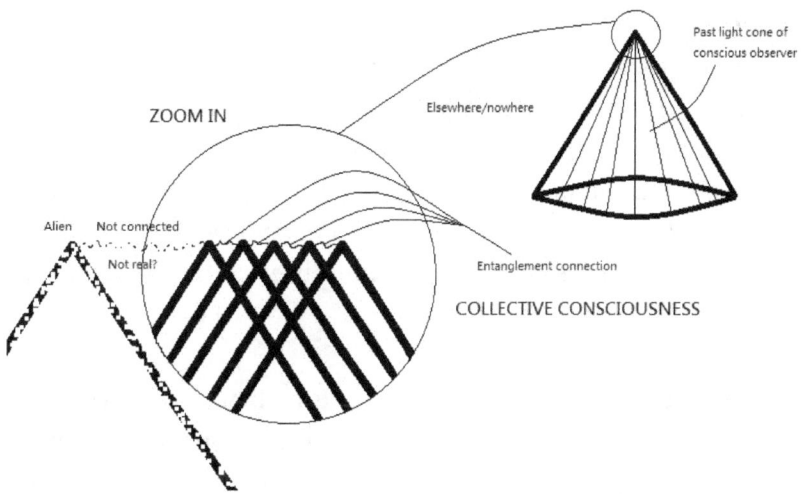

Fig. 14. Entanglement is of crucial importance for the unity of reality. Without it the orthodox interpretation of quantum mechanics would lead to *solipsism* (the belief that only oneself is real).

Eugen Wigner himself didn't find a solution for this apparent paradox and therefore gave up the orthodox consciousness based interpretation.

Let us now continue by asking some questions. For mere illustrative reasons we will keep the space-time notion alive for just one more moment:

1. **Where in 'space-time' must collective consciousness be located to create quantum effects as we observe them?**
On a slice perpendicular to the *ict*-axis (the time axis) and moving forwards in time (in the direction of growing black holes and expanding space).*

2. **What is the speed of this motion?**
Since there is no external reference system, there is no sense to this question. Time, same as space, can only be measured relatively to a reference system.*

*Of course these questions don't come up anymore after we give up the space-time picture.

3. Which parts of 'space-time' have certain quantum states?

Only those which lie on the past light cone of a conscious observer are in a certain quantum state. Hypothetical other observers who lie on the past light cone of a conscious observer, for example aliens, cannot be conscious, because they are not located on the conscious now-slice of space-time. If they were, we would end up with endless many conscious planes in space-time, which would result in a certain quantum state for the whole space-time. <u>That would wipe out quantum effects and bring us back to a classical space-time.</u>

4. How does a particle manage to exist in a superposition of many mutually exclusive states?

The different states are located in different co-existing and presently indistinguishable spaces (or 'spacetimes'), and it is our consciousness which is in all of them at once. Every measurement leads to the collapsing of a wave-function, and therefore to the disappearance of a number of space-times from the group of consciousness inhabited space-times, the *equivalence group*. The measurement does not have to be accompanied by any real knowledge of nature. It can be a pure observation, as made by conscious animals.

5. How come particles can violate local causality by having mysterious connections over huge distances known as quantum entanglement?

These connections do not appear mysterious at all from the similar worlds perspective. Here all the classical worlds have orthogonal spins for each pair of entangled particles. Since all the constituent states of a superposition are states of the whole universe, it is no wonder that spins are related over distance. What has caused philosophers and physicists headaches for almost a century appears so trivial in this new approach that it seems not even worth mentioning.

6. What was before the appearance of conscious life on Earth?

This question refers to a mental time (or *quantum time*, in opposition to the static physical time of relativity) before the emergence of consciousness, a time which does not exist. It therefore does not really make sense to ask such a question. We can say that the whole past of the universe was decided at the very moment that consciousness emerged. This is not a completely new concept. The Copenhagen interpretation, which is the very first interpretation of quantum-mechanics, leads to the same conclusion.

7. What would happen if conscious life became extinct in the whole universe?

The universe would fall back into any arbitrary quantum state which includes consciousness. Only conscious quantum states are quantum states at all. Everything else is nonexistence. Everything on the past light cone of a conscious observer is existing, because it all influences his or her actual state of mind and therefore his or her reality.[3] External reality exists only in correlation to states of mind and can be seen as the common experience between individual internal worlds (minds).

8. Is the past physically 'out there'?

It isn't. We may be tempted to count things on our past light cones to the past and imagine that those things somehow 'go on' (or simply 'exist') simultaneously. However, there isn't really anything forcing us to adopt such a space-time approach to physical reality. Quite opposite, *similar worlds theory* forces us to give up the space-time paradigm: if the quantum forces come into existence only through the interplay between the constituents of the equivalence group, then those forces can not exist on single space-times. From this new perspective the word 'space-time', as fancy as it has become today, has to be viewed as misleading and should be given up. Two things such as space and time being able to influence each other doesn't mean we have to view them as one entity (space-time).

That however doesn't mean the universe has no 'memory'. In the chapters 5.1 and 5.5 we will see that it indeed has, just not in form of a physical dimension but rather in form of experiences from different conscious individuals.

In the Copenhagen (or Von Neumann-Wigner) interpretation as well as in the similar-worlds interpretation, it is consciousness which leads to the choice between states (the reduction of the wave function), but the description of superpositions is very different. The Copenhagen interpretation does not explain what it means for a particle to be at many places at the same time or to be smudged over space and time. In the similar-worlds interpretation it is our consciousness which is in many presently indistinguishable spaces at once (fuzzy observer principle). This can explain what Einstein called the 'spooky action at a distance' and what is now referred to as *quantum entanglement*. We have orthogonal spins at each of the co-existing spaces (times), it is therefore not strange to find two distant particles to be orthogonal each time we measure them.

The Copenhagen interpretation also did not make any statements about the nature or the emergence of consciousness. In contrast to that the similar worlds approach was developed together with a detailed theory on consciousness (presented in chapter 5.1) that can tell exactly what things are conscious and therefore able to cause a collapse of the wave function.

From their names, similar-worlds theory and many-worlds theory sound alike. However, the many-worlds theory separates superpositions only at the macroscopic level where they would not appear anyway according to the orthodox view. So this kind of splitting up into worlds does here not help reducing the paradox nature of superpositions. There is furthermore no requirement for those worlds to be similar, because they do not even influence each other (!). This might make the reader wonder how this is operationally any different from a collapse of the wavefunction.

The answer is that it is not, and in fact physicists should care more about where 'the cut' is to be placed instead of what name to call it ('collapse/reduction of the wavefunction' or 'splitting of worlds').

We will come back to this issue in chapter 2.3 where we will examine the different interpretations of quantum mechanics in greater detail. This will mainly serve the purpose of pointing out the shortcomings of present theories and is not really required for the understanding of the theory presented in this book. Also it is not necessary for giving support to the similar-worlds interpretation. In chapter 4.2 I will provide conclusive evidence for this interpretation; evidence that rules out all other interpretations.

The concept of a bunch of co-existing 'space-times' can make the notion of a superposition in quantum-mechanics plausible. It can also explain the mysterious phenomenon of quantum entanglement, because every 'space-time' has a pair of particles with correlated steady spins. It does however not explain why the probability density of a particle is *wavelike* and not just a single moving mountain.

The wave nature can be understood as a <u>movement in another complex dimension of space</u>, which makes the particle undetectable on its main path of movement for a moment. *This continuous disappearance and reappearance seems to us like a wavelike movement* (correlated to the spin of the particle).[4]

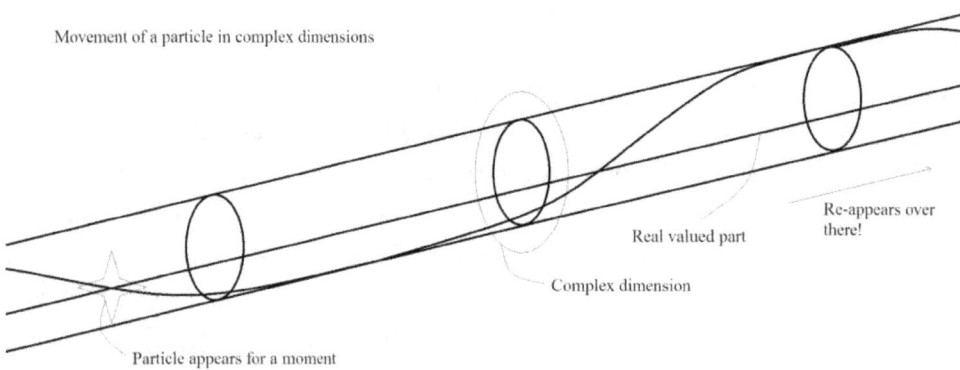

Fig. 15. The wave nature of particles can be visualized as a periodical movement in a complex-valued dimension.
[Note: This is a simplification. We will get familiar with the accurate picture later.]

We will see how this works in detail when we discuss the nature of space in the next chapter.

Notes:
1. The reader might wonder how the particle chooses which of its uncertainties are greater when nobody is measuring it purposefully. The answer is that a measurement does not have to involve any measurement device. Which of the uncertainties Δx or Δp is larger is decided by the environment which plays the role of the measurement device.
2. A term coined by the Greek neuro-physicist Doros Kiriakoulis (2016; personal correspondence). He used it to refer to the way the similar worlds approach describes superpositions and entanglement.

3. The reader might be tempted to try maintaining the space-time picture of the world by talking about different degrees of uncertainty within (not on) the past light cone of the observer. However, one has to notice that quantum effects and quantum forces only come to existence through the interplay of the whole equivalence group. There is no world line of the observer and thus no physically existing past. That however doesn't means the universe has no 'memory'. We will see that it indeed has, just not in form of a physical dimension but rather in form of collective memories.

4. Fig. 15 is showing a highly simplified picture where it looks as if the real part is just a thin line. The reality is that it is not a particle that is orbiting its elementary space, it is a complex number z that orbits the center of the complex plane and which represents the energy of the particle or the surface area of the elementary space. The complex plane is not physically 'out there', only the elementary space and its complex-valued surface is.

1.3 SPACE PARTICLE DUALISM THEORY [03.06.2004 (minimal mass for black holes with Hawking radiation [see 3.1]); 03.08.2005 (gravity shields itself off [see 3.2]); 10.08.2005 (elementary spaces and gravity; background independence); 22.08.2005 (complex elementary spaces and primordial dark matter [see 3.3]); 06.09.2005 (complex elementary spaces and twistor theory [see 3.7]); 04.10.2005 (holographic principle); 07.11.2006 (link to the wave function); 06.04.2007 (flatness problem); 18.11.2017 (glowing night-sky argument invalid)]

What is it that is bending and twisting?

What is space? According to general relativity space can bend and twist. It has a shape. Bodies moving in space follow this shape and that makes space look like a gravitational field. But what is it that is bending and twisting? It must be something; it can't be nothing. Could it be that space and particles are somehow different aspects of one thing?

In the standard model of particle physics elementary particles are theorized to be *point-like*. The electron for example has a certain mass, but it has no extention whatsoever; its size is thought of to be 'zero'.

Now, according to general relativity every mass has its so called *critical circumference*. If a body shrinks under this critical circumference, it becomes a *black hole*. Wouldn't that mean that all basic quantum particles are black holes? Well, if they are point-like then they must be black holes. Though some people would doubt that it makes sense to talk about a black hole if it is smaller than the Planck length, or, and which is equal, if it has a mass smaller than the Planck mass. Ordinary black holes are theorized to have a very subtle radiation called *Hawking radiation*.[1] It is a very weak radiation for big black holes that becomes very intense for microscopic black holes, but disappears completely for black holes smaller than the Planck mass. This is because a black hole smaller than the Planck length would have to emit radiation with wave lengths so short that the corresponding energy would have to exceed the black hole's entire mass energy.

That is why it stops radiating (see chapter 3.1). And that is good, because otherwise our elementary particles would just evaporate away.

Gravity is a very weak force. One cannot really talk about a gravitational field when it comes to single elementary particles. What if we regard the tiny little surfaces of these elementary black holes as mere mathematical entities which only gain physical reality when their diameter grows over the Planck length? What if we use these mathematical entities to construct a 3-dimensional space? A pattern of overlapping 2-dimensional sphere surfaces would form a 3-dimensional space and maybe our space is constructed in exactly this way. Would that make any sense?

Yes, it would! The surfaces we are talking about are directly proportional to the mass of their generating particles. More mass would mean more surfaces and therefore more space. More paths would lead through such areas. The result would be a seemingly bended space (!). And indeed it is somehow bended, because it consists of sphere surfaces, but it is not bended in the way Einstein imagined it to be.

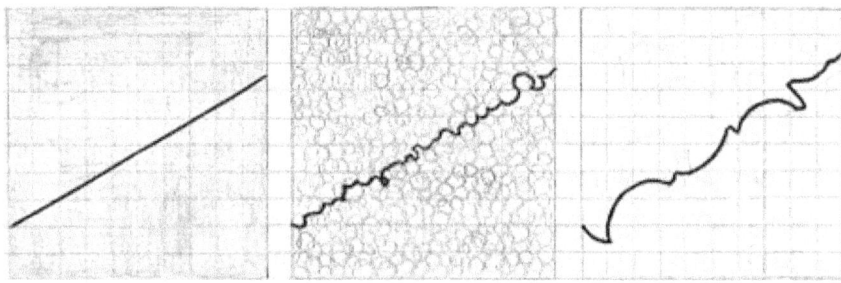

Fig. 16. Seemingly straight path of a particle in flat space.[2]
[sketch from the 10th August 2005]

At first this thought sounds adventurous. How could those tiny little event horizons of even tinier particles ever form a gapless pattern that looks anyway like our well-known 3-dimensional space? And indeed real particles could never form such a patchwork, otherwise there would be no empty space in the universe. The patchwork mainly consists of virtual particles, which are part of *vacuum fluctuations*. One tiny little part of space, as small as a nutshell, already contains more energy in the form of vacuum fluctuations than there is real mass energy in the whole universe (!). That seems more than enough for getting a gapless patchwork. Knowing this we don't have to worry about leaks or even walls in our pattern. The vacuum energy is always the same regardless of the expansion of the universe. Therefore the expansion cannot come to an end, because of such a thing as a lack of space.

However, those elementary spaces have to be much bigger than black hole surfaces in order to give us a gapless patchwork of space, because entities smaller than the Planck length could hardly influence physical reality. So, the analogy between black holes and elementary particles doesn't go very far. We can use the same formula $R = 2\,G\,M/c^2 = 2\,G\,E/c^4$ both for the radius of black holes and elementary spaces, but the G in the formula for elementary spaces must be exchanged by another constant G_E related to the strength of the other forces at the fundamental level (they merge at high energies and small scales). So the radius of elementary spaces is given by:

$$R_E \; = \; \frac{2\,G_E\,E}{c^4}$$

That means both the extended nature of elementary spaces and gravity are side effects of those other forces. In this scheme the strength of gravity depends on the vacuum energy density. The higher this density is, the weaker is the emergent force gravity. We will come back to this in chapter 3.7 where we will examine the structure of empty space.

Vacuum fluctuations have the same density everywhere in the universe, so they can only give us the arena; they cannot cause gravity themselves. If the elementary spaces of the real particles were supposed to cause all the gravity alone, then gravity would be unable to reach beyond the gravitational body itself. What actually happens is that every charge creates a stream of virtual particles. This stream becomes denser and denser near the gravitational source. The higher density of elementary spaces in the area surrounding a massive body is then causing gravity.

If gravity is a side effect of the other forces, then *rest mass* must be a side effect of charge. This has far-reaching consequences, turning upside-down almost all we thought to know about gravity. Many aspects of this radical new view will be kept for the last chapter (chapter 6).

As every virtual particle contributes equally to the fabric of space and to the differences in the density of space, the introduction of a graviton particle becomes just as surplus as the introduction of gravity as a force. Same as in general relativity gravity is regarded to arise from geometric properties of space itself, with the main difference being the type of geometry that is based.

That means gravity is not a fundamental force, but merely an emergent property that appears only at large scales.

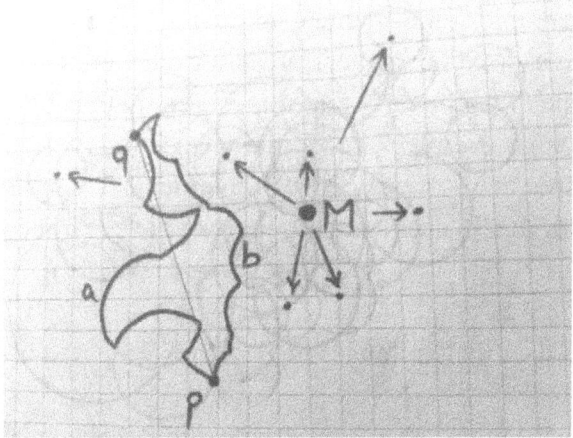

Fig. 17. A bended space at a fundamental level. The shortest connection between two points in space p and q, is a curved line, because the granular nature of space forces

a particle to follow many zig-zag paths, even if it is seemingly going straight. The picture shown is strongly exaggerated to make the effect more clear.
[sketch from the 10th August 2005]

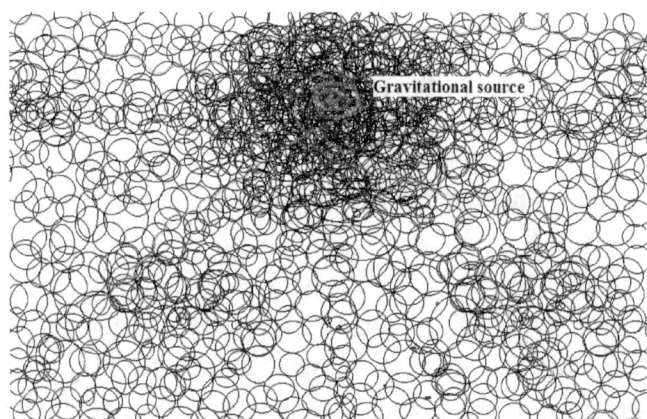

Fig. 18. The density of elementary spaces is higher near a gravitational body. More paths lead to such a body then away from it. This results in what we call gravity.

To summarize: according to this theory, which we will from now on refer to as 'space particle dualism theory', all elementary particles carry their own little quantum of space. This quantum of space has the shape of a sphere attached to the particle which has a radius proportional to the mass energy of the particle. The elementary spaces of virtual particles from the quantum vacuum overlap and combine to form a patchwork that looks like the 3-dimensional space we are familiar with. Yet since all movements of particles occur on those 2-dimensional surfaces called elementary spaces, the 3-dimensional space we are so familiar with is in fact 2-dimensional on its smallest scales. Only the way the spaces are connected is 3-dimensional. In this model gravity could be understood as a 'more' of space. Places with more mass, would have more 'space', and therefore more possible paths through them. On large scales that would look like gravity.

As mentioned already in the last chapter, these elementary spaces are complex-valued surfaces. The space they form is therefore divided into parts of plus-plus-complex (matter), plus-minus-complex (anti-matter), minus-minus-complex (exotic anti-matter) and minus-plus-complex (exotic matter) areas. A side effect of this is an elegant explanation for the type of *dark matter* that was present in the early universe and which evolved into the supermassive black holes that we see today. It should be noted here that this is not the matter that is missing in astronomical star surveys. We will come to terms with the origin of that in chapter 4.14, after we have learned about the actual age of the universe.

How much primordial dark matter that leads to will be discussed in the chapters 3.6 and 6.7.
Having complex-valued elementary spaces can not only explain primordial dark matter, but it can also account for the wave nature of the probability distributions in quantum mechanics. According to this theory, every particle is an elementary surface with a

complex-valued area. The complex number quantifying this area orbits the center of the complex plane, repeatedly leaving and re-enter the (real & imaginary) positive part of it, making the particle continuously disappear and re-appear on its path of movement. This shapes its location probability function into a wave.

Space particle dualism theory changes gravity on all scales making it different both from Newtonian gravity and general relativity. While the deviations from general relativity are rather subtle at small scales (see chapter 6.2 – 6.4), they become enormous at cosmological scales (see chapter 4). In this theory gravity does not influence the expansion of the universe and can not lead to a contraction.

What do other theories say about gravity on intergalactic scales?
Newton's gravity could not make the universe collapse as a whole. Newton assumed the universe to be static; not expanding, nor contracting. He explained this stability by assuming it to have no borders, no centre and endless many stars. In such a universe the gravity of endless many stars would pull from all directions, resulting in no overall movement at all.
In *Einstein's theory of gravity* mass can bend space and this bending of space is what then causes gravity. Here gravity doesn't depend on differences in mass density, but on the absolute value of the mass energy in a region of space. Therefore gravity from opposite directions bends space as well; allowing space even to bend into a 3-dimensional sphere surface (the surface of a 4-dimensional ball).
A universe with such a curvature is called *closed universe*. It is a universe which has no other choice but to start collapsing at some point and to end in a so called 'big crunch'.

Many people liked the idea of a closed universe, because it avoids irritating implications of an eternally expanding universe, which becomes colder and colder, having an endless future without life and in complete darkness.
However, in 1998 Saul Perlmutter, Brian P. Schmidt, and Adam G. Riess discovered that the universe's expansion is not slowing down but *accelerating*! That was a shock, because everyone was expecting the universe to slow down under the gravity of all the mass in it. No one could figure out what could make the universe expand even faster now. Many physicists started using the term *dark energy* for that which caused the acceleration.
How is this issue treated in space-particle dualism?
In this theory space cannot curve. What is interpreted as curvature of *space-time* in relativity theory are different densities of elementary spaces here. Therefore there is no chance of getting a closed universe. This solves the *flatness problem* and takes away the *fine tuning* one would need in general relativity to get a flat universe; the fine balance between gravity and expansion speed. In the standard model of cosmology the early universe would need to have a density very close to the critical density; departing from it by only one part in 10^{62} or less. In the mainstream this is solved rather artificially by inflation theory. We will come back to it in chapter 4.4.

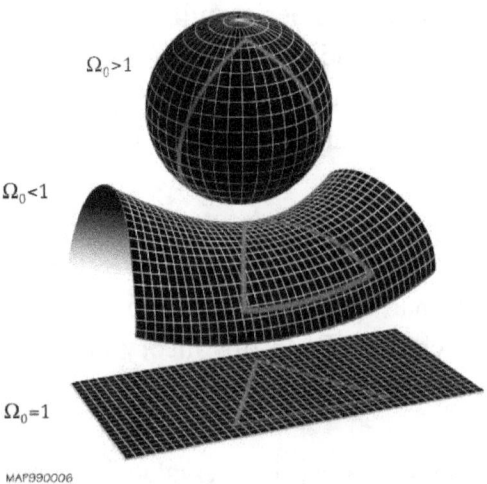

$\Omega_0 > 1$

$\Omega_0 < 1$

$\Omega_0 = 1$

MAP990006

Fig. 19. Flatness problem: in Einstein's gravity theory the chance for getting a flat universe is only $1 : 10^{62}$, incredibly tiny, yet our universe is flat as far as we can see. In space particle dualism theory however the flat universe is the only possible geometry.

Gravity in space particle dualism theory is always <u>local</u> and <u>never global</u>. Therefore the universe could <u>never collapse</u> under the gravity of the mass it contains, as it had to according to Einstein's general relativity. In light of the fact that there was never any observational evidence for the universe's expansion to have been slowing down in any period of its history, the theory seems pretty much on the track here.

This however doesn't mean that Einstein's field equations for gravity are completely wrong, but simply that they are not describing the curvature of space-time but the density of elementary spaces (similar to the resolution of a picture) and the influence of this density on the flow of time.

The fact that the density of space has an influence on the flow of time is here not taken as evidence for supporting a picture where space and time are merged into an entity called 'space-time'. Quite the opposite, the theory presented is refusing Minkowski's *block universe paradigm*, as it is in gross conflict with the principles of quantum mechanics. Thus Einstein was very right in his first judgement of Minkowski's idea: he thought of it as being merely a fancy mathematical trick which isn't really necessary.

Now if the universe is unable to collapse, then what made it expand in the first place? Almost all processes in physics are reversible, at least theoretically. Do we here face the first completely *irreversible process*? What does the speed of the expansion depend upon? Thermodynamics is the only realm of physics where we are used to and expect irreversible processes. Could it be that the expansion of the universe is somehow linked to thermodynamics?

Let's approach this by asking some more general questions:

1. Why is there an expansion at all?
As explained previously there is no existence without measurements by conscious

observers. If the universe was always inhabited by conscious life in mental time, the time we feel flowing, then why does it need to create itself a past in physical time which starts with a Big Bang followed by a rapid expansion? Or in other words: why does the past have to be limited?

What would it be like if it wasn't limited?
If we lived in a static universe with an endless physical past, then it seems we would have to see light from all regions of the universe regardless of how far away they are, simply because there would have been enough time for light from all regions of space to reach our location. This would make every spot of our night sky become as bright as the glowing sun.
Yet this argument scratches only on the surface of the paradoxy inherent to such a universe: the very fact that stars are still around and didn't yet burn out already shows that we must live in a universe with a finite past.[3]
In a static universe with an infinite past stars would have burned out long time ago; to be precise: an infinite amount of time ago.

This rules out universes that exist forever. But what about static universes that are created somehow in their entirety and exist till life in them gets extinct? In such a universe there couldn't be light carrying any information about anything that happened before the moment of creation. Yet one would think that such a universe that is created into the middle of its evolution would have some virtual past built into it, expecially into the photons that were already around. The only way to avoid photons carrying pictures of some form of past, virtual or real, seems inevitable. It would require the universe to start off with no photons whatsoever between stars.
If it is inevitable that the universe contains images of a past, then this past can't be an infinite past. Yet the only way to have a finite past is there to be some border region where densities become too high to allow for causal records. That sounds like a *heat death* right at the beginning of time. How can the beginning of time be so much like the end of time in classical thermodynamics? Could it be that there are two sides of the second law thermodynamics? Could it be that the expanding universe itself gives rise to the second law and thereby the arrow of time?

All the past of the universe is layed out spacially in the present as information we get from photons. In this way of sense the Big Bang is simply the border of the universe.
For very remote object measuring the photons they emit doesn't collapse their own wavefunction, but just that of those photons. For the objects themselves all the measurement of their photons does is determining an assembly of potential present states. Their whole existence is mere potentiality.

In this line of thinking the remote past of the universe we see represents merely a logical backwards continuation of what we see in our immediate environment. This 'continuation' is created automatically with each measurement, resulting in a kind of 'emergent universe'. This follows directly from the orthodox interpretation of quantum mechanics when we apply it to the universe as a whole. But do we really live in such an emergent universe? In chapter 4.2 I will present strong evidence that will prove beyond any reasonable doubt that our universe is indeed of such nature.

2. Why is the expansion of the universe accelerating right now?

My suggestion is that the expansion of the universe is directly dictated by the thermodynamic arrow of time and is therefore irreversible. There will be no reversal of the expansion – no collapse. From this perspective a collapse would mean a reversal of time. Stephen Hawking also once suggested that a collapse would reverse time, but he arrived at this conclusion from his no-boundary proposal; a speculative idea about the nature of the Big Bang singularity.[4] He later changed his mind and called this his 'biggest blunder'. It however shows that it is not too unconventional to link the time arrow to the expansion of the universe. In contrast to Hawking, we don't have to worry about time going backwards, because in this scheme the universe cannot collapse.

How does thermodynamics work at a universal scale? How can we talk about entropy in a borderless universe? Isn't it true that we can only talk about the density of entropy when it comes to the universe? So what is happening with the density of entropy? Is it rising? Is it dropping?

This is not at all an easy question. Entropy is growing locally at places with a high matter density, but because of the expansion of space, this doesn't seem to be true for the overall entropy density. What if the expansion of space levels out every local entropy increase, so that the entropy density is always constant?

If this is true, it would mean that the expansion of space is a thermodynamic process. What is the main source of entropy increase in the universe? It is black holes.

An interesting characteristic of the entropy growth in black holes is that is exponential. That is because the entropy of a black hole is proportional to its surface and thus it grows with the square of the black hole's mass.

It is this exponential entropy growth in black holes which causes the accelerated expansion of the universe. If the entropy in black holes grew linear, the universe would expand linear too.

To summarize: the expansion of space in a region is exactly proportional to the local entropy increase in this region. The acceleration of this expansion is due to the exponential growth of entropy in black holes, which are the objects in the universe with the highest entropy.

We will come back to this 'entropic expansion' in chapter 4, where we will analyse and test it in detail.

Notes:
1. However, as we will see in chapter 4.3, Hawking radiation doesn't exist in space particle dualism theory. It was first speculated that there is an analogous evaporation process due to 'thermal tunneling', but it turned out that this isn't the case.

2. When looking at this illustration (or the cover of this book) which shows how particles are supposed to move through space according to this theory, we may be asking ourselves what prevents particles from getting off-course (or space-time). We have to remind ourselves that these elementary spaces do not cause any friction. They are not solid objects but just space itself. Therefore, they cannot change vectors in space, just like a bended space-time cannot change vectors. It is pretty much like with a marble rolling down a vertical labyrinth. It might move a bit to the left and a bit to the right, here and there, but it won't change its general direction of propagation. That would indeed violate the preservation of energy.

3. This glowing night-sky argument is used by Hawking in chapter 3 of his popular book "The universe in a nutshell" (2001). It is oversimplifying things by not taking account of the limited lifetime of stars. However, before the Big Bang theory became prevalent, there were steady state models of there universe where new matter was theorized to appear out of nothing, compensating the expansion induced density drop. In such models one could theoretically imagine stars to form forever. Within the scope of the limited and highly questionable logical consistency of such models the glowing night-sky argument may make some sense.

4. The *no-boundary proposal* belongs to the cosmological Hartle–Hawking theory. It claims that before the Big Bang the universe was in a state such as that time becomes a space-like dimension. Basically speaking, the different histories or worlds of many worlds interpretation are lined up and form another 'direction of time', which he calls *imaginary time*. This imaginary time is supposed to influence the history of the universe in real time. However, there is no interference between those universes; they are totally disconnected. The statistical chance for us to be in a particular universe among other universes can not be called influence. At least it is not a causal influence. It doesn't make sense to postulate a dimension which is not accessible by anything, and if it is accessible, then we get something like the *many worlds immortality theory* which has its own problems (see chapter 2.3). To my knowledge Hawking is the only one who uses this concept of imaginary time.

2. IMPLICATIONS FOR QUANTUM MECHANICS

2.1 The measurement problem in detail [11.11.2005 (cross wiring); 24.08.2013 (wrong notion of many observers/entanglement restricting reality); 4.3.2016 (wrong notion of large objects remaining in superposition for half sec.); 13.4.2016 (large objects remain in superposition for $1/20$ sec.)); 10.03.2017 (weakness of micro-PK explained by required timing precision); 23.12.2017 (shift to the frequency of subconsciousness, namely a $1/40$ sec.)]

As already mentioned in chapter 1.2, Gödel's incompleteness theorem proves that consciousness can not be described by any formal system. In other words: it is non-algorithmic and can thus not be simulated on a computer. To me this expels all mechanical explanations, even those which are based on hypothetical non-algorithmic laws of nature (such as suggested by Penrose (1994)). The access to truths and meanings, which is one of the main properties of consciousness, can only be based on some sort of mind-body interaction, in which the randomness in quantum mechanics is replaced by synchronicity and meaningful choices (free will). For that to be possible, large superpositions are all what is required.

Superpositions don't simply stop to exist when quantum states reach the macrocosm. In many cases the states under consideration are not visible, either because they are states of invisible gases or because they are hidden behind walls or within a larger physical object.

One also has to look at when the measurement reaches the conscious part of the brain: longer pathways lead to larger superpositions. At first glance that seems to explain the cross wiring for all senses in the brain (except smell). However, there is no noticeable growth in the superposition without an increase in the number of alternatives to be chosen. Therefore only passing synapses really enlarges the superposition. Large is here not referring to the spacial distribution of the superposition, but to the diversity of different states it contains.[1]

One main reason why Einstein and many others had great difficulties to accept quantum mechanics as the ultimate truth was that it seemed to violate Leibniz's <u>principle of the sufficient reason</u>, which states that everything has to have a cause (a reason to be) (Gottfried Leibniz; *La Monadologie*, 1714). If we accept the natural role of consciousness in quantum mechanics, then there is no conflict with this principle anymore, since randomness can be replaced by non-computable determinism.[2]

Many of those who implemented a role for consciousness in the reduction of the wave function did not consider Gödel's theorem nor mind-body interactions. Von Neumann, after whom the explicit orthodox form of the Copenhagen interpretation, the Von Neumann-Wigner interpretation, is named, was regarding the mind to be fully describable in terms of computation. Even Gödel himself had great difficulties to accept the consequences of his theorem for consciousness. He preferred to assume that our mind is based on a *principally incomprehensible algorithm*, which would have to be so complicated that it could only have been implanted into the brain by god himself. The fact that artificial intelligence is impossible is commonly acknowledged among mathematicians familiar with formal logic.[3] Books discussing Gödel's incompleteness theorem usually have a section or at least a note, stating that the theorem also implies that logical reasoning cannot be automatized, thus there can be no robot successfully imitating a human (for an example see chapter 8 of "Gödel's Proof", Ernest Nagel & James R. Newman; 1958). However, taking full account of this fact would require a theory of consciousness, and developing such a theory is commonly referred to as 'the hard problem'. Hard, because philosophers have been thinking about how to solve this problem for about two millennia without much progress. Many even regard it as a principally unsolvable problem (Karl Popper; 1974).

I will suggest a solution for this problem at the end of this book (chapter 5). This solution is closely related to ideas from Platon, Frege, Popper and Penrose.

People who held consciousness responsible for the collapse of the wavefunction include some of the greatest minds in physics of the last century, such as John Von Neumann (1932), Fritz London (1939), Werner Heisenberg (1957), Eugene Wigner (1967) and John A. Wheeler (1998). Eugene Wigner, however, changed his mind in his old days, while Wheeler insisted and propagated consciousness as the driving force in the universe; especially with his slogans 'its from bits' and 'everything is information', as well as his notion of the 'participatory anthropological principle'.

None of them, however, tried to develop a theory of consciousness. So their understanding of the orthodox interpretation (concrete form of the Copenhagen interpretation), was that although consciousness is causing the collapse, it should not have any influence on its outcome. This sounds reasonable at first sight, because we are not aware of having any direct influence on quantum processes we observe, but it cannot be true for the brain, because we know that some kind of mind matter interaction must exist at least there. If consciousness is able to interact with a material object like the human brain, and replace the quantum uncertainty in it by its *free will*, then why can't it also influence the 'quantum decisions' around it? After all perception and thought/imagination lie mostly on the same neurons.

In fact consciousness does influence those decisions. While we observe the world, we always choose between different realities – different quantum states. So why aren't we like powerful wizards controlling everything that happens around us?

This is a good question and the answer is not that obvious.

We might first take note of the fact, that we are not choosing alone. I might want to have sunny weather, while others may wish for rain. In a first attempt we may think that this is all what differentiates reality from dreams. Yet even if one is alone in a desert or in a forest, nature doesn't appear to follow ones wishes.

We might then consider entanglement and hope to get some restrictions on reality from it. However, entanglement only leads to some constraints between the properties of certain quantum numbers such as spin; it can not give any restrictions on position and impulse.

The longer superpositions have time to evolve before the measurement takes place, the more different choices can be made. The time for anything to become conscious is a half second, and we may be tempted to believe that any wavefunction in the outside world has a half second time to evolve – plenty of time for chaotic systems to evolve into various states all kept in superposition. However, the brain measures its environment with the frequency of 40 Hertz. So, although the wavefunction of the individual photons of an object and the subsequent individual signals in the brain remain in superposition for a half second, the object itself can only remain in superposition for about 1/40 of a second.

Many think that 'collapse of the wavefunction' is something very violent, which happens very suddenly and instantaneously all over space. Although that is true for the wavefunction of a single particle, the wavefunction of a larger system doesn't really collapse. What happens is that it is reduced in one place and grows in another place. Although the 'collapse' or 'reduction' happens everywhere at the same time, it does so only for things which are entangled.[4]

Our consciousness, which is ourselves, controls only our brain, but since all wavefunctions collapse in the brain of conscious observers, the control of the sensor means also a control of the outside world. A measurement is only a measurement if it reaches a conscious observer. So, all 'quantum decisions' are made at the very moment of observation. Even events that took place millions of years ago, far away in the cosmos, become real just in the very moment when their light reaches an observer (see chapter 4.2). It is the *act of observation* which leads to the choice of realities.

Macroscopic objects are constantly measured and also consist of a large number of particles, so it seems reasonable that influencing them is hard. But what about single quantum particles? Theoretically the observer should have full control over them.

Since many decades parapsychologists conduct experiments with quantum randomness generators gaining positive results. Test subjects, who are mostly common people, are asked to observe a random row of ones and zeros with a strong inner expectation or wish to get more ones or more zeros. Although these are single quantum decisions at the quantum level, the participants fail to reach a success rate of 100%. In single trials the success rate may reach 60% or even 70%, but when averaging over thousands of participants and millions of trials we find a success rate of only 51% (for details see "The Conscious Universe", Dean Radin; 1997). That is puzzling. If it is consciousness which leads to the choice, then why is it so hard to influence even a single quantum decision? Considering that ones and zeros have the same likelihood, it seems odd that we have such a weak influence on this choice. It is a common believe among researchers in parapsychology that it is subconsciousness which has the power to influence quantum states.

Physicists sometimes suggest this too: in a book about quantum philosophy written by Lothar Arendes, he noticed "... It is also strange that we are not aware that and how we conduct this reduction. It would therefore have to be a performance of the subconscious" ("Is physics giving us knowledge about nature?"; Lothar Arendes; 1992; page 78).

However, this assumption doesn't seem to help us much when trying to understand the weakness of the effect. Many researchers may feel that the subconscious is simply not convinced enough that it has a real influence on reality, and that is why the effect is so small.

Similarly many people with an idealist believe system think that it is only a lack of believe into idealism among the majority of people which prevents everyone else to develop superpowers (Werner Hendrik, Ludovic Krundel; personal correspondence; 2016).

Having to control ones subconsciousness does make mind matter interaction more difficult, but that is not enough to explain the weakness of the effect.

The correct answer has to do with the singular nature of those quantum decisions. While dealing with single binary decisions with a 50% chance for either result allows the observer to influence the decision without reducing entropy (without choosing an unlikely state), it is still difficult, because he has to wish at a certain moment in time, with a very narrow time window of only a 40th second. Any such influence must therefore be linked to *gamma brain waves*.

There are other forms of mind over matter interaction with different degrees of difficulty and efficiency. We will come back to this in chapter 5.3.

Another question we have to ask in connection to the wavefunction collapse is if it happens directly when electromagnetic waves reach the brain, or not until the signal of the detected event reaches consciousness. Even before it reaches consciousness, changes in the electromagnetic field may already cause a sufficient difference in the brain, big enough to make a difference for the conscious state of mind. Yet if that would be true, the wavefunction would collapse all the time, making an efficient mind brain interaction impossible. Also quantum erasers suggest that the measurement can be undone at least shortly after it happens, and that means the wavefunction can not be collapsing that quickly. In principle experiments could be conducted to find for how long after a measurement the measurement still can be undone (see chapter 2.5).

As we know through experiments first conducted by Benjamin Libet, everything needs about half a second to reach the conscious level.[5] That means every signal coming into the brain remains in a superposition of multiple states for half a second.

However, because consciousness sees 20 pictures per second and subconsciousness 40 pictures per second (only half of the sensory information makes it to the conscious level), that means wavefunctions collapse in the brain 40 times a second. Since we have an integrated consciousness of all our environment in each moment, the wavefunction must be collapsing in a synchronized fashion.

The Penrose-Hameroff model

Roger Penrose proposed a criterion for the wavefunction collapse, called OR (orchestrated reduction), where a difference in the gravitational field of more than the Planck length leads to it. Together with Stuard Hameroff he created a model of the brain in which the wavefunction collapses in a synchronized/orchestrated fashion every 40th

second. Basically what they did was to artificially calculate the wavefunction collapse in way that would lead to the observed brain frequency (in the perception realm).

However, although this seems to be a nice explanation for both the holistic nature of consciousness and the frequency of the brain, it does not allow a single neuron to fire in superposition. Therefore all superpositions in the Penrose-Hameroff model are superpositions of totally identical brain states that differ only in the molecule distribution inside microtubules.

That means we have a superposition covering the whole brain and every break down of this superposition is a conscious moment, but according to this model the break-downs (the reductions of the wave function) are not caused by consciousness, but produce consciousness.

According to Hameroff even an electron flying around alone for some thousand years would have a conscious moment, just by moving far enough to create sufficiently different space-time geometries. To me it seems that the whole objective reduction thing is ruling consciousness completely out, instead of explaining it.

Also it requires unrealistic environmental isolation for Hameroff's microtubules. With consciousness itself as the cause for the wave-function collapse, nothing of that is needed.

Originally Penrose had considered superpositions of neurons firing and not firing (Penrose; 1989). When he realized that OR (gravity induced wavefunction collapse) doesn't allow for such superpositions, the only way out was Hamerhoff's suggestion to base consciousness on hypothetic computation inside microtubules.

We have to here differentiate between the time single brain signals are in superposition and the time the whole brain is in superposition.

The half second needed for the collapse is referring to the time needed for a signal to travel from the entrance through the external senses, until the reduction of its wave function in the subconscious and conscious parts of the brain. In the Penrose-Hameroff model such a distinction doesn't exist. According to their model the wave function is reduced already shortly after the signal entered the brain. Long time before the signal reaches the conscious level and even the subconscious level! All superpositions left for consciousness to work upon would be superpositions of macroscopically totally identical states of the brain, only differing in the matter distribution within the microtubules of neurons. For any real influence to the brain activity it would have to push these matter distributions between every single quick collapse of the wave function. Not only would that take incredibly long time, it would be impossible, because consciousness could not really choose between states of the mind which are identical to it and begin to differ only after dozens of further collapses took place.

The reader might be confused to hear that all those single wavefunction collapses in the Penrose-Hameroff model correlate to choices between totally identical states. Shouldn't every single of those 40 snapshots of the brain show different brain activity? The answer is, yes, they do, but in this model that brain activity is never in superposition, all what is in superposition is the molecule distribution within microtubules and that is all what can be chosen.

So if we were to say that this is a model of free will, it would have free will based on something that is totally unconscious, namely activity in microtubules.

Also if we believe consciousness is based on 'calculations' within such microtubules (also suggested by Hameroff), then it seems hard to still link it to noncomputability, even if we say that the choice between superpositioned calculations is made in a noncomputable fashion.

That would mean that we have 40 or even more 'choices' each second, all of them producing consciousness or subconsciousness, but none of them being real choices between real alternatives. That would make no sense. It would make it impossible to influence quantum decisions not only outside the brain, but inside as well.

The subconscious

If subconsciousness is responsible for the collapse of the wave function of states outside the brain, then it must be of the same non-computable nature as consciousness. Is there evidence for this? As Libet describes in his book "mind time", most ideas, even creative ideas, come from the subconscious. Consciousness is just working upon these ideas. In "the emperor's new mind" Penrose describes many mathematicians and other scientists having experienced complete solutions to problems in their fields suddenly popping into their minds. Solutions for new problems cannot be found just by applying algorithms. The subconscious must therefore be non-computable as well.

Other evidence for the subconscious to be of a similar nature as the conscious is that the area most associated with the generation of feelings, the *amygdaloid nucleus*, does not lie on the conscious cerebral cortex, but in the subconscious *diencephalon*. It is also responsible for the filtering of information approaching the conscious cerebral cortex. Exactly this filtering is what is separating the conscious from the subconscious. Any kind of isolation within the brain can increase the number of consciousnesses within the brain. So called split brain patients, which have their *corpus callosum*, the connection between the two hemispheres of the brain, cut apart, have two consciousnesses in one and the same brain. Same as with the subconscious, these patients do not feel their second consciousness, and only under some special experimental conditions the existence of the second consciousness becomes apparent.[6]

The neuroscientist John Eccles tried to bypass the irritating philosophical implications of a self which is split up into several separate consciousnesses by claiming that only the side which is able to express itself verbally can count as truely conscious. This language centrism had been criticized by Penrose, who argued that mathematicians do not use any language at all when thinking about mathematical problems (Penrose; 1989). Indeed, verbalizing thoughts is just a way to memorize them easier and to communicate them with others. In the same year that Popper and Eccles published their collaborative work on consciousness, claiming that consciousness requires language and that only the left hemisphere is conscious; Donald Wilson and his colleagues investigated a split-brain patient with both hemispheres being able to perfectly understand spoken language. After some time the right hemisphere even learned how to speak (!). From then on, both hemispheres could clarify independently that they are conscious. They even had different preferences and dislikes. It should now be clear that consciousness does not require language. Could it then be that the subconscious is an independent person, just like the conscious? I imagine the subconscious as a loose collection of thoughts, feelings and spontaneous desires. To ask who is having these thoughts, feelings, and desires, is to misunderstand the nature of these things. We are not having thoughts, feelings and desires, we are these thoughts, feelings and desires. For them being what they are it is not necessary to be accompanied by any abstract notion of a self. Different from the double-consciousness of the split-brain patients, the conscious and the subconscious have always roughly the same information content and differ only in the grade of detailedness.

The subconscious can however not know that it is the subconscious. It has no means of communicating independently from the conscious, so it will not develop the feeling of a separate self. Any of those feelings which are part of the subconscious can potentially become part of the conscious. These feelings can cause the performance of various gestures and somatical movements, without becoming fully conscious. In other words: we do a lot of things, without thinking about them.

The superior consciousness is responsible for permitting actions, rethinking ideas, changing strategies, verbalizing and memorizing things which spread out from the subconscious. The subconscious must always be in a dreamlike state of mind. That is a state of chaos and disorder, but also a state of creativeness. Only a part of our subconscious thoughts reaches consciousness. Most subconscious ideas are taken into action without consciousness being asked for permission.

If sub-consciousness is, as suggested here, really of the same non-algorithmic nature as consciousness then it is fully capable of reducing a wave function, and thereby choosing between realities. That means, the state reduction (choice of realities), must take place earlier than a half sec. after sensing (the time for an observation to become conscious). The choice between external realities is a subconscious choice, while consciousness only chooses between superpositioned states of the mind. We might be able to give our subconsciousness certain tendencies for its choice between realities, but since normal macroscopic objects are measured 40 times a second, they stay in superposition for only 1/40 seconds. Only the individual photons reaching our brain and the subsequent signals in our brain remain in superposition for a half second.

The time external states remain in superposition depends upon the frequency of the brain. The fact that the brain has an overall frequency with which it measures the external world shows us that there are structures in the brain which must be in a *quantum coherent state*. These structures are most likely the *microtubules* in brain cells. They are a precondition for the *holistic nature* of consciousness, namely for the fact that we have a consciousness of whole pictures, while each neuron is only responding to certain parts of this picture. It was Penrose and Hameroff who first put forward a model of consciousness based on microtubule. However, although accepting the microtubule model, I do not think it has to be based upon Penrose's OR theory.

Quasicrystals

Another means to test when exactly the wavefunction collapse takes place could be to analyse quasicrystals, which have a 5-fold symmetry that is regarded to be not obtainable by any classical local growth pattern. As Penrose first suggested in "The emperors new mind", these kinds of crystals must grow in superposition and that is what allows them to find the lowest energy level state in the very moment the wavefunction collapses. There might be irregularities in the patterns of these crystals, which are of a size proportional to the superposition they have been in.
It will be much easier for these crystals to grow, if they can remain in a superposition until a measurement takes place. Penrose's one-graviton criteria (OR) would again make it very hard for them to grow. Interestingly Penrose was the first one to analyse this 5-fold symmetry, many years before it was found in nature.

To prove our capability of choosing between realities, we would have to look at a rather isolated quantum system, such as a newly generated pair of entangled photons. For getting our sub consciousness to influence the quantum choice in a systematic way, we might have to evaluate the different choices with certain emotions or ideas. For example, two states A and B might be represented by the detector as a smiling face and a sad face respectively. Other representations which subconsciousness is more sensitive to might be considered as well. If consciousness is what causes the state reduction, then it should have some influence on it. It would be strange if we can influence the reduction of mind states for thoughts, but not for sensory information, while both lie on the same neurons and are superpositioned just in the same way.

Parapsychology

And indeed we can, and that has been shown consistently in countless experiments (see Dean Radin; 1997). The effects can be weak or strong, depending on various aspects such as entropy, timing flexibility and training. *Microkinesis* (the influencing of single quantum decisions) is weak because it has to happen at a particular moment in time. *Macrokinesis* (the shifting of large bodies) is weak, or better, hard to achieve, because of the sheer number of particles involved (huge drop in entropy required). Microkinesis is often ignored by the scientific community because it can only be proven statistically, and many non-experts in the field believe it is possible to explain the rather small effect away by *conformation bias*. Macrokinesis on the other hand is ignored even more commonly simply due to the fact that only very few people are able to demonstrate it and those who did demonstrate it did so mostly under non-laboratory conditions. Therefore many scientists simply don't know that these phenomena have been observed under laboratory conditions as well. There is a general tendency to turn a blind eye on everything consciousness related. That is not so much because scientists love to be materialists, but more because there is no theory of consciousness yet which could tell what is conscious and what not. That is also why the orthodox interpretation of quantum mechanics is so unpopular nowadays.

As consciousness is very central to the theory described in this book, solving the mind body problem is an absolute necessity. However, since the proposed solution doesn't have much influence on the actual physics of the universe in general, it is kept for chapter 5.1 and 5.2. The influence of the observer on external states will be further analysed in chapter 5.3.

Notes and references:
1. If microtubule really perform calculations as suggested by Hameroff, then the purpose of the cross wiring in the brain could be a gain in microtubule length.
2. The notion 'non-computable determinism' is strongly influenced by Penrose [1] - [3]. However, there is an inherent vagueness in Penrose's description of what he means by this. Sometimes he suggests an interaction with the Platonic world (see [1], chapter 10.8 "In contact with the Platonic world"), and sometimes non-algorithmic 'laws' of nature (see [2], chapter 7.9 and 7.10). In chapter 4.11 of this book, I will give some criticism of the later view.
3. Gödel's Proof (1958), Ernest Nagel & Ernest Nagel & James R. Newman ([6]).
4. The fact that a quantum wave-function collapses simultaneously on every spot in the universe, implies that we need a concept of 'now' which is beyond relativity (see chapter 1.2). Physical time as described by Einstein's theory of relativity denies the existence of a present, and treats the whole space-time as equally real. Both our considerations on consciousness and quantum theory lead us to the conclusion that the present we see and feel is more real than the past and the future. Our feeling of a flowing time is not an illusion, contrary to what relativists suggest.
5. Mind Time - The Temporal Factor in Consciousness (2004); Benjamin Libet ([18]).
6. "The Emperor's New Mind"; chapter "Split-brain experiments" and "Non-verbality of thinking"; Penrose, 1989).

2.2 On the nature of reality [19.05.2013 (entropic expansion hypothesis); 10.03.2015 (entropy as number of worlds)]

Noncomputable determinism

Using the Von Neumann-Wigner interpretation in combination with interactionalism we arrive at the conclusion that real randomness doesn't exist. What we use to call quantum random is according to this scheme fully determined by conscious and subconscious choices. We are not aware that we are choosing, but we do and there is always a reason why a quantum random event is decided this way or another.

The reason why most people are not able to use this freedom of choice in order to make objects shift or change is that such shifts and changes require to purposely coordinate each and every particle in order for the system as a whole to reach the desired state. The main obstacle in achieving this is thermodynamics (see chapter 5.3).

That means hidden variables do exist, but they are of psychological nature, meaning that they reside in the observer himself. Therefore the description of nature as a system full of randomness is just an approximation only valid for the lowest degree of intention within observers.

We may feel that consciousness itself is pretty random, yet if it really was, humans wouldn't be able to construct mathematical proofs and also psychology couldn't be a science then. As Roger Penrose already pointed out correctly consciousness is both deterministic and noncomputable (see chapter 1.2).

Productive logical reasoning requires access to truth and meaning and can therefore not be done a machine – it is non-computable. Nevertheless it is deterministic, because its outcome is predestined by logic. It is already there; forming a part of eternal mathematical truth, accessible to consciousness in principle.

A world without Qualia

Meanings and symbols exist only in our minds. Yet a world without them would not be a world. A world only consisting of point-like particles with positions and impulses, would not be visible. Even if a god could look at it, he would not see more than moving points. Without a body and without subjective experience, he would have no Qualia (mental representation of physical entities), and therefore no representation of the outside world within himself. As we will see in chapter 5.1 the way things are represented in our consciousness depends upon our experience with those things. We describe red as a warm colour and blue as a cold colour, because many things which emit light in the frequency of red are hot things or burning things, while blue things tend to be cold. We can say that Qualia are like symbols for the things in the external world. Without these symbols there could be no representation of the external world within our consciousness. One needs symbols to express meanings. A world without symbols and meanings would be like an unexpressed equation or an empty paper. It would simply not be there. So the physical world needs the mental world. The physical world is the space, the tie, between different minds.

The same is true for the Platonic world of mathematical notions and idealized objects: it can only exist when it is at least sometimes expressed and represented in nature, or in the mind. None of these three worlds can exist separately.

Net spread by minds

Considering the strong dependence of the external world on consciousness, we might even say that the external world exists only as a net spread out by minds. The minds are the knots in this net. Just as the net is dependent on its knots, the world is dependent on our minds. The space in between is just space between minds, not objects, and it is not filled with matter, but just with information. What lies between minds is seen by many minds, and thereby becomes reality, following physical laws. These laws might ultimately be merely the laws of logic. How could it be possible to reduce the laws of nature to mere logic?

We have to recall that quantum mechanics as it is presented in the similar worlds approach can be derived from Leibniz's identity of indiscernibles, which can be seen as part of formal logic. This further validates Einstein's hope that there will be only one logically consistent theory that describes the entirety of physics.

Nature can never know more about itself than any observer could know about it. If it is impossible to know if an object turned left or right, then the object will turn both left and right and might even interfere with itself. If our epistemological borders are at the same time the borders of nature; if nature does not know more about itself than we can possibly know, then it is to conclude that WE are nature. Nature can exist only as a network between minds. A nature without minds would have no colours, no sounds, no smells, no tastes, no meanings, and no symbols. In short: It would have nothing of what we see in it, and what we identify with it. Even positions would not exist, since there would be no distinct points (observers) to define them upon. Time wouldn't exist either: time in relativity is static; just there; not different from space.

In quantum theory things change whenever we make a measurement. Without us, without a measurement, nothing would happen; nothing would change. There would be no time, no space; nothing.

Universes always begin with very small (primordial) black holes and a high density, and then expand towards infinity. They do this till no one can survive in them. So when we are gone, and all other life forms are gone, because the universe got too cold, then it must enter an infinite Omnium-like superposition and from there it can only collapse into another conscious inhabited state. If quantum random choices tend to be meaningful, then the newly created universe would be similar to the last universe, not only in its initial state but in its evolution as well (we will come back to this in chapter 4.10).

Why does it expand?

Why does the universe need to expand? Why does it even expand faster and faster? What drives it? Ever since the discovery of the laws of thermodynamics, it was always regarded as a precondition for the flow of time to have a system beginning in a low entropy state. That means the universe was always thought of as to have evolved from a very special state (low entropy) to a rather random state (high entropy). Yet that would make our universe very special and fine-tuned. What if we, instead suggest that entropy itself is the driving force behind the expansion? On first sight this seems to lead to an entropy minimum for the beginning of physical time, which would fulfill the common belief that the universe started with a low entropy state, but if we think about it more we see that it is hard to talk about entropy when it comes to the universe as a whole. The universe might be endless, and, at least the visible part of the universe (the cosmological event horizon) is growing with the speed of light. On a cosmological scale it makes more sense to talk about the entropy density instead. If every local entropy growth leads to an expansion of space, then the average entropy density must always be constant. That means the 'Big Bang' did not have low entropy and was instead full of primordial black holes. In other words: the beginning of physical time was not special or fine-tuned at all.

What is entropy?

But what exactly is entropy? It is a measure for the potential of heat to flow from hot spots to cold spots. Its value (without the Boltzmann constant k) can be interpreted as the number of microscopic constellations of particles which correspond to the same macroscopic appearance. Since the border between microscopic and macroscopic is highly dependent on the size of conscious observers, this is a rather imprecise and random definition. On a quantum level, entropy can be given a different and more precise but very similar definition. Here it can be a direct measure for our knowledge about nature, as represented by the wave function. In the similar-worlds interpretation we can identify the logarithm of the number of similar world with the value of entropy. It is often suggested that entropy is somehow a measure for 'information'. We can interpret it this way, but it is then a negative measure for information: the higher the entropy, the less we know about the system. So entropy can be both a measure for chaos and for the loss of information. Of course decoherence theorists look at it different, because for them a collapse of the wavefunction would mean that 'quantum information' is lost. That is also why for them black holes seem to violate 'information conservation', no matter if they evaporate or not (information is inaccessible in both cases).[1]

Penrose's CCC model

Roger Penrose speculated that the universe might 'forget' about its size, when all black holes evaporated through the Hawking process, and all particles decayed, so that nothing else than light remains. For light the universe has no size, and distances do not exist. So it could restart with a Big Bang, since arguably that is a similar state. He describes this new view at length in his book "Cycles of Time" (2010). This approach depends crucially on both the instability of electrons and protons and the existence of Hawking radiation. Both are predictions of the standard model of physics, but there is no experimental evidence for either of them.

If black hole radiation exists and is perfectly thermal, which is indeed Penrose's position, then all the otherwise stable particles could be 'shredded' (turned into pure light) by black holes. Yet this would work only if really each and every rest mass particle in the universes gets at some point swallowed by a black hole. In an acceleratedly expanding universe it is not at all sure if that is possible.

As we shall see at the end of chapter 4.3, there is no Hawking radiation and as we will see in chapter 4.8 electrons and neutrons are indeed stable.

The here presented model

According to space particle dualism the universe 'forgets' about its size much earlier, namely at the moment no observers are left. It then enters an Omnium-like superposition of all possible states and from there it collapses into the first state that contains conscious observers. Of course the thereby formed new universe would have a dense initial state (what is referred to as the 'Big Bang' in the mainstream), just like the last universe (or 'eon'), but this dense initial state is, and always was something in the physical past, something on our past light cones. It was never something in a 'present moment', because only moments which are experienced by someone can be regarded as 'real moments', everything else is merely light rays showing us a logically plausible past.

Conclusion

Einstein always ignored consciousness in his attempt to find a final theory of everything. He could also not accept the randomness introduced by the measurement process in quantum mechanics. He did not see that quantum theory has as much logical and philosophical justification as arguably relativity theory has. Quantum mechanics was not part of his intuition, because his intuition was based on a materialistic world view.[2] Science always tried to be objective, to leave the mind out of everything. That might work for big things, but fails completely for the small.

Note:
1. If black holes don't emit Hawking radiation, then the information that is trapped inside them can't get out. However it would still be accessible to somebody who decided to commit suicide by entering a black hole.
2. This is however not entirely true: Einstein had some rudimentary interest in parapsychology and even wrote a preface to a book on the subject. Similar to David Bohm later who tested psychics like Uri Geller, Einstein seems to have been a determinist and realist who at the same time believed in the 'power of the mind' (psi). This apparent contradiction in the worldview of both men maybe shows that they didn't mistake their scientific models with reality.

2.3 The different interpretations of quantum mechanics [26.08.2013 (rough analysis); 04.05.2016 (categorisation)]

There are various interpretations of quantum mechanics, and on first sight it is hard to differentiate them clearly. Not all of them have very clear-cut definitions. For the purpose of this book it makes sense to categorize them according to the following scheme:

Interpretations with a collapse/change of the wave function, we can call '*choice theories*' (an alternative term would be '*collapse theories*', which is actually the term in common usage). The many world-like approaches we can call '*no-choice theories*'. Then we can differ between those where the choice is thought to happen roughly at a macroscopic scale and those where consciousness has to be involved. The former can be called '*scale choice theories*', and the later '*consciousness choice theories*' (usually called 'consciousness cause collapse'). Then we can further differ between active and passive 'consciousness choice theories'. '*Active consciousness choice theories*' are those where consciousness is thought to have influence on this choice. Theories that to some degree take account of relativity theory we can call 'relativistic'.

(1) Copenhagen interpretation (Category: *Consciousness scale choice theory*; Proponents: Niels Bohr (1927); Werner Heisenberg (1957); Note: According to Heisenberg *scale* is here the macroscopic scale and defined by the community of observers being able to distinguish states.)

According to this interpretation a measurement causes the reduction of the wave function. It is not made clear what it is that has to measure. It seems that for Bohr it was enough if a measurement apparatus measures, because it is macroscopic. There is no clear-cut definition of what has to count as macroscopic and what as microscopic. It has often been suggested that a measurement has to be done by a conscious observer to count as such. That is why the term Copenhagen interpretation can stand for the Von Neumann-Wigner interpretation as well.

In this interpretation it wouldn't be possible for consciousness to interact with the brain, because the choice between alternatives is made during the measurement, before any signal reaches the brain, and in the brain any neuron firing can also be regard as macroscopic and would therefore lead to a choice/collapse.[1] Therefore the standard Copenhagen interpretation doesn't allow quantum coherence in the brain. In this interpretation it is only relevant that a difference between states is visible (macroscopic), and not whether it has already been observed or not.

Hence, a mind brain interaction is impossible, because no superposition would be large enough to represent alternative realities. A superposition could here only represent a single measurement event or a single firing of a neuron. Superpositions would have to stop here.

It is important to note that for Heisenberg the wavefunction was representing the accessible knowledge of the observer. He saw the wavefunction changing in two steps: the first step being a state causing macroscopic differences and thereby being available to the community of observers, and the second, the measurement result arriving in the consciousness of an observer.

The split into two steps probably serves the purpose of avoiding the possibility for the observer to influence the result of the measurement using his consciousness (psychokinesis). For Heisenberg the result of the measurement is decided when it arrives at a macroscopic level and is made accessible to the community of observers. However, it is unclear how this two step evolution of the wavefunction could work. Is the wavefunction supposed to freeze on its way from the measurement apparatus to the brain of the observer? How could the quantum state be decided already, if the wavefunction doesn't change before it is registered in the consciousness of the observer? I think that Heisenberg simply did not involve the brain in his considerations. The observer cannot measure his own neurons. Therefore it must have appeared to him, that the wavefunction could not possibly contain information about states of the observer itself.

From his writings it is very clear that he required a conscious observer to be looking at the measurement in order for it to count as such and to lead to the reduction of the wave function. We can see this from what he says in his book "Physics and Philosophy" (1958)[2]:

"The measuring device deserves this name only if it is in close contact with the rest of the world, if there is an interaction between the device and the observer. ... If the measuring device would be isolated from the rest of the world, it would be neither a measuring device nor could it be described in the terms of classical physics at all."

So for Heisenberg it was clearly the observer who causes the collapse. The main reason why he stated that the outcome of the experiment is already decided at the measurement device is that he did not expect the wavefunction to further evolve or change on its path through the brain of the observer. And when we talk about a wavefunction as a tool for predictions used by a physicist during experiments, then he is right. But now we are able to make more sophisticated experiments which include both particles and the observer. For both Bohr and Heisenberg the wavefunction was always a wavefunction of some small system of consideration. Never had it included the observer itself or the universe. Heisenberg said:

"To begin with, it is important to remember that in natural science we are not interested in the universe as a whole, including ourselves, but we direct our attention to some part of the universe and make that the object of our studies. In atomic physics this part is usually a very small object, an atomic particle or a group of such particles, sometimes much larger - the size does not matter; but it is important that a large part of the universe, including ourselves, does not belong to the object."

It was John Von Neumann who first considered quantum mechanics for the universe as a whole, including observers (Von-Neumann; 1932). Therefore the Von Neumann or Von Neumann-Wigner interpretation, which states that it is consciousness which causes the reductions of the wavefunction is just the Copenhagen interpretation applied to the universe as a whole. Heisenberg probably never thought that it would be possible to decide where and when between measurement apparatus and the consciousness of the observer the reduction of the wavefunction actually happens (as we will see in chapter

2.5, this is now possible using the quantum eraser experiment). For Heisenberg the wavefunction was always something used as a tool to predict the outcome of small experiments never including the observers or the universe.

It was Werner Heisenberg who coined the name 'Copenhagen interpretation', so it is reasonable to use his definition. The reason why so many people use the Copenhagen interpretation without assuming a relation to consciousness is that they do not study how Bohr and Heisenberg defined a measurement device. Since it is the measurement device where the wavefunction is reduced in the Copenhagen interpretation, one tends to think that it is a matter of size, scale or number of particles when and where the wavefunction collapses. But in superconductors we have coherent superpositions of about 10^{20} electrons. It can therefore not be a question of the size. Furthermore both Bohr and Heisenberg made clear that the measurement device does not cause the reduction of the wavefunction because it is big, but because an observer is watching it. So the only thing which differentiates the Copenhagen interpretation from the Von Neumann-Wigner interpretation is its field of application. The Copenhagen interpretation is only concerned about the wavefunction of closed systems not including the observers themselves. The Von Neumann-Wigner interpretation is basically the same interpretation, but concerns more the universe as a whole, including observers. By excluding observers and the universe the Copenhagen interpretation does not have to worry about how to develop a quantum theory of the brain when the brain would have to be regarded as a system unceasingly measuring itself. Secondly, it doesn't have to worry about who measures the observer. Such considerations are often regarded to lead to solipsism (Eugene Wigner; 1982). And finally, it also doesn't have to worry about who measures distant areas of the universe, something which leads to *cosmological solipsism* as considerations in chapter chapter 4.2 will show.[3]

However, to many the Copenhagen interpretation stands for the view that a measurement already causes the collapse fully independent from conscious observers. This viewpoint is so common, that I find it necessary to comment on it here: such a scale dependent approach would not only be entirely arbitrary due to the fact that there is nothing telling us at which scale the reduction is supposed to happen, but it can also be regarded as disproven by various experiments and phenomena, superconductivity being only one of them.

Another piece of evidence against it is the *quantum eraser experiment*, which shows us that before measurement information reaches the brain, the measurement can still be undone (see chapter 2.5). <u>That means the wavefunction can not have collapsed, although it reached a macroscopic level</u>. We therefore have to assume that it doesn't collapse till the conscious state of the observer is influenced.

One would assume that scale dependent approaches should be very unpopular, because of their randomness. But one of the most popular interpretations, the *many worlds interpretation*, is also a scale dependent theory, which from an instrumentalist point of view is equal to the scale dependent version of the Copenhagen view, predicting the same amount of quantum coherence.

In my opinion there are no different interpretations of one theory. Every meaningful change in interpretation, should lead to falsifiable experiments.

Therefore the scale dependent version of the Copenhagen view is neither a theory nor an interpretation but simply an anti-philosophical ad hoc standpoint.

The historically correct consciousness dependent version of the Copenhagen interpretation on the other hand is too limited to account for a large enough range of experiments. It seems that if we say that the result of the measurement is decided in the moment the measurement device, which has to be supervised by a conscious observer, detects the particle, that is equal to saying that the wavefunction is reduced in the moment light from the measurement device reaches the eye of the observer.

In fact there is a half up to one second time to make the measurement undone, because that is the time information needs to become conscious in the brain (Benjamin Libet; 1985), and that is why quantum erasers work. If the measurement result was decided already when the measurement apparatus detected the particle, then quantum erasers could not work.

That however, doesn't mean we have to say the Copenhagen interpretation is wrong. It is simply an approximation not taking the observer itself into account. When extending its principles to the observer, we get the Von Neumann-Wigner interpretation. We get there without adding new premises, but only by reducing limitations on the range of application.

STATUS: Correct but incomplete. The Von Neumann-Wigner interpretation is the complete version.

NOTE: The scale dependent version of this interpretation is simply wrong, as superconductivity and quantum erasers prove. In addition it is not defined, because there is no distinct scale to choose.

(2) Orthodox interpretation or Von Neumann-Wigner interpretation (Category: *Passive consciousness choice theory (consciousness causes collapse)*; Proponents: John von Neumann (1932); F. London and E. Bauer (1939); Eugene Wigner (1967); Barrow and Tipler (1986); John Archibald Wheeler (1990); Anton Zeilinger)

This interpretation can be understood as a more clearly defined version of the Copenhagen interpretation. While in the Copenhagen interpretation there is a certain degree of confusion about what constitutes a measurement, this interpretation states clearly that measuring is about gaining information, and that information can only exist through consciousness. Therefore it must be the observer's consciousness which causes the collapse/change of the wavefunction. In this interpretation macroscopic objects are in definite states, because they are detected all the time through the electromagnetic waves they emit and reflect, while isolated quantum particles remain in superposition for long time. The human body has a lot of senses dispersed all over the body. Every detection event is represented by the firing of a certain neuron in the brain, yet there is no other neuron or nerve that could detect the firing of the neuron itself. Therefore the neuron remains in superposition until the activity reaches a conscious level (if subconscious is also conscious, then a subconscious level is enough).

Penrose criticized this approach for its circularity. He pointed out that "… Observation in the usual sense is something much more subtle (than OR), and it is unclear how to start developing a quantum mechanical description of the brain, if it had to be regarded as something unceasingly measuring itself (!)".

I see the difficulty, but I don't think it would be easier to explain consciousness if the reduction of the wave function would be something happening all over in the universe, without consciousness being involved. That would be very far from the original philosophy of quantum mechanics as its founders understood it (especially Werner Heisenberg). In my opinion that is to misunderstand both the essence of quantum theory and of consciousness. Also it would most probably lead to panpsychism. Although Penrose claims that a theory of everything would have to describe consciousness as well, he still hopes that consciousness is not necessary to understand the collapse of the wave function. He uses something we understand even less than consciousness, namely quantum gravity, to solve this problem. Although his approach sounds sophisticated, it does not work when we think it till the end.

Others have criticized the subjectivity in this approach. It had been argued that even if one observer did a measurement here, it still counts as unmeasured for other observers, so that we are in danger of ending up in solipsism basing this view. I don't see the difficulty. Gained knowledge does not have to be gained by each observer separately in order to count as such. We can say that a state has been measured by the community of observers. That's also how Fritz London and Edmond Bauer argued in a paper from 1939.[4] Eugene Wigner argued similarly that the first observer would collapse the wavefunction, so that it would be already settled for a second observer. He expressed this using a thought experiment now known as 'Wigner's Friend'. It is an extension of the well known 'Schrödinger's cat' thought experiment. Here the protagonist, which is Wigner, doesn't check if the cat is alive by himself, but lets his friend do it. He then asks his friend about the result. The questions then is, if the state of the cat (death or alive) was decided before or after Wigner asked the question. Wigner considered superpositions of different states of mind as impossible and therefore assumed that the first observer collapses the state.

Of course the example with the cat is serving only the purpose to make the whole thing more dramatic. Cats are sentient beings which are surely conscious and thus they should be well able to 'measure' themselves. Taking account of the cat's consciousness, the 'Schrödinger's cat' thought experiment is already involving two observers and thus is equivalent to the 'Wigner's friend' case. However, for the sake of this thought experiment we may imagine the cat being a zombie without consciousness (it is a quite wide spread misconception among some intellectuals that animals have no consciousness, because they do not have a sophisticated language to express their thoughts).

Unfortunately Wigner gave up his consciousness based interpretation of quantum theory. In a speech he gave in German (not his native language) in 1982 he mentioned that other observers would have to lie on his past light cone, because light needs time to travel from here to there, so that he is measuring only their state as it was in the past.[5] The present state of those observers around him would then be an undecided superposition. The flaw in this argument is that it doesn't account for the <u>entanglement between the observers</u>, which allows them all to be part of a <u>collective wavefunction</u>. Entanglement leads to the connection of all individual wavefunctions. The individual parts of this joint wavefunction grow (**U**) and shrink (**R**) as time goes by, but the *joint wavefunction* itself doesn't really grow much in size, because the entropy density of the universe remains always the same (see chapter 4.1 & 4.7).[6] However, there is still growth associated with the growth of the cosmological event horizon, but this growth is not of any local relevance.

Premise: The result of a measurement is not decided before signals arrive in the brain AND become conscious (which happens with half a second delay).

Conclusion-1: This allows larger superpositions then the standard Copenhagen approach, hence quantum coherence in the brain is possible.

Conclusion-2: We need to assume the existence of collective consciousness (community of observers) here in order to avoid solipsism.

Conclusion-3: The result of the measurement is decided in the brain of the observer, which makes this theory potentially interactionistic.

Conclusion-4: Information can not be destroyed by macroscopic interference as it could occur at least in principle in decoherence theory.

Conclusion-5: It potentially fulfills the requirements of the Penrose-Lucas argument (non-algorithmic consciousness) when allowing consciousness to influence the choice – which would be natural, if the choice is made in the brain.[7]

Evidence-1: The quantum eraser experiment proves that measurement information can be erased before arriving in the consciousness of an observer. Thereby it disproves the scale dependent version of the Copenhagen interpretation, and gives strong support for the Von Neumann-Wigner interpretation.

Evidence-2: The quantum coherence found in microtubule is an example for macroscopic wide range superpositions, which contradicts the standard Copenhagen approach and gives support for the Von Neumann-Wigner interpretation or for OR (Anirban Bandyopadhyay; 2014).

STATUS: Basically correct and potentially including interactionalism, but never fully developed in its implications, mainly because it did not implement ideas from relativity or neuroscience.

(3) Interactionalism (Category: *Active consciousness choice theory*; Proponents: John Eccles, Dean Radin; John Lucas; Henry Stapp; Note: Variation of the Von-Neumann interpretation)

The Von-Neumann interpretation can be further split up into two versions, one where the observer causes the change of the wave function without any influence on the outcome, and one in which he in addition has influence on the outcome (choice between realities). We can call this interpretation 'interactionalism', and put it into the category '*active consciousness choice theory*'. Only interpretations carrying the attribute '*active*' can possibly take Gödel's incompleteness theorem and its implications for consciousness into account.

Some people have based interactionalism on theories which do not make consciousness responsible for the choice (like OR) or even on *no-choice theories* (many worlds immortality theory). It is mainly due to these theories, that I introduced the categories *active* and *passive*.

STATUS: The most correct preexisting view, taking into account proven properties of consciousness. However, in most situations it is enough to refer to it as the Von Neumann-Wigner interpretation.

(4) Decoherence/Many-minds interpretation (Category: *Passive consciousness no-choice theory*; Proponents: Dieter H. Zeh; 1970)

All possibilities come into existence equally. There is no reduction at all and thus nothing left to do for consciousness. It is a deterministic theory, which explains the randomness we see as due to our restricted perspective, because we are part of the superposition and therefore unable to see the other states/worlds. In this theory nothing ever really happens. The world is a static whole, which never changes, very much like the space-time in Einstein's relativity theory.

A prediction of this theory is macroscopic interference. However, if it really exists, it would destroy information and could eventually even delete memories. Many states could thus merge back into one state. Nonconservation of information is a direct result of refusing the role of the wavefunction as representing the maximal accessible knowledge of the observer. However, nowadays so called *quantum information* is valued more than real information. During Heisenberg's time, a reduction of the wavefunction was a gain in information, because the wavefunction was representing the knowledge of the observer. The more measurements the observer makes, the more certainty he gains about a system, and as a result of that the wavefunction shrinks. Now with decoherence theory gaining popularity, the wavefunction has lost its original meaning. In this new twisted view a measurement is regarded as a loss in information, because it entangles oneself with a certain measurement result and thereby hides the supposed reality of the Omnium – a world where everything happens at once, without a flow of time.

Eugene Wigner had good reason to refuse decoherence theory and many worlds theory back in the 60's. He refused the notion of a wavefunction of the universe which is central to many worlds theory as senseless.[8] This viewpoint is equal to mine: A state which is a mix of all possible states is not any different from nothingness.

Decoherence theory depends crucially on having certain restrictions on the wavefunction of the universe; properties which can not be in superposition, such as the topology of the universe or its expansion speed; things having to do with the fine-tuning of our universe. However, once we discover that supposedly free parameters as the initial expansion speed of the universe depend on what happens inside the universe, then the so called wavefunction of the universe dissolves in front of our eyes. Without any restricts to its form and without any reduction to ever have taken place at any time during the 'history' of the universe, there would be no probabilities anymore. Every event would have the same likelihood. Nothing would really happen. The Omnium would remain to be an Omnium. It is not possible to get a reality out of such an everything-and-nothing.

STATUS: It violates information conservation; it does not account for the flow of time; it depends on the notion of an undefined wavefunction of the universe; it negates the existence of information; it doesn't account for any requirements the nature of consciousness imposes upon reality; it doesn't allow anything to happen more likely than anything else.

(5) Many-worlds interpretation (Category: *Relativistic scale no-choice theory*; Proponents: Hugh Everett; 1957, Stephan Hawking; Status: Scale (macroscopic) is not an absolute measure. This interpretation can therefore be only an ad hoc explanation.)

The difference of this interpretation to decoherence theory is not so obvious. Same as in decoherence theory we have endless many-worlds, all fulfilling different alternative realities. The main difference is that here every world has its own space-time. Whenever a new alternative appears, the world splits up again. It becomes like a huge tree with many branches. The main problem with this interpretation is that it just transformed the measurement problem instead of solving it. It continues to exist as the problem of when the world has to split up exactly. Since the observer is not given any particular role in this theory, the splitting can only arbitrarily depend on the reached scale, leaving the whole theory more or less undefined.

Although I regard the many-worlds interpretation as highly misleading, I took some inspiration from it for my own theory. On first sight it seems plausible to have the dead cat and the living belonging to different alternative realities. Yet the particles of the cat would still be in fuzzy superpositions in each of the states. If consciousness is not to play any important role here, then there seem to be no reason to split up macroscopic states, but not microscopic states.[9] Of course splitting them up would prevent interference on a microscopic level, and that would contradict our observations.

Similar to the Copenhagen interpretation the many worlds interpretation is a fuzzy defined version of another interpretation: the 'many-minds interpretation'.

Another problem has to do with the lack of macroscopic interference. If quantum states really split up into states on different space-times when reaching a macroscopic level (what level that may ever be), macroscopic interference would be impossible. The other parallel states could never influence us, so from an experimentalist point of view, there seems to be no difference to a collapse theory. And indeed I think we can regard the splitting into different disconnected space-times as 'a collapse of the wave function'. Even in the Copenhagen interpretation we could imagine that other alternative choices might be realized in another separate reality, yet no one of those with the 'Copenhagen spirit' would be interested in such 'realities'.

Although no interference is possible, some scientists, like for example Stephan Hawking, still claim that we could find ourselves in any of those parallel universes, so that some kind of 'sum over histories' is still important here.

Now it appears that we are confronted with two sorts of sums over histories. Those lying on the same space-time, being basically 'sums over particle histories' based on real interference, and those lying on different space-times which are disconnected, being 'sums over universe histories' based on some kind of Copernican principle for universes (!).

I have my difficulties to buy in statistical arguments using the Copernican principle. Same as for Richard Gott's *future predictions based on the Copernican principle*, it appears to me that I would not be myself if I were placed in another world.[10] I can therefore not give any probability for being placed in this or that world. In contrast to that giving probabilities for a certain galactic neighbourhood is much more acceptable. However, even if we allow such arguments, their predictive power goes down to zero when we consider more and more different kinds of universes, as is the present tendency of many 'meta-cosmologists'.

Because the split into different space-times is here depending on the scale, the whole approach is subject to the same criticism as the scale dependent version of the Copenhagen interpretation.

STATUS: From an instrumentalist point of view identical to the scale dependent version of the Copenhagen interpretation, which is simply disproven by the existence of superconductivity and quantum erasers. In addition it is not defined, because there is no distinct scale to choose.

NOTE: Those other worlds are also ruled out by a principle of Leibniz which states that only things that can mutually influence each other can be co-existing (a result of Leibnizian relationism). It was part of my own philosophy long before I heard that Leibniz had said that. I regard it as a natural extension of Berkley's 'to be is to be perceived' principle.

(6) **Mny-worlds immortality theory** (Category: *Active conscious no choice theory*; Proponents: Hans Moravec (1987); Bruno Marchal (1988); Max Tegmark (1998))

It sounds paradoxical to have a theory where consciousness is actively doing something, while nothing really happens, but yet this is taken as a serious possibility in this theory. In this variety of the many worlds theory, the consciousness of the observer is believed to be choosing its path through different worlds. It is thought to do this independently from all the other observers. So it would be possible for consciousness to always choose a path through the different worlds where it never dies, no matter how unlikely that would be otherwise. To me it seems that such a switching of space-times requires that the other space-times are not already occupied by other consciousnesses (or souls). We would end up with a rather ridiculous model, where all consciousnesses start together in the same space-time, for then choosing all different paths, which separate them from each other. Arrived in those new space-times the conscious observer would then be surrounded by 'unconscious zombies'. Of course this view is not reaching the 'Gödelian requirements', because unconscious zombies could not simulate conscious beings (taking the non-algorithmic nature of consciousness into account).

So this theory assumes that we have free will in respect to our fate, but we do not have free will in our own brain, since human beings are here regarded to be basically like robots (completely algorithmic). Or maybe the proponents of this theory are only suggesting that consciousness is passively reading along the different 'histories', very much like following a rope, so that it would naturally always follow a path in which it survives. However, if this is the only freedom we have in this theory, then there is no reason to ever assume that consciousness is choosing a 'path' through many worlds. It would then only be required to be located in a space-time where it is immortal. So would all other consciousnesses, all separated from each other and surrounded by emotionless zombies who, for whatever reason, pretend to be sentient beings.

Another problem with this theory is that it would predict no free will in our brains, but unlimited psychic powers. If we are free to choose our path through parallel space-time, we would be able to do telekinesis on epic dimensions and free from all limitations. Some people with a less deep understanding of the measurement problem, indeed believe that we have potentially unlimited powers both to determine our fate and even to practice extreme forms of psychokinesis, simply because of the existence of those other worlds.

One advantage of this theory in comparison with the standard many worlds theory is that here the other worlds do potentially influence 'reality', because they represent different choices, even without the possibility of interference. These choices are however unphysical, because they are not part of any obtainable wave function, but just of a hypothetic wave function of the universe (Omnium wave function).

Denying the collapse of the wave function does only work if we also deny free will. Free will without a collapse of the wave function would give any observer god-like omnipotence. And this would not be an omnipotence that can be controlled: it would simply wipe out any difference between a dream and reality. Not only because the missing state reduction, but also because the observer is supposed to be choosing his path independent from all other observers. That is exactly what a dream is like.

Same as the standard many worlds theory, this interpretation can be regard as <u>disproven by the quantum eraser experiment</u> (path information can be deleted after the screen is hit).

STATUS: This approach would lead to omnipotence of every conscious observer and thus contradicts reality. If we do not require free will, but only the immortality of every consciousness, we basically get something more or less equal to the standard many-worlds theory, which is proven wrong already for other reasons.

(7) Similar-worlds interpretation (Category: *Relativistic consciousness choice theory*; Proponents: The author (Sky Darmos (2003))

Consciousness is located in all presently indistinguishable worlds at once. That is why particles seem to be located at many places at once. The wave nature of the probability function can be explained by combining this interpretation with space particle dualism theory that was developed roughly a year later.

As described in the last chapters this approach is not a mere interpretation, because it works only by changing the structure of quantum theory. Its main influences are the Von Neumann-Wigner interpretation, the many-worlds approach (to a very limited extend), and the notion of a phase space. As I mentioned before, it is somehow plausible to separate different quantum states into different worlds, especially if one wants to try to include general relativity, but if we continue this to the microcosm we lose interference. However that is not the case if we refrain from insisting that every world would have a separate consciousness. It is not really plausible to talk about different consciousnesses if the state of the brain is still the same. So even if we have many classical worlds, they would appear quantum-like to consciousness, because for consciousness only the state of the brain matters. The split of the world into macroscopic and microscopic is also due to consciousness; and even the notion of entropy does only make sense with consciousness, since entropy can be understood as the logarithm of the number of states which do not change the macroscopic appearance of an object. In this interpretation we even have a more exact interpretation of entropy. We can define the entropy of the universe as the number of constituting similar worlds.

STATUS: This is a whole new interpretation of the wavefunction, which produces superposition by having a conscious cloud in phase space. It links quantum mechanics and relativity closely together and is therefore not just another interpretation but part of a new theory. Yet there is still some justification to call it an interpretation, because the new theory (space particle dualism) it is part of, does not introduce any changes in the dynamics of quantum physics. The changes appear mainly on the relativistic side. So we can speak of a relativistic extension of the Copenhagen and the Von Neumann-Wigner interpretation. Another feature is more clarity on how to use the principles of quantum mechanics in the human brain as well as in cosmology.

8) **Objective Reduction** (Category: *Active scale choice theory*; Proponents: Roger Penrose (1989); Stuard Hameroff)

A state is chosen automatically after the space-time geometry of different states differs more than a Planck-length. This leads to a picture were we have different space-times for each moment, which is very similar to the picture of an everchanging equivalence group in similar worlds theory. However, the criteria for this objective reduction (short OR) would have to be applied only to the present now-slice of space-time. Any small differences between two space-times in one now-slice lead to huge differences in the past and future of this slice. If we accept the whole space-time to be real, it seems to make no sense to apply this criterion only to the present slice of space-time. It seems also to depend on the angle of the now-slice. I don't see how we are supposed to choose a certain now-slice if we don't involve conscious observers. Maybe the whole idea is based on a projective approach to space-time as suggested by twistor theory, but Penrose doesn't seem to have specify that.

In this theory consciousness could only exist, if superpositions are kept up in the brain by extreme environmental isolation. Such high isolation could exists only in microtubule. That is why Penrose had to switch from superposition of neurons firing and not firing ([1]) to superpositions in microtubule only ([2]). There is conclusive evidence for quantum coherence in microtubule, found by Anirban Bandyopadhyay (2014). However, one can get much more quantum coherence basing the Von Neumann-Wigner interpretation.

The main motivation for this approach is to somehow link quantum theory and relativity together. Penrose seldom mentions that, but it was initially inspired by some properties of his twistor theory. The idea has some justification and seems to be the best way to gain back some of the realism taken away by quantum theory. Both the Von Neumann-Wigner interpretation and the many-worlds approach lead to pictures of the world which appear highly unreal and irritating.

Yet it ignores all the beautiful and meaningful philosophical background of quantum theory. For some it might be just a random description of what we observe, but for me quantum theory has much more philosophical justification as say relativity theory.

Superpositions cannot be seen. We know about them through the interference between different superpositioned states. Big massive objects, like the objects we can see easily with our eyes, have an infinitesimally small wave and therefore no measurable interference occurs. Without Consciousness the question about if or when a state reduction occurs would not be of great importance. Our observations would be still the same, even without any state reduction (choice of realities).

Penrose's proposal for an objective reduction eliminates what he is trying to explain: consciousness. The Von Neumann-Wigner interpretation gives consciousness an important place in the physical world and one is not going to explain consciousness by eliminating this role. The whole philosophy of quantum theory is based on the idea that observation creates reality. Without the role of consciousness, quantum theory becomes a entirely arbitrary theory without any core statement. Thus eliminating consciousness is like eliminating the essence of quantum mechanics – a philosophically very significant essence!

It is unclear yet, if the quantum eraser experiment already disproves OR or not. That depends on how much mass is moved during the measurement which is undone.

STATUS: Leads to panpsychism. It contradicts the fact that even in the brain consciousness exists only in certain areas. Its whole attraction comes from its potential explanation of the brain frequency (40 Hertz). It can not account for the fact that all conscious experiences are delayed in time for a half second.[11]

Furthermore the collapse criterion here is a hypothetical prediction of a yet not existing theory of quantum gravity. OR can therefore not be regard an interpretation. As an interpretation it would be a *scale choice theory* and would be subject to the criticism brought forward against the scale dependent version of the Copenhagen interpretation. It is also not a theory, because it consists of nothing more than the hypothesis itself. It is however unsure if superconductivity or the quantum eraser experiment disprove this hypothesis yet already or not.

Notes and references:

1. Heisenberg always used the word '*change*' instead of '*collapse*'. 'Collapse' can here be understood as a derogatory term. John Von-Neumann was the first to strictly differ between the unitary evolution **U**, and the collapse of the wave function **R**. This differentiation loses is meaning when we talk about a 'change of the wave function' instead. So, instead of talking about 'collapse theories', we can talk about 'choice theories', 'change theories' or 'non-static/dynamic theories'. And instead of 'consciousness causes collapse', we can say, 'consciousness causes change/choice'. Denying that the wave function changes is denying the existence of time – a very popular way of thinking in our times (!).

2. *Physics and Philosophy* (Werner Heisenberg; 1958); Chapters 2 (History), 3 (Copenhagen interpretation) and 5 (HPS).

3. The standard Von Neumann-Wigner interpretation does also lead to *Cosmological solipsism*. The reason why this has never been pointed out by anyone has to do with the fact, that there is a lot of confusion on how to analyse a dynamic theory (having a dynamic time evolution) within the framework of special relativity using light cones in space-time. Von Neumann never made such considerations. Therefore we can say that the Von Neumann-Wigner interpretation is not *relativistic*.

4. F. London and E. Bauer, "La théorie de l'observation en mécanique quantique" (1939), English translation in Quantum Theory and Measurement, edited by J.A. Wheeler and W.H. Zurek, Princeton University, Princeton, 1983, pp. 217–259.

5. Eugene Wigner - The Quantum Mechanical Meaning of the Concept of Reality (1982, German Presentation). Available under:
[https://www.youtube.com/watch?v=T3gsGjw2WrY].

6. I find the common argument that this would violate the probabilities in quantum mechanics as invalid. Probabilities are random. Each quantum property for itself is allowed to take any value within a certain range. To allow consciousness to replace this randomness by meaningful choice is not a violation of logic. In opposite, it fulfills Leibnitz's principle of the sufficient reason.

7. Michael Esfeld, (1999), Essay Review: Wigner's View of Physical Reality, published in Studies in History and Philosophy of Modern Physics, 30B, pp. 145–154, Elsevier Science Ltd.

8. I wanted to avoid the term 'universal wavefunction', because it is a term in common usage for proponents of decoherence theory and many worlds theory.

9. One may well, also imagine a many-worlds theory without a split into different space-times. However, it seems impossible to then account for the fact that we do not see the other parts of the wavefunction, without assuming consciousness to be responsible for the split. From all the presented no-choice theories, the many-minds

interpretation appears to be the only one which has at least some degree of inner consistency.

10. Mentioned in the last chapter of: J. Richard Gott, Time Travel in Einstein's Universe: The Physical Possibilities of Travel Through Time, 2002, Houghton Mifflin Books, ISBN 0-618-25735-7.

11. For a detailed analysis see "Mind Time - The Temporal Factor in Consciousness", by Benjamin Libet (2004).

2.4 Using quantum erasers to test animal consciousness [01.10.2015 (test if measurement by an animal can be undone); 27.04.2017 (delaying the erasing to see where the collapse happens); 29.06.2017 (testing robot consciousness)*; 06.08.2017 (Faraday cage); 27.08.2017 (fiber loop entry and exit); 24.11.2017 (four mirrors; mirror sphere; mirror cube); 29.01.2018 (fiber loop is better); 11.08.2018 (using a Sodium-2 gas)]

*Asked to do by Piotr Boltuc (Poland) and Ludovic Krundel (France).

Also published in: APA Newsletter on Computers and Philosophy (Volume 17, number 2) (spring 2018); pages: 22 – 28 (without the Sodium-2 gas-update).

John A. Wheeler was a strong supporter of 'consciousness causes collapse' and one of the first to apply this principle to the universe as a whole, saying:

"We are not only participators in creating the here and near, but also the far away and long ago."

How did he come to this conclusion? In the 70's and 80's he suggested a number of experiments aiming to test if particles decide to behave like waves or particles, right when they are emitted or sometime later. For example, one could change the experimental constellation in respect to measuring the path information (polarizations at the slits) or the impulse (interference pattern) after the particle has already been emitted. When the experiments were done many years later, it turned out that what particles do before they are measured isn't decided until after they are measured. This led to Wheeler concluding:

"Quantum phenomena are neither waves nor particles but are intrinsically undefined until the moment they are measured. In a sense, the British philosopher Bishop Berkeley was right when he asserted two centuries ago 'to be is to be perceived'."

But many others preferred to rather believe that information partially travels to the past, than to believe that reality is entirely created by the mind. Therefore Wheeler brought the experiment to an extreme by suggesting to conduct it on light emitted from remote galaxies. The experiments showed Wheeler to be right again. The universe indeed materializes in a retrospective fashion.[1]

Later in the 90's new experiments were suggested to test other temporal aspects of quantum mechanics. The so called *quantum eraser experiment* was also about changing one's mind on whether to measure position (particle) or impulse (wave), but here the decision was <u>not delayed but undone</u> by erasing the path information.

The erasing is usually not done by deleting data in a measurement apparatus, but simply by undoing the polarization of the entangled partner of a given photon. Polarization doesn't require absorbing a particle. It is therefore no measurement, and the result wouldn't really be introducing much more than Wheeler's delayed choice experiment already did, but there is a special case, namely undoing the polarization of the entangled partner after the examined photon arrived at the screen already. That is indeed possible, which means the screen itself, although being macroscopic, can be in superposition, at least for short periods of time. <u>This proves that the screen didn't make the wavefunction collapse.</u> If we can already prove this, then there must be a way of finding out where exactly the wavefunction collapses.

Using quantum erasers to test consciousness

Polarizers can be used to mark through which of two given slits A or B a photon went, while its entangled partner is sent to another detector. The interference pattern disappears in this situation, but it can be restored if the entangled partner passes another polarizer C, which can undo the marking, resulting in the restoring of the interference pattern. This deleting can be done after the photon arrived at the detector screen, but not long time after. Arguably, it is the signal's arrival at the consciousness of the observer that sets the time limit for the deleting.

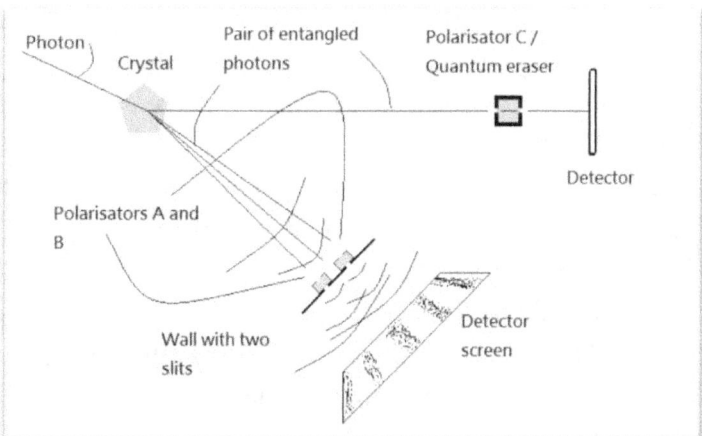

Fig. 20. Interference pattern reappears when the quantum eraser is used. That happens even if the quantum eraser is placed in a larger distance to the crystal then the screen.

If decoherence theory (or Bohr's scale dependent version of the Copenhagen interpretation) was right, then the screen should have measured the photon and thereby should have destroyed any chance for interference, simply because it is 'macroscopic' (no quantum behavior). Yet that is hard to say, because if one doesn't believe in the collapse of the wavefunction (decoherence theory is a no-collapse theory), then interference, and therefore information loss (erasing), may occur at any moment after the measurement (see chapter 2.3). [2, 3]

In the Von Neumann-Wigner interpretation it is said that a measurement has to reach a conscious observer in order for the wavefunction to collapse. Yet, if the wavefunction collapsed right in the eye of the observer, there wouldn't be much time for erasing the measurement. Light signals from the measurement arrive almost instantaneously at the eye of the observer (at the speed of light). Thus we can exclude the possibility that the eye ball of the observer causes the collapse of the wavefunction. [4, 5]

In chapter 2.1 I suggested that one could try to delay the erasing more and more in order to figure out in which moment in time and where in the brain the wavefunction collapses. It may collapse at a subconscious level already (single projection to the *cerebral cortex* taking less than a half second), or at a conscious level (double projection to the cerebral cortex taking a half second).
It is sometimes suggested that if it is the subconscious which is responsible for the collapse of the wavefunction, then that could explain why we seem to have almost no influence on into which state it collapses. [6]

If erasing the measurement is possible till half a second after the measurement, then consciousness causes the collapse. If this time is slightly shorter, let's say a 1/3 second, then subconsciousness causes the collapse. We can know this because the temporal aspects of consciousness have been studied quite excessively by the neuroscientist Benjamin Libet. [7]

If we now replace the human by a robot, we would have to place all humans very far away, in order to avoid having them collapsing the wavefunction. Yet, as soon as the measurement reaches the macrocosm, changes in all fields reach the human with light speed. And for the wavefunction to collapse, no real knowledge of quantum states needs to be present in the consciousness of an observer. All that is needed is different quantum states to lead to distinguishable states of the mind.

Another technicality is that although the wavefunctions of macroscopic objects around us collapse every 40th of a second (the frequency of our brain in the perception realm), the single photons and subsequent brain signals remain in superposition for almost half a second (see chapter 2.1).
When looking at mind over matter interactions, which are mostly about influencing macroscopic systems, the 40th second is crucial (see chapter 5.3), whereas for quantum erasers, which are about single photons, it is the half second which is crucial.
After testing humans one can go on and test animals with different brain structure. In some animals the subconscious/conscious level could be reached earlier or later, and that should affect the time limit for the quantum eraser.

Of course, when there is a way to check experimentally if something has consciousness, one can do that for all kinds of things, even robots, cameras, stones and so forth. It is my belief that something totally algorithmic can't be conscious, simply because such a consciousness wouldn't affect the systems behaviour. Only a system which is quantum random can have a consciousness that actually affects the system.

Obviously opinions deviate strongly here, but the good thing is that we don't need to solely rely on beliefs or formal arguments anymore; we can actually go on and experimentally test it.

What we can do is this: assume that a robot would become aware of things very fast, much faster than the half second it takes for humans. One can then go on and test that by putting the robot in front of the experimental device together with a human. If the robot makes quantum erasing impossible already before the signals reach human consciousness, then the robot is conscious.

Of course, this doesn't account for the possibility that robot consciousness, if existent, is slower than human consciousness (humans experience everything a half second delayed in time!).

Some people think that replacing the human observer by a camera and seeing that the wavefunction still collapses already proves Von Neumann wrong (Paris Weir; personal correspondence; 2017). They miss the point that the quantum state reached the macrocosm already when entering the camera. From there it continues spreading out very fast. According to the Von Neumann view the first time the wavefunction collapsed was after the emergence of life, yet that doesn't have any obvious impact on the world. In Everett's many worlds interpretation the wavefunction never collapses, and again there are no obvious implications. That means only if we try to rapidly erase the measurement, can we hope to learn something about where the wavefunction collapses.

In decoherence theory, decoherence replaces the wavefunction collapse. In this theory objects can be treated classically as soon as interference is lost. Calculating when interference is lost is relatively easy: for any macroscopic object it is 'lost' almost instantaneously. Yet this doesn't tell us when a measurement becomes irreversible. The issue of irreversibility is independent from decoherence (losing of interference), and looking at the ontology of decoherence theory, one would have to assume that erasing a measurement should always be possible. Some took this literally, which led to the creation of rather bizarre theories, such as the 'Mandela-effect' were the past is not regarded unchangeable anymore and the universe becomes 'forgetful'.

According to Max Tegmark decoherence theory may even lead to a bizarre form of solipsism where consciousness 'reads' the many worlds always in a sequential order which leads to its succession – its survival. That is expressed in his thought experiment 'quantum suicide'. Rather surprisingly Tegmark doesn't use this to make a case against decoherence theory, but rather wants to show how 'thrilling' it is.

Schrödinger's cat in real

For entities that have a consciousness which is faster than human consciousness, one can easily test that looking at how much the time window for the quantum eraser is shortened. However, accounting for entities with a slower consciousness we have to try to isolate the whole system from humans and all other potentially conscious animals. This could be done by moving the whole experiment into a *Faraday cage* and/or placing it deep beneath the surface of earth and far away from human observers. Nothing that happens inside this Faraday cage should be able to influence anything on the outside.

If the experiment is really perfectly isolated, then the erasing of the which-path information could be delayed further and further. All one would have to do is to let the entangled partner photon continue its travel, for example by making it travel circular inside *optical fibers*. Yet, if the delayed erasing is to be successful, the entangled partner has to finally hit the third polarizer before the Faraday cage is opened.

Considering how far photons travel in a half second (about 150 thousand km), some way to store them without measuring them must be found. Photons travel slower inside optical fiber, reducing the distance travelled in a half second to only 104,927 km, but that is still by far too long for a distance to be travelled in a laboratory. One way to slow them down further could be to let them enter some sort of glass fiber loop. Trapping photons inside mirror spheres or mirror cubes, similar to the 'light clocks' in Einstein's thought experiments, is probably not feasible. That is mainly because in such mirror cages photons are often reflected frontal (in a 90° angle), and that is when the likelihood of a photon to be absorbed by the mirror is highest (the worst choice here being a mirror sphere[8]). Ordinary mirrors reflect only about half of the photons that hit them. Even the best laser mirrors, so called supermirrors[9], made exclusively for certain frequencies reflect only 99.9999% of the light and with many reflections (inside an *optical cavity* made of such supermirrors) a single photon would certainly be lost in a tiny fraction of a second. That doesn't happen in a glass fiber wire because there reflection angles are always very flat. [10]

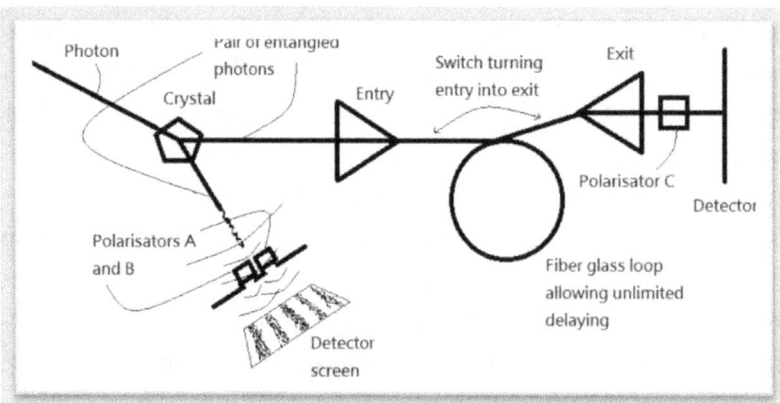

Fig. 21. Using a fiber glass loop with an entry that can turn into an exit the erasing of the which-path information can be delayed as much as wished by the experimenter.

It might prove itself to be very difficult to get the photons in and out of the loop, but even more difficult it seems to get them entering the glass fiber wire in the first place, after they are created together with their entangled partners at the crystal. An option could be to make the glass fiber wire wider at the one end which is used as the entry. One could also guide the photons into the wire by using a focusing lense or a series of guiding mirrors. The first glass fiber wire would lead the photons to the fiber loop. At the place of entry into the loop, the first fiber wire has to be almost parallel to the loop. If the photons always travel into the same direction, they won't ever leave the loop in this case. After sufficient delaying time is gained, the photons have to be taken out and be directed to the third polarizer. That could be achieved if the direction of the entrance fiber wire could be switched, so that the entrance becomes an exit. This exit could then be made pointing into the direction of the third polarizer.

In some sense, this experiment would be the first real "Schrödinger's cat" experiment, because just like in Erwin Schrödinger's thought experiment, an animal is put inside a box, here a Faraday cage, and it is theorized about if the animal is in superposition (indicating unconsciousness) or in a certain state (indicating consciousness). But here we have an experimental constellation, which allows us to actually check if the animal was in a superposition or not. As for "Schrödinger's cat" in his original thought experiment, one could either just find the cat alive or dead after opening the box. There wasn't any way to tell if the cat had been dead or alive from the beginning, or if it was in a superposition of both states (alive and dead).

Alternative set-up

Using such fiber-glass loops to store light is very difficult, and we would do good to look for a better alternative.
The speed of light is not always the same. In air light is slightly slower than in a vacuum and in water it is even a bit slower.
Is there a medium that can slow down photons so drastically that a significant quantum erasing delay can be created?
Apparently there is: using a technique called electromagnetically induced transparency (EIT) one can slow down photons to incredibly low speeds.
The first person to use this technique was Stephen Harris at Stanford University in 1991. In 1999 Lene Hau and a team from Harvard University used a Bose-Einstein condensate of *ultra cold sodium atoms* to slowed down light to a factor of 30 million. In 2001 she was even able to stop a beam completely.[11]

If we need a half second delay, then going with a 30 millionth the speed of light, we would need an ultra cold sodium chamber only 5 meters in diameter.

(Unconcious) robot in a Faraday cage

For cats we can be pretty sure that they are conscious, so we can't really make them enter a superposition of being alive and dead at the same time. For robots that's different: we can be pretty sure that they are unconscious. So, if we want to dramatize the experiment, we could have the robot destroying itself when it 'sees' an interference pattern.[12] The destruction of the robot (as well as the interference pattern on the screen) could then be erased/undone (!) by the third polarizer. Of course all this has to happen before the Faraday cage is opened. This basically means that the whole past of what happened inside the Faraday cage is decided when it is opened.

However, this is much different from Schrödinger's cat, and maybe much more dramatic. Instead of being in a superposition of destroyed and not destroyed, the robot would 'experience' a state of having been definitely destroyed and then a state of never having been destroyed. Of course that can't be 'experienced', and it is just our way of talking about things as if they were real without us looking at them ('looking' here stands for any form of influence to the observer).

A less paradoxical way of talking about this robot, is to say that if he destroys himself in the past depends on whether the interference pattern is restored in the future.

Conclusion

Why hasn't this experiment been proposed before? One reason is that delaying the erasing for more than just tiny fractions of a second is rather difficult (photons are just too fast). The other reason is that very few physicists are proponents of the Von Neumann-Wigner interpretation and even less are familiar enough with concepts in neurobiology in order to link them to things in physics.

And finally, there is the general misconception, that choosing different interpretations doesn't influence predictions on experimental results.

We can categorize interpretations of quantum mechanics into scale dependent and consciousness dependent approaches (see chapter 2.3). Most interpretations exist in both variations. We therefore shouldn't really care if there is a wavefunction collapse or a splitting of worlds, because operationally they are the same. All what operationally matters is where the cut is to be placed: is it scale dependent, or consciousness dependent?

It is my opinion that the present results of quantum eraser experiments already disprove scale dependent approaches. Some, such as Penrose's gravity induced wavefunction collapse theory, might be fine with a detector screen being in superposition for short periods of time. Further delaying the erasing will however make it increasingly difficult for any scale dependent theory to survive.

In my opinion the interpretation and ontology of a theory is just as important as its mathematical structure. Without a proper interpretation it is not possible to correctly apply the mathematical formalism in all situations. That is just as true for relativity theory. Only by correctly interpreting both theories unification can be conceived.

I hold that *pure interpretations* don't exist and that philosophy, correctly done, always leads to hard science.

Note: This is not only an experiment, but can also be turned into a device/product for testing consciousness. The applications would be broad. It could for example measure when consciousness is delayed because of drug use.

One who would be perfect for conducting the experiment is the Austrian quantum experimentalist Anton Zeilinger. That is because he is most skilled and renowned in working with interferometers. He could also be good for giving advices on how to conduct the experiment. However, considering the stigma that lies on consciousness, it is more likely that Dean Radin is going to do it instead.

Notes and references:
1. Retrospective here doesn't mean that something travels into the past, but that the past is created at the moment of measurement.
2. Though they would claim that information is not something that must be accessible to individuals, but it can be something like the wavefunction of the universe, which is thought of to be out there without being accessible to any particular observer. In this line of thinking no information is really lost.
3. As already mentioned in chapter 2.3 decoherence theory can lead to issues with information conservation. If interference is always allowed, then it will happen even with vanishing wavelengths. Within a universe that never experienced a collapse of the wavefunction, quantum probabilities might get lost totally. If the universe is in all possible states right now, then those states should, arguably, all have the same likelihood. In such a world there would be no reason for an observer to experience a certain succession of states more likely than another.
4. Von Neumann's original paper was discussing the question at which place in the brain of the observer the wavefunction might be collapsing.
5. Unless the extra distance travelled by photon B is not much longer than the distance of the observer to the measurement device for photon A.
6. Lothar Arendes (1992) [Gibt die Physik Wissen über die Natur? - Das Realismusproblem in der Quantenmechanik].
7. Benjamin Libet, *Mind time: The temporal factor in consciousness, Perspectives in Cognitive Neuroscience*. Harvard University Press, 2004. ISBN 0-674-01320-4.
8. Video on the behavior of light in a spherical mirror: [https://www.youtube.com/watch?v=zRP82omMX0g].
9. Entry on supermirrors in an encyclopedia of optics: [https://www.rp-photonics.com/supermirrors.html].
10. A helpful discussion on trapping photons between mirrors can be found here: [https://www.physicsforums.com/threads/light-in-a-mirrored-sphere.90267/].
11. Goodsell, Anne; Ristroph, Trygve; Golovchenko, J. A; Hau, Lene Vestergaard (2010). "Physics – Ionizing atoms with a nanotube". *Physical Review Letters*. 104 (13). arXiv:1004.2644 Freely accessible. Bibcode:2010PhRvL.104m3002G. doi:10.1103/PhysRevLett.104.133002.
12. Of course an interference pattern involves many particles. If only one particle pair

is used, then there would be no real pattern, but still particle A wouldn't arrive at the two possible positions corresponding to straight paths through the slits. That indicates that it interfered with itself. It doesn't really make a difference for the experiment if it is just one pair or many in a row. The erasing works in both cases.

2.5 Nonlocal perturbation as a means of probing the wavefunction collapse
[20.03.2018 (Radin's new experiment is ordinary PK); 21.03.2018 (retrocausality experiments show superpositions can be preserved for very long times)]

Also published in: APA Newsletter on Computers and Philosophy (Volume 17, number 2) (spring 2018); pages: 22 – 28 (without the Sodium-2 gas-update).

Dean Radin:
"Experimental test of the Von Neumann-Wigner interpretation"

In 2016 at the 'The Science of Consciousness Conference' (TSC) in Tucson Dean Radin held a lecture which was titled 'Experimental test of the Von Neumann-Wigner interpretation'[1]. Although that was not the name of the associated paper[2], the experiments he had conducted were basically presented as evidence for consciousness collapsing wavefunctions. That has been indeed shown by Radin, yet the way the experiment was described can be somewhat misleading as to what was really happening. It was a double slit experiment involving participants 'observing' the double slits and thereby altering the *interferometric visibility* of the interference pattern. These human observers were not literally watching the double slits with their eyes. They were not staring at the slits to look through which slit the photons passed. If they did so the photons would go into their eyes and thus we wouldn't have a chance to analyse how the interference pattern was altered. What they did instead is they focused on the slits with their mind. The way Radin puts it, the observers tried to look at the double slits with their 'inner eye', in an ESP sort of way. This would be remote viewing; yet one can only remote view things that already exist. A photon that is flying through a double slit does not have a position yet, so the position of the photon is not existing information at that stage.

Therefore in this experiment the wavefunction is not collapsing any time earlier than usually. It doesn't collapse at the double slit, not even for some of the photons. The wavefunction still collapses only when the photons are registered at the screen and the picture of the screen arrived at the conscious part of the observer's brain.

This experiment is in its essence not different from any other micro-PK experiment. Any form of psychokinesis (PK) is a proof that something is in superposition; that the wavefunction hasn't collapsed. If somebody can perform PK on lets say a cup, it means that the whole cup is in superposition (for a 40th second). Yet if the target object is a single quantum event we speak about micro-PK and all that we can be sure to have been in superposition is the associated quantum particle. However, the observer having an effect on it makes it at least plausible that its quantum state did collapse somewhere in the brain of the observer.

In this sense all nonlocal perturbation experiments can be seen as evidence for consciousness based interpretations of quantum mechanics. Yet, having to deal with so many different interpretations with several of them being related to consciousness, it is obviously not enough to demonstrate the observer effect in order to prove that the orthodox interpretation is the only option.

For some reasons the psi-effect Radin found at the double slits was much stronger then what he and others usually find using other set ups such as random number generators (RNG). His result had sigma-5 significance. Maybe the more interesting set up is the main reason for this.

In parapsychology the physical world view a researcher subscribes to can have a significant impact on how data is interpreted. If someone, in spite of quantum mechanics, believes reality to be based on a time-symmetric spacetime block universe for example, he is likely to interpret nonlocal perturbation as precognition.

While I believe the observers were conducting usual micro-PK on the photons, Dean Radin believes the photons were 'measured' by remote viewing and thereby the interference pattern was altered. Without going beyond the conventional quantum theory that is afflicted in ambiguity it will be hard to convince Radin that it was actually micro-PK and that he should have asked his participants not to mentally 'look', but to 'wish'. A similar debate I have with him about his precognition experiments which I interpret as to represent cases of micro-PK as well (the future picture is selected by a RNG).

He showed that people can react to quantum randomly selected pictures in advance.[3] For me this is a form of PK: the picture that aligns most with the state the mind & body was in before is chosen, as a sort of path of least resistance among quantum alternatives. For him however it is precognition. From a general relativity perspective his opinion makes more sense. From a quantum perspective PK is the more plausible explanation. The same also works backwards in time: various researchers have shown that when one uses a computer to record random bits produced by a RNG which are left un-observed for hours, days and in some cases even for a half year, one still can go and influence the outcome. Looking at this from a spacetime perspective one might suggest that the record in the past was influenced by the observation in the future – an example for retrocausality. And indeed both Dean Radin and Stephan A. Schwartz argue that way (Stephan A. Schwartz; personal correspondence; 2017). However, from a quantum perspective it is more plausible to assume that the record was in superposition all the time before it was played.

An argument against this view by Schwartz is that the success rates are somewhat higher for these retrospective experiments than for ordinary RNG experiments.

Summarizing we can say that Dean Radin's double slit observer effect experiment can't determine when and where the wavefunction collapses. It is a regular double slit experiment and that is a thing a regular double slit experiment just can't do.

Therefore it is not a test of the Von Neumann-Wigner interpretation to any extends beyond the usual micro-PK experiments.

All we can infer from it is that the observers influenced the outcome. When this influence manifested we can't know from it. For instance, it doesn't disprove Roger Penrose's gravity induced wavefunction collapse (OR). What Roger Penrose believes is that it is gravity that induces the collapse, but that it somehow gives rise to consciousness. Others like Max Tegmark believe that consciousness is choosing its path through an Omnium-like universe of all possible states – an example for this idea is the aforementioned 'quantum suicide' thought experiment. These are all examples for theories that don't link the wavefunction collapse to consciousness, but still hold that consciousness has influence over it.

So when testing interpretations of quantum mechanics there are two aspects to consider:

1. Does the observer have an influence on quantum states?
2. When and where does the wavefunction collapse?

Dean Radin's 50 years of research answer (1) with a definite yes, but for answering (2) we need to do do the quantum delayed eraser experiment I described in the last chapter. Fortunately Radin has just recently expressed interest in conducting that experiment in his lab in the near future (Dean Radin; personal correspondence; 2018).

Ludovic Krundel: Delayed choice double slit experiment observed by a robot

Beginning in 2013 Ludovic Krundel had been promoting an experiment where a robot is looking at a double slit set-up with humans staying as far away as possible.[4] He suggested that if the robot is unconscious then checking through which slit the photons goes shouldn't destroy the interference pattern.

There are several problems with this: firstly, an unconscious robot isn't any different from a normal measurement device, and our experience with measurements is that we can never both obtain the path information and the impulse information (interference).

Secondly, any measurement by the robot would bring the quantum states into the macrocosm and from there it is just a matter of time till the observer's state is influenced.

The way he described it, it was a delayed choice experiment. Presumably that was influenced by the pre-Wheeler notion of a particle deciding to travel as a wave or a particle before taking off. While accepting the reality of delayed choices one might think that they cannot happen when the measurement is done by an unconscious robot. It is not too obvious that even when using the Von Neumann criteria of measurement (consciousness induced collapse of the wavefunction), a measurement doesn't have to be directly displayed to a human in order to count as such. Even in the physicist community people still sometimes misunderstand the Von Neumann-Wigner interpretation in this essential way (Paris Weir; personal correspondence; 2017). This is on the one hand because pondering about the interpretation problem isn't very much encouraged in general; and on the other hand because Von Neumann himself did not spend much time formulating his interpretation in detail. A clarification that different quantum states only need to lead to different brain states in order to count as measured, without the requirement of any concrete knowledge of these states, would have been very useful. It is this lack of clarity that led to a lot of confusion on if and how to apply quantum mechanics to the macroscopic world.

Possibly (but not necessarily) inspired by Dean Radin's double slit experiment Ludovic Krundel furthermore suggested that different 'degrees of consciousness' may lead to differences in the *interferometric visibility* within the interference pattern.
Not knowing about Radin's experiment, I criticized that, arguing that consciousness is either there or not and that one can't measure levels of it, especially not with a quantum experiment. When I then first heard of it, it almost seemed as if Krundel was right. However, looking at how it was done it became evident that the *interferometric visibility* in Radin's experiment was simply a measure for how strong the mind over matter effect was. If we took that as a measure for the degree of consciousness, then we we would be deemed to be regard as totally unconscious most of the time.

Note:
1. TIC 2016 TUCSON, page 194. A video of the lecture can be found here: [https://www.youtube.com/watch?v=uSWY6WhHl_M].
2. Radin, D. Michel, L., Delorme, A. (2016). "Psychophysical modulation of fringe visibility in a distant double-slit optical system". Physics Essays.29(1), 14-22.
3. Dean Radin (2000). "Time-reversed human experience: Experimental evidence and implications". Boundary Institute, Los Altos, California.
4. Actually Ludovic briefly showed me the slides of his presentation in summer 2015. On the last pages they were describing a *delayed choice experiment* designed to detect consciousness. I didn't really understand them, and totally forgot about it. When I saw them again in 2017 in an actual presentation of him, it was as if I had seen them for the first time. However, it is possible that my own idea on using *quantum erasers* was unconsciously inspired by his slides.

2.6 The rate of wavefunction collapse [04.10.2020 (smallest meaningful frame resolution)]

Sponsored by: Lin Yi Song (China).

The main test Roger Penrose and Stuart Hameroff have to offer for their model of consciousness is demonstrating a correspondence between the rate of gravitationally induced wavefunction collapse in the brain and the frequency of gamma brain waves. As mentioned in earlier chapters, gravitationally induced wavefunction collapse is what Penrose speculates to be part of a yet unknown quantum gravity theory.

In space particle dualism theory it is instead the projection of information to the neocortex and back which leads to the collapse of the wavefunction. In this chapter we shall explore how the frequency of gamma brain waves, often taken to be the frequency of the subconscious, can be explained in this framework.

The rate of wavefunction collapse, or wavefunction change, anywhere in the world, depends on the frequency of the brainwaves of conscious observers. This frequency depends ultimately on the firing rate of neurons.

There are many different frequencies present in the brain. The type that is most relevant to the collapse of outside wave functions are gamma brain waves. Those have a frequency of 40 Hertz, which means that our subconscious sees 40 pictures a second. The subconscious level requires only one projection to the neocortex, while the conscious level requires two. Only about half of the incoming information is projected to the neocortex twice. That is why our conscious mind sees only 20 pictures per second.

Why is this the frequency of our brain? We know that flies see more pictures per second. Why is that? It is plausible to assume, that due to the speed with which they move through their environment, implying a field of vision changes more rapidly, they need to have more snapshots per second.

Let's assume that most moving objects we look at are about 5 meters away from us, and most of them are other humans, moving at a standard walking speed of 5 kg/h.

If we see 40 pictures per second, then we see such a walking speed person move 35 mm between each picture. At a distance D of 0.7 meters, which is the length of our arms, and which can be assumed to be our average engaging distance, that corresponds to an angular diameter δ of

$$\delta = 2 \tan^{-1}\left(\frac{d}{2\,D}\right) = 0.049989 \, \text{mm} \approx 0.05 \, \text{mm}$$

The smallest objects visible to the naked eye have a diameter of about 0.058 − 0.072 mm. This is strikingly close to the result we have obtained above.

So instead of trying to explain the frequency of the subconscious, the gamma brain waves, with something as sophisticated as mass-energy displacements that lead to the emission of one graviton, as proposed by Roger Penrose and Stuart Hameroff, the present proposal is suggesting that it simply depends on the average firing rate.

Sometimes when we look around, we do not see anything moving. However, even when turning the head, our field of vision changes at roughly the same rate as if someone was walking by.

When we just stare at an unmoving object, and there is also nothing else moving, we do not seem to need a frame resolution of 20 (conscious) or 40 pictures (subconscious) a second, yet aside from our visual input we also have a mental picture, and that is usually about things we have seen, or things we conceptualize. Those pictures usually require the same frame resolution as what we usually see around us.

References:
1. Yanoff, Myron; Duker, Jay S. (2009). Ophthalmology 3rd Edition. MOSBY Elsevier. p. 54. ISBN 978-0444511416.

3. IMPLICATIONS FOR PARTICLE PHYSICS

3.1 **Minimal mass for black holes with Hawking radiation** [03.06.2004 (minimal mass); 14.08.2016 (not a precondition anymore)]

Note: In the early days of space particle dualism theory the analogy between elementary particles and black holes was taken more literally, so that having the Hawking radiation stopping at a certain scale was seen as a precondition for the consistency of the theory. Later it turned out that the elementary spaces of particles are much larger than their Schwarzschild radius, which means that elementary particles are not black holes and cannot become black holes.

However, outside the framework of space particle dualism this minimal mass still has some relevance, because it firstly shows that gravity cannot make point particles evaporate away, and secondly that primordial black holes wouldn't reveal their existence by powerful explosions at the end of their lifetime.
It is also historically important, because it led up to the discovery of space particle dualism.

The average photon energy of a black hole's Hawking radiation is given by

$$E = \frac{h \, c^3}{16 \, \pi \, G \, M}$$

From this we can see that the smaller the mass of the black hole becomes, the more intense is this radiation. Yet this can't go on forever, because at some point the energy of the radiation photons must exceed the mass energy of the black hole itself.
That means there must be a minimal mass for black holes beyond which they cannot continue to evaporate any further. There are always two particles involved in the Hawking-process: a virtual particle and its corresponding anti-particle. The tidal force of the black hole pulls them apart, leading to one of them escaping and the other falling into it. This process transfers some energy from the black hole to the escaping particle which thereby turns from a virtual particle into a real particle.
Yet if the black hole had exactly the right mass for putting all its mass energy into those two virtual particles, it could disappears in an instant, so that none of these two particles fall into it, and both escape as real particles.
For finding the minimal mass of a black hole, we have to look at exactly this special case, so we have to set the mass energy E of the black hole equal to that of the two escaping particles. Because there are two of them, each must have the energy $E/2$. According to $E = mc^2$ we can now replace M by $E/2c^2$ and get:

$$E = \frac{h \, c^3}{16 \, \pi \, G \, \left(\frac{E}{2 \, c^2}\right)}$$

$$E = \frac{2\,h\,c^5}{16\,\pi\,G\,E}$$

$$E = \sqrt{\frac{h\,c^5}{8\,\pi\,G}}$$

Dividing this by c^2 we get 1.08825×10^{-8} kg. That is very close to the Planck mass, which is

$$m_P = \sqrt{\frac{\hbar\,c}{G}} = 2.17651 \times 10^{-8}\ \text{kg}$$

It is to conclude that black holes smaller than the Planck length do not emit Hawking radiation.

This is not a completely new idea: many experts expect some changes for the laws of physics at the Planck scale, and most would also bet that the radiation has to stop for black holes smaller than the Planck scale. One can come to this conclusion from various angles: some arrive there by realizing that the radiation time would have to become shorter than the Planck time; others instead argue that nothing could reasonably be smaller than the Planck length.
It is quite obvious that point particles couldn't evaporate away, if their radiation would have to by far exceed their own mass energy. Yet it is somewhat less obvious that black holes that started off with masses above the Planck mass can't evaporate away without leaving back Planck size remnants. After all it might be possible for big black holes to evaporate down to the Planck scale and to then disappear completely by emitting a last pair of high energetic photons. Yet for that to happen the last two emitted photons would have to have have a mass energy precisely equals that of the black hole, which is quite unlikely.

It may seem that the above can readily explain why we haven't witnessed powerful explosions from primordial black holes left over from the beginning of physical time. Yet thinking about it again it wouldn't make much of a difference for an outside observer if the black hole evaporates away completely or if it leaves a little rest mass.

The minimal mass we calculated here did not take account of the extended nature of elementary spaces. Taking account of this leads to a much higher minimal mass (see chapter 4.10), turning off Hawking radiation long time before it becomes really intense.
[1] One may think that this is a perfect explanation for why we haven't found evidence for this form of radiation; however as we will see already in chapter 4.3 Hawking radiation doesn't exist in space particle dualism theory, nor does any analogous form of

thermal or non-thermal radiation associated with event horizons. It is only outside the framework of this theory that the unobserved Hawking radiation poses a problem for the notion of primordial black holes.

Note:
1. Aside of that it might be that if primordial black holes have to start off with a rather high initial mass, they are not able to lose weight* in the dense primordial soup of the young universe; and even if they managed to lose weight, they would eventually be swallowed by larger black holes at some point. Considering the huge amount of primordial dark matter that all condenses into black holes it seems quite unlikely that any small primordial black holes could be left over till this day.

*Of course the whole argument here becomes surplus once we realize that black holes can't radiate their mass away (see chapter 4.3).

3.2 Gravity shields itself off [03.08.2005 (idea); 20.03.2014 (calculation); 10.11.2017 (same calculation but using the elementary space radius instead)]

Note: This part is again more of historical value and belongs to a time when the analogy between particles and black holes was taken more literally. However, outside the framework of space particle dualism it still remains a valid way of how to prevent infinities in point particle theories.

In every point particle theory that doesn't take gravity into account the mass energy of charged particles seems to go against infinity. That is because they are assumed to have a radius of zero, so that there are no opposite sides which could repel upon each other. But in order to let a spherical charge shrink to the size of zero, one would need an infinite amount of energy. In theories which do not include gravity, one has no other choice than adjusting these infinities through renormalization to the observed values. Now, if we look at the gravitational field of say an electron there is a chance that gravity itself shields away the infinite mass.
If one lets a spherical charge shrink against the size of zero, an event horizon will form around it when a certain critical circumference is reached (the Schwarzschild circumference). The sphere might continue to shrink after that, but any information about a further increase of self-energy within the horizon cannot reach beyond the horizon. So the problem of an infinite self-energy does not really come up when including gravity.
What mass-energy would we get by letting a spherical charge equal to the charge of an electron shrink to its Schwarzschild radius? The energy of a spherical charge is given by

$$E = \frac{3\,Q^2}{20\,\pi\,\varepsilon_0\,R}$$

with Q being the charge, here the charge of an electron e. The Schwarzschild radius is given by $R = 2GM/c^2$. Using the energy equivalent instead that is $R = 2GE/c^4$. If we put the two equations together, we get

$$E = \sqrt{\frac{3\,e^2\,c^4}{20\,\pi\,\varepsilon_0\,2\,G}}$$

Dividing by c^2 we get the mass equivalent, which is 1.017×10^{-9} kg. This is quite close to the Planck mass and thus much bigger than the real mass of an electron (9.1×10^{-31} kg). Although this proves that point particle theories are free of infinities when accounting for gravity, it didn't explain the observed value for the electron mass.

When using the radius of the particle's elementary spaces which will be introduced in chapter 3.7, then the above formula results in a mass of 3.22×10^{-28} kg, which is somewhat close to the mass of the proton (1.6726×10^{-27} kg).[1]

Anyway, as will also be shown in chapter 3.7 gravity is a side effect of charge. That means it doesn't make much sense to try to derive the mass of particles from anything else than their charge. How this can work we will see in chapter 3.11.

Note:
1. One might be tempted to speculate that the running of coupling can account for this difference and that this mass represents the naked mass of the proton or even the electron (the mass it would have without the dense cloud of virtual positrons around it). However, the naked mass would always have to be smaller than the observed mass, which rules the electron out. Associating it with the proton wouldn't make sense either, because the electromagnetic charge is mainly related to the electron, while the proton's bigger mass is related to the strong force (see chapter 3.11).

3.3 The old source-surface model [22.08.2005 (complex mass/energy; oscillating mass/energy); 06.03.2007 (anti-matter is matter moving on negative space); 04.09.2015 (revise; corrected primordial dark matter prediction); 09.12.2019 (corrected mistake in old model; symmetry)*]

*In collaboration with: Hải My Thị Nguyễn (Vietnam).

Note: The scheme for the correspondence between particles and elementary spaces presented in this chapter is the scheme the theory used from August 2005 until August 2019. It is now replaced with a more advanced scheme, which will be introduced in chapter 3.6. The old model has however still historic value, and should be introduced, in order to better understand the benefits of the new model.

In space particle dualism theory, elementary particles have complex-valued masses and mass energies. That leads to a complex valued space and a universe which is split up into four parts: a plus-plus (matter), a plus-minus (antimatter), a minus-minus (exotic antimatter) and a minus-plus (exotic matter) part. We can't directly interact with those parts that have a negative real part, except of through gravity, and it is very tempting to think that those must therefore represent what we refer to as 'dark matter'. As we will learn later, while those other parts of our universe indeed have the properties we ascribe to dark matter, they are not to be identified with the missing matter in galaxy surveys. Actual dark matter, which is those parts with a negative real component, must have condensed into primordial black holes right at the beginning of physical time.[1] Those primordial black holes must have by now all evolved into supermassive black holes, which are all fairly well observable.

How much primordial dark matter do we have to expect?
Treating elementary spaces similar to black holes, we can imagine that different complex masses, or energies, generate different types of complex elementary spaces, analogeous to the horizons of black holes.

An elementary space can be quite different from its source mass or source energy in its complex number composition. We can see that if we enter different complex numbers into the formula for the elementary space. The surface area of a black hole is given by:

$$ A = \frac{16 \pi G^2 m^2}{c^2} = \frac{16 \pi G^2 E^2}{c^6} $$

For elementary spaces, G has to be replaced by another constant G_E which will be introduced in the chapters 3.7 & 3.8. However, for the present analysis this isn't important and can be ignored for the moment.

If the mass m, or more general, the energy E, is a complex number, then it has the form $a + bi$. So E^2 is

$$(a + bi)^2 = a^2 + 2abi + (bi)^2$$

If a becomes 0, then only $(bi)^2$ is left, and all non-zero values for b lead to negative values for the surface (because b^2 can only be positive and i^2 is -1). If b becomes 0, then only a^2 is left. In this case all values for a lead to positive values for the surface. We can make a list of all possible values and outcomes for a and b. Elementary spaces with no real component as well as elementary spaces which are non-complex are marked gray and represent primordial dark matter:

Elementary Space	$a \in \mathbb{R}^+$ $b \in \mathbb{R}^+$	$a \in \mathbb{R}^-$ $b \in \mathbb{R}^-$	$a \in \mathbb{R}^+$ $b \in \mathbb{R}^-$	$a \in \mathbb{R}^-$ $b \in \mathbb{R}^+$
$|a| = |b|$	$\mathbb{I}^+$	$\mathbb{I}^+$	$\mathbb{I}^-$	$\mathbb{I}^-$
$|a| > |b|$	$\mathbb{C}^{++}$	$\mathbb{C}^{++}$	$\mathbb{C}^{+-}$	$\mathbb{C}^{+-}$
$|a| < |b|$	$\mathbb{C}^{-+}$	$\mathbb{C}^{-+}$	$\mathbb{C}^{--}$	$\mathbb{C}^{--}$
$|a| = 0$	$\mathbb{R}^-$	$\mathbb{R}^-$		
$|b| = 0$	$\mathbb{R}^+$	$\mathbb{R}^+$		

If we treat all of these spaces equally we can conclude that there is $1 - 4/16 = 75\%$ primordial dark matter. Ignoring the non-complex elementary spaces on the other hand gives 50%. Counting $\mathbb{R}^+$ to the observable part would lead to $1 - 6/16 = 62.5\%$, which is close to the 63% that is claimed to be the dark matter percentage in the early universe.[2] However, this percentage is derived from the cosmic microwave background using standard general relativity cosmology, not space particle dualism theory.

Even if that figure was right, decoupling (the universe becoming transparent) is only close to the border of the past (the 'Big Bang' in standard model terminology) according to general relativity, while according to space particle dualism the universe has spent half of its lifetime in the non-transparent pre-decoupling state (see chapter 4.15).

We can therefore ignore this figure. We will come back to the issue of determining the initial matter-dark matter ratio in chapter 3.6.

If a and b can have different relative values, then what could they be representing respectively? One hypothesis is that of *oscillating mass* or *oscillating energy*, namely that between measurements the energy of a particle has all the different complex values that amount to the same absolute value. In such a scheme the measured value would only representing the most likely one.

What if instead a and b need to be always equal? This seems like an unnatural requirement for a complex number, but we have to keep in mind that we are talking about a complex surface area and not about a point on a complex plane or on a Riemann sphere; at least not one that is out there in any straight forward physical sense. We will examine this and other possibilities in chapter 3.6. As for now we will explore the idea of oscillating mass/energy first.

Let's see of which spaces the four different subuniverses consist of according to the above scheme:

Positive part of the universe:

$$U^{\mathbb{R}^+} = \{\, \mathbb{C}^{++}(++);\ \mathbb{C}^{++}(--);\ \mathbb{C}^{+-}(+-);\ \mathbb{C}^{+-}(-+);\ \mathbb{R}^+(+);\ \mathbb{R}^+(-)\}$$

Negative part of the universe:

$$U^{\mathbb{R}^-} = \{\mathbb{C}^{-+}(++);\ \mathbb{C}^{-+}(--);\ \mathbb{C}^{--}(+-);\ \mathbb{C}^{--}(-+);\ \mathbb{R}^-(+);\ \mathbb{R}^-(-)\}$$

Imaginary part of the universe:

$$U^{\mathbb{I}^+} = \{\mathbb{C}^{++}(++);\ \mathbb{C}^{++}(--);\ \mathbb{C}^{-+}(++);\ \mathbb{C}^{-+}(--);\ \mathbb{I}^+(++);\ \mathbb{I}^+(--)\}$$

Negative imaginary part of the universe:

$$U^{\mathbb{I}^-} = \{\mathbb{C}^{+-}(+-);\ \mathbb{C}^{+-}(-+);\ \mathbb{C}^{--}(+-);\ \mathbb{C}^{--}(-+);\ \mathbb{I}^-(+-);\ \mathbb{I}^-(-+)\}$$

Anti-particles are sometimes described as particles which run backwards in time. Running backwards in time equals moving in a negative space. In the here presented *energy-surface model* of space particle dualism anti-particles can be interpreted as particles with negative mass energy, which generate a positive elementary space. If their elementary space is negative as well, they become primordial dark matter. The statement that they move on negative space is referring to their impulse, which is a movement in the accumulative entirety of extended space. It is not referring to their orbital movement around the elementary space they are attached to.

In this picture $\mathbb{C}^{++}(--)$ would be the anti-particle of $\mathbb{C}^{++}(++)$ and $\mathbb{C}^{+-}(-+)$ the anti-particle for $\mathbb{C}^{+-}(+-)$. The $\mathbb{R}^+$-particles would, although belonging to our sector of the universe, be completely dark, because they cannot have a wave function without being complex. We can anyway regard $\mathbb{R}^+(-\emptyset)$ as the anti-particle of $\mathbb{R}^+(+\emptyset)$.

Since measurements always take place in the positive sector, their movement is always dictated by the forces in this sector. Without conscious observers there is no measurement, and no quantum forces. Particles which are not part of this positive sector, can therefore only respond to gravity.

The four particles $\mathbb{I}^+(++)$, $\mathbb{I}^+(--)$, $\mathbb{I}^-(+-)$ and $\mathbb{I}^-(-+)$ are not in any way connected to our sector. Formally they have to be treated as tachyonic. Just as anti-particles can be treated as particles with negative mass moving on negative space, tachyons can be treated as particles with imaginary mass, moving on imaginary space. They would however not really behave like tachyons, because in space particle dualism time can only arise through measurements. Thus they cannot use our space as their time. Also they cannot react to any forces, except of the emergent force of gravity.

The above scheme produces only two different non-dark particles together with their anti-particles. Switching the signs of the source mass without changing the signs of the surface corresponds to turning matter into antimatter. Dark matter is when the real part of the surface (irregardless of the source mass) is negative.

We have four non-gray areas in the table, but only two non-dark categories, particles and anti-particles. The definition of anti-matter used in this model is rather complicated and very different from that of (primordial) dark matter, which doesn't care about the sign of the source mass. If matter and anti-matter differ only in the signs of their source mass while having the same types of surfaces, then what does a change in the sign of the imaginary component of the surface mean?
This is left unclear in this scheme and as we will see in the next chapter, the diversity of non-dark elementary particles (without anti-particles) arises from the symmetries of elementary spaces alone, and doesn't require switching any signs.

Switching of signs should be associated with anti-matter and exotic matter (primordial dark matter) alone. In chapter 3.6 we will see how that can be realised.

Notes and references:
1. For that to be possible, primordial dark matter has to be 'cold'. There are theoretical reasons to assume that it is indeed cold and we will come to those in chapter 4.10.
2. "Content of the Universe – Pie Chart". Wilkinson Microwave Anisotropy Probe. National Aeronautics and Space Administration. Retrieved 9 January 2018.

3.4 Symmetry groups [22.12.2005 (U(1) × SU(2)); 01.02.2006 (SU(3))]

All particles of the standard model can be described basing the symmetry group

$$U(1) \times SU(2) \times SU(3)$$

The common view is that this broken symmetry has to be obtained by some sort of symmetry break from a larger symmetry, namely one of the symmetries $SU(5)$, $SU(8)$, $SO(10)$ or $O(16)$. Each of these larger symmetries would bring us a whole zoo of not yet encountered particles. Those are mainly the supersymmetry partners of well-known particles. In this chapter we will see that the symmetry of the standard model follows naturally from the structure of space in space particle dualism theory, so that no symmetry break is necessary.

The just mentioned symmetry scheme has to be replicated for each of the three known generations of matter. Such as multilayer symmetry isn't very natural and therefore I suggest that the different generations of particles can be understood if we account for the possibility that elementary spaces can merge. We will investigate this possibility in the chapters 3.9 and 3.10.

In pragmatic particle physics theories symmetry patterns are usually only utilised to bring some order into the zoo of particles. In theories which claim to be really fundamental, such as string theory, these symmetries are at the same time statements about the structure of space itself.[1] The number of compactified extradimensions in string theory is determined by constancy requirements, such as avoiding infinities in calculations. The resulting number of dimensions however, 26, 10 or 11, has a direct impact on the symmetry of the resulting particle physics. Those resulting symmetries are all too large, and have to artificially be reduced through symmetry break. Space particle dualism refuses this type of artificial solution.

What is $SU(2)$ standing for? It stands for the symmetries of a complex 2-sphere and is the complex equivalent of $SO(2)$. The S stands for special, and means that only spheres of the radius 1 are considered.
This reminds us strongly of elementary spaces: they too are complex 2-spheres.

$SU(3)$ is the symmetry group of rotations in a 3-dimensional complex space. The space of space particle dualism is indeed a 3-dimensional complex space, yet the third dimension is only an emergent property arising from the connections of elementary spaces. We can call it a granular or a 'fractional' dimension. The more particles we involve, the closer it gets to the value 3. In this context a highly curved space in relativity corresponds to a space with huge differences in the value of this granular dimensionality. However, because of the omnipresence of virtual particles we always have a dimensionality very close to 3 and therefore a very weak gravity. We will come back to the weakness of gravity in chapter 3.7.

What is the dimensionality of space particle dualism at the most fundamental level? This level would be the level of a single elementary space and therefore the answer must be 'two', but it is a complex 'two'. That corresponds to $SU(2)$. Quite opposite to string theory, space particle dualism has fewer dimensions at the fundamental level than at the macroscopic level. Therefore we can say that it realizes the holographic principle.

All three symmetries are complex. That shows us that the structure of all particles is based on their appearance in the whole complex space, and not only on how they look like on the real half of it.

What about $U(1)$? It corresponds to a 1-dimensional complex space. Maybe that is indicating that elementary spaces can lose one of their dimensions. Even when losing one dimension they could still have a wave function, because the wave function depends mainly upon the complex value. We will further investigate on this possibility in the next chapter.

What if they lose a or b instead? They would have no wave function anymore, and could not interact by any means, except of gravity. They would be very strange particles, having only mountain-like probabilities, instead of the usual wave-like probabilities; and they would have the spin 0. This kind of particle can only be a prediction of space particle dualism, because only here the wave function is the result of a deeper underlying structure. This gives us more degrees of freedom, and the possibility to consider objects with very different probability densities; such which do not obey Schrödinger's equation. Particles with only a probability mountain instead of a probability wave could only be explained within the framework of space particle dualism. Those are the particles $\mathbb{R}^+$, $\mathbb{R}^-$ and $\mathbb{I}^+$ we encountered in the last chapter. Together with particles just mentioned, we would then have the following four main classifications of spaces or particles:

2 complex dimensions (ordinary and dark fermions) – SU(2)
1 complex dimension (ordinary and dark bosons) – U(1)
2 real/imaginary dimensions (non-quantum dark matter) – SO(2)
1 real/imaginary dimension (non-quantum dark matter) – O(1)

Different gauge bosons are associeted with different symmetry groups. Electromagnetism is mediated by virtual photons, and as we will learn in the next chapter they have less degrees of freedom in their spin, which suggests that they have only one complex dimension, hence their symmetry is U(1).
The weak force is mediated by massive particles, which suggest 2 dimensions, hence SU(2).
The strong force arises from the interplay of multiple particles separated in 3 dimensional complex space, hence SU(3).

Note:
1. 11-dimensional stringtheory or M-theory has 3 extended dimensions and 8 compactified ones, with one of them being 'short' instead of compactified. That leads to 16 supercharges and corresponds to either SO(32) or $E_8 \times E_8$ depending on the details of the model. E_8 is also popular in other approaches to quantum gravity, such as Garret Lisi's "Exceptionally Simple Theory of Everything" or Klee Irwin's "Emergent Gravity". These other theories don't talk of extradimensions explicitly, but they assume there is some type of hyperdimensional 'crystal' at each Planck volume of space. That is essentially the same as extra-dimensions, with the only difference being that they are not deformed like in string theory.

3.5 The bizarre nature of quantum spin [15.10.2006 (basic idea); 04.09.2015 (revise; corrected dark matter prediction); 10.10.2015 (difference between fermions & bosons); 10.02.2020 (boson-fermion difference arises from two and fourfold morphing of complex oscillating mass)]

We just saw that complex numbers give us some structure to generate particles. Yet these particles don't seem to be enough to assemble the standard model of particle physics. How can we enrich this structure? Rising the number of dimensions, like in string theory, is not allowed here, because space particle dualism does not work with *compactification* (hiding dimensions by rolling them up). One more dimension at the fundamental level would mean 2 more dimensions on the macroscopic level: if overlapping 2-spheres generates a 3-dimensional space, then overlapping 3-spheres would create a 4-dimensional space. The only way to get more dimensions in this theory would be to switch to a more complex type of numbers. There is another kind of numbers more complex than complex numbers, called *quaternions*. Anyway, quaternions have three different imaginary constants i, j, k, which means there would be way too many mirror symmetries, while what we know of is really just the matter-antimatter symmetry and the matter-dark matter/exotic matter symmetry. Thus quaternions seem to not fit our purpose.

What we will be looking at now is a much simpler way to create more complexity. In the last chapter we mentioned the symmetry U(1) and the possibility of elementary spaces to lose one of their dimensions. As mentioned before they would still have normal wave-like probability densities. What I want to suggest here is that it is the dimensionality which makes fermions and bosons different. What is the main difference between these two types of particles? Fermions have half numbers for their spin $(1/2, 3/2, 5/2, ...)$, while bosons have whole numbers for their spins $(0, 1, 2, ...)$. The spin angular momentum of a particle is

$$S = \frac{h}{2\pi}\sqrt{s(s+1)}, \qquad s = \frac{n}{2}$$

If we insert different natural numbers for n, we get $0, 0.5, 1, 1.5, 2, ...$ How can we interpret these different spins?

Fermions always have a spin of $1/2$, if they are fundamental particles. That actually means they need to rotate by $720°$ in order to become identical with themselves. Same as with quantum entanglement, space particle dualism is here again the only way to come to a real understanding of quantum properties.

Very generally speaking fermions are charged particles which can access all directions in space for their spin. Bosons on the other hand don't have charge and always spin perpendicular to their direction of movement. This lack in degrees of freedom suggests that bosons have less dimensions than fermions.

The complex number z representing their mass or energy orbits the centre of the complex plane in the same way for both particle types, but the effects are different. Both go from particle to anti-particle over to exotic antiparticle, then to exotic particle, for then arriving back to the original particle state. In each of those four steps there is a sphere growing and shrinking back again. However, for bosons the four spheres are perceived as only two, because bosons are usually their own antiparticles.[1]

While the complex number goes around, a complex surface with a positive real part appears and grows. After it reaches its maximal extent, the imaginary part becomes negative, but because bosons are their own antiparticles, that goes unnoticed. Then the surface shrinks again, until the real part becomes negative. In that moment the particle becomes undetectable. The corresponding wavefunction is now a valley.

In effect we deal with one positive surface and one negative.

With fermions things are very different. Here the switch to imaginary-negative does not go unnoticed. It is perceived as a whole new surface. This means the oscillating mass/energy of fermions is perceived as a fourfold transformation.

Instead of twice growing and shrinking, from $0 + i$ over $1 + 0i$ to $0 - i$, and then from $0 - i$ over $-1 + 0i$ back to $0 + i$, where the positive real part can be thought of like the frontside of an object and the negative real part as sort of like the backside, or as simply as two states, we have four sides now, corresponding to four states.

Using an everyday life analogy, we can think of bosons being like a vertically rotating bottle. The space it occupies in our field of sight grows and shrinks as it rotates; and there are two sides to it.

Because the imaginary part doesn't make a difference for bosons, we need to rotate our bottle only once. Each side, the positive (matter) and the negative (exotic matter) grows and shrinks once. For fermions we have the antimatter states, so here we have four growing and shrinking sides, corresponding to a bottle which we need to rotate $360°$ degrees twice for bringing it back into its original state.

The oscillating mass-energy and surface of an elementary space can be described by a complex number z, with a rotating angle θ, so that we get

$$z = e^{i\theta} = \cos\theta\, a + \sin\theta\, b\, i$$

If it is the unity circle, with $|z| = 1$, we can simply leave out the factors a and b.

Regardless of particle type, the time for one orbit is always

$$t = \frac{\lambda}{c}$$

There is a conceptual confusion one tends to undergo when considering the implications of this. Since the size of an elementary space is always proportional to the mass energy of the attached particle, namely

$$R_E = \frac{2\,G_E\,M}{c^2} = \frac{2\,G_E\,E}{c^4},$$

One tends to see a contradiction here: larger complex numbers should take longer to go around the centre of the complex plane or the Riemann sphere. This is mistaking the Riemann sphere with the elementary space. The Riemann sphere can be used to represent the complex number which quantifies the surface area of the elementary space. It is not the elementary space itself. Consequently there is no energy associated with the orbiting of the complex number z around its Riemann sphere. All the energy of a particle lies within the surface area of the resulting elementary space and the wavefunction its oscillations create.

On first sight elementary spaces may seem similar to the compactified dimensions of string theory. This has led to suggestions to place vibrating strings onto them (Bernhard Umlauf; 2015; personal correspondence). However, this would be completely in vain, because the wave function, and therefore the energy of the particle, depends solely upon the continuous appearance and disappearance of these complex surfaces. Any local vibration or tension of a string would not have any effect on the measureable energy.[2]

Space particle dualism doesn't allow the existence of spin-2 particles, which rules out the graviton is a potential intermediating particle of gravity. As mentioned before gravity is here believed to arise from density differences in the quantum vacuum, which implies that it doesn't require an intermediating particle.

Another particle with a strange spin-value is the Higgs-boson. According to space particle dualism the only way for a particle to have a spin of zero is either for it to not be fundamental, or to have a purely real or purely imaginary mass, which would prevent its mass from oscillating.

In the second case it would have no normal wave function but just a probability mountain. That would work for a primordial dark matter particle, but it might be problematic for the Higgs boson, because it can't be 'dark'. Its heaviness could be a hint that it isn't fundamental, and in that case the spin of 0 wouldn't be a problem, because it would be merely an accumulative spin composed of non-zero spin particles.[3]

As mentioned above, bosons can only spin orthogonal to their direction of movement, which only leaves the two choices left-handed and right-handed.
Over quite some period of time it was thought that the Riemann sphere with the complex number z on it is physically real and somehow identical to the elementary space itself. This idea was clearly imported from twistor theory, where two Riemann spheres determine spin and relativistic impulse of a particle.

Today it is clear that the projective properties of the twistor space are not necessary in order to explain relativisitic impulse and special relativity in general. In chapter 3.13

we will see that speed dependent time dilation and length contraction are a result of the uniform growth of elementary spaces.

Quite a few things can be done with the Riemann sphere as a mathematical tool, but that can be said of almost any basic mathematical concept. We can represent relativistic transformations of the sight field of observers on the Riemann sphere. We can also represent quantum spin on it. In space particle dualism we represent the mass of a particle using the Riemann sphere, and we can do the same with the spin and with the impulse, but that doesn't mean every particle carries dozens of Riemann spheres with it. To think so, would be to make the same mistake as those who mistake spacetime diagrams for real physical entities we live in.

A complex number is somewhat richer in structure than a pure real number. That makes us think of it in highly geometrical ways, but that doesn't mean that this geometry is a physical object. If we want, we can project all real numbers onto a single ring, including the number ∞, but that doesn't turn every real number we 'see' into a physical ring.

If an elementary space was the same as a twistor, or a Riemann sphere, then it could hardly appear and disappear, because all complex numbers are present on it at every time. If we then say the number z is in fact at the same time our particle that physically orbits its elementary space, then we are basically saying that an elementary space consists of plus-plus, plus-minus, minus-minus and minus-plus junks of surface, while in fact at every time the whole surface of an elementary space has only one single complex value for its surface area.

Both general relativity and twistor theory were mistaking mathematical objects for physical objects. A physicist has to always be cautious about committing this mistake.

Note:
1. Those who really think there is anything to gain from string theory based approaches may try to identify the rest mass of a particle with the string vibration. However, this way photon-strings wouldn't be allowed to vibriate at all.
2. It is yet still unclear what role the Higgs field and in particular the Higgs mechanism plays in space particle dualism theory. The Higgs mechanism besides being widely accepted doesn't allow for the prediction of any particle masses. One should think that explaining mass should result in the prediction of at least some particle masses.
3. The only exception from this rule is the W-boson.

3.6 Dark symmetry and great unification [03.09.2005 (negative frequencies); 19.08.2019 (reducing degrees of freedom); 08.09.2019 (same absolute values for a and b); 30.09.2019 (oscillating surface)*; 24.12.2019 (Necati Demiroglu's expression for complex mass); 08.02.2020 (Demiroglu is right)**]

*In collaboration with: David Chester (US); Lyudmila Alexeyevna (Kazakhstan).
**In collaboration with: Necati Demiroğlu (Turkey).

As we saw in chapter 3.4, space particle dualism nicely reproduces the symmetries of the standard model. However, at the same time it seems to have too many additional degrees of freedom: distinguishing between different types of elementary spaces by the complex factors 'a' and 'b' of the surface area should be enough; distinguishing between a source energy (or mass) for that surface and the surface itself seems to violate strict *geometrism* or relationism. If energy itself is merely a result of the interaction of wavefunctions, and if the orbiting of the complex number z around the Riemann sphere or the complex plane which generates this wavefunction has no energy in itself (see chapter 3.5), then it is obvious that energy is purely geometric.

Historically it is unclear if allowing different values for a and b was aiming at creating some type of oscillating mass/energy, or if a and b were more meant to represent different particle properties, such as mass and impulse.

How much primordial dark matter do we get if we don't distinguish between source energy and surface area?
In that case we simply change the sign of the imaginary part to get anti-particles and we change the sign of the real part to get exotic matter particles, the constituents of primordial dark matter. However, elementary spaces with no imaginary component or no real component are possible too, and those would also be 'dark'. As mentioned in chapter 3.3, those have no wavefunctions. We could therefore call them classitrons, for 'classic particles'.

We can create a new chart with reduced degrees of freedom and with the corresponding symmetry groups instead of the names of particles:

Matter	Anti matter	Dark matter	Anti-dark matter
$\widehat{\mathbb{C}}^2_{++}$	$\widehat{\mathbb{C}}^2_{+-}$	$\widehat{\mathbb{C}}^2_{-+}$	$\widehat{\mathbb{C}}^2_{--}$
SU(2)	Anti-SU(2)	Dark-SU(2)	Anti-dark-SU(2)
$\widehat{\mathbb{C}}^1_{++}$	$\widehat{\mathbb{C}}^1_{+-}$	$\widehat{\mathbb{C}}^1_{-+}$	$\widehat{\mathbb{C}}^1_{--}$
U(1)	Anti-U(1)	Dark-U(1)	Anti-dark-U(1)
$\mathbb{C}^3_{++}$	$\mathbb{C}^3_{+-}$	$\mathbb{C}^3_{-+}$	$\mathbb{C}^3_{--}$
SU(3)	Anti-SU(3)	Dark-SU(3)	Anti-dark-SU(3)

The different rotational symmetries of these different types of spheres produce the different particles of the associated symmetry groups named in the above chart.

The SU(3) symmetry associated with the strong force is then about the elementary spaces orientation in relation to the overall six-dimensional space.

$a = 0$ or $b = 0$ gives us the also dark 'classitrons'. Those are:

Classitrons	Anti-classitrons
$\mathbb{R}^2_+$ Spheron	$\widehat{\mathbb{R}}^2_-$ Anti-spheron
$\mathbb{R}^1_+$ Circlon	$\widehat{\mathbb{R}}^1_-$ Anti-circlon
$\widehat{\mathbb{I}}^2_+$ Tachyon	$\widehat{\mathbb{I}}^2_-$ Anti-tachyon
$\widehat{\mathbb{I}}^1_+$ Tachyophoton	$\widehat{\mathbb{I}}^1_-$ Anti-Tachyophoton

We realize that if U(1) has to correspond to the electromagnetic force, then we have to fit both photons and electrons into it. One main difference between a photon and an electron is the degrees of freedom for their spins.

Roger Penrose speculated that massive particles could be described by bundles of twistors. Analogously we could describe electrons as merged photon pairs.

This could be understood in a very visual way: in ordinary general relativity the only way to stop a photon would be to capture it in a black hole. According to space particle dualism black holes are only quasi-black holes, so they can't stop photons (see chapter 4.3). However, elementary spaces are in many aspects more similar to mathematical black holes. They grow according to their mass energy, and they can fully absorb things onto their surface. Combining the elementary spaces of two photons could therefore both stop them and give them one additional dimension and thereby more degrees of freedom.

If we assume that exotic matter is to a certain extend analogous to ordinary matter, then we must assume that the anti-matter counterparts of exotic matter only appear in matter-antimatter pair generation. Consequently we can count 3 elementar space types for matter and 7 for primordial dark matter; leaving out their dark counterparts. If we treat all of those equally, we arrive at the conclusion that there should have been 70% primordial dark matter. However, as already suggested in chapter 3.3, it is unlikely that non-complex elementary spaces (classitrons) contribute to gravity. If they don't, we arrive at a more symmetric prediction of 50% primordial dark matter.

The problem is we don't know if dark matter exerts just as much gravity as ordinary matter. It might be that the dark sector has only dark photons, and no dark gluons, since it is not much connected to the emergent SU(3)-space. In that case primordial black holes would have been relying on rather weak gravity, analogous to the usually unmeasurable charge gravity of electrons (it is dwarfed by the hypercharge gravity of protons and neutrons), which we will come to in the chapters 6.6 and 6.7.

The above scheme has the very strong advantage of explaining the so called missing negative frequencies. The equation for the energy of a photon obviously allows both positive as well as negative values for E:

$$E = f \times h$$

Both space particle dualism and twistor theory have convincing and in fact very similar ways to deal with these negative frequencies. In space particle dualism they are dark or 'exotic' photons, generating and moving on negative space. In twistor theory they are light rays directed to the past.

The more symmetric relation between matter and primordial dark matter that is central to the new scheme introduced in this chapter could be helpful when it comes to great unification of the three forces. They do meet exactly if we give them some form of partner particles. Those don't have to be supersymmetric; any symmetry does the job. A symmetry involving exotic matter, the constituents of dark matter, we could refer to as 'exotic symmetry' or 'dark symmetry'.

When we were looking at complex mass or energy in chapter 3.3 and 3.5, we were looking at different relative values of a and b. In a paper titled 'electric charge as a form of imaginary energy', the theoretical physicist Tianxi Zhang also proposed that the energy of particles is complex.[1] He identified a with mass and b with electric charge. This is not a bad idea, because it allows to easily create anti-particles by changing the sign of the imaginary part. If we wanted to change the sign of the real component instead, then the particle would turn 'dark'.

However, as we will see in chapter 3.12, mass is merely a side effect of charge, so treating them as separate entities can be problematic. Less problematic it is to say that the sign of the imaginary component determines the sign of the charge.

What about identifying a with charge and b with impulse?
That does not seem to work: if a was the charge or mass, then photons, which are massless, would have no real component, and that would turn them into classitrons. Also if increase in impulse didn't affect the real component, then there would be no Lorentz transformation (see chapter 3.13).

What if a and b are instead the same?
Obviously to get quantum mechanics, they still have to oscillate, so assuming a and b to be equal is about the measured value only, or the original state, if you will.

In such a setting the only variation we can have would be change of signs. b turning negative corresponds to turning a particle into an antiparticle, while a turning negative corresponds to turning a particle into an exotic particle.

In such a scheme it seems a bit unnatural to also allow setting $a = 0$ or $b = 0$, and if we don't allow that, then we don't have classitrons, and get a total symmetry between matter and exotic matter. That might indeed be necessary for achieving the unification of all forces at high energies.

Surface-energy dualism

In the early days of space particle dualism there was still a distinction between the particle and the elementary surface. In fact that was simply confusing the number z that orbits the Riemann sphere for a particle that orbits its elementary space.
When we understand that there is no particle, but only an elementary surface or space, then we understand that distinguishing between mass and surface is misguided.

We have to identify the rest-surface with rest-mass, and the grown surface with relativistic mass or energy.

Having a point-particle makes it easier to visualize a path through granular space.[2] The trouble starts when trying to imagine interactions. It is unclear how two point particles are supposed to interact, when they could not possibly ever overlap or touch each other, because after all a point is infinitely small.
In the early days it was assumed that an interaction can be regarded as having taken place when one point-particle has entered the elementary space of another point-particle. It helped to imagine things, but it was provoking the question of why not to simply use merging surfaces instead of point-particles on surfaces that behave exactly like merging surfaces.

Oscillating surface

Oscillating mass or energy and surface is what creates the wavefunction. Dropping the distinction between source mass/energy and surface simplifies things a lot. The signs of the mass or energy are at the same time the signs of the surface area.
That means we do not enter a complex number for the mass or the energy when we calculate the surface area A_E of an elementary space, but rather we use an equation that already has a real and an imaginary component. That can be done by simply multiplying our usual equation by $(1 + i)$:

$$R_E = \frac{2\,G_E\,E_{\mathbb{R}}}{c^4}(1 + i) = \frac{2\,G_E\,m_{\mathbb{R}}}{c^2}(1 + i)$$

$$A_E = \frac{16\,\pi\,G_E^2\,E_{\mathbb{R}}^2}{c^8}\,(1+i) = \frac{16\,\pi\,G_E^2\,m_{\mathbb{R}}^2}{c^4}\,(1+i)$$

Mathematically speaking we can express the idea of oscillating mass/energy and oscillating surface as:

$$E_{\mathbb{C}} = |a + bi| = \sqrt{a^2 + b^2} = E\,e^{i\,\Phi}$$

$$A_{\mathbb{C}} = |a + bi| = \sqrt{a^2 + b^2} = E\,e^{i\,\Phi}$$

How unusual is the concept of complex mass? What about complex charge? In fact complex charge is not a strange concept at all. Charge is a result of oscillating electric and magnetic fields. The wave itself is described using complex numbers. We could indeed say that in between two measurements the charge of a particle oscillates from negative to positive and back, and not only that, but also imaginary positive and imaginary negative. As we will learn in chapter 3.12, mass is simply a side effect of charge. If that is the case, and if charge is already oscillating, then so should mass and mass energy.

Also, when we look at the Schrödinger equation, we see that what is oscillating or waving there is the energy of the particle itself:

$$\hat{H}\,|\Psi(t)\rangle = i\,\hbar\,\frac{d}{dt}\,|\Psi(t)\rangle$$

$$\hat{H} = \hat{T} + \hat{V}$$

With the Hamiltonian operator $\hat{H}$ consisting of the kinetic energy operator $\hat{T}$ and the potential energy operator $\hat{V}$.

Although it is not the observed energy that is oscillating, the mathematical description clearly points to what represents the energy being oscillating, and if that thing is the surface of an elementary space, then it is plausible that this surface is oscillating too.

How can we imagine the trajectory of a particle through granular space, if it is in fact not a point, but a surface? Its impulse is still unambiguous pointing into one direction and so we could simply take the centre of our real surface to be the point that travels through granular space.

Demiroglu's complex mass expression

According to space particle dualism the energy of a particle is a complex number z which orbits the complex plane or the Riemann sphere. In this process the particle oscillates from matter, to antimatter, then to exotic antimatter, then to exotic matter, for then to arrive back to the matter-state. Only when it is in its matter state, can we measure it.

Above we mentioned the complex version of the elementary space radius. What is the relativistic version of it? For that we can add the relativistic correction term γ into it:

$$R_E = \frac{2\,G_E\,m_{\mathbb{R}}}{c^2\sqrt{1-\dfrac{v^2}{c^2}}}\,(1+i)$$

As the complex number z orbits the center of the complex plane, the real part and imaginary part increase and decrease. Only in four points do the real and the imaginary component have the same absolute value. It seems unnatural to assume that the measured value of the complex mass should always be such that the real and the imaginary componant have the same absolute value.
Do we know anything about the behaviour of imaginary mass?
Usually the only particles theorized to have an imaginary mass are tachyons. That is because by definition tachyons, which are only hypothetic particles, travel always at a speed higher than the speed of light.

The assumption that they must have an imaginary mass comes from what happens when when we enter a speed greater than the speed of light in the relativistic correction term γ:

$$v > c \Leftrightarrow \gamma = \sqrt{1-\frac{v^2}{c^2}} \in \mathbb{I}$$

The term under the root becomes negative, which turns the correction term into an imaginary number. Now that is usually taken to mean that the mass of the particle is imaginary, but this interpretation is very problematic, because if we divide an imaginary impulse by this correction term, we get a real impulse:

$$v > c \Leftrightarrow p = \frac{m_{\mathbb{I}}\,v}{\sqrt{1-\dfrac{v^2}{c^2}}} \in \mathbb{R}$$

People argue instead that properties like impulse and energy should be real numbers, and that therefore mass should be imaginary. In some more recent formulations however the mass is regarded to be real.[3, 4, 5]

The Turkish physicist Necati Demiroğlu has specialized in the physics of complex mass. He believes that all particles have complex masses, and he has developed a relativistic expression for the complex mass of particles.[6, 7, 8, 9] According to his research the complex impulse of any rest-mass particle is given by:

$$p = \frac{m_{\mathbb{R}} \, v}{\sqrt{1 - \frac{v^2}{c^2}}} + \frac{m_{\mathbb{R}} \, v}{\sqrt{1 + \frac{v^2}{c^2}}} \, i$$

According to this equation, the imaginary component of the impulse of a particle decreases whenever the real component is increasing. This is exactly what happens with our complex number z in space particle dualism.

Using Demiroglu's equation we can formulate the full relativistic elementary space as:

$$R_E = \frac{2 \, G_E \, m_{\mathbb{R}}}{c^2 \sqrt{1 - \frac{v^2}{c^2}}} + \frac{2 \, G_E \, m_{\mathbb{R}}}{c^2 \sqrt{1 + \frac{v^2}{c^2}}} \, i$$

$$A_E = \frac{16 \, \pi \, G_E^2 \, m_{\mathbb{R}}^2}{c^2 \sqrt{1 - \frac{v^2}{c^2}}} + \frac{16 \, \pi \, G_E^2 \, m_{\mathbb{R}}^2}{c^2 \sqrt{1 + \frac{v^2}{c^2}}} \, i$$

This does not mean that the imaginary component eventually disappears completely at high speeds. At higher and higher speeds it approaches the value:

$$\lim_{v \to c} R_{\mathbb{I}} = \frac{16 \, \pi \, G_E^2 \, m_{\mathbb{R}}^2}{c^2 \sqrt{1 + \frac{v^2}{c^2}}} \, i = \frac{16 \, \pi \, G_E^2 \, m_{\mathbb{R}}^2}{c^2 \sqrt{2}}$$

Note:
1. "Electric Charge as a Form of Imaginary Energy"; Tianxi Zhang (2008); *Department of Physics, Alabama A & M University, Normal, Alabama, USA*.
2. As we will see in chapter 3.15, using point particles leads to orbital speeds that greatly exceed the speed of light. Interestingly that happens very close to the frequency at which the elementary space grows over the wavelength of the particle (see chapter 3.9).

This is fine if we don't talk about a movement in ordinary space but in complex space, because the cosmic speed limit exists only in ordinary space, yet still we may be introducing unnecessary complexity by distinguishing between two types of movement, one in ordinary space and one in complex elementary space.

3. Recami, E. (2007-10-16). "Classical tachyons and possible applications". Rivista del Nuovo Cimento. 9 (6): 1–178. Bibcode:1986NCimR...9e...1R. doi:10.1007/BF02724327. ISSN 1826-9850.

4. Vieira, R. S. (2011). "An introduction to the theory of tachyons". Rev. Bras. Ens. Fis. 34 (3). arXiv:1112.4187. Bibcode:2011arXiv1112.4187V.

5. Hill, James M.; Cox, Barry J. (2012-12-08). "Einstein's special relativity beyond the speed of light". Proc. R. Soc. A. 468 (2148): 4174–4192. Bibcode:2012RSPSA.468.4174H. doi:10.1098/rspa.2012.0340. ISSN 1364-5021.

6. Necati Demiroglu; Yalcin O. & Ozum S.. "A Simple Methodology for Quantum Mechanical Theory of Tardyons andTachyons".

7. Necati Demiroglu; Kuantum Mekanigi & Yeni Metotlar. 2013.

8. Jesse Timron Brown, Necati Demiroglu. "Forms of Time: Fields and Particles", 2013.

9. Necati Demiroglu. "Fields and Particles" (15.08.2019). DOI: 10.21276/sjet.2019.7.8.1.

3.7 The density of space [22.09.2005 (fractional dimensionality); 06.09.2015 (improved formula for granular dim.; vacuum energy calculation); 14.08.2016 (solution for the hierarchy problem); 21.11.2017 (corrected granular dimensionality value); 02.04.2018 (corrected value for G_E)]

Note-1: This section had to be changed a lot after it was found that the radius of the elementary spaces of particles is by far larger than their Schwarzschild radius (2016).

Note-2: A second change had to be introduced after it was found that G_E has to be $G \times 10^{36}$ and not $G \times 10^{37}$ as previously thought (2018). For details on that see chapter 3.12. The precise formula for G_E can be found in chapter 3.8.

The problem with most approaches to quantum gravity is they work with entities at the Planck scale and are therefore almost impossible to verify experimentally. Before August 2016 the real size of elementary spaces was unknown. Although space particle dualism had much more predictive power than those theories already back then, regarding its basic entities it was even more inaccessible to direct experimentation. At that time elementary spaces were theorized to be even smaller than the Planck length. Yet, the only way to calculate a finite value for the vacuum energy without using the methods of renormalization is to make use of the Planck length. Vacuum fluctuations should stop at the Planck length, and wavelengths should appear always as multiples of the Planck length. Therefore our elementary spaces must be larger than the Planck length, because otherwise it will be impossible to build a gapless patchwork of space with them.

In both quantum electrodynamics (QED) and stochastic electrodynamics (SED),. consistency with the principle of Lorentz covariance and with the magnitude of the Planck constant require the energy of the quantum vacuum to be 10^{113} Joules per cubic meter.[1,2] This mass energy alone would, if it wasn't part of a homogeneous background energy, already produce a black hole with a radius of 10^{65} light years (Euclidian radius according to the circumference), which is by many orders of magnitude larger than the universe! It sounds as if this was enough horizon surface to give a gapless patchwork of space. Yet we have to bear in mind how small elementary spaces would be if they really corresponded to the critical circumference of their particles. If we imagine this vacuum energy to be consisting of only virtual electrons, then we would need 10^{127} electrons ($10^{113}/10^{-14}$ J) per cubic meter. If we multiply this by their Schwarzschild volume, which is $10^{-171} \, m^3$, we can fill up only $10^{-44} \, m^3$ of space with them. This shows again in yet another way that we need elementary spaces much larger than the Schwarzschild radius of a particles.

How much larger? Well, as already mentioned in chapter 1.3 gravity is a side effect of the other forces; an emergent property that shows up only on large scales. If that is the case, then we don't need the G in the formula for the size of elementary spaces to be in any way related to the strength of gravity. Instead we need to exchange it by a new constant that determines only the size of elementary spaces. We shall soon see what constant that is.

The exact value for the vacuum energy is a delicate issue. One can get different results, basing different assumptions. John Wheeler for example, calculated another value, which was $10^{106} \, J/m^3$, assuming the fluctuations to stop at the Planck scale. Yet still all these values are the values one gets after applying a method called renormalization. Without the application of this method, we could get only infinite results. Why is that? That is simply because vacuum fluctuations can have any value for their energy. The more energetic they are, the shorter is their life span. This life span can be calculated as

$$\Delta E \times \Delta t = \frac{\hbar}{2}$$

Since E and t can be any real numbers, there are endless many possible combinations for their values.

The only way to directly measure the vacuum energy is the *Casimir experiment* where two uncharged metallic plates are put a few nanometers apart in order to measure how the restriction for the wave lengths of the vacuum fluctuations between the plates (only a whole number of waves is allowed) is creating a negative pressure which lightly pushes the two plates together. However, this experiment is only measuring differences in energy levels and not absolute values.

If gravity depended on absolute mass energy values instead of differences in mass, then vacuum energy could influence the universe's expansion and the expansion rate could be a measure for such values, but it is not, otherwise this energy would have already destroyed the universe long ago.

A first approach of creating finite values for vacuum energy could be to allow only multiples of the Planck time, which is

$$t_P = \sqrt{\frac{\hbar\, G}{c^5}}$$

That is 5×10^{-44} seconds. The values for energy could analogously be restricted by allowing only multiples of the Planck length for the corresponding wave:

$$E = \frac{h\, c}{\lambda}, \qquad \lambda = n \times \sqrt{\frac{\hbar\, G}{c^3}}$$

This does not give us a fixed smallest possible energy, because it does not set a limit to the length of a wave (the longer the wave the smaller the amount of energy it carries). Furthermore, a wave-length difference of the Planck length corresponds to a larger difference in energy for highly energetic particles than for low energy particles.
However for our purpose of getting limited values for the vacuum energy it is for the most part already enough to quantise Δt, because ΔE and Δt are here dependent up on each other.[2]
Still the problem of a lower limit for ΔE remains. A solution for this problem will be introduced in the next chapter. There we will see that there is indeed a longest wavelength at least for vacuum fluctuations. In the present chapter we will work with the hypothesis that wavelengths of one meter and above are too weak to make a difference.
This seems reasonable for conceivable finite sums of weaker and weaker photons, but such sums don't necessarily converge to a finite value when we continue them to infinity. Therefore only the next chapter can show if our simplification here was a valid approximation.

In this chapter we will choose to quantize energy levels in the quantum vacuum through the quantization of time. However, this is only the most convenient approach, not the most correct one. There is a number of problems with it:

1. It doesn't give us an energy minimum.
2. It may include sequences of photons with practically the same energy but only slightly different lifetimes.
3. It is the energy which determines the lifetime, not the other way around.

4. This form of quantization can't be used for anything else than the quantum vacuum.

The quantization of energy levels outside an atom belongs to what we may call 'second quantization'. Same as the various Planck units it is based upon it doesn't belong to quantum theory itself but to 'quantum gravity'.

If we want to measure the energy of a particle we look at its wavelength and we can't possibly measure differences in wavelength smaller than the Planck length. Thus quantization through the wavelength should be preferred over the rather indirect quantization through the lifetime, expecially because it is appliable universally and not only to vacuum fluctuations. The quantization through the lifetime will only serve as an approximation that will be substituted by a more precise calculation in the next chapter.

Without applying normalization, but only this second form of quantization, we will probably get a value higher than 10^{113} J/m^3. Only with the exact value it will be possible to calculate the density of space in the absence of gravity. Seemingly bended space has a density higher than that, yet the apparent bending is caused solely by the difference in density, not its absolute value.

The density of space tells us about the number of connections each elementary space has with other elementary spaces and that determines its *granular dimensionality*.

In a two dimensional universe we could define granular dimensionality as

$$D = 1 + \left(1 - \frac{1}{n}\right)$$

where n is the number of connections each elementary space has with other elementary spaces. In this 2-dimensional analogy we are tempted to use $2n$ instead of n, because two intersecting circles always have two points of intersection. In the 3-dimensional case we have circles of intersection. Yet we can not know the position of a particle on an elementary space. Directions in 3-dimensional space are already fully described by the possible connections of one elementary space to another. The number of accessible directions is dictated by the number of accessible other elementary spaces. It doesn't matter on which spot our particle passes over to the next elementary space. Therefore sphere diagrams can be equal to net diagrams in practice (see fig. 22).

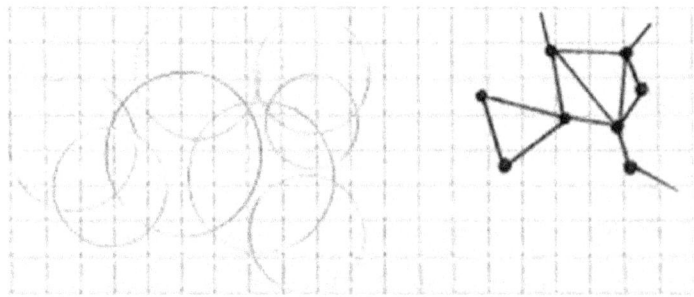

Fig. 22. Transformation of a sphere diagram into a net diagram.
[sketch from the first Februar 2006]

$n = 10$ would here already result in a dimensionality of $D = 2.9$, which means that flatlanders inhabiting such a world would already feel pretty 2-dimensional. The number of sphere surfaces we need to connect in order to get a feeling for 3-dimensionality is much higher.

We can define the granular dimensionality of a space that is formed by overlapping spheres as:

$$D = 2 + \left(1 - \frac{1}{\sqrt{n}}\right)$$

The root is due to the increasing number of connections needed in higher dimensions. Using different values for n, we get

$$n = 1, 2, \ldots 10 \rightarrow 2; \ 2.4142; \ \ldots 2.6837$$

$$n = 100 \rightarrow 2.9$$

$$n = 1000 \rightarrow 2.9683$$

$$\ldots$$

$$n = \infty \rightarrow 3$$

This scheme works for spaces build up by spheres of any dimensionality d $(d = D_E)$. In general:

$$D = d + \left(1 - \frac{1}{\sqrt[d]{n}}\right)$$

If we include the invisible complex part of our universe, we have to set $d = 4$, and get a dimensionality close to 6. This is not including the time dimension, which is here not regarded to be a dimension.[3]

How many degrees of freedom do we have for a virtual particle in one cubic meter of space? That would be a Planck cubic meter, namely

$$1.616 \times 10^{-105} \ m^3$$

multiplied by 10^{105}, which means that we have 10^{105} degrees of freedom in a cubic meter of space. What about the degrees of freedom for energy? We know that Δt is not allowed to be shorter than the Planck time. That gives us an energy maximum, which is the Planck energy:

$$E_{PL} = \sqrt{\frac{\hbar c^5}{G}}$$

In this first approximation we will ignore waves that are longer than one meter. For $\lambda = 1$ we get

$$E_{min} = h \times c$$

That is 1.986×10^{-25} J. Vacuum fluctuations with this energy can exist for 2.653×10^{-10} sec (approximately $1/c$). That is 4.922×10^{33} times the Planck time. These are our degrees of freedom for time and the sum of all different allowed energies is

$$\sum_{i=1}^{n=10^{33}} E = \frac{\hbar}{2\,n \times t_P}$$

$$\int_{i=1}^{n=10^{33}} E = \frac{1}{n} + r \times 978{,}056{,}249.4$$

$$\int_{i=1}^{n=10^{33}} E = 7.4882 \times 10^{10} \text{ J}$$

If we calculate this,[4] we get 7.5×10^{10} J. If we now multiply this by the above mentioned factor of 10^{105}, the degrees of freedom for the position, we get 7.5×10^{115} J/m^3. This is higher than the 10^{113} J/m^3 we had before, and we can even get a bit more than that by including other elementary particles.

However, as aforementioned only if we use elementary spaces larger than the Planck length, we can have a gapless patchwork of space, because the Planck-length is the minimal distance of vacuum fluctuations that are supposed to have different positions, and if the elementary spaces of these vacuum fluctuations are smaller than the Planck length, then one can hardly overlap with the other.

In *Grand Unifying Theories* (GUT) the electromagnetic force, the weak force and the strong force are, at very small distances or high energies, united into one single force.[5] That means their power increases not with the square of the distance but with a higher rate and this is because when approaching particles at short distances one dives into a thick cloud of virtual particles. This cloud is created by the polarizing effect charged particles have on the quantum vacuum around them.

We might be tempted to assume that something like this could also apply to gravity and that this way gravity could be so strong at short distances that it makes elementary spaces big enough to give a gapless patchwork of space.

Following this line of thought we would arrive at the hypothesis that gravity is originally as strong as the united GUT force and that it appears weak only at large distances. Yet arguing this way we would be forgetting that gravity is not a force but an emergent phenomenon. Also it wouldn't make sense that the other forces are only weakened a little, while gravity is weakened so much.

Gravity is based on space (or 'space-time') geometry, while the other three forces are based on probability waves and force transmitting particles. Therefore the very commonly raised idea that gravity could merge with those other forces in strength is misguided. Gravity can't both be based on space geometry and on force transmitting particles called 'gravitons'. It may seem that the energy of gravitational waves needs to be carried by some sort of particle, but just as density differences in a gas can travel as sound waves, distortions in the density of space can travel as gravitational waves, their whole energy lying in the distortion itself.

After all if gravitons existed they would have to behave like 'dark matter', interacting only through gravity. Both this and the incredible weakness of gravity would make them undetectable forever.

What effect does vacuum energy have on gravity?

If gravity was mediated by only one type of particle, let's say virtual 'gravitons' (if they existed), the differences in granular dimensionality would be huge, resulting in a very strong gravitational force. However, as already mentioned in chapter 1.3 all virtual particles in the quantum vacuum contribute equally to the value for the granular dimensionality.

The strength of the gravitational pull between two locations in space depends on the ratio between the granular dimensionality of those two locations, which is

$$F_g \propto \frac{2 + \left(1 - \dfrac{1}{\sqrt{n_V + n_1}}\right)}{2 + \left(1 - \dfrac{1}{\sqrt{n_V + n_2}}\right)}$$

with n_1 and n_2 being the number of connections of one elementary space to other neighbouring elementary spaces, and n_V being the minimal number of connections due to the omnipresence of vacuum fluctuations. If gravity was yet just another quantum force, it would be similar in strength as the other forces, depending solely on the density of 'gravitons' around the gravitational body. One may think that having all the different force transmitting particles contribute equally to the strength of gravity should make gravity stronger, as there are more particles intermediating it.

One may then also think that adding all those virtual particles from the quantum vacuum wouldn't change much of this, since vacuum energy is everywhere the same. But looking at the terms, we see that adding a fixed high value representing the vacuum energy to both n_1 and n_2 immediately brings the value for D very close to 3. Therefore the differences in the granular dimensionality of different spots in a gravitational field are usually extremely tiny. That is why gravity is so weak. Things need to be as big as moons or planets to have noticeable gravity. This is known as the *hierarchy problem*.

How much weaker than the other forces is gravity? It is 10^{38} times weaker than the strong force, 10^{36} times weaker than the electromagnetic force, and 10^{26} times weaker than the weak force. If the proportionality factor for the size of elementary spaces corresponds to the strength of the united GUT force, it would bring us close to the right magnitude we need for not only having a gapless patchwork of elementary spaces but having so many layers of them that gravity becomes very weak.

Using the most general estimate of 10^{113} J/m^3 for the vacuum energy density and small elementary spaces based on a small G, we saw that the vacuum fluctuations within 1 m^3 give us only enough fabric of space for a tiny volume of 10^{-44} m^3. Basing our own higher estimate for the vacuum energy density (10^{116} J/m^3) we can fill up 10^{-41} m^3 (again basing a small G).

If the proportionality factor for the size of elementary spaces corresponds to the united GUT force, then elementary spaces could be 10^{36} or even 10^{38} times bigger than the usual value for G suggest. The different quantum forces meet at a strength of about 10^{37} times the strength of gravity. With a proportionality factor that corresponds to a force 10^{37} times stronger than gravity, we would get elementary spaces which are 10^{37} times larger.

It seems if we want to use a single proportionality factor which is force neutral, then we have to use 10^{37}. This assumption is reinforced by the fact that if we could approach an electron so close that its virtual particles cloud disappeared, then the charge we would find would correspond to this strength of force. This seems like a solid argument and for long time it was believed that 10^{37} indeed is the right proportionality factor (August 2016 – April 2018), but a thorough analysis revealed that using the strength of the 'naked charge' (without the virtual particles cloud) in combination with the observed electron mass (not the 'naked mass') is inconsequent and if we used the naked mass instead the proportionality between the strength of electromagnetism and gravity wouldn't change despite the running of coupling (see chapter 3.12). The reached conclusion was that 10^{36} is instead the right proportionality factor to use.

When using G as the proportionality factor the elementary space of an unmoving electron would have a radius of only 1.35268×10^{-57} m. With this new constant, which we will from now on call G_E and which shall only be applied to elementary spaces, we get a size for electrons of 10^{-21} m. That is 10^{-63} m³. If we imagine the whole vacuum energy to exist in form of rest mass electrons, we get 9×10^{128} electrons ($7.5 \times 10^{115}/8.1 \times 10^{-14}$ J), which we could round up to 10^{129}. If every electron has an elementary space with a volume of 10^{-63} m³, then we get close to 10^{66} layers of elementary spaces (129 minus 63), which means that there are about 10^{66} connection from one elementary space to other elementary spaces. In this case the granular dimensionality of empty space would be about

$$D = 2 + \left(1 - \frac{1}{\sqrt{10^{66}}}\right) = 3 - 10^{-33}$$

That would mean differences in granular dimensionality can only show up 34th digits after the decimal point. This would make gravity 10^{34} times weaker than the other forces. However, we know it is at least 10^{37} times weaker than the GUT force and 10^{36} weaker than electromagnetism.

Both gravity and electromagnetism are forces with an unlimited reach in distance. That makes it very much conceivable that gravity is a side effect of electromagnetism. From there it is not a big step to the conclusion that it must be a side effect of not only electromagnetism but all the quantum forces.

However, in our calculation of the size of elementary spaces we used the low energy value for G_E and consequently we have to stick to the same when it comes to granular dimensionality. Thus $3 - 10^{-35}$ should be the true value for granular dimensionality, representing a gravity that is 10^{36} weaker than electromagnetism.

The value of 7.5×10^{115} J/m³ was an approximation counting in only the contribution from one kind of particles. We have 17 fundamental particles in the standard model, which do all contribute to the value of the vacuum energy density. Yet these make up only 40% of all matter with the rest being dark. Counting in all these we get 3.85×10^{130} electron-like elementary spaces ($3.1 \times 10^{117}/8.1 \times 10^{-14}$ J), which leads to a dimensionality of

$$D = 2 + \left(1 - \frac{1}{\sqrt{10^{67}}}\right) \approx 3 - 10^{-34}$$

$$D \approx 2.999999999999999999999999999999991$$

That would make gravity 10^{35} times weaker than electromagnetism. That is quite close to 10^{36}, but the difference is significant. We will see in the next chapter if using a more accurate quantization will bring us closer to the actual value.

In space particle dualism theory a high value for the energy of the vacuum is nothing harmful. It is needed to get a fully developed space. In standard physics things are very different: there even a homogeneously dispersed energy like the vacuum energy has a gravitational effect, which is in this case repulsive, leading to the immediate destruction of the universe; a problem known as the *vacuum catastrophe*.

That doesn't happen here, because in space particle dualism theory gravity is always acting local and never global. In this picture only differences in mass density cause gravity, not absolute values of mass energy density. Thus a substance that is dispersed totally homogeneously, such as the vacuum energy, could not have any gravitational effect.

From the perspective of space particle dualism, we have to regard the acceleration of the universe's expansion as unrelated to vacuum energy. But what does control the expansion of the universe then?

The flow of time needs a *timelike limited universe* that shrinks when viewed backwards in time and that reaches a maximal density and temperature the moment primordial black holes disappear (a state that is analogous to the 'Big Bang' of the mainstream). That has to do with entropy. I suggested that it is entropy itself which forces the universe to expand. We will come back to this *entropic expansion hypothesis* in chapter 4.1.

Notes and references:

1. Peter W. Milonni – "The Quantum Vacuum".
2. De la Pena and Cetto "The Quantum Dice: An Introduction to Stochastic Electrodynamics".
3. Although moving in space can influence the flow of time, time is not treated as a dimension in space particle dualism. From the perspective of this theory space-time diagrams are merely visualizations of relativistic impulses. When we represent impulses by little arrows with certain lengths pointing in certain directions in space, we do not assign any reality to the arrow itself. In space-time diagrams the values of impulses are no longer represented by the length of vectors, but by the inclination of worldlines against the time axis. Replacing the vector length by an additional dimension doesn't make it any more real than it was before. It is merely of illustrative nature.
4. r is here the Euler-Mascheroni constant, which is the limiting difference between the harmonic series and the natural algorithm. It must here be employed as a correction term for very large sums and its value is 0.577218.
5. Their strengths do not meet exactly. Ways to make the match exact are supersymmetry (not without manually adjusting the masses of supersymmetric partners) or so called mirror particles.

3.8   Frequencies for probing space [03.09.2016 (critical photon frequency); 05.09.2016 (critical electron impulse); 08.09.2016 (largest relevant wavelength in the quantum vacuum); 21.11.2017 (new vacuum energy value through direct quantization of energy & corrected granular dimensionality); 02.04.2018 (corrected value for G_E); 27.05.2018 (G_E can be known independently from G); 05.07.2018 (deriving n from G_S); 17.01.2019 (using slits); 19.01.2019 (higher frequencies allow using wider slits); 22.01.2019 (needs to be done in outer space); 24.11.2019 (using monoatomic lattice to function as slits); 04.02.2020 (critical wavelength for photons and electrons almost exactly the same); 07.02.2020 (photon measures most precise of all particles)*; 08.02.2020 (corrected SI unit of G_E)]

*In collaboration with Necati Demiroğlu (Turkey); David Chester (UCLA & MIT; US); Smita Yadav (India); Finn (Germany); Anwar Jawad (Iran).

In the last chapter we found that the proportionality constant for the size of elementary spaces is not the gravitational constant G, but a new G scaled up to the strength of electromagnetism. The only function of this new constant is to determine the size of elementary spaces. Yet the fact that it is applied to inertia mass makes it pretty similar to the gravitational constant of general relativity. That is why we chose to call it G_E, which can stand for 'gravitational constant of elementary spaces' or simply 'elementary space constant'. Its value is given by G multiplied with the ratio between the coupling constant of the electromagnetic force, also called the fine structure constant α, and the coupling constant of gravity[1], denoted as α_G:

$$G_E = G \times \frac{\alpha}{\alpha_G} = G \times \frac{k_e\, e^2}{G\, m_p^2} = G \times [\, 1.236348099 \times 10^{36}\,]$$

From this we can not only see why electromagnetism is about 10^{36} times stronger than gravity, but we also realize that we actually don't need G in order to know G_E, because G is canceled out, so that

$$G_E = \frac{k_e\, e^2}{m_p^2} = 8.246441821 \times 10^{25}\ \mathrm{m^3\ s^{-2}\ kg^{-1}}$$

In the last chapter we further realized that gravity is only a side effect of the other forces and that there is no 'graviton'. If gravity is a side effect of the other forces that means that mass is also just a side effect of the various charges. Thus in space particle dualism the usual gravitational constant G is applied very different from general relativity. It applies only to rest mass and not to the overall mass energy, simply because only charges can give rise to the emergent phenomenon gravity. We will study the implications of this radical new view in detail in the chapters 6.2 till 6.6.

Using this value for G_E, an electron which would otherwise have an elementary space radius of 10^{-57} m is thereby increased to the size of 10^{-21} m, being only three orders of magnitude below the smallest probed distance of 10^{-18} m.

But what happens when the wavelength of a particle becomes shorter than the diameter of its elementary space? If the analogy between black holes and elementary spaces were to be taken serious we would have to assume that a maximal frequency would be reached when the elementary space becomes equal in diameter to the wavelength, but now that we have elementary spaces depending on G_E and black holes on G, black holes and elementary spaces become very different.

Therefore there is currently no reason to assume a particle would reach a maximal frequency or energy when its wavelength becomes shorter than its elementary space diameter. Furthermore, if the elementary space of a particle has always a bigger radius than its Schwarzschild radius, then a single particle can never become a black hole, even when exceeding the Planck mass. Although not being the maximal frequency, it is still a highly significant frequency, marking the energy at which the nature of space reveals itself. We can calculate this frequency by taking the elementary space diameter, replacing M by E/c^2 and setting it equal to the wavelength of a photon:

$$\frac{4\,G_E\,E}{c^4} = \frac{h\,c}{E}$$

$$E_{crit_\gamma} = \sqrt{\frac{h\,c^5}{4\,G_E}}$$

$$E_{crit_\gamma} = 2.2 \times 10^{-9}\,\text{J}$$

This corresponds to a frequency of 3.3×10^{24} Hz. For comparison: visible light has a frequency of $430 - 750$ THz, one THz being 10^{12} Hz, and gamma rays have a typical frequency of 10^{19} Hz. The wavelength is:

$$\lambda_{crit_\gamma} = 9.006586757 \times 10^{-17}\,\text{m}$$

Very-high-energy gamma rays have frequencies from 2.42×10^{25} to 2.42×10^{28} Hz. Does that mean we can we use existing gamma ray detectors to probing the hyper-spherical nature of photons? Possibly yes, but only with space-based gamma ray detectors, because the earth-based ones detect gamma rays only indirectly, through the Cherenkov radiation they produce in the atmosphere.

Electron microscopes can be used to see much smaller things than microscopes based on visible light. Of course we do not have to rely on visible light only, and so electrons do not necessarily reach shorter distances (higher resolution) than photons. In order to find out which of the two reaches down to smaller scales, we will now calculate the critical wavelength of electrons.

The relativistic de Broglie wavelength of an electron is given by

$$\lambda = \frac{h}{p} = \frac{h\sqrt{1-(v/c)^2}}{m_0\,v}$$

If we set this equal to the elementary space diameter of a relativistic mass, we get

$$\frac{4\,G_E\,m_0}{c^2\sqrt{1-(v/c)^2}} = \frac{h\sqrt{1-(v/c)^2}}{m_0\,v}$$

Now we bring this into the general form for quadratic equations:

$$\frac{4\,G_E\,m_0^2}{h\,c^2} = \frac{1-(v/c)^2}{v}$$

$$\frac{4\,G_E\,m_0^2}{h\,c^2} = \frac{1}{v} - \frac{v}{c^2}$$

$$c^2 - v^2 = \frac{4\,G_E\,m_0^2}{h\,c^2}\,c^2\,v$$

$$c^2 - v^2 - \frac{4\,G_E\,m_0^2}{h\,c^2}\,c^2\,v = 0$$

$$-v^2 - \frac{4\,G_E\,m_0^2}{h\,c^2}\,c^2\,v + c^2 = 0$$

And solve it for v:

$$v_{1,2} = \frac{-c^2\,\dfrac{4\,G_E\,m_0^2}{h\,c^2} \pm \sqrt{c^4\left(\dfrac{4\,G_E\,m_0^2}{h\,c^2}\right)^2 + 4\,c^2}}{2}$$

Since negative speeds can be ruled out we can ignore one of the solutions and write:

$$v = \frac{-\dfrac{4\,G_E\,m_0^2}{h} + \sqrt{\dfrac{16\,G_E^2\,m_0^4}{h^2} + 4\,c^2}}{2}$$

Inserting the electrons mass m_e for m_0, we get

$$v_{e^-_{crit}} = 299{,}792{,}457.793 \text{ m/s}$$

This is 99.9999999% of lightspeed and corresponds to a de Broglie wavelength of $9.0065857 \times 10^{-17}$ m. Now let's compare that to the critical wavelength of photons:

$$\lambda_{crit_\varepsilon} = 9.0065857 \times 10^{-17} \text{ m}$$

$$\lambda_{crit_\gamma} = 9.006586757 \times 10^{-17} \text{ m}$$

That is a 6 digits agreement. The alignment of these two figures is so neat that it makes us think that higher precision calculations would bring it further and further into agreement.

The constants involved were not the same. In the first calculation we had h, c and G_E, while the second calculation involved m_e, besides the aforementioned. Nevertheless we got almost the same results. While electrons are better for probing small distances in the low energy realm, they become equal in efficiency in the high energy sector. One has to keep in mind that the electron's mass is just another form of energy. It is hard to produce and utilize photons with the same wavelength as electrons, while the energy expenditure and feasibility become equal when approaching very high energies and small distances.
However, even at high energies electrons and photons are not the same and the fact that their critical energies are almost equal is probably due to the fact that the rest mass energy of the electron becomes very small compared with its overall mass energy when it approaches light speed.

The above similarity in the critical wavelengths of photons and electrons is pretty stricking however, and at least judging from this low precision calculation (only a few digits were calculated), it seems like the electron is reaching down to lower distances than the photon.

What about other rest-mass particles?
The heaviest elementary particle that can appear isolated is the Higgs-boson. It has a mass of $2.228328369933 \times 10^{-25}$ kg. When we repeat our above calculation with this mass, we arrive at a much lower critical speed and a larger critical wavelength, namely:

$$v_{crit} = 3{,}635{,}369.78004 \text{ m/s}$$

$$\lambda_{crit} = 8.1789265 \times 10^{-16} \text{ m}$$

If smaller masses can probe shorter distances, does that mean we can surpass the precision of photons and electron when using neutrinos, the particles with the smallest known mass?

In order to answer that question, we want to first find out if the electron really reaches or possibly even surpasses the photon when it comes to measuring small distances.

For finding that, we want to explore if there is a mass m at which the critical wavelength equals exactly the critical wavelength of the photon. We therefore state:

$$\frac{4\,G_E\,m}{c^2\,\sqrt{1-(v/c)^2}} = \frac{h\,\sqrt{1-(v/c)^2}}{m\,v} = \frac{h\,c}{\sqrt{\dfrac{h\,c^5}{4\,G_E}}}$$

First we extract the value of the term under the root from the left part of this threefold equation:

$$\frac{h\,\sqrt{1-(v/c)^2}}{m\,v} = \frac{4\,G_E\,m}{c^2\,\sqrt{1-(v/c)^2}}$$

$$\frac{h\,(1-(v/c)^2)}{m\,v} = \frac{4\,G_E\,m}{c^2}$$

$$1-(v/c)^2 = \frac{4\,G_E\,m^2\,v}{h\,c^2}$$

Now we take the second equation-pair and try to isolate the term under the root:

$$\frac{h\,\sqrt{1-(v/c)^2}}{m\,v} = \frac{h\,c}{\sqrt{\dfrac{h\,c^5}{4\,G_E}}}$$

$$\sqrt{(1-(v/c)^2)\,\frac{h\,c^5}{4\,G_E}} = m\,v\,c$$

$$(1-(v/c)^2)\,\frac{h\,c^5}{4\,G_E} = m^2\,v^2\,c^2$$

Now we can put in our previous equation for the term under the root:

$$\frac{4\,G_E\,m^2\,v}{h\,c^2}\,\frac{h\,c^5}{4\,G_E} = m^2\,v^2\,c^2$$

$$v = c$$

From this we conclude that for a particle to match the measurement precision of the photon precisely, it would have to travel with the same speed as the photon. Mathematically expessed that is:

$$\lim_{\substack{v \to c \\ m \to 0}} \frac{h\,\sqrt{1 - (v/c)^2}}{m\,v} = \frac{h\,c}{\sqrt{\dfrac{h\,c^5}{4\,G_E}}}$$

To be absolutely sure we additionally solve the above equation for m.

Therefore we first square it and then multiply it by $m^2\,v^2/h^2$:

$$1 - \frac{v^2}{c^2} = \frac{4\,G_E\,m^2\,v^2}{h\,c^3}$$

Now we introduce a dimensionless constant $\alpha = 2\,G_E\,m^2/h\,c$ and rewrite the above in dimensionless quantities, which yields:

$$1 - \frac{v^2}{c^2} = 2\,\alpha\,\frac{v^2}{c^2}$$

We then use this same constant α to rewrite our formula for v, which yields:

$$\frac{v}{c} = -\,\alpha + \sqrt{\alpha^2 + 1}$$

Solving for v/c in the first equation yields:

$$\frac{v}{c} = \frac{1}{\sqrt{1 + 2\,\alpha}}$$

Setting this equal to the second equation to remove v/c gives:

$$\frac{1}{\sqrt{1 + 2\,\alpha}} = -\alpha + \sqrt{\alpha^2 + 1}$$

This equation only has a solution for $\alpha = 0$, implying $m = 0$.

Particles with mass can come infinetely close to the speed of light, but they can never reach it. For measuring distances smaller than the smallest distance photons can measure, a particle would have to exceed the speed of light, which is impossible.

That shows that although electrons come close, no particle with rest mass can reach the photon in its measurement precision.

9×10^{-17} m seems to be the limit for the direct measurement of distances using only one particle. However, indirect methods of distance measurement can reach deeper. The gravitational wave detector LIGO in Livingston measures distance differences in the magnitude of 10^{-18} m , using laser interferometers. When measuring the interference between waves one is not longer dependent on the wavelength for reaching smaller distances.

These critical wavelengths are not minimal wavelengths (maximal frequencies). Although the particle cannot be used to directly probe shorter distances, it still can approach higher frequencies corresponding to shorter wavelengths, the shortest wavelength probably being the Planck-length.

What about maximal wavelengths? In our calculation of the vacuum energy in the last chapter we thought that it is not important to set an upper limit for the wavelength of particles, because we assumed that a sum of smaller and smaller energies would converge to a finite result – that is why we stopped at one meter.
If the Planck length maintains its role as the shortest physically relevant length, then the lowest frequency might be reached when the elementary space diameter shrinks down to the Planck length. By setting the elementary space diameter equal to the Planck length we can find this lowest energy:

$$\frac{4\,G_E\,E}{c^4} = \sqrt{\frac{G\,\hbar}{c^3}}$$

$$E_{min} = \frac{c^4\,\sqrt{G\,\hbar/c^3}}{4\,G_E}$$

$$E_{min} = 3.9578 \times 10^{-28}\ \text{J}$$

This corresponds to a wavelength of 502 m. Yet this is below the longest detected wavelength which is about 3,000,000 km. That would mean there is no upper limit for wavelengths. However, elementary spaces with a surface area smaller than a Planck surface would probably not influence the granular dimensionality of space. This will again become important in chapter 6.6 when we examin the range of the different forces.

In any quantum theory of gravity space has to be build up by quanta of space which are carried by particles, mostly virtual particles, and there cannot be an infinite amount of them. In other approaches to quantum gravity, like loop-quantum gravity, the atoms of space always remain at the same scale, the Planck scale, regardless of their energy.
When we are concerned about the energy of the vacuum we mainly want to know the value for the granular dimensionality of empty space, and since virtual photons with energies of less then 3.9578×10^{-28} J do not contribute to this granular dimensionality, we can ignore them. That doesn't mean they don't exist in the quantum vacuum. In fact they must exist, because quantum forces are dependent on virtual photons of all wavelengths, including such which have less than this minimal energy (see chapter 6.6).
Real photons can also have energies smaller than this, but they have to be measured at one point in order to have existed, and for that they have to cause some detectable macroscopic distortion (see chapter 4.3). There is yet no known maximum for a wavelength that can still be detected.

So let's repeat the calculation for the vacuum energy from the last chapter by using this value for the lowest relevant wavelength. As a virtual particle this lowest relevant energy photon would have a lifetime of

$$\Delta t = \frac{\hbar}{2\, E_{min}} = 1.332253621 \times 10^{-7} \text{ sec}$$

That corresponds to 2×10^{36} Planck times, which gives us a new value for n in the calculation of vacuum energy:

$$\sum_{i=1}^{n=2\times10^{36}} E = \frac{\hbar}{2n \times t_{PL}}$$

$$\int_{i=1}^{n=2\times10^{36}} E = \frac{1}{n} + r \times 978,056,180.7$$

$$\int_{i=1}^{n=2\times10^{36}} E = 8.2317 \times 10^{10} \text{ J}$$

This is only slightly higher than the energy calculated in the last chapter, and leads to the same overall vacuum energy density of about 10^{116} J/m^3 (8.4568×10^{115} J/m^3).

In the last chapter it was mentioned that quantizing the energy through the lifetime of a virtual particle is only an approximation. So let's now get a more precise value by quantizing the energy directly by using natural multiples of the Planck length for the wavelength. The energy of a photon is given by $E = h\,c$ and as mentioned before the minimal energy for virtual particles we calculated corresponds to a wavelength of 406 m, which equals 2.5×10^{37} Planck lengths. Thus we can re-calculate our sum as:

$$\sum_{i=1}^{n=2.5\times10^{37}} E = \frac{h\,c}{n \times l_{PL}}$$

$$\int_{1}^{2.5\times10^{37}} E = \frac{1}{n} + r \times (1.229061821 \times 10^{10})$$

$$\int_{1}^{2.5\times10^{37}} E = 6.109096188 \times 10^{11} \text{ J}$$

Multiplying this with the degrees of freedom for space (10^{105}) yields $6.109096188 \times 10^{116}$ J/m^3, which is 8 times higher than the value we got using time based quantization.
We can be pretty sure that this is indeed the true value for the vacuum energy, when counting in virtual photons only.

Now, let's repeat the calculation of granular dimensionality using these improved values. The number of electron-like elementary spaces is:

$$n_e = \frac{6.109096188 \times 10^{116} \text{ J/m}^3 \times 17 \times 2.5}{8.187105651 \times 10^{-14}}$$

$$n_e = 3.171286643 \times 10^{131}$$

If each of them has a volume of 10^{-63} m^3, then that yields 10^{68} connections and a granular dimensionality of

$$D = 2 + \left(1 - \frac{1}{\sqrt{10^{68}}}\right) \approx 3 - 10^{-34}$$

$$D \approx 2.99999999999999999999999999999999991$$

That means any difference in granular dimensionality could only show up at the 35th digit after the decimal. That roughly corresponds to the weakness of gravity. Gravity is 10^{36} times weaker than the electromagnetic force.

Although this calculation was more precise than the calculation before, it still simplified things a lot by using same size elementary spaces regardless of the different energies that the particles have. However, we can be pretty sure that 10^{-21} m is a good average for the radius of elementary spaces.[3]

What is not so sure is if single particles in the quantum vacuum can really reach up to the Planck energy. After all using G_E the elementary space of a Planck particle would be 40 meters in radius. If such an object could exist, it would look pretty much like a 'giant atom'. Yet of course as a virtual particle it would exist only for a single Planck time (9×10^{-44} seconds), which makes it less scary. We will come back to this issue in chapter 3.15.

Can we go the other way around and derive the value for granular dimensionality from the value of G?
First we have to clarify that what the above calculation was trying to find was the strength of the gravity arising from the strong force, and not the usual gravity measured without taking account of mass defect. This goes deep into issues that will be discussed in chapter 6. For now we shall simply refer to the true value of G as G_S (the gravitational constant of the strong force). As we shall see in chapter 6.5, its value is $7.110645989 \times 10^{-11}$ m^3 kg^{-1} s^{-2}.

If gravity was as strong as electromagnetism it would have the value of G_E. The value of $1/\sqrt{n}$ divided by 10 then tells us how weak gravity is.[4] Thus G_S is given by:

$$G_S = \frac{k_e\, e^2}{m_p^2} \times \frac{1}{10\,\sqrt{n}}$$

We can solve this for n, which yields:

$$n = \left(\frac{k_e\, e^2}{10\, G_S\, m_p^2}\right)^2$$

$$n = 1.344977701 \times 10^{70}$$

And that leads to a granular dimensionality of:

$$D = 3 - 8.62268375 \times 10^{36}$$

The true value for n we obtained here is two orders of magnitude higher than what we calculated above and in the last chapter. That is probably due to the fact that we didn't take account of changing elementary space radii. A more precise calculation taking account of this can be found in chapter 3.15.

Probing elementary spaces with a multi-slit graphene plate

How can we most directly find evidence for the extended nature of elementary particles as predicted by space particle dualism theory?

The most direct test would be to let a photon above the critical frequency travel through a slit that has the critical wavelength as its diameter. That would be a tiny 9.006×10^{-17} m. For electrons it would be 8.861×10^{-17} m. Both is far beyond of what is possible with today's technology.

Higher energies means larger sizes for elementary spaces. How small needs to be our slit if we sent ultra-high frequency photons through them? Ultra high frequency photons are photons with a frequency of more than 2.42×10^{28} Hertz. Our slit would have to have a diameter of:

$$d = \frac{4\, G_E\, h\, f}{c^4} = 6.548097488 \times 10^{-13} \text{ m}$$

Such a slit would still be very hard to build.

The highest frequency of gamma rays that has been detected is 10^{30} Hertz. That corresponds to a slit width of $1.352912881 \times 10^{-11}$ m.

Nanoengineering professor Darren Lipomi has succeeded creating ultra-narrow slits between two nanostructures by employing a graphene spacer, which is then etched away to create the gap.[5]

For probing elementary spaces, a large array of slits would have to be cut into graphene plates and the plates could then be placed on top of a gamma ray detector. A large number of such slits is necessary in order to ensure that a rare ultra-high frequency photon gets stuck in one of them.

Earth based observatories for gamma ray bursts only measure the Cherenkov radiation that is created when gamma rays hit the atmosphere. This way the wavelength of the gamma rays can only be inferred from the amount of Cherenkov radiation that is detected at the ground.

However, gamma ray bursts are typically observed with satellite based telescopes which detect the gamma rays directly, not through subsequent *Cherenkov radiation*, because there is no atmosphere in which it could emerge.

Is the highest ever measured frequency of 10^{30} Hertz directly measured, or only indirectly inferred from Cherenkov radiation?

Since we are not interested in the source of the gamma ray bursts, but only in single high frequency photons within it, one could assume it would be enough to have a few of them making it to the ground. However, having a small number of them making it to the ground is not enough, because for there being any chance of some hitting or passing through the slits, there would have to be a large number of them in the first place. That means the experiment can only be done in outer space.

In order to increase the likelihood of a detection it should not be a single slit, but preferably a grid of many atom-diameter-wide slits.
If it is a single slit, then it can be two very sharp edged plates that are moved closer and closer together, observing when particles cease to be able to traverse the slit.
However, with many slits in a larger plate, one would have to be able to cut very precisely to have them have only one atom distance from one side of the gap to the other. Maybe if one makes it thinner and thinner close to the slit, one can push single atoms out of the structure using a gentle laser.

Graphene is easy to work with and it should be possible to create many slits in a single graphene plate. Such plates have a diameter d of 3×10^{-10} m and since it is a single atom layer structure, any slits in it would have that diameter too. In order to probe elementary spaces photons with a frequency of

$$f = \frac{d_{gap}\, c^4}{4\, G_E\, h} = 1.108718906 \times 10^{31} \text{ Hertz}$$

Would have to be sent through such slits. This is a very high frequency and it is unclear if there are enough photons of such a high frequency even within gamma rays, to make such a test feasible.

Researchers at the Department of Energy's Lawrence Berkeley National Laboratory and their collaborators at the University of Frankfurt, Germany have managed to conduct double slit experiments on single hydrogen nuclei.[6]
If something like that could be done with ultra-high frequency photons, then that would open up the possibility to probe elementary spaces with slightly lower frequencies. Using single hydrogen nuclei as slits would allow narrowing down the slit width to $7.4084809162 \times 10^{-11}$ m (1.4 atomic units), which brings down the required photon frequency to:

$$f = 2.737974287 \times 10^{30} \text{ Hertz}$$

It is very questionable if we can manage to get such rare ultra high frequency photons to scatter on specific nuclei. It is a very unrealistic scenario.
Similarly unrealistic it is to cut many slits in an existing structure and hope that there are enough of them so that a photon of exactly the right frequency might pass through

one of them. This becomes even more illusive when we think about how photon frequencies are measured. The most common method is to use a spectrometer, and when doing so, all the relevant photons would be pretty much at the same area of our graphene plate.

If the width of our slit must be roughly the size of an atom and even slightly below, then can't we simply take a single layer atomic structure and let the short wavelength photons travel through between the atoms?
The problem is that the structure would be destroyed long before the frequencies we are interested in could have a chance to pass through without causing damage.
We are inclined to not want to use a many layer structure, because we tend to think that if there are many layers, then surely any photon must get stuck in between.

This is however not the case for monoatomic lattice structures. Every element of the periodic table has three distinct lattice constants A, B and C. These are usually between 200 pm and 2500 pm.[7] As we can see from this small sample below, the lattice constants increase with the atomic number:

Element	A	B	C
H (1)	228.58 pm	228.58 pm	358.43 pm
He (2)	246.4 pm	246.4 pm	671.1 pm
Li (3)	250.71 pm	250.71 pm	406.95 pm
...	...	...	...

Obviously for getting the narrowest possible slits, we need to go for the lightest element in the periodic table. That is hydrogen. In order for hydrogen to form a lattice structure, it needs to be in solid form, and for that it needs to be kept below its melting point, which is 14.01 K (corresponding to $-259.14\,°C$), which is fairly easy to achieve.
It means that our new slit would have a diameter of:

$$d_{gap} = A = B = 2.2858 \times 10^{-10} \text{ m}$$

Now the required frequency is:

$$f = \frac{d_{gap}\, c^4}{4\, G_E\, h} = 8.4476989 \times 10^{30} \text{ Hertz}$$

If photons get stuck in atomic lattice with a lattice constant of 2.2858×10^{-10} meter, while according to ordinary quantum mechanics they should be able to pass any gap larger than 10^{-22} meter, then this would be direct evidence that the radius space particle dualism gives for elementary particles is correct.
If elementary spaces were any larger than the elementary space constant suggests, then elementary spaces would get stuck between atoms at even lower energies.

So the experimental constellation is a spectrometer in outer space with a solid hydrogen plate behind it and in front of a photon detector.

It is quite an important experiment to do, because as we will learn in chapter 6.8 there is quite some doubt about the factor 2 used in the formula for elementary spaces. If elementary spaces are derived from pure units, instead of the a contrast to gravity, then that would suggest that the factor of 2 which is there because we simply scaled up the Schwarzschild radius is not justified. In that case the above calculated frequency has to be doubled.

The experiment is very feasable and not too costly, but due to the fact that space particle dualism is not very widly known yet, it will probably take quite a few years until there is a space agency that attempts to carry it out.

Notes and references:
1. Here I used a coupling constant for gravity that is defined basing the proton mass. Elsewhere it is often defined using the electron mass, but since G is here regard to arise as a side effect of hypercharge, we have to use the proton, because the electron has no hypercharge. The reader may want to suggest that we should better use a single quark, because it is more fundamental, but a quark doesn't have an integer charge and is therefore unfit for comparing gravity to electromagnetism.
2. Using the energy momentum relation $E^2 = (p\,c)^2 + (m_0\,c^2)^2$.
3. We might be tempted to take h as the lowest energy, but that would make h dependent on the definition of the second or the meter.
4. The division through 10 is necessary because the potency is showing how many nines there are already, while the first digit where differences in granular dimensionality can show up is one order of magnitude below that.
5. "Using the Thickness of Graphene to Template Lateral Subnanometer Gaps between Gold Nanostructures"; 2015; Aliaksandr V. Zaretski; Brandon C. Marin; Herad Moetazedi; Tyler J. DillLiban Jibril; Casey Kong; Andrea R. Tao; Darren J. Lipomi.
6. "The Simplest Double Slit: Interference and Entanglement in Double Photoionization of H2"; D. Akoury, K. Kreidi, T. Jahnke, Th. Weber, A. Staudte, M. Schöffler, N. Neumann, J. Titze, L. Ph. H. Schmidt, A. Czasch, O. Jagutzki, R. A. Costa Fraga, R. E. Grisenti, R. Díez Muiño, N. A. Cherepkov, S. K. Semenov, P. Ranitovic, C. L. Cocke, T. Osipov, H. Adaniya, J. C. Thompson, M. H. Prior, A. Belkacem, A. L. Landers, H. Schmidt-Böcking, and R. Dörner.
7. [https://periodictable.com/Properties/A/LatticeConstants.st.html?fbclid=IwAR0N 0cdASIjQ2dGmDlJlgQboh2P4IF3mosEoi-UVbm_eFPXdAVwZxd3Fp-o].

3.9 Mass out of light? [20.11.2005 (basic idea); 19.08.2019 (there is more to this idea than previously thought)]

Note-1: This chapter was written before the true origin of mass and the real size of elementary spaces was found. Therefore many of the expressed thoughts are outdated from todays perspective. It is however still of historic value because it was leading up to the *lepton conglomerate hypothesis* which is the topic of chapter 3.10.

Note-2: New developments described already in chapter 3.6 show that the here discussed idea is not entirely wrong after all, namely when one replaces merging black holes with merging elementary spaces.

In 2005 the different types of spaces were associated only with mass-less particles, because those were regarded as more fundamental than particles with a rest mass. This was clearly inspired by Penrose's twistor theory, where massive particles are also believed to be secondary. But even outside of twistor theory the general tendency is to do so: in the most generally accepted view particles are believed to gain their masses through the interaction with the so called Higgs field.

Later in 2015, the difference between massive and massless particles was modelled as a difference in dimensionality, massless particles having only one dimension, while massive particles having two; the evidence for that being the increased degrees of freedom in spin for mass-carrying particles (see chapter 3.5). In 2019 however, it became clear that if complex 2-spheres with their $SU(2)$ symmetry are to correspond to the weak force, and complex 1-spheres with their $U(1)$ symmetry to the electromagnetic force (see chapter 3.4), then both types of elementary spaces must give rise to both mass-carrying and massless particles. As already mentioned in chapter 3.6, that can be achieved by the merging of two massless particles.

Back in 2005 the black hole particle analogy had led to the hypothesis that one could get rest mass particles by letting bundles of massless particles form tiny black holes. This seemed to be an elegant idea, because in a world without atoms, event horizons would be the only thing that can stop photons. A black hole made out of photons would look just like a black hole made out of ordinary mass; at least that is the case in standard physics. However, in today's space particle dualism theory photons could not form black holes, because their elementary spaces are much larger than their Schwarzschild radii, and in addition to that, they don't even have Schwarzschild radii, because in space particle dualism only particles with a charge or hypercharge exert gravity.[1]

Even if that wasn't the case, general arguments we will encounter in chapter 4.3 show that black hole event horizons trap only rest mass particles. We will also learn that the size of a black hole is only proportional to the number of quarks it contains.

Considering these details, it turns out that elementary spaces are very similar to classical black holes, just different in size, while actual black holes are in fact just quasi-black holes, deviating significantly from our traditional model of black holes.

While it is hard to see how to get fixed constant rest masses from thinking of electrons as bundels of light rays that got trapped in an event horizon, there is actually hope to get that when thinking of them as pairs of 1-dimensional photon-elementary spaces. As we learned in chapter 3.7, there is a certain critical frequency at which the elementary space diameter grows larger than the wavelength of the particle.

But even from within the standard model of physics it is hard to see how to get fixed constant sizes for such horizons.

If we instead take the known charges of rest mass particles and let them shrink in size till event horizons form, we do get fixed constant masses, but these masses are near the Planck mass, which is far too heavy (see chapter 3.2). Alternatively shrinking them down to only the critical wavelength (chapter 3.7), gets us close to the proton mass.

We know that photons have no electrical charge, and therefore black holes created out of photons wouldn't have charge either. That is different for elementary spaces created in that manner, because it is the structure of elementary spaces which generates charge and hypercharge.

A universe filled with only massless particles would have no size and scale and therefore hardly any basis for talking about conditions that allow for the formation of event horizons. Therefore massless particles and mass-carrying particles must be relying on each other; after all, without scale there can't be wavelength either.

As we will see in chapter 3.12 it is possible to link the mass of the electron to the strength of electromagnetism. Before we do so we will take the electron mass as a basic fundamental constant and see what we can do with it.

There are three generations of elementary particles which do all share the same properties and differ only in their masses. Among these three generations only the particles of the first generation are stable. The other particles appear only as products of high energy collisions of particles and dissolve into stable particles very quickly. Our world would be totally the same even if they didn't exist. So why was nature so wasteful to create them in the first place? This mystery has been puzzling particle physicists ever since the first discovery of second and third generation particles.

When trying to figure out how to represent the different generations of particles using elementary spaces, one is quickly reminded of the different energy levels in an atom. Those energy levels exist due to the nature of the wave function, which allows only whole number-values for the number of waves in every orbit. Could anything similar apply to elementary spaces?

It might turn out that the stable particles are the only really fundamental particles, and that other particles can be understood as elementary spaces with more than one particle on them. The simplest picture of that kind would be to think of the muon as consisting of two electrons and one positron, and the tauon as three electrons and two positrons, all on the same elementary space. The opposite electric charges of electrons and positrons would level each other out, so that the charge always remains to be 1. As electrons are fermions, which obey the exclusion principle, they cannot stay close to each other for long time. Also bringing them together requires big amounts of energy. I suggest the big discrepancy between their masses, for example the muon being 207 times heavier than the electron, to be due to the large amount of energy which is required to bring them together.

Whatever the true origin of the three generations is, it seems rather obvious to me that that elementary particles are quite too simple structures for having to be explained by something as complicated as vibrating strings in 11 dimensions[2] with some extraordinary topology.

If one takes a closer look at them, it is not hard to see that their masses and charges are associated. That will be shown in detail in chapter 3.11.

If string theory was true, and everything depended on randomly formed extra dimensions, then none such patterns or symmetries were to be expected. A complete mess would be much more likely then.

Also the infinities of point particle theories are more intuitively and simply resolved by using spheres instead of strings, especially because spheres can form space itself, while strings do not have such appealing properties.

Notes:

1. This shows that in space particle dualism theory the event horizon is not a perfect measure for the entropy of a black hole.

2. Now that supersymmetry is proven wrong (the Higgs boson's supersymmetric partner was not found), string theory has to go back to the usage of 26 dimensions – the number of dimensions it had before 1984 (!).

3.10 Muons and tauons as lepton conglomerates [16.07.2014 (mass increase due to degeneracy pressure)*; 07.08.2015 (positron shields half away); 19.08.2015 (mass increase due to electromagnetic repulsion); 21.02.2016 (virtual particles cloud); 09.04.2018 (repeated all calculations with the corrected value for G_E and the critical wavelength)]

*In collaboration with Bernhard Umlauf (Germany).

Note: The masses calculated here are inertia masses. According to space particle dualism the gravitational mass of the muon and the tauon equal that of the electron.

If we take the model of muons and tauons consisting of electrons and positrons (or electrons only) on single elementary spaces seriously, we will have to calculate the mass energy increase which comes along with bringing these particles together that close. For this there are two forces we need to take into account. According to the Pauli exclusion principle identical fermions cannot occupy the same quantum state simultaneously. In systems made up of many fermions, let's say electrons, this leads to a so called degeneracy pressure, when the density of the system reaches a certain level. That is because they try to be different in impulse, while being forced to take same positions. That means they move faster. The result is an increase in the pressure and the temperature of the system. Pressure and temperature are thermodynamic features and can only apply to systems made out of many particles. For a system of two electrons and one positron it is enough to apply the uncertainty principle to find the increase in impulse.

First of all we have to find out how large the uncertainty of impulse becomes when pressing particles together into a space which is smaller than anything that has been probed with today's technology.

According to the uncertainty principle the product of the uncertainty in position Δx and the uncertainty in impulse Δp is equal to half of the reduced Planck constant $\hbar$.

$$\Delta x \times \Delta p = \frac{\hbar}{2}$$

Let's first look at the case where we know Δx with maximal accuracy, which means to take the Planck length for Δx. That would be:

$$\sqrt{\frac{h\,G}{c^3}} \times \frac{m_0\,\Delta v}{\sqrt{1 - \frac{\Delta v^2}{c^2}}} = \frac{\hbar}{2}$$

If we solve this for Δv, we get:

$$\Delta v = \sqrt{\dfrac{1}{\dfrac{4\,G\,m_0{}^2}{c^3} + \dfrac{1}{c^2}}}$$

The term on the down left is so small that Δv must reach the maximum, namely $\Delta v \approx$ c. That means in this case the electron would be in a superposition of all possible speeds, from the speed of zero up to speeds arbitrarily close to the speed of light.

In order to get the resulting mass energy increase, we need to find the sum of all those endless many different impulses and divide it by the number of terms (also an infinite number).

We get the following sum:

$$m = \frac{m_0}{n} \sum_{k=1}^{n-1} \left(\sqrt{1 - \frac{(c - v_k)^2}{c^2}} \right)^{-1}$$

Whereby

$$v_k = k \times \frac{c}{n}$$

This can be transformed into:

$$m = \frac{m_0}{n} \sum_{k=1}^{n-1} \left(\sqrt{\frac{k}{n}\left(2 - \frac{k}{n}\right)} \right)^{-1}$$

By reordering the summands we obtain:

$$m = \frac{m_0}{n} \sum_{k=1}^{n-1} \left(\sqrt{1 - \frac{v_k^2}{c^2}} \right)^{-1}$$

Or

$$m = \frac{m_0}{n} \sum_{k=1}^{n-1} \left(\sqrt{1 - \frac{k^2}{n^2}} \right)^{-1}$$

When we take the limit $n \to \infty$ we get:

$$m = m_0 \int_0^1 \left(\sqrt{1 - \beta^2}\right)^{-1} d\beta = m_0 \times \frac{\pi}{2}$$

$\pi/2$ is approximately a factor of 1.57. Two electrons and one positron – that is 3 times the mass of a single electron. With the factor of $\pi/2$ we get about 4.7 times the mass of an electron.

However, elementary spaces are much larger than the Planck length, so the uncertainty in impulse might be well below light speed. In order to find that out we can input the diameter of an elementary space together with the relativistic impulse into our uncertainty equation:

$$\frac{4 G_E m_0}{c^2} \times \frac{m_0 \Delta v}{\sqrt{1 - \frac{\Delta v^2}{c^2}}} = \frac{\hbar}{2}$$

If we solve this for Δv, we get:

$$\Delta v = \frac{h c^2}{\sqrt{64 G_E^2 m_0^4 + h^2 c^2}}$$

Here the term $64 G_E^2 m_0^4$ is again so small that Δv equals c. That means the above result $\pi/2$ holds true.

Next we have to calculate the mass increase coming from the energy needed to overcome the magnetic repulsion, which is only partly cancelled out by the positron in between (it would cancel out half of the charge).[1]

So how much is this energy? If we were to press an electric charge into an infinitely small area we get 1.017×10^{-9} kg, which is close to the Planck mass (see chapter 3.2). However, we know well that elementary spaces are much larger than the Planck length. If we replace the G in the formula from chapter 3.2 by G_E (the elementary space constant), we get

$$E = \sqrt{\frac{3 e^2 c^4}{20 \pi \varepsilon_0 2 G_E}}$$

$$E = 9.1552 \times 10^{-11} \text{ J}$$

This corresponds to a mass of $1.01865891 \times 10^{-27}$ kg, which is 1,118 times the mass of the electron. Twice the charge of an electron gives us twice this mass, which would be 2,236 times the mass of the electron. That is already far more than the mass of the muon.[2]

So it can't be that the energy increases till the particles arrive at each other's elementary space. Till when then? Till the distance is less than the wave length of the electron? This wave length is about 10^{-4} m. That is quite large and too easy to reach. In particle physics there is another length or area for the electron (the cross section σ), which tells us when it is likely to interact with other particles; it is called the *classical electron radius*, and has the value

$$r_e = \frac{1}{4\,\pi\,\varepsilon_0} \times \frac{e^2}{m_e\,c^2} = 2.8179403227 \times 10^{-15}\,\text{m}$$

The energy required for shrinking a charge to a certain size is

$$E = \frac{3\,Q^2}{20\,\pi\,\varepsilon_0\,R}$$

Because we are concerned with single electrons, Q is here given in natural multiples of the electron charge:

$$Q = n \times e$$

If we divide by c^2 (in order to get a mass) and add the original mass $(n \times m_e)$ multiplied by the aforementioned factor $\pi/2$, we get

$$M_l = \left(n \times m_e \times \frac{\pi}{2}\right) + \frac{3\,(n \times e)^2}{20\,\pi\,\varepsilon_0\,r_e\,c^2}$$

For two electrons that is a mass of 5.048×10^{-30} kg, which is 5.54 times the mass of the electron.[3] That is obviously too less mass increase for creating a muon. What if elementary spaces need to grow larger than the particle's wavelength in order to melt together and give a new particle? In that case we need to use the elementary space diameter linked to the critical electron impulse found in chapter 3.8, which is 8.862×10^{-17} m. We need a radius and not a diameter, so we take half of that, which is $R_{crit} = 4.431 \times 10^{-17}$ m. We can now calculate the muon (or tauon) mass as

$$M_l = \left(n \times m_e \times \frac{\pi}{2}\right) + \frac{3\,(n \times e)^2}{20\,\pi\,\varepsilon_0\,R_{crit}\,c^2}$$

The resulting mass is 1.419×10^{-28} kg, which is 155.77 times the mass of an electron. If a positron is involved we have partial neutralization and the formula changes to

$$M_l = \left(n \times m_e \times \frac{\pi}{2}\right) + \frac{3\,Q^2}{20\,\pi\,\varepsilon_0\,R_{crit}\,c^2}$$

$$Q = \left(n_{e^-} - \frac{n_{e^+}}{2}\right) \times e$$

That results in a mass of 8.25×10^{-29} kg, which is 90.57 times the mass of the electron. Three electron charges would give a mass increase of 348.13 times the electron mass. With partial neutralization by two positrons this is 160.48 times.

The factors we need are 207 and $3{,}477$, because a muon is 207 times and a tauon $3{,}477$ times heavier than an electron.

Are there other factors which could bring further mass increase?

An electron attracts virtual positrons and thereby polarizes the vacuum around it. Therefore it constantly carries a cloud of virtual positrons with it. This cloud of virtual positrons reduces the observed charge of the electron. It is not yet known what the bare charge of the electron is, and naive calculations even end up with the result ∞, but in Roger Penrose's "The Road to Reality" we find an interesting speculation on this. He writes:

"In terms of natural units the electron's charge is about 0.0854, and it is tempting to imagine that the bare value should be 1, say. That would correspond to a scaling up factor of 11.7062 or about $\sqrt{137}$ instead of ∞."

For calculating the mass increase through the growth of this virtual particles cloud we just need to know that it is much heavier than the electron itself. A stronger charge would produce a bigger virtual particles cloud. If we double the charge, we also double the radius of the cloud. A two times bigger radius means an 8 times bigger volume, since $(2\,r)^3/r^3 = 8$. Accordingly the lepton (muon or tauon) mass would be given by

$$M_l = \left(n \times m_e \times \frac{\pi}{2} \times 8\right) + \frac{3\,(n \times e)^2}{20\,\pi\,\varepsilon_0\,R_{crit}\,c^2}$$

For two electrons that is 177.76 times the mass of the electron; for three it is 381.12. If we have a real positron reducing the charge, we then have only a volume increase of 3.375, because $(1.5\,r)^3/r^3 = 3.375$. We calculate:

$$M_l = \left(n \times m_e \times \frac{\pi}{2} \times 3.375\right) + \frac{3\,Q^2}{20\,\pi\,\varepsilon_0\,R_{crit}\,c^2}$$

$$Q = \left(n_{e^-} - \frac{n_{e^+}}{2}\right) \times e$$

For two electrons and one positron we get $101.76\,m_e$, for three electrons and two positrons we get $179.14\, m_e$.

In our above considerations we looked both at electron-only systems as well as electron-positron systems in which the overall charge is always $1\, e$. We know that the muon and the tauon have the same charge as the electron. Were the electron-only cases all unphysical?

Leaving the positrons out we get a higher overall mass, but wouldn't that give us a bigger electric charge, while the muon is well known to have the same charge as the electron?

Where does electric charge come from; and why is it that elementary particles can only have the charges -1, $+1$ or 0, taking the charge of the electron as a unit? Even quarks, which have fractional charges like $+2/3$ and $-1/3$ miraculously come together to always form charges of -1, $+1$ or 0; they don't appear alone.

The charge of an electron is a property of its wave function or more precisely of its spin. An elementary space with more than one particle on it would still only generate one wave function. We might therefore not need to have a positron to neutralize any extra charge.

But what would happen with the charge of these electrons? While pressing them together, they would have a combined charge, and their virtual particles cloud would grow, but once united onto one elementary space, they would share one wave function, and lose the additional charge. That would be a charge conservation violation, if we imagined to really produce muons this way. What happens in reality is that muons are created together with anti-muons out of the vacuum when enough energy is supplied. If a muon were to really be created by the fusion of existing electrons, extra electrons would have to be created for not violating the conservation of charge.

After the electrons are captured on one single elementary space, the grown virtual particles cloud would shrink back to its normal size, with the muon taking over the thereby lost mass.

The here presented scheme did not yet provide us with correct values for the masses of the muon and the tauon, and it is unsure if corrections and more precise calculations will do better, but it can in anyway be understood as an example of how one can, in the framework of this theory, gain a ladder of excessively increasing mass values out of a given value (the mass of the electron).

Notes:
1. Different combinations of distances for the three particles give us different degrees of neutralization, e.g. 0.9, 0.8, 0.7, ... It is to assume that all these exist in superposition, so that we have to take the sum of them to get the actual shielding. Since the sum $1 + 0.9 + 0.8 + 0.7 \ldots + 0.1 + 0$, divided by the number of its summands (10) equals 0.5, we have a shielding of 0.5.
2. The problem becomes even more severe when we use the naked charge, which according to an educated guess of Penrose (mentioned later in the text) is to a factor

of about $\sqrt{137}$ stronger than the charge we observe. However, this factor is also hidden inside G_E, since it's the virtual particles cloud which leads to the so called running of coupling.

3. Using the naked charge we get a factor of 329!

3.11 Predicting the masses of the electron, the neutrino and the proton [15.07.2017 (neutrino mass); 17.07.2017 (corrected value); 22.06.2018 (new neutrino mass); 23.06.2018 (Klaus Lux's electron mass calculation)*; 24.06.2018 (the mass of both the electron and the neutrino are sourced entirely from their force fields); 30.06.2018 (the mass of the proton is also entirely sourced from its force field); 04.07.2018 (neutron mass related to beta decay); 24.09.2019 (gluon mass); 25.09.2019 (mass of W- & Z-bosons possibly related to critical photon & electron energy)]

*In collaboration with Klaus Lux (Germany).

Also published in: Advances in Theoretical and Computational Physics (Volume 2, Issue 4) (winter 2019). www.doi.org/10.33140/ATCP.02.04.03.

According to space particle dualism, gravity is a side effect of charge. Could the same apply to inertia?

If that is the case, then one could try to predict the masses of different particles by simply looking at their charge.

The proton has the same electric charge as the positron, but besides electric charge (mediated by virtual photons), it also has a strong 'charge' or hypercharge, often referred to as 'colour-charge' (mediated by gluons), and a weak charge (mediated by W- and Z-bosons). That could be what makes it heavier than an electron.

The neutrino which has only a weak charge has almost no rest mass.

And then we have all the massless particles which happen to be chargeless.

Elementary particle	Strength of force	Mass
Electron	$\alpha = \dfrac{k_e\, e^2}{\hbar c} \approx 1/137 = 7.2 \times 10^{-3}$	$0.5109989461 \dfrac{\text{MeV}}{c^2}$ $= 9.109\,383\,56(11) \times 10^{-31}$ kg
Neutrino (sum of three 'flavors')	$\alpha_w = \sqrt{\dfrac{\tau_\Delta}{\tau_\Sigma}} = \sqrt{\dfrac{6 \times 10^{-24}\text{ s}}{8 \times 10^{-11}\text{ s}}}$ $= 2.738612788 \times 10^{-7}$	$0.12 \dfrac{\text{eV}}{c^2} =$ $= 2.138958468 \times 10^{-37}$ kg

Up-quark	$\alpha_s = 1$	$2.4\dfrac{\text{MeV}}{c^2} = 4.277916936$ $\times 10^{-30}$ kg
Down-quark	$\alpha_s = 1$	$4.8\dfrac{\text{MeV}}{c^2} = 8.555833871$ $\times 10^{-30}$ kg

The neutrino is 4,258,794 times lighter than the electron. This is quite a small mass and fits well with the weak charge of the neutrino, but the factors don't really match: the difference in strength of force is merely a factor of 26,646.
The reason might be the short range of the weak force, causing it to polarize much smaller regions of vacuum than the electromagnetic force which has no distance limit.

The quarks are a bit tricky: the rest mass values presented here are for isolated quarks without the gluon cloud around them, which do not exist in reality. It is a bit like an electron without virtual photons around it. Such an electron wouldn't have an electric charge. In the same way a quark without its gluons wouldn't have a hypercharge. That is why here it seems to be just 9.39 times heavier than an electron.

In reality quarks are combined in pairs of three, assembled to protons and neutrons. Here the proton is 1,836 times heavier and the neutron 1,838 times heavier than an electron. A third of that is 612 and 612.6 respectively. This is very roughly close to the difference between the strength of the electromagnetic force (represented by the electron) and the strong force (represented by the up- and down-quarks), which is also 2 orders of magnitude – at a distance of 10^{-15} m it is 137 times stronger than electromagnetism.

For the neutrino the comparison between charge and (inertia) mass was off by a factor of 160, while for the proton it was a much smaller factor of 4.5. That already shows that the relationship between charge and inertia isn't as simple as that between charge and gravity.

How could inertia rest mass be explained then?
The German physicist Klaus Lux suggested that the mass of particles might be equivalent to the energy of their force fields and he demonstrated that on the example of the electron (personal correspondence; 2018). In his words:

"... For that I assumed the electron to be a small spherical charged object with a field that is oriented perpendicular to its surface and that reaches out into all directions and extends to infinity, corresponding to the field of a spherical capacitor whose inner sphere is filled with a charge and whose outer sphere has an infinite diameter ..."

[Disclaimer: The following calculation of the electron mass is the intellectual property of Klaus Lux who gave me permission to use it in my book (personal correspondence;

2018).]

The capacity of a spherical capacitor in a vacuum is given by:

$$C = \frac{4\pi\varepsilon_0\, r\, R}{R - r}$$

With:

$$\varepsilon_0 = \frac{1}{\mu_0\, c^2} \approx \frac{10^7}{4\pi c^2} = 8.854\,187\,817 \times 10^{-12}\ \text{F}\cdot\text{m}^{-1}$$

When $R \to \infty$, the capacity is given by:

$$C = \frac{10^7\, r}{c^2}$$

The capacity of a condensator is given by $C = Q/U$, and the energy of a charged condensator by:

$$W = \frac{1}{2}\, C\, U^2$$

Solving $C = Q/U$ for U yields $U = Q/C$ and inserting that into the equation for W yields:

$$W = \frac{Q^2}{2\, C}$$

Now we insert the equation for C and get:

$$W = \frac{Q^2\, c^2}{2 \times 10^7\, r}$$

Inserting the electron charge e for Q, the classical electron radius r_e for r and dividing through c^2 yields:

$$m_e = \frac{e^2}{2 \times 10^7\, r_e}$$

However, the true value for the electron mass is twice this value and thus given by the simpler equation:

$$m_e = \frac{e^2}{10^7 \, r_e}$$

Which yields:

$$m_e = 9.109382913 \times 10^{-31} \text{ kg}$$

With the measured value being: $9.109383561 \times 10^{-31}$ kg.

At this point it has to be admitted, that m_e was already part of r_e, which is given by:

$$r_e = \frac{e^2}{4 \, \pi \, \varepsilon_0 \, m_e \, c^2}$$

This may seem like circular reasoning, but it is in fact very desirable: only a theory in which all constants depend upon each other can be a theory devoid of free parameters.

Klaus Lux's idea is not totally new. The idea of an electromagnetic explanation for inertia mass goes back to J. J. Tomson who first conceived it in 1881.[1] Others worked it out in greater detail between 1882 and 1904. One of the most engaged among them was Hendrik Lorentz (1892; 1904).[2, 3] They calculated the electromagnetic energy and mass of the electron as:

$$E_e = \frac{1}{2} \frac{e^2}{r_e}; \; m_e = \frac{2}{3} \frac{e^2}{r_e \, c^2}$$

Which implied:

$$m_e = \frac{4}{3} \frac{E_e}{c^2}$$

Wilhelm Wien (1900)[4] and Max Abraham (1902)[5] then concluded that the total mass of bodies is identical to their electromagnetic mass.

The $4/3$ factor violates special relativity and over the years there were many physicists showing how it disappears in a strictly relativistic treatment of the matter.[6, 7, 8] In 2011 Valery Morozov showed that a moving charged sphere has a flux of non-electromagnetic energy and this flux has an impulse that is exactly $1/3$ of the sphere's electromagnetic impulse.[9, 10]

The remaining factor of $1/2$ in Klaus Lux's calculation can probably be explained in a similar way.

For unclear reasons electromagnetic explanations of mass have been rejected in recent decades, giving way to the *Higgs mechanism*. Unlike the above explanations the Higgs mechanism doesn't provide us with any predictions on the masses of different particles.

It is puzzling how the energy in the electromagnetic field can be ignored as a source of mass by so many.

If the above considerations are correct, then the inertia mass of a particle is proportional to the energy stored in the polarized quantum vacuum around it.

We could now try to do the same calculations for the neutrino.
First we have to determine the 'classical radius of the neutrino'. Therefore we take the formula for the classical electron radius and scale down the charge e by multiplying it with the coupling constant of the weak force α_W. Furthermore we have to replace the electron mass m_e by the neutrino mass m_ν and the electric permeability constant of the vacuum ε_0 by another constant ε_{0_W} that is related to the weak force. This yields:

$$ r_\nu = \frac{(e\,\alpha_W)^2}{4\,\pi\,\varepsilon_{0_W}\,m_\nu\,c^2} $$

$$ \alpha_W = \sqrt{\frac{\tau_\Delta}{\tau_\Sigma}} = 2.738612788 \times 10^{-7} $$

ε_{0_W} can be determined by taking the formula for ε_0 and replacing α by α_W. e is of course again replaced by $e\,\alpha_W$. This yields:

$$ \varepsilon_{0w} = \frac{2\,\alpha_W}{(e\,\alpha_W)^2}\,\frac{h}{c} = 628.801983 \text{ F m}^{-1} $$

$$ r_\nu = 1.267402294 \times 10^{-35} \text{ m} $$

For C we take the equation with a finite radius R in order to account for the finite range of the weak force and set $R_W = 10^{-15}$ m. Replacing all the constants as above yields:

$$ C = \frac{4\,\pi\,\varepsilon_{0w}\,r_\nu\,R_W}{R_W - r_\nu} $$

Inserting this into the equation for W yields:

$$ W = \frac{Q^2\,(R_W - r_\nu)}{8\,\pi\,\varepsilon_{0w}\,r_\nu\,R_W} $$

Inserting $e\,\alpha_W$ for Q and dividing through c^2 yields:

$$m_v = \frac{(e\,\alpha_W)^2\,(R_W - r_v)}{8\,\pi\,\varepsilon_{0_W}\,r_v\,R_W\,c^2}$$

$$m_v = 1.069479234 \times 10^{-37}\ \text{kg}$$

Again same as with the electron the true mass is twice this value, so it is given by:

$$m_v = \frac{(e\,\alpha_W)^2\,(R_W - r_v)}{4\,\pi\,\varepsilon_{0_W}\,r_v\,R_W\,c^2}$$

$$m_v = 2.138958468 \times 10^{-37}\ \text{kg}$$

With the measured value being: $2.138958468 \times 10^{-37}$ kg – exactly the same (!).

$R_W - r_v$ actually reduces to R_W, because r_v is too small to make a difference when subtracted.
It is obvious that the whole equation depends crucially on R_W, the range of the weak force, a number which depends on the mass of the W- and Z-bosons. The heavier they are, the lighter is the neutrino, but the neutrino mass is not used when calculating R_W.

From the above it should be clear that the electron's and the neutrino's mass are both sourced from their electromagnetic energy. The electron's mass is its electromagnetic mass and the neutrinos mass is its 'weak mass'.

Often people doubt that accuracy of the weak force coupling constant, because it is derived by comparing two different particle decays, and this may seem a bit arbitrary to many. The above calculation shows that it is an accurate measure for the strength of the weak force.

The fact that we used different formulas for the electron and the neutrino, one for forces with unlimited range and one for forces with a limited range R shows that at least for the neutrino we don't just get the mass we inserted when calculating the 'classical radius'. It is this limited range that explains the original discrepancy to a factor of 160. Had we used the formula for unlimited range, we would have been again off by this factor.

Will this scheme work for the quarks, protons and neutrons as well?
Most of the proton's mass is sourced from the gluon field in between the constituent quarks. We can therefore hardly talk about the mass of a single quark. Rather we have to look at the mass of the proton, or more precisely a third of this mass.
Nevertheless we will be talking about the 'classical radius of the quark' and not of the proton, because the proton is not an elementary particle. We again take the formula for

the classical electron radius and this time we scale up the charge e by dividing it through the fine structure constant α. Furthermore we have to replace the electron mass m_e by a third of the proton mass $m_p/3$ and the electric permeability constant of the vacuum ε_0 by another constant ε_{0_s} that is related to the strong force. This yields:

$$r_q = \frac{3\,(e/\alpha)^2}{4\,\pi\,\varepsilon_{0_s}\,m_p\,c^2}$$

$$\alpha = \frac{k_e\,e^2}{\hbar\,c} = 7.297352571 \times 10^{-3}$$

ε_{0_s} can be determined by taking the formula for ε_0 and replacing α by α^{-1}. e is of course again replaced by e/α. This yields:

$$\varepsilon_{0s} = \frac{2}{\alpha\,(e/\alpha)^2}\,\frac{h}{c} = 1.256637061 \times 10^{-6}\ \text{F}\,\text{m}^{-1}$$

$$r_q = 6.091895339 \times 10^{-19}\ \text{m}$$

For C we take the equation with a finite radius R in order to account for the finite range of the weak force and set $R_S = 3 \times 10^{-15}$ m. Replacing all the constants as above yields:

$$C = \frac{4\,\pi\,\varepsilon_{0s}\,r_q\,R_S}{R_S - r_q}$$

Inserting this into the equation for W yields:

$$W = \frac{Q^2\,(R_S - r_q)}{8\,\pi\,\varepsilon_{0s}\,r_q\,R_S}$$

Inserting e/α for Q and dividing through c^2 yields:

$$m_q = \frac{(e/\alpha)^2\,(R_S - r_q)}{8\,\pi\,\varepsilon_{0s}\,r_q\,R_S\,c^2}$$

$$m_q = 2.787137084 \times 10^{-28}\ \text{kg}$$

Again same as with the electron and the neutrino, the true mass is twice this value, so it is given by:

$$m_q = \frac{(e/\alpha)^2 \left(R_S - r_q\right)}{4\,\pi\,\varepsilon_{0_S}\,r_q\,R_S\,c^2}$$

$$m_q = 5.574274167 \times 10^{-28} \text{ kg}$$

A proton contains three quarks. If we multiply the above quark mass by three, we get:

$$m_p = 1.67228225 \times 10^{-27} \text{ kg}$$

With the measured value being: $1.672621898(21) \times 10^{-27}$ kg – extremely close. Even higher precision could be reached by using more precise values for R_S.

However, the proton does not only have a strong charge, but electric charge as well. The corresponding mass (the mass of the electron) has to be added to the above calculated mass. Doing so yields:

$$m_p = 1.673193188 \times 10^{-27} \text{ kg}$$

Above we used $R_S = 3 \times 10^{-15}$ m as the range of the strong force. In fact this is the range of the residual strong force, caused by the little rest charge that remains due to spacial inhomogeneities in the distribution of the different color charges inside the proton. The strong force is the strongest at a distance of 1.1×10^{-15} m. Beyond this distance, which is the typical distance between nuclei, it drops off rather rapidly (see fig. 23).

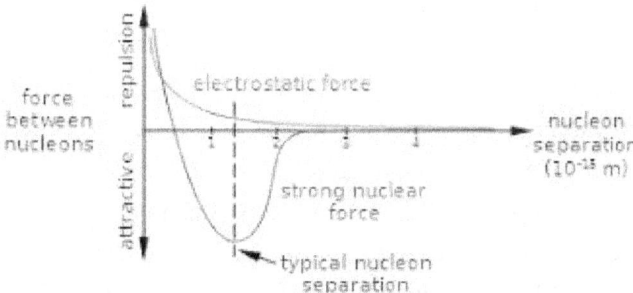

Fig. 23. The strength of the strong force in dependence to the distance.

If we set $R_S = 1.1 \times 10^{-15}$ m instead and then add the electron mass to our result, we get:

$$m_p = 1.672606524 \times 10^{-27} \text{ kg}$$

This is a 5 digits match with the measured value. For the electron we had a 6 digits match and for the neutrino a 10 digits match. Considering the rather complicated nature of the strong force, this is a result much better than expected.

What about the neutron? Why is it 2.5 electron masses heavier than the proton?

When the neutron is outside the nucleus, it decays with a half life of 10.5 min, into a proton, an electron and an anti-neutrino. This is known as beta decay and while the opposite, namely a proton turning into a neutron, while emitting a positron and a neutrino, is possible too, it never happens outside the nucleus, because there the proton doesn't have any extra energy that could allow it to turn into a particle heavier than itself.

For turning into a proton, an electron and an anti-neutrino, the neutron certainly needs some extra mass energy. Although a neutrino has a very small rest mass, in beta decays it can easily have a bigger mass energy than the electron. If the available energy was very small, such an equal distribution of impulse among electron and anti-neutrino would be impossible.

It is reasonable that the electron needs a bit more energy than just the energy associated with its rest mass, say 1.25 of it, and for enabling an equal distribution of impulse, the same amount of energy should be available for the neutron as well, adding up to 2.5 m_e.

This is by no means a rigorous derivation of the neutron mass, but being just a small multiple of the electron mass heavier than the proton, shows that it is most likely related to the beta decay.

Of course the question remains, why is it the neutron which is heavier and not the proton? Maybe a world in which protons could decay in an out of themselves, would be way too fragile. While neutron decay increases the number of positively charged protons and negatively charged electrons, proton decay would lead to both protons and electrons disappearing; creating a world without charge.

Something that remains unexplained in this model is the fact that the W- and Z-bosons have such huge masses, apparently regardless of their charge.

The Higgs mechanism which is now regarded to be the explanation for the existence of rest mass by most physicists, was originally conceived as an explanation for the masses of the W- and Z-bosons only.

The calculations in this chapter should have made clear now, that the Higgs mechanism should not have been extended beyond its original purpose. Doing so means to both overcomplicate things and to introduce arbitrary parameters into the theory.

After all the Higgs mechanism involves a field too, similar to the force fields we discussed above. It might well be that the masses of the W- and Z-bosons are somehow related to the energy of this field. If they are the only particles receiving their mass through the Higgs mechanism, then they must be the only particles interacting with the Higgs-boson, and thus their mass should be related to that of the Higgs-boson. This raises the question where the Higgs-boson itself receives its mass from.

It is as if the weak force was some strange mutation of the electromagnetic force where all the mass or charge of the electrons went into the photons. And indeed, the heavier the W- and Z-bosons are, the weaker is the mass and hypercharge of the neutrino. If the different forces are just variations of the same underlying geometrical structure, then such inversions are indeed to be expected.

If the weak force is such a variation of the electromagnetic force, then its inverse could be the strong force. The transmitter particles of the strong force, the gluons, different from the massless photons of the electromagnetic force, do have a charge, namely the strong hypercharge. As mentioned before the self-interaction of gluons makes it very hard to figure out how much energy individual gluons carry, let alone know if they have any non-zero mass. If the strong force is some form of inverse analogy to the weak force, then we could expect the gluon to have a mass that stands in the same relation as the mass of the neutrinos to the mass of the W- and Z-bosons; since the W-bosons have electric charge, we will be looking at the Z-boson only.
This would lead to the following mass for gluons:

$$\frac{m_g}{m_q} = \frac{m_v}{m_z}$$

$$m_g = \frac{m_v\, m_q}{m_z}$$

Using the average between the up and the down quark for m_q yields:

$$m_g = 8.096495577 \times 10^{-42} \text{ kg}$$

It is hard to isolate the quark mass strictly. Using a third of the proton mass instead yields:

$$m_g = \frac{m_v\, m_p}{3\, m_z}$$

$$m_g = 7.339822104 \times 10^{-40} \text{ kg}$$

Obviously such a small mass can't slow down gluons much. Similar to neutrinos, which are estimated to have a mass of 2.14×10^{-37} kg (average of three flavors), gluons would move very close to the speed of light.

Our above hypothesis established a mass relation between fermions and their corresponding gauge bosons. It wasn't able to predict the mass of the W- and Z-bosons. Is there any clue as to how it could be derived?

When we translate the critical photon energy from chapter 3.9 into a mass we get

$2.447830123 \times 10^{-26}$ kg. This is somewhat close to the mass of the W- and Z-bosons, which is at about 4×10^{-27} kg.

According to space particle dualism elementary spaces can merge as soon as they exceed the critical energy; the energy which makes them grow larger than their wavelength. The discrepancy of about one order of magnitude could then possibly be due to the running of coupling.

Note:

1. Thomson, Joseph John (1881), "On the Electric and Magnetic Effects produced by the Motion of Electrified Bodies", *Philosophical Magazine*, 5, 11 (68): 229 – 249, doi:10.1080/14786448108627008.

2. Lorentz, Hendrik Antoon (1892a), "La Theorie electromagnetique de Maxwell et son application aux corps mouvants", *Archives neerlandaises des sciences exactes et naturelles*, 25: 363 - 552.

3. Lorentz, Hendrik Antoon (1904), "Electromagnetic phenomena in a system moving with any velocity smaller than that of light", *Proceedings of the Royal Netherlands Academy of Arts and Sciences*, 6: 809 – 831.

4. Wien, Wilhelm (1900), "Über die Möglichkeit einer elektromagnetischen Begründung der Mechanik" [On the Possibilitz of an Electromagnetic Foundation of Mechanics], *Annalen der Physik*, 310 (7): 501 – 513, Bibcode:1901AnP...310..501W, doi:10.1002/andp.19013100703.

5. Fermi, Enrico (1922), "Über einen Widerspruch zwischen der elektrodznamischen und relativistischen Theorie der elektromagnetischen Masse" [Concerning a Contradiction between the Electrodynamic and Relativistic Theory of Electromagnetic Mass], *Physikalische Zeitschrift*, 23: 340 – 344.

6. Dirac, Paul (1938), "Classical Theory of Radiating Electrons", *Proceedings of the Royal Society of London A*, 167 (929): 148 – 169, Bibcode:1938RSPSA.167..148D, doi:10.1098/rspa/1938.0124.

7. Rohrlich, Fritz (1960), "Self-Energy and Stability of the Classical Electron", American Journal of Physics, 28 (7): 639-169, Bibcode:1960AmPh..28..639R, doi:10.1119/1.1935924.

8. Swinger, Julian (1983), "Electromagnetic mass revisited", Foundations of Physics, 13 (3): 373 – 383, Bibcode:1983FoPh///13//373S, doi:10.1007/BF01906185.

9. Morozov, Valery B. (2011). "On the question of the electromagnetic momentum of a charged body". Physics Uspekhi. 54 (4): 371–374. Bibcode:2011PhyU...54..371M. doi:10.3367/UFNe.0181.201104c.0389.

10. Janssen, Michel; Mecklenburg, Matthew (2007). "From classical to relativistic mechanics: Electromagnetic models of the electron". In Hendricks, V.F.; et al. (eds.). Interactions: Mathematics, Physics and Philosophy. Dordrecht: Springer. pp. 65–134.

3.12 Linear superpositions of particle types [20.08.2004 (SWI explains neutral particle mixing); 04.09.2020 (lifetime for particle type superpositions)]

Quantum mechanics explains how superpositioned particle properties spread and evolve over time, yet it does not predict which types of properties can be in superposition. For example, an electron can be in a superposition of having an up-spin and having a down-spin; it can also be in a superposition of going left and going right, however, it cannot be in a superposition of being an electron and being a positron.
It would be straight forward if that was the case for all particles, but it is not: neutrinos are always in superposition of being electron-neutrinos, muon-neutrinos and tauon-neutrinos, and there are neutral mesons that are linear superpositions of different quark compositions.

Soon after the similar worlds interpretation was proposed in 2004, it was realized that it could explain why certain electrically neutral particles exist in a superposition of different particle types.

This phenomenon is often referred to as 'flavor mixing'. In the standard model it is not quite clear why it occurs.

A notable example for flavor mixing are neutral K-mesons, also called Kaons. Their quark composition is a superposition of $d\bar{s}$ (down quark & anti-strange quark) and $s\bar{d}$ (strange quark & anti-down quark). The two different particles are referred to as K_S^0 and K_L^0, with the 0 representing a net charge of 0, and S and L standing for 'short' and 'long' respectively, referring to the different decay times of the two particles. They do not exist as separate entities, but instead always as a linear superposition of both types. Their quark composition can be denoted as:

$$K_S^0 = \frac{d\bar{s} - s\bar{d}}{\sqrt{2}}$$

$$K_L^0 = \frac{d\bar{s} + s\bar{d}}{\sqrt{2}}$$

Another example are so called 'neutrino oscillations'. There are three types of neutrinos, electron neutrinos, muon neutrinos and tau neutrinos. Every neutrino is a linear superposition of all three types. They have slightly different masses, and therefore also different wavelengths. The different wavelengths of the different types interfere with each other, and that is how one type of neutrino can 'shape shift' into another type of neutrino. The typical oscillation period being proportional to L/E, with L being the distance travelled, and E being the total energy of the neutrino.

According to the similar worlds interpretation, on which space particle dualism is based upon, we co-exist in all worlds which are temporarily indistinguishable. That typically includes worlds in which the positions and impulses of particles are slightly different, but because anything that is temporarily indistinguishable goes, it also allows for particles to have different identities in different worlds.

The number of charges however is typically the same in all worlds, while rest masses do not need to be the same in different worlds. We can simply notice this and accept it as reflecting the conservation of charge and the convertibility of energy respectively, but we can also think one level deeper and try to explain it.

According to the time energy uncertainty relation, the time needed for an energy difference ΔE to be noticed is shorter the greater the energy is. We can therefore calculate the time for which two states, or worlds, are indistinguishable, according to:

$$\Delta t = \frac{\hbar}{2\,\Delta E}$$

When it comes to virtual particles, ΔE represents their whole mass energy, as a virtual particle is a particle that has to come into existence first from nothing.
When it comes to flavor oscillations, ΔE represents only the energy of the particle property that was not there before.
For a charged particle this energy is the energy kept in its force field. In chapter 3.11 we saw that this energy is equal to the energy equivalent of the rest mass of a particle.

For how long could a superposition between a proton and a neutron be maintained?
For answering this question we have to see how much energy there is in the electromagnetic force field of the proton and compare that to the neutron, which has no electromagnetic force field. It is important to note here, that while the neutron has electromagnetic charges, which give it its mass, they do neutralize outside of the neutron, and therefore they do not influence the environment in a way that makes similar worlds distinguishable. The time after which switching a neutron for a proton would become noticeable is thus given by:

$$\Delta t = \frac{\hbar}{2\,m_p\,c^2}$$

$$\Delta t = 2.2038744 \times 10^{-24} \text{ s}$$

At room temperature (293.15 Kelvin) the most likely velocity of a proton is:

$$v_{th} = \sqrt{\frac{2\,k\,T}{m_p}}$$

$$v_{th} = 2{,}200\,\frac{m}{s}$$

At this velocity it has an energy of:

$$E = \sqrt{(p\,c)^2 + \left(m_p\,c\right)^2}$$

$$E = 1.5032776 \times 10^{-10}\,J$$

If we calculate our hypothetic proton-neutron oscillations using the same equation as for neutrino flavor oscillations, we arrive at:

$$f_{FO} = \frac{\Delta m^2\,c^3\,v}{4\,\hbar\,E} = 2.6152006 \times 10^{18}\,Hz$$

That is one oscillation every $3.8237984 \times 10^{-19}$ seconds. For this kind of proton-neutron oscillations to really occur, this oscillation time would have to be smaller than the above calculated lifetime Δt of the proton-neutron superposition, which was 2.2×10^{-24} seconds. We can write this condition as:

$$f_{FO} > \frac{1}{\Delta t} \rightarrow mix$$

For a proton travelling near the speed of light ($v = c$), what energy would it have to reach to fulfil the above inequality relation? We can find that out by turn the above inequation into an equation:

$$\frac{4\,\hbar\,E}{\Delta m^2\,c^4} = \frac{\hbar}{2\,m_p\,c^2}$$

$$E = \frac{m_p\,c^2}{8}$$

That is less than its mass energy $E = m_p\,c^2$, and can therefore never be reached. This shows that charge conservation cannot be violated; or at least it cannot be violated for long enough to be noticed.

There is no difference between the different neutrino flavors, aside from their mass. We can explain the difference in mass by theorizing about merging elementary spaces (see chapter 3.10), but that is irrelevant to the outside observer.

For neutrinos, the left side of our above inequation does not change. For Δm we simply enter the difference in mass between the neutrino flavors. The right hand side however changes completely, because neutrinos do not have charge, which would allow for easy detection. Their weak charge does not do much, aside from giving them their mass.

In fact neutrinos interact so weakly with their environment that even a one light year long lead block would stop only half of the neutrinos flying through it.

Without interaction, there can be no measurement, and therefore our Δt should be a rather large number.
How large? There are 100 trillion neutrinos passing through our bodies every second, and yet the chance that even just one of them will interact with any atom in our body within a lifetime, is only 25%.[1,2]

That means in every second the chance for a neutrino to interact is only 1 in 10^{24}. The age of the post-decoupling universe in seconds is about 10^{21} (see chapter 4.12). This shows why for most neutrinos their wavefunction never collapses.

If their wavefunction never collapses, then the linear superposition of different neutrino types also never collapses, and therefore any oscillation rate can do the job of 'shape shifting' the neutrinos. No oscillation could be long enough to take longer than the age of the universe.

Mathematically expressed that is:

$$f_{FO} > \frac{1}{\Delta t} \rightarrow mix$$

$$\frac{\Delta m^2 c^4}{4 \hbar E} > \frac{1}{10^{24} \, [s]}$$

The right side of this inequation is essentially saying that the wavefunction of a neutrino collapses 10^{-24} times every second, which equals saying that it collapses once every 10^{24} seconds.

We shall end this chapter by calculating a typical flavor oscillation rate for a neutrino:

Neutrinos from the sun, which are generated in proton-proton collisions typically have energies of about 400 keV, which corresponds to 6×10^{-14} J. The difference between the masses of the different neutrinos are not known with any precision, but taking a third of the mean neutrino mass, which is the sum of three flavors, must be a well enough approximation. The neutrino mass is 2.14×10^{-37} kg. This yields:

$$\frac{\Delta m^2 \, c^4}{4 \, \hbar \, E} = 1{,}520{,}412.09421 \text{ Hz}$$

That is 1.5 million oscillation a second.

Note:
1. "Ghostly Cosmic Neutrinos Are Stopped Cold by Planet Earth", Calla Cofield. Space.com. November. 22, 2017.
2. Link: [symmetrymagazine.org/2010/02/15/just-how-often-are-you-hit-by-a-neutrino].

3.13 Unitary rest surface, relativistic mass and special relativity [29.03.2018 (wrongly assumed that relativistic mass arises from spherical nature of elementary spaces); 30.03.2018 (wrongly assumed different scale factors for different forces); 31.03.2018 (relativity of motion); 02.04.2018 (elementary spaces slightly smaller; same size for different particles)]

Note: Writing this chapter was not an easy task, because the main conclusions were reached during the writing process and a lot of things had to be revised.

Unitary rest surface for fermions

When the constant G_E was introduced in earlier versions of this book, it was defined as G times the difference in strength between gravity and the GUT force. Why was this definition chosen?

When the three different fundamental forces meet at high energies they become same in strength and that strength is about 10^{37} times the strength of gravity. The idea of using this number as the scaling up factor for G_E was grounded in the notion that if one could zoom in onto elementary spaces, this would be the strength of the force one would find.

The idea of using this very same scaling factor G_E even for electrons and photons at low energies came from thinking about the difference in force strength as just being due to the virtual positron cloud that surrounds every electron. Yet if we want to use the 'naked charge', then we should also be using the 'naked mass', which is unknown. If Penrose's 'educated guess' of attributing the running of coupling entirely to this vacuum polarisation effect is correct, then we have to expect that the real (naked) mass of the electron is smaller to the same amount as the real charge is bigger.

Thus we shouldn't adjust G_E to the 'more true' value $G \times 10^{37}$, because then we will have to use a smaller mass m_e, and at the end we will have changed nothing at all.

However, if GUT is true and all forces meet in strength at high energies, then the 'naked masses' of all particles must essentially be the same.
A tempting idea had been to attribute the different inertia mass of particles to some form of *vacuum pressure* different size elementary space surfaces might exert, yet if the above is true, then the surface area of elementary spaces must be the same for different types of particles, rendering such approaches unfeasible.

In this case, the different inertia masses could be arising from differently strong interactions with the quantum vacuum. If the different forces are same in strength at the fundamental level, then it must be the different form of interaction which makes them different on large scales. The different masses must then simply correspond to differently strong vacuum polarisation effects. This is in accordance with with the well accepted notion of the *running of coupling*.

If these considerations are correct, then the elementary space 'rest surface' is all the same for the different particles, and can be calculated as

$$A_E = \frac{16 \pi G_E^2 m_e^2}{c^4}$$

$$G_E = \frac{k_e e^2}{m_p^2}$$

Elementary space of massless particles

What about the size of photons? It makes sense to adopt the same G_E that we use for electrons, simply because electrons and photons are both particles associated with electromagnetism, but what about other massless particles? Photons are very different from gluons; is it justified to use the same G_E for both particles?
What do we know about gluons? Do gluons and photons of the same wavelength carry equal amounts of energy?
Unfortunately there is no way of finding that out experimentally, because gluons are always bound in the nucleus, and even theoretically there is no value that could be derived in any straight forward way from quantum chromodynamics (QCD).
However, it is to assume that the energy carried by a force transmitting particle is proportional to the strength of the force. That is why in theories assuming the existence of a gravity particle called 'graviton' nobody wonders why it hasn't been observed: giving the extraordinary weakness of gravity, such a particle would have to carry incredibly small amounts of energy.[1]
If the scale factor G_E is indeed equal regardless of the associated force, then the circumference of a boson's elementary space is

$$C_{E_B} = \frac{4\pi G_E h f}{c^4}$$

Due to the lower dimensionality of bosons they do not have a real number surface, and thus we can give only a circumference as their 'real surface'.

Their full complex valued surface is however geometrically still a sphere, because one complex dimension corresponds to two real dimensions.

For getting a feeling for the hidden parts of our universe, we shall take a look at the full complex surfaces of both fermions and bosons:

The full complex surface of fermions and bosons

The full complex surface area of a boson's elementary space is given by

$$A_{\mathbb{C}_B} = \frac{16\pi G_E^2 h^2 f^2}{c^8} (1 + i)$$

It might be that only the real part of an elementary spaces' surface can generate inertia, and with only one complex dimension there is only one real dimension and therefore no inertia.[2]

For the electron the surface of the full complex elementary space is given by

$$A_{\mathbb{C}_B} = \frac{(2\pi)^{5/2}}{\Gamma(5/2)} \times \left(\frac{2 G_E^2 m_e^2}{c^2}\right)^4 (1 + i)$$

Where Γ is the gamma function, for which it is known that $\Gamma(1/2) = \sqrt{\pi}$ and $\Gamma(x + 1) = x\Gamma(x)$. Applying this yields

$$A_{\mathbb{C}_B} = \frac{4(2\pi)^{5/2}}{\sqrt{\pi}} \times \left(\frac{2 G_E^2 m_e^2}{c^2}\right)^4 (1 + i)$$

Since for inertia only the real part of this complex manifold is relevant, we can put aside this equation till when it may become of use for other purposes.

Unitary growth of elementary spaces leads to Lorentz transformation

Lorentz contraction is what makes moving electric charges generate a magnetic field. That is because it makes otherwise evenly dispersed charges appear uneven, which causes electric repulsion or attraction. From the resting frame of reference this looks like a magnetic force.[3] That is why with increasing velocities the energy of the magnetic field increases to the same amount that length is contracted, time slows down and mass-energy increases. All that is given by the same Lorentz factor γ, which is

$$\gamma = \frac{1}{\sqrt{1 - \frac{v^2}{c^2}}}$$

If mass is just a side effect of charge and relativistic mass-energy increase is simply to identify with the increase in magnetic energy, then inertia mass must arise from relativity and electromagnetism alone.

This increase in magnetic energy leads to a relativistic mass-energy increase and that is the typical mass-energy increase that takes place when charged particles approach higher and higher impulses. Along with the mass-energy, the size of the elementary space grows too. Does that violate the relativity of motion?

That seems not to be the case: in a reference system in which the elementary spaces of all particles are uniformly grown in size, the new size would be perceived as the rest mass and all other constants would change in a way that would make the difference unrecognizable from within the reference system.
Yet if the elementary spaces appear as grown when 'viewed'[3] from the outside reference system, then a *Lorentz transformation* is just what would appear natural here. That is to say that the changed constants inside the moving reference frame would be perceived as time dilation and length contraction from the outside.
Does that mean special relativity can be derived in its entirety from the Galilean equivalence principle applied to elementary spaces? That might indeed be the case, though it would have to be shown in a more rigorous way.
Elementary spaces grow into all directions, but length contraction takes place only in the direction of movement. Why is that?

The size of elementary space depends on the constants G_E, m_e and c, which themselves again depend upon G and e. Changing these all together will not be recognizable from within the frame of reference, but from outside of it. Furthermore one would recognize it when looking from within the moving frame of reference to the outside world. But does this Galilean transformation of the reference system really imply all the constants to change together? Or is it just some?

If one wanted to make elementary spaces larger by changing fundamental constants, then the most direct way would be to change the denominator, which is c^2. Making c smaller would make elementary spaces bigger.
Since all constants are related, from within the reference system the change would not be felt, but from outside this slowing down of the speed of light would be perceived as a slowing down of time itself, so that the speed of light becomes just the same as in a resting frame of reference.
The length contraction in the direction of movement is then a consequence of this

slowing down of time (or of the 'speed of light'), because from within the moving frame of reference nothing changes and light has still to travel the same distances it does usually. That is why from outside things in the moving frame of reference must appear contracted in length.

Instead of saying *time* slows down in order to allow the speed of light to be constant, we said *light* slows down in order to appear constant in all frames of reference. This may sound like a tautology, and it might be one, but it makes sense in this context, because if it can be shown that special relativity is a consequence of applying the Galilean equivalence principle to elementary spaces, that would mean that not only general relativity, but also the more basic special theory of relativity can be derived from space particle dualism.

Notes:

1. Of course if we assumed the existence of a graviton particle, we would think of its energy to be a result of a renormalization process as well, so that the 'true' energy is equal to the transmitting particles of the other forces. After all, judging from the reference system of bosons themselves, scale doesn't exist, and since they have no rest mass and, from their own perspective, no energy as well, their elementary spaces are scaleless, or zero, too.

2. It is however unclear if the two real dimensions really have something to do with mass and inertia, or if it has more to do with having a charge and with polarizing the vacuum.

3. A popular account on how special relativity explains electromagnets can be found here: [https://www.youtube.com/watch?v=1TKSfAkWWN0] (Veritasium (2013); "How Special Relativity Makes Magnets Work").

4. Before the critical frequency or impulse is reached, the size of elementary spaces can't be observed directly, but only indirectly through measuring the energy of the involved particles. The assumption of this 'thought experiment' is that measuring the frequency can be equated to measuring the elementary space size itself.

3.14 Probing granularity of space with photons [30.06.2017 (delay according to the mainstream figure for the age of the CMB); 26.04.2019 (using the correct age of the CMB according to SPD); 29.06.2019 (using refined age of the CMB)]

Note: Many researchers think that space quantization approaches have been disproven by the missing delay of photons from the borders of our universe. Indeed that is at odds with most space quantization approaches, yet not with space particle dualism theory, because here the quanta of space are allowed to overlap.

The granularity of space leads to a certain degree of zig-zagging in the paths of particles.[1] This makes even otherwise straight paths slightly longer. One could imagine that this effect becomes measurable for photons that have travelled through the whole visible universe.

The deviation from a straight path is proportional to the granular dimensionality. That means, for a photon which travelled one meter, the deviation is only about 10^{-37} m, or more precisely:

$$\Delta x_{del}(x) = \frac{x\,G}{G_E} = \frac{x\,G\,m_p^2}{k_e\,e^2}$$

$$\Delta x_{del}(x) = 8.08833694 \times 10^{-37}\ \text{m}$$

After $1.236348099 \times 10^{36}$ m this deviation becomes 1 m.
Using the age the photons from the CMB and basing the mainstream figure for the age of the universe, we get an overall delay that corresponds to a distance of only $3.520018353 \times 10^{-19}$ m.

In chapter 4.12 we will learn that the true age of the CMB is in fact $4.2130109541636 \times 10^{13}$ yr.
Using this figure we obtain a delay distance of $1.074628687 \times 10^{-15}$ m.
However, the most distant observed galaxy is 439,441,351,917 light years away, and that gives a delay distance of $1.120899727 \times 10^{-15}$ m.

If the effect was bigger we could hope to find evidence for it in anomalous redshifts for objects near the border of the past, but its incredible smallness makes it untestable.

Notes:
1. This chapter was inspired by attempts of other theorists who tried to find evidence for loop-QG in the delay of intergalactic low energy photons. In 2007 a team claimed to have found the opposite, namely high energy photons being slower. This result was however not confirmed later on. If the effect exists at all, it might be due to different slowing down rates when intergalactic photons pass through various gases that may lie between us and their place of origin.

3.15 Granular dimensionality revisited [15.07.2018 (quantizing velocity; equation for connectivity in granular space); 17.07.2018 (solution)*; 19.07.2018 (result hints at further restrictions on vacuum fluctuations); 28.07.2018 (contribution from fermions); 29.07.2018 (contribution from photons & particle count)]

*In collaboration with Hal 9000 (Germany).

Note: this chapter is using the Planck-length, which contains G, a constant which is dismissed in SPD. However, as we will see in chapter 6.7, there is an alternative formulation of the Planck-length without G.

Also published in: Advances in Theoretical and Computational Physics (Volume 2, Issue 4) (winter 2019). www.doi.org/10.33140/ATCP.02.04.02.

In the chapters 3.7 and 3.8 we tried to derive the granular dimensionality of empty space from the vacuum energy density. The approach used there was extremely simplified, because it ignored differences in the size of elementary spaces.
Furthermore it was looking at virtual photons only, and generalized the result to other particles.

We shall now look at the degree of connectivity n of elementary spaces provided by electrons and other fermions.

The formula for the tightest packing of spheres can provide us with the elementary space density required for a continuous space, and that is:

$$\rho = \frac{\pi}{3\sqrt{2}}$$

We want to know how many times a volume of space can be filled up with different sized elementary spaces.
The volume of each elementary space is:

$$V = \frac{32\,\pi\,G_E{}^3}{3\,c^6}\left[\frac{m_e}{\sqrt{1-\frac{v^2}{c^2}}}\right]^3$$

There are $1/V_P$ of them in a cubic meter of space and they need to arrive at the density ρ in order to fill up the cubic meter once. So the contribution from one particular relativistic mass of an electron is:

$$n_E = \frac{96\ G_E{}^3\ \sqrt{2}}{V_P\ 3\ c^6} \left[\frac{m_e}{\sqrt{1 - \dfrac{v^2}{c^2}}}\right]^3$$

We need to quantize the mass-energy of these virtual electrons, and therefore we must look at their wavelength[1], which is:

$$\lambda = n \times l_P = \frac{h\ \sqrt{1 - \dfrac{v^2}{c^2}}}{m_e\ v}$$

Solving this for v yields:

$$\frac{\lambda\ m_e\ v}{h} = \sqrt{1 - \frac{v^2}{c^2}}$$

$$\frac{\lambda^2\ m_e^2\ v^2}{h^2} + \frac{v^2}{c^2} = 1$$

$$v^2 \left(\frac{\lambda^2\ m_e^2}{h^2} + \frac{1}{c^2}\right) = 1$$

$$v^2 \left(\frac{\lambda^2\ m_e^2\ c^2\ +\ h^2}{h^2\ c^2}\right) = 1$$

$$v^2 = \frac{h^2\ c^2}{\lambda^2\ m_e^2\ c^2\ +\ h^2}$$

$$v = \frac{h\ c}{\sqrt{\lambda^2\ m_e^2\ c^2\ +\ h^2}}$$

Now we can substitute λ^2 by $n^2\ l_P^2$ and enter this term into the original equation for the number of connections n_E on elementary spaces, which yields:

$$n_E = \frac{96\ G_E{}^3\ \sqrt{2}}{V_P\ 3\ c^6} \left[m_e : \sqrt{1 - \frac{h^2}{n^2\ l_P^2\ m_e^2\ c^2\ +\ h^2}}\right]^3$$

The next step is to take the sum of all different values for n, up to the value that corresponds to the 'maximal relevant wavelength' we found in chapter 3.8 ($n = 2.5 \times 10^{37}$):

$$\sum_{i=1}^{n=2.5\times10^{37}} n_E = \frac{96 \, G_E{}^3 \, \sqrt{2}}{V_P \, 3 \, c^6} \left[m_e : \sqrt{1 - \frac{h^2}{n^2 \, l_P^2 \, m_e^2 \, c^2 + h^2}} \right]^3$$

For the sake of simplicity we will now replace the individual groups of constants by the factors A, B, C and D:

$$\sum_{i=1}^{n=2.5\times10^{37}} n_E = A \left[B : \sqrt{1 - \frac{C}{D \, n^2 + C}} \right]^3$$

$$A = \frac{96 \, G_E{}^3 \, \sqrt{2}}{V_P \, 3 \, c^6} = 8.280023422 \times 10^{118}$$

$$B = m_e = 9.1093835611 \times 10^{-31} \text{ kg}$$

$$C = h^2 = 4.390479797 \times 10^{-67}$$

$$D = l_P^2 \, m_e^2 \, c^2 = 1.948163391 \times 10^{-113}$$

This can be further simplified into:

$$\sum_{i=1}^{n=2.5\times10^{37}} n_E = A \left[B : \sqrt{1 - \frac{1}{D' \, n^2 + 1}} \right]^3$$

With:

$$D' = \frac{D}{C} = 4.43724486 \times 10^{-47}$$

The order of magnitude of D is 10^{-113}. When we divide this by C, we get an order of magnitude of 10^{-47}. When we square the maximal value for n, we get 6.25×10^{74}.

In the following we will use a method which only applies to $D' \, n^2 \ll 1$. The terms for which $D' \, n^2 \gg 1$ we will calculate separately.

Accordingly our first sum goes only to $n = 10^{23}$, which is about the half of 47. After isolating the factors A and B and replacing the root, we get:

$$n_{E_1} = A\,B^3 \sum_{i=1}^{n=10^{23}} \left(1 - \frac{1}{D'\,n^2 + 1}\right)^{-\frac{3}{2}}$$

In this case $D'\,n^2$ is much smaller than 1, and that allows the following simplification:

$$1 - \frac{1}{D'\,n^2 + 1} \approx 1 - (1 - D'n^2) = D'\,n^2$$

Thus the sum is reduced to:

$$n_{E_1} = A\,B^3 D'^{-\frac{3}{2}} \sum_{i=1}^{n=10^{23}} n^{-3} = A\,B^3 D'^{-\frac{3}{2}}\,\zeta(3)$$

With $\zeta(3)$ being Apéry's constant, which has the value:

$$1.202056903159594285399 \ldots$$

Putting the physical constants back into our equation we get:

$$n_{E_1} = \frac{96\,G_E{}^3\,\sqrt{2}\,m_e^3\,(n^2\,l_P^2\,m_e^2\,c)^{-\frac{3}{2}}\,\zeta(3)}{V_P\,3\,c^6}$$

And that is:

$$n_{E_1} = 4.889092274 \times 10^{212}$$

This value is far higher than what we expected, namely the $n_E = 10^{70}$ we found in chapter 3.8.
Back in chapter 3.7 and 3.8 we didn't take account of the varying elementary space size, and thus we got a much lower value for n_E there, although the amount of vacuum energy we assumed to exist was the same.
This might be a hint that vacuum fluctuations equivalent to the Planck energy in fact don't exist. Such a fluctuation would have an elementary space with the radius of ridiculous 40 meters (!), making it look like the macroscopic atoms the ancient Greek theorized about.

The shorter the lifetime of a quantum fluctuation is, the more energetic it is. Shouldn't the same apply to extension in space? For particles with wavelengths larger than their own diameter that is certainly true. Fluctuations with huge energies are usually confined to smaller volumes of space. Large elementary spaces would violate this rule.

If this line of reasoning is correct, then the critical energy from chapter 3.8 should at the same time be the maximal energy of vacuum fluctuations. Particles with larger energies do exist, but presumably only as real particles, not as virtual particles in the quantum vacuum.

As we saw in chapter 3.8, the critical energy corresponds to a wavelength of 8.862×10^{-17} m and that equals $5.483132598 \times 10^{18}$ Planck lengths.
That means a new calculation should start with $n = 5 \times 10^{18}$ and end with $n = 2 \times 10^{37}$.

Again we note that:

$$1 - \frac{1}{1 + D' \, n^2} < 1 - \frac{1 - D'^2 \, n^4}{1 + D' \, n^2} = 1 - (1 - D' \, n^2) = D' \, n^2$$

And that this difference becomes neglectable for 'small' values of n, namely whenever:

$$n \ll \frac{1}{\sqrt{D'}} \approx 10^{24}$$

So whenever n is smaller than 10^{24}, the whole term simplifies into $D' \, n^2$.

For very big values of n on the other hand, we have the simple estimate:

$$1 - \frac{1}{1 + D' \, n^2} < 1$$

Overall we can assess that:

$$\sum_{n=n_1}^{n_2} \left(1 - \frac{1}{1 + D' \, n^2}\right)^{-\frac{3}{2}} > \sum_{n=n_1}^{n_2} \left(\min(D' \, n^2)\right)^{-\frac{3}{2}} = \sum_{n=n_1}^{n_2} \max\left(D'^{-\frac{3}{2}} n^{-3}, 1\right)$$

Whereas

$$n_1 = 5.483132598 \times 10^{18} \ll \frac{1}{\sqrt{D'}}$$

and

$$n_2 = 2.512019674 \times 10^{37} \gg \frac{1}{\sqrt{D'}}$$

That means we have to separate the sum there. For n_1 we have:

$$\sum_{n=n_1}^{\frac{1}{\sqrt{D'}}} \max\left(D'^{-\frac{3}{2}} n^{-3}, 1\right) = D'^{-\frac{3}{2}} \sum_{n=n_1}^{\frac{1}{\sqrt{D'}}} n^{-3} \approx D'^{-\frac{3}{2}} \int_{n_1}^{\frac{1}{\sqrt{D'}}} x^{-3} \, dx$$

Adding back in A and B and solving the integral yields:

$$n_{E_1} = A B^3 D'^{-\frac{3}{2}} \left[-\frac{1}{2} x^{-2}\right]_{x=n_1}^{\frac{1}{\sqrt{D'}}} \approx \frac{1}{2} A B^3 D'^{-\frac{3}{2}} n_1^{-2}$$

$$n_{E_1} = 3.521607475 \times 10^{60}$$

And for n_2 that is simply:

$$\sum_{n=\frac{1}{\sqrt{D'}}}^{n_2} \max\left(D'^{-\frac{3}{2}} n^{-3}, 1\right) = \sum_{n=\frac{1}{\sqrt{D'}}}^{n_2} 1 \approx n_2$$

$$n_{E_2} = A B^3 n_2$$

$$A B^3 = 6.258907484 \times 10^{28}$$

$$n_{E_2} = 1.572249874 \times 10^{66}$$

Adding together the two yields:

$$n_{E_\varepsilon} = n_{E_1} + n_{E_2} = 1.572253396 \times 10^{66}$$

As we saw in chapter 3.8 the connectivity (number of connections) of elementary spaces in empty space is:

$$n_E = \left(\frac{k_e e^2}{10 \, G_S \, m_p^2}\right)^2$$

$$n_E = 1.344977701 \times 10^{70}$$

It is good that our result here did not exceed this value. If it did, the theory would be in serious trouble.

Now we have to calculate the contribution from massless bosons like the photon.

Out of $m = E/c^2$ and $E = h\,c/\lambda$ as well as the equation we used before for the contribution of electrons follows that:

$$n_{E_\gamma} = \frac{96\,G_E{}^3\,\sqrt{2}}{V_P\,3\,c^6} \left[\frac{h}{n\,l_P\,c} \right]^3$$

$$n_{E_\gamma} = \frac{1}{n^3} \times \frac{96\,G_E{}^3\,\sqrt{2}\,h^3}{V_P\,3\,c^9\,l_P^3} = \frac{1}{n^3} \times \frac{A\,h^3}{c^3\,l_P^3}$$

For finding the contribution from all different wavelengths we take the sum from the shortest wavelength up to the longest wavelength. For photons these correspond to:

$$n_1 = 5.572591903 \times 10^{18}$$

$$n_2 = 2.512019674 \times 10^{37}$$

Thus the photon contribution to the connectivity of elementary spaces is given by:

$$n_{E_\gamma} = \sum_{n_1}^{n_2} \frac{1}{n^3} \times \frac{96\,G_E{}^3\,\sqrt{2}\,h^3}{V_P\,3\,c^9\,l_P^3}$$

The constant cluster on the right has the value:

$$A' = \frac{A\,h^3}{c^3\,l_P^3} = 2.11752448 \times 10^{98}$$

We can now approach this sum using an integral:

$$A' \sum_{n=n_1}^{n=n_2} n^{-3} \approx A' \int_{n_1}^{n_2} x^{-3}\,dx$$

$$n_{E_\gamma} = A' \left[-\frac{1}{2} x^{-2} \right]_{x=n_1}^{n_2} \approx \frac{1}{2} A' n_1^{-2}$$

This yields:

$$n_{E_\gamma} = 3.409447167 \times 10^{60}$$

This is far below the overall value of 10^{70} for the connectivity of elementary spaces in empty space. This calculation shows that rest mass particles like electrons contribute by far more to n_E than massless particles like photons.

The question now is if the diversity of particles in the quantum vacuum can explain the gap between our before calculated value of 10^{66} and the expected overall value of about 10^{70}.

In chapter 3.7 we multiplied our results there by 17, for 17 fundamental particles, and then by 1.5, to account for all the dark matter particles.

However, it is to be doubted if dark matter particles really contribute much to the quantum vacuum, because as we will see in chapter 4.9 they do not have impulse, so there would be no different energy levels which could pile up and significantly increase the overall vacuum energy.

Furthermore we should be counting in all subatomic particles, not only the fundamental ones. Including the not yet observed ones that gives a total of 226 baryons, 196 mesons, 6 leptons and 4 vector bosons and 1 scalar boson. That gives a total of 433 subatomic particles. Accounting for all their anti-particles we have to multiply that by 2 and arrive at 866. Two of these particles, the photon and the gluon, do not have rest mass, and do therefore contribute much less to the connectivity of granular space. That leaves us with a factor of 862, which yields:

$$n_E = 862 \times n_{E_\varepsilon} = 862 \times (1.572249874 \times 10^{66})$$

$$n_E = 1.355279391 \times 10^{69}$$

Here I didn't calculate the contribution from different rest mass particles separately. That is because we know the only thing that is different with these particles is the mass, and since mass is a side effect of charge, differences in mass are presumably only due to the running of coupling.

However, a vast majority of the particles we counted were subatomic compound particles (mesons and baryons) composed of 2 or 3 quarks.

Three quarks correspond to three elementary spaces, thus each baryon counts as three particles. This gives us a new particle count, which yields:

$$2 \times [(226 \times 3) + (196 \times 2) + 6 + 2 + 1] = 2{,}158$$

Using this particle count we arrive at:

$$n_E = 2{,}158 \times n_{E_\varepsilon}$$

$$n_E = 3.392915228 \times 10^{69}$$

Of course, on earth we have a much higher vacuum energy density due to all the force transmitting virtual particles that are emitted by all the charges and hypercharges around. In chapter 6.3 we will look at the possibility that gravity is slightly stronger in absolutely empty space. However, the current research is hinting that a slightly stronger gravity would have an effect on all other constants as well, rendering the difference impossible to observe. At the end of chapter 6.3 I will instead suggest that differences in granular dimensionality show up as what we know as gravitational time dilation.

Even if it is impossible to measure a higher G even far away from any gravitational source, the number of connections between elementary spaces must definitely be lower there, although we might not be able to measure that, because it changes the way we perceive space and time.
If this line of reasoning is correct, the departure between the actual value of G and the predicted one, can tell us how much time dilation we have here on earth relative to a hypothetical observer infinitely far away from any gravitational source.

Note:
1. One could consider a quantization through the size of elementary spaces, however, the wavelength is more directly measurable and should thus be preferred.

4. IMPLICATIONS FOR COSMOLOGY

4.1 **The entropic expansion principle** [19.05.2013 (hypothesis); 28.11.2013 (erroneous first calculation); 28.02.2014 (revised calculation); 25.10.2015 (acceleration only due to m^2); 02.11.2015 (improved calculation); 11.11.2015 (matching curves); 01.11.2018 (comparing the curves of dark energy and entropic expansion)*; 12.05.2019 (using the new age of the universe/the CMB from chapter 4.12); 29.06.2019 (using refined age of CMB; Hubble value within the Milky Way); 15.07.2019 (corrected mistake in the 2014-calculation; corrected local Hubble); 20.07.2019 (average SBH mass derived local H_0; local average SBH mass derived local H_0)]

*In collaboration with Ludovic Krundel (France).

In chapter 1.3 it was stated that:

"The expansion of space in a region is exactly proportional to the local entropy increase in this region. The acceleration of this expansion is due to the exponential growth of entropy in black holes (they have a maximum of entropy)."

I want to reformulate this into a new, *fourth law of thermodynamics*:

In a sufficiently large range (the range of galaxy clusters), the entropy density is <u>not increasing, nor decreasing</u>. For guaranteeing this form of entropy conservation, the universe is expanding.

Or simpler:

§ The average entropy density in the universe is always constant.

Because of the second law of thermodynamics, the total entropy in closed systems has to increase. But in an open system like the universe things are different: here any local increase of entropy is levelled out by the expansion of space. Since stars will continue to burn and black holes will continue to grow for still quite a long time, the expansion of space will not end before life ends.

This forth law of thermodynamics is only possible because of black holes. It makes thermodynamics independent from any low entropy border condition.

While searching for a mechanism for the expansion of space I used the analogy of a gas that slowly spreads out in space. However, the spreading of a gas increases entropy, while the expansion of space itself keeps the overall entropy constant. Yet this analogy helped me to link the expansion of the universe to entropy increase. I asked the question: what if the expansion of space is itself a thermodynamic process?

That would indeed solve the mystery of why the universe appears to have started off with a low entropy. The answer would be both simple and surprising: it didn't!

In the primordial universe black holes were smaller and had a lower entropy, but the density of black holes and matter in general was higher, so that the overall entropy density was exactly the same as today.

Can such a proportionality between entropy increase and the expansion of space be established using concrete observational data?

Before the formation of stars, primordial black holes would have been the only means of expansion.

However, the dropping matter density ρ caused those primordial black holes to consume less matter and to grow slower.

This changed with the formation of galaxies and stars. By then the primordial black holes had accumulated enough matter around them to be independent from the average matter density ρ in their growth.

The first stars appeared about 200 million years after the universe became transparent (photon decoupling). Short living big stars burn out after only 10,000 years. Smaller stars like our sun can burn for up to ten billion years. Some of these very big stars formed black holes after their death.

In the early universe there were many quasars. We might suspect the universe to have had a faster entropy increase back then, but we would ignore that there were less and smaller black holes.[1] The following table shows the acceleration of the expansion in detail:

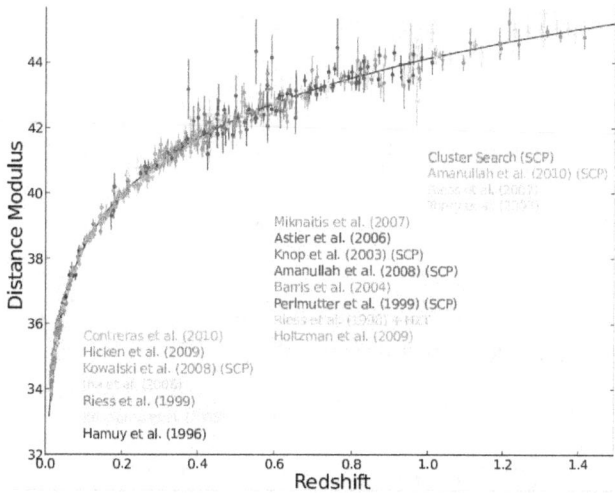

Fig. 24. This table is the result of data collection conducted by the Lawrence Berkeley National Laboratory and available at [http://supernova.lbl.gov/Union/]. It shows a clear acceleration for z < 0.6.

It shows that the universe was never decelerating. It also reveals that the expansion is a quite simple exponential function in time, just as it should if it depends on black hole entropy.

We will come back to this later. For now, let's first examine the local expansion in our galactic neighbourhood.

In order to prove that the expansion of the universe is indeed linked to its entropy, we want to first calculate the entropy increase of our galaxy, the Milky Way, and then the distance increase to our neighbour galaxy, the Andromeda galaxy. We thereby have to ignore the Andromeda galaxie's local movement towards us and think of the calculated distance increase as a growth of space only. So this distance is only a reference point. If Andromeda was not moving towards us, but standing still, the distance growth would be just as calculated.

The very most entropy is contributed by black holes. Their entropy is so big that all the rest becomes negligibly small. According to Penrose (1989) the entropy per baryon in the sun is in the magnitude of about 1, while in a black hole, it is 10^{20} (in natural units). We can therefore fully concentrate on black hole entropy only.

The Bekenstein-Hawking entropy of a black hole is given by

$$S_{BH} = \frac{A\,c^3\,k_B}{4\,G\,\hbar} = \frac{4\,\pi\,G\,k_B}{c\,\hbar}\,M^2$$

where the right term is a constant. For knowing the entropy increase factor we only need to look at M^2. It is this M^2 which gives the expansion a general tendency to accelerate.

We want to compare the entropy increase in the Milky Way in one year to the distance increase between us and a fixed point with the distance of the Andromeda galaxy in one year.

The mass of the <u>supermassive black hole</u> in the centre of our galaxy is

$$4{,}100{,}000\ M_{\odot}\ =\ 4.1 \times 10^6\ M_{\odot}\ =\ 8.2 \times 10^{36}\ \text{kg}$$

Assuming the average mass of stellar black holes to be around 10 times the mass of the sun ($M_{\odot} = 1.9 \times 10^{30}$ kg), and having about 100 million of them in the Milky Way, the <u>total mass of stellar black holes</u> in the Milky Way turns out as

$$100{,}000{,}000 \times 10\ M_{\odot}\ =\ 10^9\ M_{\odot}$$

That gives a <u>total mass of black holes</u> in the Milky Way of

$$1{,}004{,}100{,}000\ M_{\odot}$$

But we cannot treat these black holes as a single black hole, because two black holes merging into one single black hole, always lead to more entropy than the original two black holes.

There are 7 new stars every year, within the Milky Way.[2] Only one out of 1,000 is heavy enough to become a black hole. So we might not have even one new black hole in 100 years. We can therefore ignore the formation of new stellar black holes, in our calculation.

In ordinary models of galaxy evolution the Milky Way is estimated to be 200 million years younger than the universe. However, in the entropic expansion scheme described here, we need to assume that all galaxies were formed by primordial black holes, because without primordial black holes there would be no means of expansion in the very early universe when no stellar black holes were around. In this case, the Milky Way galaxy can be thought to be exactly as old as the universe, namely 13.8 billion years old. Due to the fact that the size of a galaxy is proportional to the size of its supermassive black hole, we can approximate the growth rate of Sagittarius-A by simply dividing its mass by the age of the universe, which yields:

$$0.0002971014493 \ M_\odot/\text{yr}$$

Now we can calculate the entropy increase factor as

$$X_S = \frac{S_1}{S_0} = \frac{\left(4{,}100{,}000 \ M_\odot + 0.0002971014 \ M_\odot/\text{yr}\right)^2}{\left(4{,}100{,}000 \ M_\odot\right)^2}$$

$$X_S = 1 + 1.4492753 \times 10^{-10}$$

What about the increase in distance?
Andromeda is 2.5 million light years away from us ($2.363560128 \times 10^{19}$ km). The value for the Hubble constant in the cosmic neighborhood is 74.03 km/s. That gives us a distance increase of 56.743040996 km in one year, which is an increase factor of:

$$X_{\Delta x} = \frac{\Delta x_1}{\Delta x_0} = \frac{2.363560128 \times 10^{16} \ \text{m} + 56743.040996 \ \text{m}}{2.363560128 \times 10^{16} \ \text{m}}$$

$$X_{\Delta x} = 1.000000000002837151 = 1 + 2837151 \times 10^{-12}$$

This is still far away from the value of X_S, and it doesn't look right to compare entropy in a volume, which is 3-dimensional, to something one dimensional like a distance. Therefore we should make the calculation more exact, by looking at the ratio for the increase in volume:

$$X_{\Delta V} = \frac{\Delta V_1}{\Delta V_0} = \frac{(2.363560128 \times 10^{16} \ \text{m} + 56743.040996 \ \text{m})^3}{(2.363560128 \times 10^{16} \ \text{m})^3}$$

$$X_{\Delta V} = 1.00000000002681859 = 2.681859 \times 10^{-11}$$

It doesn't make any big difference, if we look at a spherical volume, or a cubic one, like we did in this calculation. Differences occur only after the 61th digit.

That is quite close to the increase factor for entropy ($1 + 1.4492753 \times 10^{-10}$), with a difference of a factor of 0.18. This might be due to the fact that we don't know much about the entropy of the surrounding halo of our galaxy; or it may be due to uncertainties in the value of H_0.

However, as we will see in chapter 4.12, the true age of the universe since decoupling, which is required for the formation of structures such as stars and galaxies, is $4.2130109541636 \times 10^{13}$ yr. We might be tempted to use this age to estimate the growth rate of Sagitarius, but different from other structures in the universe, supermassive black holes grow since the beginning of time. Therefore we should be using the full age of the universe, which will be derived in chapter 4.15, and which we will borrow from there. It is $8.6670797702863 \times 10^{13}$ years. This yields a growth rate of:

$$4.730543745 \times 10^{-8} \, M_\odot/\text{yr}$$

Leading to an entropy increase factor of:

$$X_S = \frac{S_1}{S_0} = \frac{\left(4.1 \times 10^6 \, M_\odot + 4.730543745 \times 10^{-8} \, M_\odot/\text{yr} \right)^2}{\left(4.1 \times 10^6 \, M_\odot \right)^2}$$

$$X_S = 1.000000000000023 = 1 + 2.3 \times 10^{-14}$$

That is not at all in agreement with $X_{\Delta V}$. Does that mean the hypothesis is wrong?
Do we observe the expansion of space within the Milky Way or even within the local group? The answer is: no. In fact we can use the difference between X_S and the global Hubble derived $X_{\Delta V}$ in order to define a new local Hubble constant, which we may call the Milky Way Hubble constant H_{MW} or the Hubble constant $H_{sgr\,A}$ of Sagitarius-A:

$$H_{sgr\,A} = \frac{X_S \left(M_{sgr\,A} \right) - 1}{X_{\Delta V} \left(H_0 \right) - 1} \times H_0$$

Using the latest value of the Hubble parameter H_0 of 74.03 ± 1.42 km s^{-1} mpc^{-1} yields:

$$H_{sgr\,A} = 0.0008576141 \times H_0$$

$$H_{sgr\,A} = 0.0634891693 \text{ km s}^{-1} \text{ mpc}^{-1}$$

According to space particle dualism theory, the universe is expanding at different rates at different places in the universe. So we have different Hubble values everywhere.

The above value for the local Hubble parameter implies that in our galactic neighborhood the universe expands 1166.0256522 times slower than on average.

Why is that? As we will learn in chapter 4.12, the average mass of supermassive black holes is 5,757,999,302 $M_{\odot}$. Sagittarius A, our own supermassive black hole, is 1,404 times smaller than that. If we take the average of the mass of supermassive black holes in our immediate neighborhood the two numbers should match exactly. Mathematically speaking:

$$\frac{X_S\left(M_{sgr\,A}\right) - 1}{X_{\Delta V}\left(H_0\right) - 1} \approx \frac{M_{sgr\,A}}{M_{DH_\mu}}$$

Using this relation we can derive an even more precise value for the local Hubble parameter, namely:

$$H_{sgr\,A} = \frac{M_{DH_\mu}}{M_{sgr\,A}} \times H_0$$

$$H_{sgr\,A} = 0.0179996548 \text{ km s}^{-1} \text{ mpc}^{-1}$$

However, as we will see in chapter 4.15, black holes with low redshifts have smaller masses. Using the table of supermassive black holes from chapter 4.15, we arrive at an average mass M_{DH_μ} for supermassive black holes of:

$$M_{DH_{\mu(z<1)}} = 3,838,964,285.7$$

Which implies a local Hubble value of:

$$H_{sgr\,A} = 0.07692778 \text{ km s}^{-1} \text{ mpc}^{-1}$$

We notice that this third Hubble value is closest to the first one, which was derived independent from an average SBH mass.

This shows us that the universe is expanding in order to compensate the increase of entropy.

In the above calculation we found the growth rate of the central black hole Sagittarius-A by dividing its mass through the age of our galaxy (or of the universe, which is not much older). Thereby we assumed that the growth rate is always constant, but is it? A study published by the British astrophysicist Victor Debattist in 2013 found that the *M-Sigma* relation between the size of a supermassive black hole and its host galaxy can only be maintained if they grow together, and that implies a linear growth. [3] This shows that our assumption of a linear growth rate was correct.

The growth rate of black holes used here is much smaller than those we find in the mainstream. Is that a problem?
If the mainstream age of the universe was true, Sagittarius-A would be consuming one sun mass in 3,000 years. According to the above mentioned study by Victor Debattist, there are supermassive black holes which grow by a rate of one sun mass in 20 years, again using the mainstream age of the universe.
One constantly comes across articles that either say that supermassive black holes are 'almost incomprehensibly huge' or that they grow 'suprisingly fast'. Mainstream cosmology has a hard time explaining these growth rates.
Supermassive black holes with the sizes we observe don't make sense in a universe that is only a little bit older than the life-expectancy of a sun-like star.
The longer age of the universe we will come to in chapter 4.12 solves the mystery.

Could the relation between entropy and expansion we found above be based on mere chance? How sensibly does it depend on the volume ΔV chosen?

When testing entropic expansion in a particular region surrounding a supermassive black hole, the most sensitive and crucial factor that has to be taken care of is the volume of this region. It has to include all the low-entropy space around the supermassive black hole, because it is this space which levels out the high entropy inside the black hole. The examined region should be large enough to reach from one supermassive black hole to another.

If entropic expansion is correct, then galaxies with smaller central black holes should be surrounded by less empty space. The Andromeda galaxy is on a collision course with the Milky Way and local expansion is not powerful enough to prevent that. However, the merging of the two galaxies will certainly create more empty space around them.

This could be tested in various ways. One way would be to find out if there is really more empty space around larger galaxies with larger central black holes.

Another way is to examine if in our immediate neighbourhood galaxies with larger black holes tend to have larger redshifts. This type of survey makes of course only sense if it only includes supermassive black holes which have no other supermassive black holes between us and them, otherwise their redshift will be a sum of the redshift they cause and the redshift the black holes in front of them cause.

There is nothing in relativistic cosmology which could tie the expansion rate to the number and size of black holes. According to the presented scheme, more black holes and bigger black holes cause a faster expansion; and hence they are surrounded by more empty space. In relativistic cosmology the opposite would be expected: more black holes decelerate the expansion here.

Although the mass of black holes increases in a more or less linear way, their entropy grows exponentially, because it is proportional to the square of the mass. That is what causes the acceleration of the expansion.

The calculation above was concerning only the present expansion speed, not the acceleration over time. Both the here described entropy density conservation principle and dark energy lead to an accelerated expansion. Entropic expansion derives the speed and acceleration of the expansion from the entropy growth in black holes, while the dark energy that is used in standard cosmology is artificially set to a value that describes the observations correctly. In order for dark energy to work it has to be assumed that before the current strong acceleration phase there was a time in which the acceleration was significantly haltered by the gravity of all the mass that is contained in the universe.

We can now test how good entropic expansion describes the observed acceleration of the expansion and compare it to the expansion one would expect from dark energy without any adjustment through gravity.

The entropy of a black hole is proportional to its surface area and therefore it grows proportional to m^2. Dark energy is usually identified with vacuum energy and thus it is proportional to the volume of empty space. It therefore grows proportional to x^3, with x being the diameter of a chosen empty region of space. When comparing the two plotting them onto the both diagram we are not interested in their units but only in the slopeness of their graphs. We therefore temporally re-name m^2 into x^2, with x representing any linearly growing parameter.

In fig. 25 we see the graphs of x^2 and x^3 respectively:

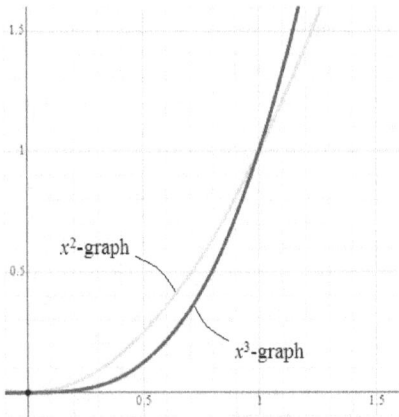

Fig. 25. The x^2-graph represents entropic expansion, while the x^3-graph represents dark energy without modification through relativistic global gravity.

In order to compare these two graphs to fig. 24, which was describing the accelerated expansion of space, we have to create $90°$-rotated mirror images of them and superimpose them with fig. 24. This can be achieved applying a little bit of trigonometry.[4] We create the image of the m^2-graph by the replacement:

$$x = s^2 \times \sin s$$
$$y = s^0 \times \sin s$$

And that of the x^3-graph by the replacement:

$$x = s^3 \times \sin s$$
$$y = s^1 \times \sin s$$

The three graphs are shown superimposed onto each other in fig. 26.

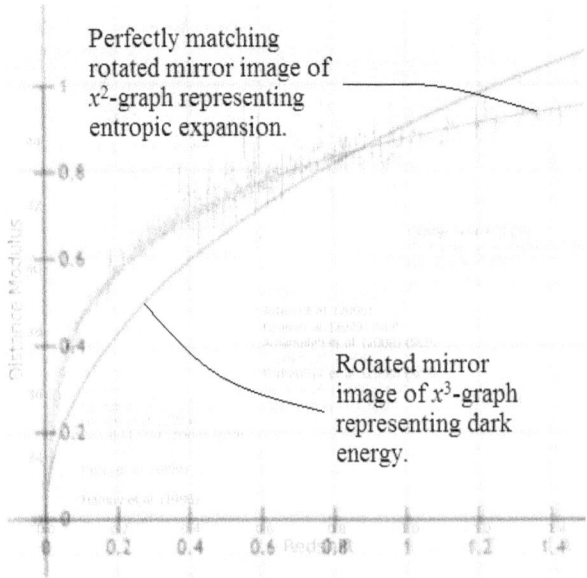

Fig. 26. Entropic expansion matches the observed expansion precisely, while dark energy fails to get even close. [Colours different from fig. 25]

The m^2-graph fits the expansion curve so well, that it becomes almost one with it, even making it hard to see, while the x^3-graph is a total mismatch. In standard cosmology this mismatch is of course 'fixed' by assuming gravity to work against the expansion and to flatten the resulting curve so much that it miraculously coincides with the entropic expansion curve.

Before the dark matter concept was introduced it was generally stated that only with dark matter there is enough stuff out there to reach the critical density ρ_c and to get the universe flat ($\Omega = \rho/\rho_c = 1$). After dark energy was introduced cosmologist went over to say that dark energy is <u>also</u> needed to have enough mass energy for a flat universe! Yet dark energy has the strange property that it becomes more and more as the universe expands, because more and more empty space with vacuum energy (supposedly equal to dark energy) is created! From this it should be clear that we never really had evidence for a density close to the critical density. Cosmologists didn't really measure the average density, but calculated the critical density and kept assuming that the average density is equal to the critical density, simply due to the fact that our universe appears to be totally flat. That means the incredible flatness of the universe has always been in gross contradiction to general relativity, because it was never successfully linked to the actual mass energy density we observe in the universe.

We also have to wonder how a geometrically flat universe can be accelerating, since only open universes with negative curvature are supposed to accelerate their expansion. From a general relativity perspective the expansion of space being so completely uncorrelated to the geometry of space is certainly puzzling. From the perspective of space particle dualism it is entirely natural, because here gravity is caused by density differences in granular space. It still is a geometrical force here, but unrelated to curvature.

Interestingly this is not the first proposal to link the accelerated expansion to entropy increase: in a work with the name "Entropic Accelerating Universe", Damien A. Easson, Paul H. Framptony and George F. Smoot linked the accelerated expansion to the entropy increase of the cosmic event horizon (2010).[3] The entropy of this horizon is, however, not related to the actual entropy inside. The radius of the cosmological horizon grows with light speed and its size depends solely on the age of the universe. That is very similar to the mass growth we assumed for our central black hole in the aspect that it is a linear growth. The exponential growth of entropy they get is compared to the accelerated expansion of the universe (see fig. 27):

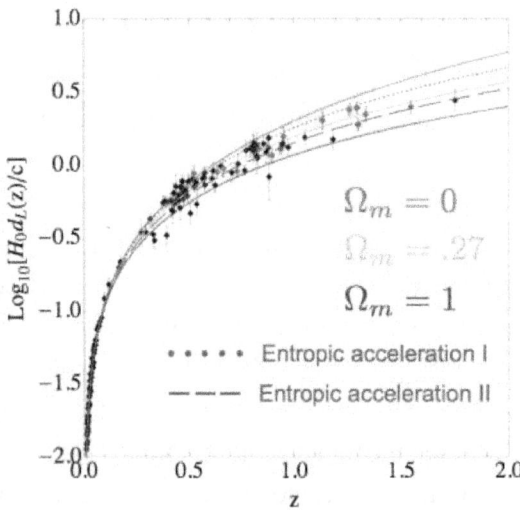

Fig. 27. Entropic acceleration resulting from linear black hole growth (linear growth of radius R or mass M).

The matter density distribution is expected to be more or less the same for most galaxies. Thus the growth of black hole mass should be quite linear over time. The inactivity of some black holes is thus levelled out by the activity of others.

Before the acceleration of the expansion was observed, most cosmologists did not even try to find a cause for the expansion itself, because they took it as an initial condition of the Big Bang that the universe is expanding. Back then, the expansion rate was thought of to have been fixed at the very moment of the Big Bang.
Therefore it was a real shock for the scientific community in 1998, when Saul Perlmutter, Brian P. Schmidt, and Adam G. Riess found that the universe's expansion is accelerating. Since then cosmologists tried to explain that with dark energy, a varying cosmological constant, vacuum energy, Higgs fields, and vice versa.
None of these explanations could predict the speed of the expansion, or make it depending on something.

Can we say that entropic expansion is an intrinsic part of space particle dualism?
This hasn't yet been shown in a rigorous way, but what we can say for sure is that it does only make sense in this theory, because an entropic expansion would be impossible in a universe with globally acting gravity.

Notes and references:
1. Indeed I was inclined to link the acceleration to an increasing formation of quasars, which seemed to to me to be backed up at least roughly by data [VII270 SDSS quasar catalog: tenth data release (Paris, 2014) (DR10Q, DR10QSO, SDSS10QSO, J/A+A/563/A54); available at [http://cdsarc.u-strasbg.fr/cats/VII.htx]], before it appeared to me that the proportionality to m^2 is all what is needed.

2. Wanjek, C. (5 January 2006). "Milky Way Churns Out Seven New Stars Per Year, Scientists Say". *Goddard Space Flight Center*. Retrieved 2008-05-08.

3. Victor P. Debattista (2013). "Disk Assembly and the M_{BH}-Sigma Relation of Supermassive Black Holes". *The Astrophysical Journal* 765 (1).

4. This trigonometric method was provided to me by my collegue Ludovic Krundel.

5. Easson; Frampton; Smoot (2010). "Entropic Accelerating Universe".Phys. Lett. B696(3):273–277.arXiv:1002.4278.Bibcode:2011PhLB..696..273E.doi:10.1016/j. physletb.2010.12.025.

4.2   Entanglement and the Fermi-paradox: Where are they? [28.02.2014 (no alien prediction); 07.07.2016 (Fermi paradox in detail); 26.09.2016 (concentric shells and consciousness); 26.10.2016 (obscured sight in spiral arm); 30.04.2017 (pattern rotation period); 14.10.2017 (sec. and third shell); 28.02.2018 (link to Filipovic's mass extinction theory & corrected rotation period; obscured sight in bulk region); 20.05.2019 (using distances according to SPD destroys the correspondence); 24.06.2019 (accounting for the oscillation of the sun over thegalactic plane restores it); 26.06.2019 (abundance of future observation hypothesis and experiment); 26.06.2019 (explaining the other three redshift periodicities); 18.07.2019 (baryonic acoustic oscillations are the cause for the strongest two periodicities); 19.09.2020 (CMB irregularities are due to micro-PK)*]

*Sponsored by Lin Yi Song (China).

Note: This chapter is mainly focusing on the cosmic microwave background (CMB) and the Sloan Digital Sky Survey (SDSS). The analysis of the former is fairly independent from the particularities of space particle dualism theory (SPD), while the later depends sensitively on it. I do not account for the SPD-specific intergalactic distances until very late in this chapter. This chapter is concentrating on the emergent universe hypothesis and the influence of so called 'abundance of future observation', but it turns out that at least one aspect of the SDSS is in fact a result of baryonic acoustic oscillations (BAO). For details on BAO, see chapter 4.16.

Since the Copernican revolution people have been asking themselves if there is life on other planets. If we are not the centre of the world, then it would be reasonable to assume that life exists in other parts of the universe as well. For long time people were speculating about life on Mars or even Venus. Probes have been sent to both Mars and Venus and didn't find any signs for life there – not even micro-organisms. Even until today there is no scientific evidence for life elsewhere in the solar system.

Is it possible to detect life on an interstellar scale? In the last 70 years we have sent television signals out into all directions of space. These signals are, at least in principle, detectable in a radius of 70 light years already. With our current technology we would not be able to detect any signals from such a distance. But if the signals are sent out and concentrated into a certain direction in space, then they become detectable even with our current technology. It would not be too much of an effort for a highly advanced civilization to send such signals into the direction of all potentially life bearing planets. How many alien civilizations would one expect to find in our galaxy? In 1961 the astronomer Frank Drake tried to approach this question with his famous equation, now known as the Drake equation. If R_* is the average rate of star formation in our galaxy, f_p the fraction of stars with planets, n_e the number of planets that potentially support life, f_l the fraction of those planets that actually develop life, f_i the fraction of those life bearing planets which develop intelligent civilized life, f_c the fraction of those civilizations which develop communication technology and L the length of time over which those technological civilizations exist, then the number of intelligent civilizations with telecommunication technology in our galaxy should be:

$$N = R_* \times f_p \times n_e \times f_l \times f_i \times f_c \times L$$

R_* is estimated to be 7 stars a year.[1] f_p was estimated $0.2 - 0.3$ back in 1961, but recently it became evident that every star has planets orbiting it, so that it is actually 1. [2, 3] A study from 2013 estimated that about 40% of stars have earth-sized planets in their habitable zone.[4, 5] So $n_e = 0.4$. If all of these planets would develop intelligent civilizations, and if these civilizations wouldn't get lost, then we would have about 40 billion intelligent civilizations in the Milky Way. One way to get less than that is to choose numbers for f_l, f_i or f_c that are smaller than 1, which is equivalent to saying that the development of life is very unlikely even on planets which potentially support life (small f_l), or that the evolution up to intelligent life is very unlikely (small f_i). If we assume that RNA, which is the predecessor of DNA, formed out of random collisions of molecules, then a small f_l would be plausible. According to estimates from Frankis Crick, one of the discoverers of the DNA structure, the chance to get such a structure out of random molecule collisions is about 1 to 10^{260} (!).[6] On the other hand life developed rather early in history. Fossils of the first simple life forms can be as old as 4.1 billion years, while earth itself is 4.5 billion years old. The time difference here is basically just the time it took for earth to cool down to a level where it could support simple cellular life. At that time there was not even any oxygen in the atmosphere – that came 1.1 billion years later. Some say, if life formed that early on earth, its emergence should be a natural law, and not an exception. This would suggest a value for f_l close to 1. Various researches suspect the formation of first RNA and then DNA to be much more likely than originally suggested by Cricks. They point out that various substances have the tendency to form double helix-like structures naturally and spontaneously in salty water.[7]

There are two classic ways to deal with the early formation of life. One approach is the hypothesis that the fabric of space-time somehow supports the formation of life. The other is that primitive life (or at least RNA) emerged in cosmic clouds and came to earth on meteoroids – this was Crick's approach. Yet the fact that there is no cellular life at all on Mars speaks against this hypothesis. Also there is not much to gain from extending the time for the formation of RNA a few billion years further into the past. That doesn't make it much more likely. But in both cases we would expect life to emerge wherever there are suitable environmental conditions.

If we now really take f_l to be 1, what can we say about f_i and f_c? It is true that only one species out of so many on earth developed high intelligence, civilization and technology? If that is the case, then it seems to be a highly improbable path for life to take. But are Homo sapiens really the sole species that developed intelligence? The Neanderthal reached a level of evolution comparable to that of Homo sapiens, but he was wiped out. What prevents today's apes from evolving into humans? It is basically only the fact that we already occupy this ecological niche. From this perspective the evolution to intelligent life does not seem unlikely at all.

There are deep mysteries in the mechanism of evolution, especially the evolution of algorithms, to which we will come back to in chapter 4.10, but for the present issue it is enough to know that f_l, f_i and f_c are probably all close to 1.

The only remaining way to reduce the number of intelligent civilizations in the Milky Way is to choose a small number for L, which equals saying that intelligent life is 'self-destructive'. An indicator of that is the 'atomic threat'. Atomic bombs present on earth now would be enough to wipe out all human life on this planet. In 1966 the astronomer Karl Sagan speculated that this may explain why we haven't received any signals from extraterrestrial civilizations.[8] Stephen Hawking also became a supporter of this view.

I don't find this convincing: technology allows humans to protect themselves from natural disasters and even impacts of meteoroids. Intelligence should allow them to survive longer or even for a potentially unlimited time. And even if self-destruction is likely, it would be hard to imagine that all 40 billion intelligent civilizations destroyed themselves shortly after developing communication technology. The life expectation of telecommunication civilizations really doesn't have to be very high to lead to big numbers for N. Assuming that f_l, f_i and f_c are all 1, we have 2.8 times the life expectancy of a typical civilization. With a ridiculously small $L = 100$ yr we already get 280 civilizations. A more realistic $L = 1000$ yr gives us 2,800 intelligent civilizations, and so on.

It would take only one out of 40 billion civilizations to be more cautious and live longer, and we would have a highly advanced civilization which could conquer the whole galaxy. Even with technology not much more advanced than ours it would be possible to send out so called Von Neumann probes which could plant life on all life supporting planets across the galaxy. A civilization could also choose to build large space ships as big as cities which could harbour generations of people, plants and animals. According to estimates by Paul Davis and John Gribbin it would take about 10 million years to colonize the galaxy using currently reachable speeds. This is a small number compared to the age of the Milky Way itself, and it becomes even smaller when better technology is used. Because of the missing air friction, momentum is always conserved in outer space, even when there is no thrust. Therefore it would be easy to accelerate a space craft to let's say 90% the speed of light, by dropping a large number of atomic bombs one after another behind it. Thereby a galaxy could be colonized in much shorter time. Even if it would still be hard to cross the whole galaxy, stellar neighbours would have no problem visiting us. At 90% the speed of light, a spacecraft could reach the next star in about 5 years. Even if the home planet of a civilization got destroyed, it could reach a life supporting planet circling another star, just by using technology only slightly more advanced than ours.

This brings us to the *Fermi paradox*: if life exists in various places across the Milky Way then it would be, considering the relative young age of earth, very hard to see why advanced civilizations didn't already form millions of years before. [9] Wouldn't such civilizations have colonized the whole Milky Way already?

When we look out to the night sky we do not see it like it is now, but how it was in the past. In relativity theory we would not worry much about this, since things exist there without being observed. But in quantum theory it is much different: here things have to be observed in order to be in certain states. As mentioned in chapter 1.2 space particle dualism can only work if there is a distinct *conscious now-slice*. Although we have certain quantum states all over the *border of the collective past light cone*, a hypothetic alien on this past light cone couldn't be conscious. It couldn't be an observer as well, because it is not on the *conscious now-slice*. Such an alien could not collapse a wavefunction. If we would allow that, we would end up with conscious light cones all over space-time, and that would at the end result in fixed quantum states for the whole of space-time. As mentioned in chapter 1.2 that would wipe out all quantum effects, and bring us back to a classical space-time.

Aside from that we have to remind ourselves that from a real quantum perspective the spacetime view on reality is of mere illustrative nature and should not be taken literally. The philosophy of quantum gravity should be based on a refusal of the spacetime paradigm.

This seems all fine, but what about the people around us? Light always needs time to reach us, even on earth. So other people should also lie on our past light cones. Yet that doesn't make them unconscious. Believing so would bring us to *solipsism*. Indeed Eugene Wigner argued that it leads to solipsism! However, we know better: the difference between people around us and a remote alien is that all particles on earth are endlessly *entangled* with each other. That makes them all belong to a single large wavefunction Ψ, which collapses in some places, but grows in other places; all at the same time. It is this oneness, or inter-entangledness, of the wavefunction that creates the conscious now-slice $S_{\circledcirc}$ (the 'eye' $\circledcirc$ being a symbol for consciousness).[10] The wavefunction of the universe can then be written as

$$|\Psi_U\rangle = |\circledcirc_1\rangle + |\circledcirc_2\rangle + \cdots + |\circledcirc_n\rangle + |A\rangle + |B\rangle + \cdots ; \; \circledcirc_1, \circledcirc_2, ..., A, B, C, ... \in I_{\circledcirc}^{-}$$

With $|\circledcirc_n\rangle$ being the various mutually entangled conscious observers and $|A\rangle, |B\rangle, ...$ various objects they observe.[11] All of them have to lie on the conscious past light cone $I_{\circledcirc}^{-}$.[12] The future light cone I^{+} on the other hand is not of particular interest for the wave function. Nothing on it is in a certain state. Things within the borders of $I_{\circledcirc}^{-}$ are partially in a certain state. The ability of observers in P or $P_{\oplus}$ (the symbol $\oplus$ representing earth) to reconstruct the past is bound to a low entropy condition on earth.

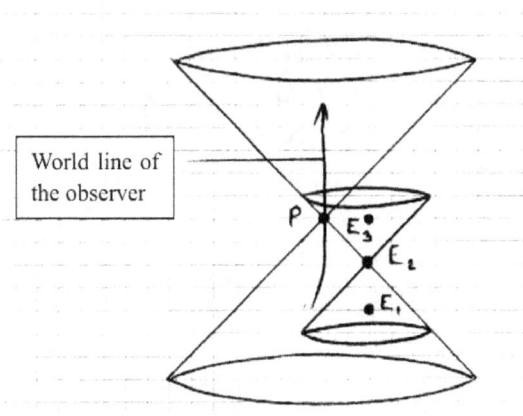

Fig. 28. Light rays which constitute the border of the future of event E_2, reach the point P of the conscious observer. That leads to the collapse of the wave function for E_2. The past of E_2, namely E_1, was measured by P before, but that was some time before. The ability to reconstruct E_1 depends sensitively on low entropy in the region P. The future of E_2, namely E_3, although being on our now-slice, is completely uncertain. It has not been measured yet. We can say that only the past of E_3, namely E_2 exists, but not the present. [sketch from the 19th May 2005]

On a cosmic scale all observers are positioned in the same point P which represents earth. And on a terrestrial scale distances are too small to require any analysis with space-time diagrams. If we take $I_{\odot}^{-}$ to be the past light cone of collective consciousness, then the point P has to be replaced by a zigzag structure 'vvvvv' when zooming in on it. As the number of conscious observers in the universe grows, the number of *entangled conscious light cones* grows.

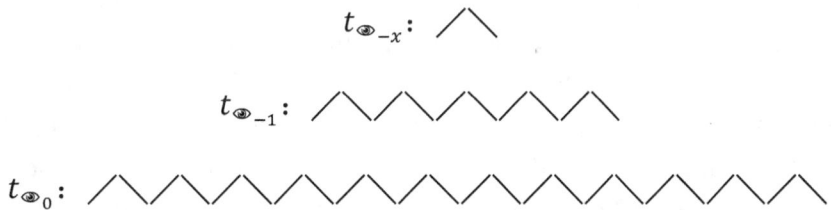

Only if the universe starts off with a single conscious light cone, all light cones can be entangled. Places in space-time beyond the light cone of a certain observer are called 'elsewhere'. If the light cone we are looking at is the joint light cone of collective consciousness, then this 'elsewhere' is just nothingness. Something which has never been measured must be in a state of maximal superposition. Such a state can be called 'Omnium' and is equal to non-existence. If we allow 'states' in the 'elsewhere' to collapse into *consciousness inhabited states*, even if it is only on the conscious now-slice, it would happen immediately all over an endless universe. That would lead to several problems:

1. Things on the past light cones of different unentangled observers in remote places of the universe would have to be in accordance with each other. Having an independent wave function, this is hard to achieve. It would mean having things outside the wavefunction influencing things inside.

2. The conscious now-slice is build by entanglement. It is hard to see how those alien observers could emerge on the same now-slice.

3. The quantum state of a *past earth* now observed by aliens in a remote place in the universe would have to be dependent both on the state of the present earth and the state which manifested through the observation of those aliens. Basically the wavefunction could no longer represent the maximal accessible knowledge, because things that have never been measured could be in certain states already.

4. Nothing could tell us how many of those other civilization would have to emerge on the now slice. Without restrictions we could have basically anything emerging from this Omnium.

Then we also have to look at how life emerges. According to the Copenhagen interpretation, the whole past of the universe was fixed in the moment consciousness emerged. If that is true, a part of the evolution on earth 'happened' completely in superposition. That brings us to point 5:

5. If it is the emergence of consciousness which teleologically determines the formation of RNA and DNA in a superpositioned past, then those other civilizations across the universe would have had to emerge all at the same time, because a planet which is unceasingly observed in its early history could not develop life.

Before we said that allowing the emergence of (unentangled) conscious observers on our past light cone would lead to endless many conscious now-slices all across space-time, and would thereby destroy all superpositions. Allowing for conscious observers emerging all across the unobserved parts of our conscious now-slice would not destroy superpositions, but it would restrict them a lot. Not only the boundary of the past $I_\oplus^-$ is in a certain state, but basically the entire space-time region enclosed by $I_\oplus^-$. However, the degree of certainty within different regions enclosed by $I_\oplus^-$ depends on local entropy conditions. The past of a black hole for example is totally uncertain. Having a universe full of conscious observers on $S_\oplus$ we would have the whole endlessly vast past region below the surface $S_\oplus$ in a fixed state, leaving just the little gap regions between the light cones undetermined.
That could be a problem for consciousness, because photons which have been measured at some time in the past don't lead to as much superpositioning in the brain as 'Omnium-photons' do.

If we are the only conscious observers in the universe, and the above strongly suggest that, then what is within our cosmological horizon should be everything that exists. Our past light cones would be all there is of the universe. Only within these cones there would be a certain quantum state. Yet even the people we see around us lie on our past light cones, because light needs time to get to us.

Different people lying on each other's past light cones can still belong to the same reality, because they are connected through entanglement, allowing their wavefunctions to change all at the same time. That means we live on a conscious now-slice of space-time which is created through quantum entanglement. This notion of a distinct present was unknown to the theory of relativity, but it becomes necessary if we introduce the quantum measurement into the picture.

Other stars and galaxies are incomparably farther away from us. Their light needs hundreds, thousands or even billion years to travel to us. When we look at the night sky, we are not looking at the universe as it is today, but as it was in the past. Our observation is therefore only measuring the past, not the present.

In our immediate environment measuring photons equals measuring the objects that emit them, because the distances the photons have to travel are tiny; yet for the photons of stars and galaxies the distances are huge, so that measuring photons does there not equal measuring the objects that emit them. Therefore the ocean of photons the earth resides in represents merely an ocean of potential existence which turns into reality just in the very moment we decide to visit those other places in the universe.

So stars do not exist on our now-slice, but only on our past light cone, while things near to us are entangled with us, and can therefore co-exist with us. Only our immediate environment, the biosphere is in a certain state, 'right now'.

Arriving here some might be tempted to search for a last resort in speculating that heat photons emitted by earth could spread entanglement into the outer cosmos. However, these photons are propagating into the future and form the border of our future I^+. Nothing on the I^+ hyper-surface is in a certain state, so there is not really anything to get entangled with.

Regions of the 'elsewhere' which are part of $S_\oplus$ and which are close to $I_\oplus^-$ have been measured very recently and stayed unmeasured for only a short time. Can they be in certain states? The answer is no. Quantum states get fuzzy very quick after measurement. Unceasing measurement is what makes this quantum world appear classic at a macroscopic scale.

Still we can not be sure how far beyond earth entanglement reaches, if at all. Surely there is entanglement between us and the pioneer probes which already left our solar system. Some properties of their constituent particles are fixed instantaneously as particles on earth that are entangled with them are measured. However, their overall state is measured hours later, since light from them needs to travel large distances to arrive to us here on earth.

Even if we had a whole planet entangled with our planet, it still wouldn't support the independent formation of life there. The formation of RNA and the subsequent formation of DNA has to happen in a vast superposition. Very much like in Wheeler's delayed choice experiment (see chapter 2.5), the future has to determine the past.

Although the above argumentation could have been derived from the standard Von-Neumann interpretation and special relativity alone, it is the concept of a distinct present, a conscious now-slice, which allows for a systematic reasoning in this realm. We can therefore say that it is a prediction of space particle dualism theory, that

§ There is no alien life. All life can only exist if connected through entanglement.

This is very much like a reversal of the Copernican revolution. Leaving out quantum mechanics, earth is in a random position in the universe, orbiting a random star, in a random galaxy, being part of a random galaxy cluster. But when considering quantum mechanics, earth is the center of this world, being the only place where things are in a certain state, and also the only place where life exists.

In space particle dualism theory, everything has to see or affect each other, otherwise it falls into non-existence. So in this theory, the world is much smaller than in theories without state reduction: no hyperspace, no colliding universes, no baby-universes born out of black holes or even just out of vacuum fluctuations at the Planck-scale, and all the other *materialistic hocus pocus*. In this idealist world view the world ends where observation ends.

Of course having no alien life forms might be pretty disappointing for many. It does also appear very unusual for a scientific theory to predict the non-existence of alien life forms. The normal scientific, and very reasonable, argument is that if life is a natural phenomenon, it must be very common all over the universe, otherwise it would look like something created by a god.

Although this prediction sounds odd, it does solve the *Fermi-paradox*. If life is common in the Milky Way, then it is strange that developed civilizations didn't form millions of years before us. Even if most destroy themselves shortly after developing the technology required for space travel and far range communication, there would still be a few which didn't. Given the enormous amount of time that already passed since the beginning of physical time, they should have already colonized most parts of the Milky Way. But they didn't. Why is that? The answer I am proposing here is: because they do not and can not exist.

What if we find alien life one day? Would that disprove space particle dualism theory? Potentially yes, but not if it could be shown that we share a common origin with them, for example in the way that they 'planted' us. However, it would make less sense to plant just the seeds for evolution and not a fully developed range of species.
I am quite confident that the SETI program will continue to find no signs of extraterrestrial life.

Besides SETI constantly failing to detect signals from (intelligent) alien life and exoplanet surveys also failing to find a single planet with an atmosphere shaped by biological organisms, is there still other evidence indicating that earth is special and that we are alone in the universe?

Indeed there is! There is overwhelming evidence that earth being the only place where wavefunctions collapse had a tremendous impact on the overall structure of the universe. This is evident from the analysis of various large scale structures in the universe. In 2005 Lawrence Krauss held a lecture where he mentioned disturbing anomalies in the cosmic microwave background (CMB); he said:

"… But when you look at the CMB map, you also see that the structure that is observed is in fact, in a weird way, correlated with the plane of the earth around the sun. Is this Copernicus coming back to haunt us? That's crazy. We are looking out at the whole universe. There's no way there should be a correlation of structure with our motion of the earth around the sun – the plane of the earth around the sun – the ecliptic. That would say we are truly at the centre of the universe."[13]

What he was talking about is generally referred to as the '*axis of evil*', a line through the microwave background which separates hotter and colder spots. The axis itself is a hot spot and less mysterious, because it is caused by the redshift due to the motion of the earth around the sun. What makes an anomaly out of it is the alignment with other structures above and below it. It turns out that if one connects the cold spots with straight lines, those lines intersect exactly where earth is. It was also noted that the sky above this axis is slightly hotter than the sky below and that the cold spots seem to be aligned in shape with this axis. However, those big cold and hot spots were not very distinct, because they were covered with an ocean of smaller irregularities, so that it was easy for the scientific community to ignore them.

Max Tegmark heard of these anomalies in the CMB and decided to analyse them with a computer program by splitting up the CMB into so called spherical harmonics. This aims to smudge out the small irregularities and leave only a predetermined number of big ones. Choosing 1, we see only the axis of evil. Choosing 2 and 3, more cold spots appear above and below it (see fig. 29 & 30).
Same as Krauss, Tegmark thought the anomalies might disappear when more exact measures of the CMB from the Planck satellite telescope are available. In 2013 they became available, but the anomalies remained.

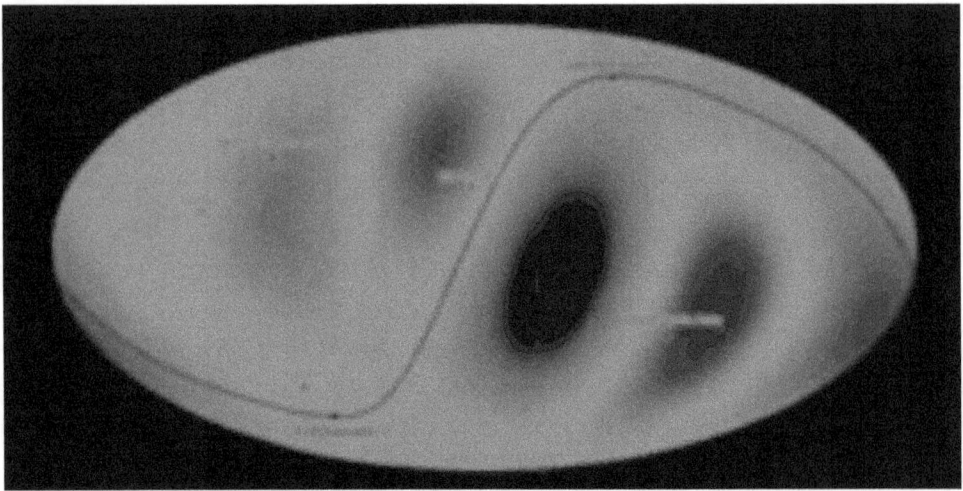

Fig. 29. The 'axis of evil' among which hot and cold spots in the microwave background are arranged on.[14]

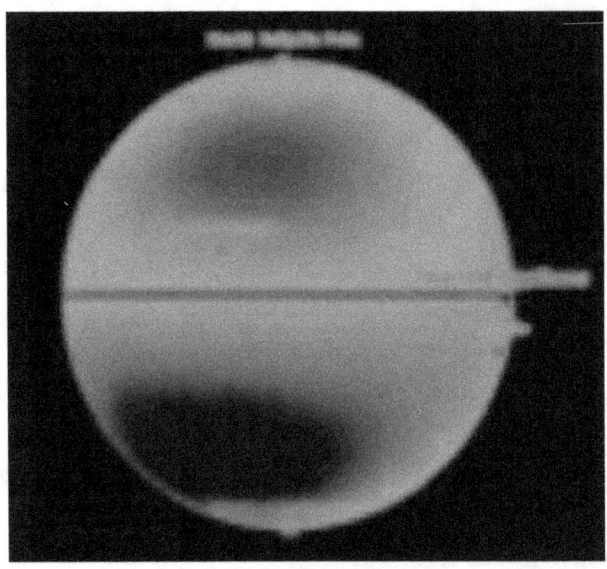

Fig. 30. If we project the microwave background onto a sphere we find that this 'axis of evil' roughly coincides with the orbit axis of the earth around the sun.

But this is not all. As Professor John Hartnett from Australia pointed out, there also seems to be a general asymmetry in the intergalactic matter distribution around us. When analyzing the distribution of photon frequencies it turns out that most photons prefer to have frequencies which are natural multiples of $\Delta z \approx 0.027$. That means a majority of galaxies is concentrated in concentric shells around us (!). These concentric shells of higher galaxy concentration lie with a distance of about 250 million light-years from each other (see fig. 31 & 32).[15]

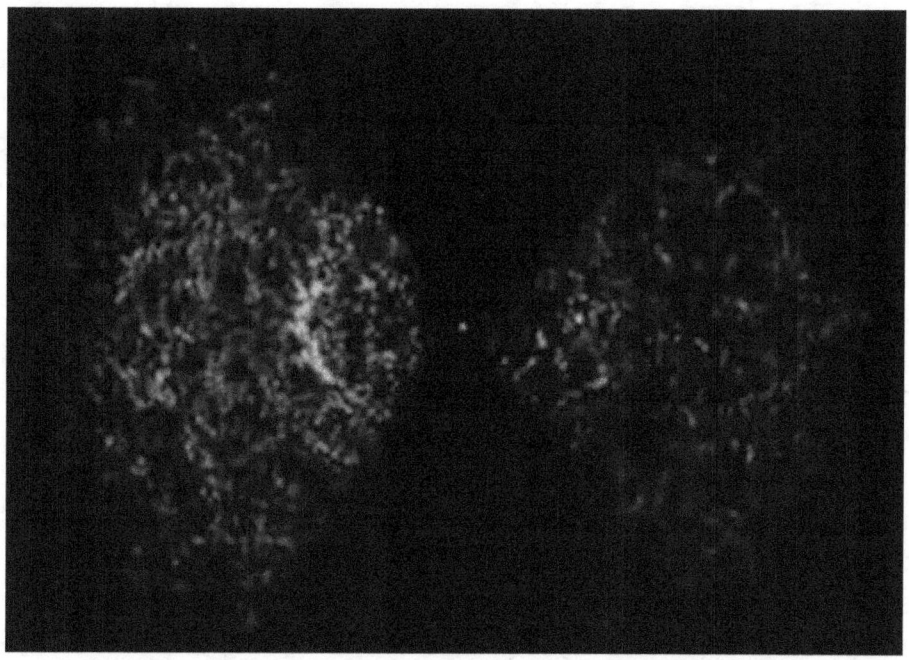

Fig. 31. Large scale distribution of galaxies in the universe according to the "Sloan Digital Sky Survey" (SDSS).

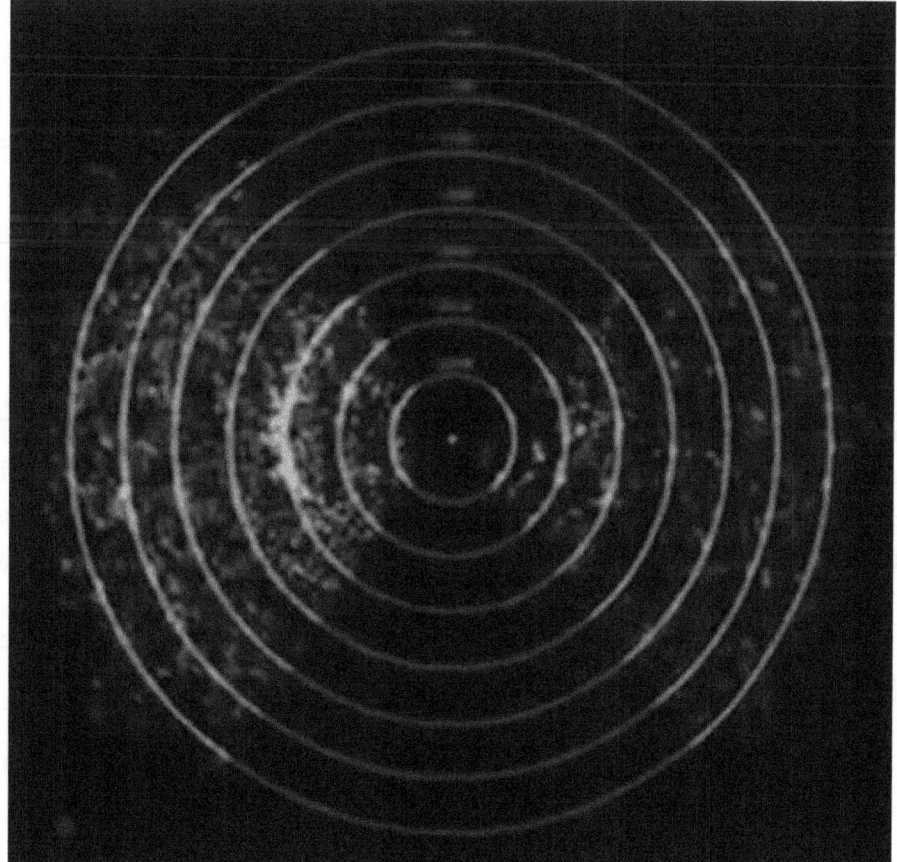

Fig. 32. The galaxies seem to lie on concentric shells with a distance of about 250 million light years.

This non-Copernican periodic structure in the galaxy distribution has a likelihood of only 1:1669 to have formed by chance and appears impossible to explain when adhering to decoherence theory where consciousness plays a rather passive role accounting only for the here merely subjective separation of realities. It can only be explained when accepting the role the Von Neumann-Wigner interpretation foresees for consciousness, having it creating reality by causing the reduction of the wavefunction all over an endlessly vast but death universe.[16]

Why this number? Why are those concentric spheres of higher galaxy density arranged in a distance of about 250 million light years from each other? When I first heard of this number, it immediately reminded me of the roughly 240 million years it takes for the sun to both orbit around the centre of our galaxy and to pass from one major spiral arm to the next and how this was linked to mass extinction events on earth due to the higher rate of asteroid impacts and supernova explosions within the spiral arms, by a team around M.D. Filipovic in 2013.[17]
What is the connection? After each mass extinction there is a new 'explosion of life' and this may lead to a higher density of galaxies, because the outer universe only slowly crystallizes out from a vast superposition into a certain state by the act of observation by conscious observers.

But this is not the only way to link the distance between these concentric spheres and the passage time of our sun from one spiral arm to the next. Even within the spiral arms of our galaxy tragic events do not occur on a regular basis. Furthermore tragic events leading to mass extinction do not only occur within the spiral arms, and there is no guaranty that the biggest such events would occur only during the passage through a spiral arm. But what is fundamentally different during the time of a spiral arm passage, is that the view to the outside universe is not as clear as in the middle between two spiral arms.[18, 19] Huge gas clouds are obscuring the sight, which leads to a less frequent measurement of quantum states beyond our galaxy.[20]

As mentioned before, the sun's galactic orbit period, which is 240 myr,[21] is also *roughly* the time the sun needs to get from one spiral arm to the next. For being related to the 250 million light years for the spacing of the concentric shells discovered by Hartnett, the travel from one spiral arm to the next must take an additional 10 million years.
Only if the travel from one spiral arm to the next takes exactly 250 million years, the here presented scheme can work.

There are several uncertainties here. The milky way galaxy was first believed to be a *barred spiral galaxy* with only two spiral arms, while today most scientists suggest a *four spiral arms structure*. Yet the testimony of the surveys conducted is ambiguous: When traced by old stars the milky way appears to have two spiral arms (see fig. 33), and when traced by gas and young stars it appears to have four spiral arms (see fig 34).[22]

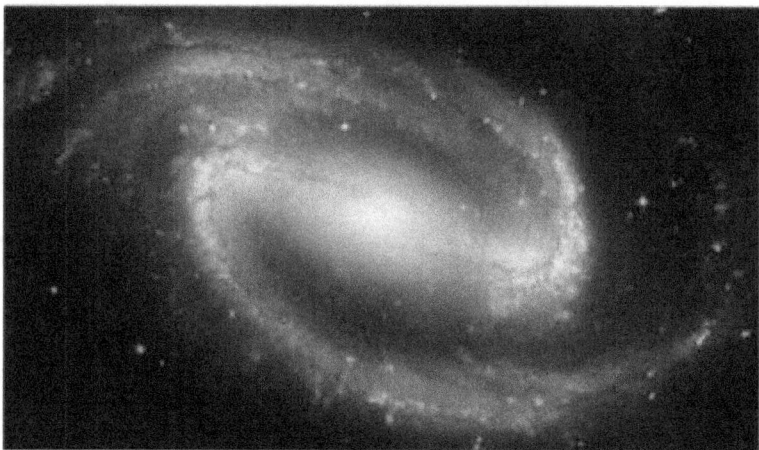

Fig. 33. NGC 3200, an example for a barred spiral galaxy with only two spiral arms.

Fig. 34. UGC 12158, an example for a galaxy believed to resemble the milky way quite closely in shape.

Furthermore, there is a big uncertainty in the spiral arm rotation period: according to recent estimates from Ortwin Gerhard from the Max Planck institute in Germany the pattern rotation period could range anywhere from 220 to 360 myr.[23]
With two spiral arms the rotation period of the milkyway galaxy has to be double the galactic orbit period of the sun for the sun to re-enter a spiral arm after one orbit:

$$240 \text{ myr} \times 2 = 480 \text{ myr}$$

This is already above the values suggested by the aforementioned recent estimates (220 to 360 myr) and therefore it seems that we have to dismiss this possibility. Furthermore the evidence of the last 15 years is increasingly pointing at a four spiral arm structure. [24, 25, 26]
With four spiral arms the galaxy must have an orbit period which is 1.25 times the galactic orbit period of the sun. That yields

$$240 \text{ myr} + \frac{240}{4} = 300 \text{ myr}$$

For this to be related to the 250 mlyr spacing of the concentric shell pattern of our universe, the re-entrance must happen 10 myr after one full orbit. Thus the rotation period of the milkyway galaxy has to be

$$240 \text{ myr} + \frac{240}{4} + (250 - 240) = 310 \text{ myr}$$

This is within the present estimates (220 to 360 myr).

The big uncertainty in the pattern rotation period arises from the spiral arms not being solid objects but merely density waves (somewhat similar to a traffic jam) which propagate at a different speed than the stars, that is, nothing moves at their orbit speed, and therefore there isn't any straight forward direct measurement one can do to determine this pattern speed.

So what can we do? What about the various theories linking spiral arm passages to things on earth, such as mass extinctions and variations in the overall temperature? They must be based on some pattern rotation speed and by examining the evidence that exists for these theories we may be able to determine the pattern speed.

Some have linked mass extinctions to something that can be easier known, namely the oscillation period in the suns galactic orbit, which is 70 million years, resulting in the sun passing through the midplane every 35 million years and being in it about 100,000 years.[27] Yet of course there must be a spiral arm where the sun passes through the midplane, otherwise there won't be a significant increase in star density.

When studying the various proposals one quickly realizes that most of them suggest periods of spiral arm passages of only about 100 million years. The cosmologist Nir J. Shaviv from the Racah Institute of Physics in Israel made an extensive research on a link between ice ages and passages through spiral arms. The passage period he suggests is 135 ± 25 million years.[28, 29] As his research shows every ice age can be linked to the passage through one of the Milky Way's various spiral arms (see fig. 35).

Furthermore he did not try to make the spiral arm pattern rotation speed fit his theory, but rather he made a very careful study of all ever proposed rotation periods and made a meticulous analysis to support the value he chose, which is $11.0 \text{ km s}^{-1} \text{ kpc}^{-1}$.

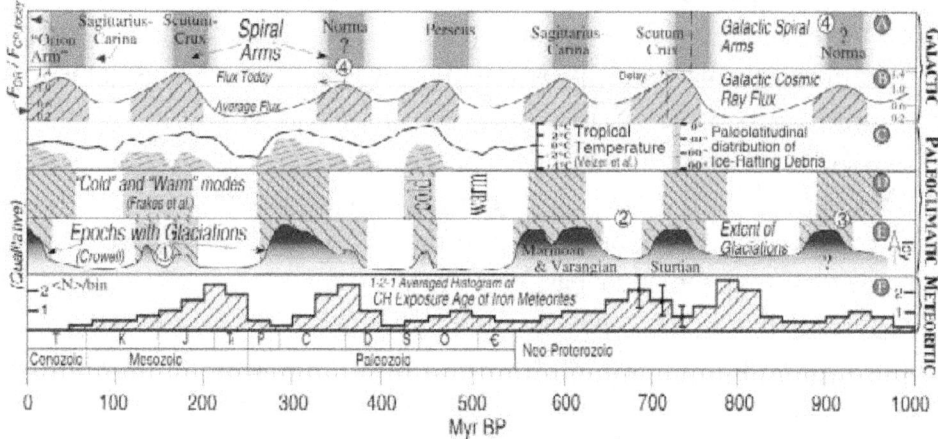

Fig. 35. Each time we enter a spiral arm there is a new ice-age. [fig. taken from Shaviv (2002)][28]

What about the initially mentioned study by M. D. Filipovic linking mass extinction to passages through spiral arms? What speed do they use? It is $11.9 \text{ km s}^{-1} \text{ kpc}^{-1}$ and based on previous research by Andrew Overholt (2009). That is very close to the value used by Shaviv, and both of them don't appear to have tuned this value in order to fit their theory. That means the more recent estimate by Ortwin Gerhard that give a great uncertainty about the pattern rotation period must be overly pessimistic.

What evidence is there for mass extinctions to be linked to passages through spiral arms? Filipovic and his team looked at six major mass extinctions and five 'lesser mass extinctions' in the past 500 million years. As shown in fig. 36 all of these eleven extinction events seem to have happened during a passage through a spiral arm.

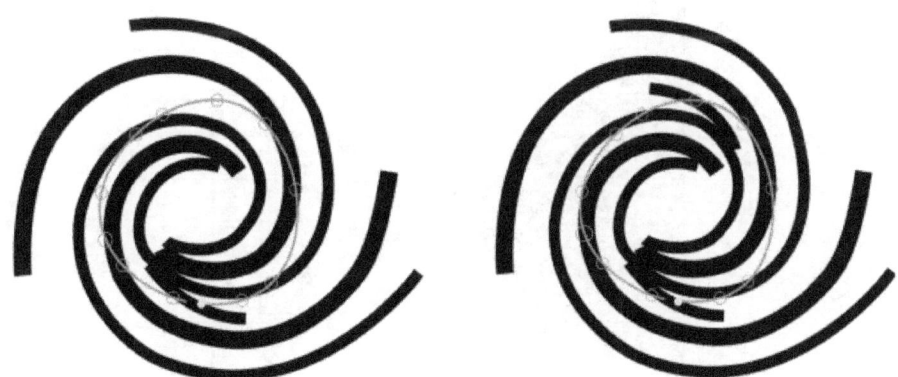

Fig. 36. The Milky Way model (left) based on Churchwell et al. (2008) and a slightly modified model with an extra sub-arm hidden on the other side of the galaxy. The blue circles represent the major mass extinctions, while the orange circles represent 'lesser mass extinctions'. [fig. taken from Filipovic (2013)][17]

According to their model we spend about 60% of the time in spiral arms and the chance for all nine events to lie within spiral arms purely accidentally is 0.36%. Both the ice-age theory and the mass extinction theory are based on almost the same pattern speed. What pattern rotation period does it correspond to? It is 500 million years! Looking again at fig. 36 we notice that although there are many thinner spiral arms all over the place, there are only two regions where all spiral arms come together and where the star density must be highest and that is besides the two sides of the central bulk. Although the thinner arms are able to cause mass extinctions and ice-ages, they probably don't significantly obscure the sight to the outside universe and therefore don't influence how the universe is build up retrospectively (with the future observation collapsing the wavefunction of the past universe).

According to Filipovic it takes the sun 500 million years to pass through all spiral arms. That means it takes the sun 250 million years to get from one side of the bulk to the other. That are the 250 million years we were looking for.

While Filipovic and his team could only be 99.64% sure that their findings are not due to mere chance, combined with the here presented theory on the emergence of the concentric shell structure, we can raise the certainty to 99.9976%.

Let's now re-examine the Sloan Digital Sky map (fig. 31) and see what other features draw our attention. Most viewers take particular note of the huge galaxy density on the left, which represents the *Sloan great wall*. It is 433 Mpc long (about 1.4 billion light years), and about one billion light years away from earth.[30] But it is not just this wall which draws the viewers attention. The galaxy density on the left side appears to be greater in general! The left side is the northern hemisphere. The *CfA2 Great Wall* and the *Hercules–Corona Borealis Great Wall* also happen to be located in the northern hemisphere. Fig. 37 shows the gross difference between the galaxy density in the northern hemisphere (left) and the southern hemisphere (right):

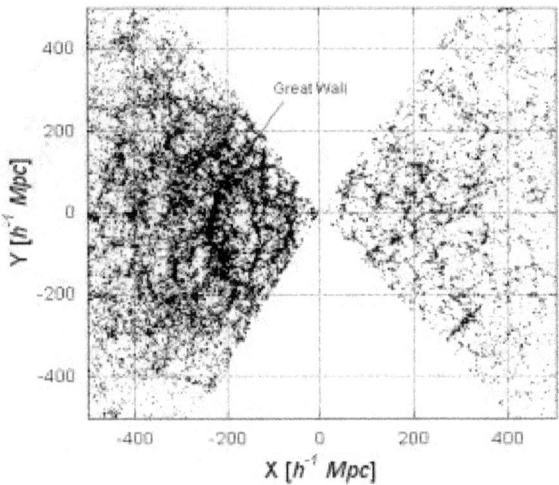

Fig. 37. Gross left right asymmetry in the galaxy distribution between the northern (left) and the southern hemisphere (right). [fig. taken from Hartnett (2008)][10]

Re-examining Fig. 31 and 32 we notice that this north-south asymmetry exists there as well. Since the galaxy formation is determined by the initial matter density distribution as reflected in the CMB, it is natural to have a correspondence in structure here. And it shows that this asymmetry can't be just due to a neater cartographing of the northern hemisphere.

Why do we have such a higher galaxy distribution density on the northern hemisphere? Using the same explanation pattern as previously, we may attribute this to the fact that earth is located above the equator plane of the milky way by about 75 to 101 light years towards the northern galactic hemisphere.[31] That makes us measuring the northern hemisphere with a slightly higher frequency. However, the layer of galactic gas orthogonal to the galactic plane is not very thick, and the difference is not big. Also during the suns orbit around the centre of the milky way it crosses the galactic equator, so that the time spend above and below the equator is more or less the same.

A more promising way of explanation is to be found on the surface of earth: 67.3% of the land mass is to be found on the northern hemisphere of the earth. Since only very few photons can reach out deep enough below the surface of the water in order to influence conscious observers there, we can, in a first simplified approach, ignore the sea, and assume that the southern hemisphere of the night sky is measured with a 32.7% lower frequency.[32]

If this model is correct, it would show that there was conscious life on earth since many spiral arm passages of the sun. When consciousness newly emerged on earth, the rest of the universe was still in maximal superposition; its structure was not settled. Although it was physically some billion years old, it was not really out there in a certain sense. It was created in a top to bottom manner by being looked at, having the habituated future determining the unhabituated past. That means it spend the first 9.6 billion years in a maximal superposition, an omnium-like state of all possible states and all possible pasts, and all its structure, including the concentric shells of higher galaxy distribution were not created before the emergence of consciousness (sentient) beings on earth.

We can imagine conscious beings sending shockwaves out to the universe causing wavefunctions to collapse. It is like dropping stones on a water surface. When one periodically stops dropping the stones into the water there will be gaps between the areas with a lot of waves. Those are the gaps between the concentric shells of higher galaxy distribution. Those gaps are caused by photons from these regions never being measured by sentient beings and therefore never having existed. Areas of the universe which happen to have sent out photons successfully detected by conscious observers would then be fuller of matter. After all only a photon which is finally measured at some point is a real photon. Its whole being together with its past can only come into existence by the process of measurement by a conscious observer.

If 7 is the total number of these concentric shells of higher galaxy distribution, then conscious life on earth could be about 1.75 billion years old. Yet multicellular life started 1.5 billion years ago. If we assume that the emergence of multicellular life marks the emergence of consciousness, then there can't be more than 6 concentric shells of higher galaxy distribution.

It may well be that the seventh concentric shell can be dismissed as statistically not significant enough, or that multicellular life started earlier. Further research has to show if one of these two possibilities is true.

Let's now have a closer look at the distribution of galaxies and compare it with the biological evolution. In fig. 38 we have the Sloan digital sky map laid besides a table on the history of life on earth.

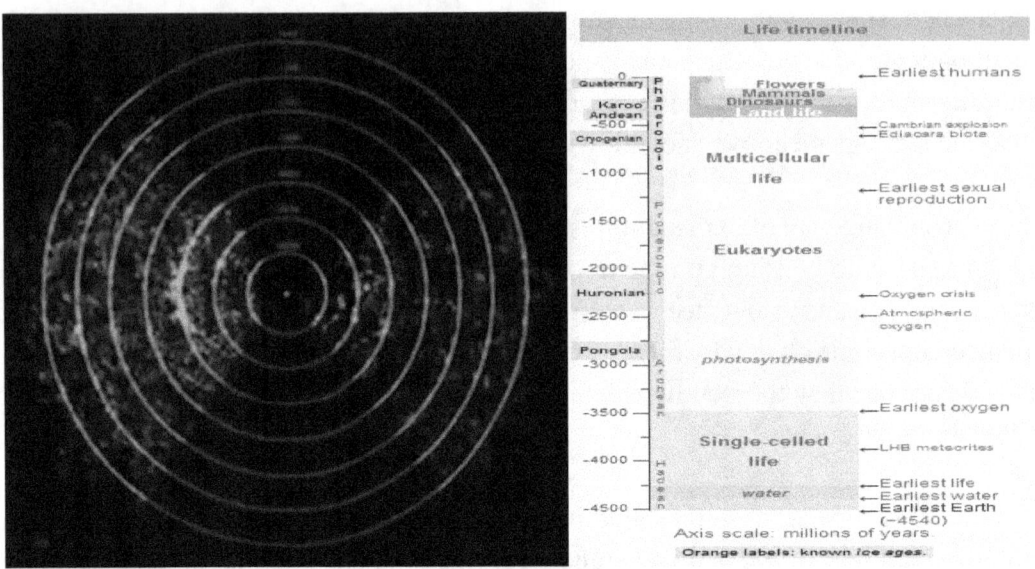

Fig. 38. A comparison between patterns in the distribution of large scale structures in the universe on the left, and events in the evolution of life on earth on the right. One concentric shell corresponds to 250 million years.
Note: As we will see later, SPD has a different scaling.

Now let's look at where the galaxy distribution is highest.

It is on the third shell, at a distance of about 750 million light years. On the table on the right we have the *ediacara biota*, which lasted from 635 myr ago till 542 myr ago. It peaked in the *Avalon explosion* 575 myr ago. Then we have the *Cambrian explosion*, which began 541 myra and lasted 20 or 25 million years. If we take the two together, we have a time span from 635 myra till 516 myra (or 521 myra). Including all of the Cambrian period, which lasted for 55.6 myr, we have a span from 635 myra till 485.4 myra. The highest concentration of galaxies we find between the third and the second shell, which corresponds to 750 myra till 500 myra. As we see, there is a rough correspondence here. We have to keep in mind that there is no chance to have a peak of galaxy distribution density at a distance of let's say 635 mlyr, because that would be in between two shells, corresponding to a time when earth was hidden in a spiral arm. However, as the map shows, there seems to have been a lot of conscious activity on earth even beyond 750 mlyr. The sponge-like animal *Otavia antiqua* emerged 760 myra in the *Tonian period* (1000 myra till 720 myra).[33] The first *metazoans* (animals by biologic definition) are believed to have appeared 800 myra. That matches the super dense third shell pretty well. We may now look at all the known megastructures within the *conscious sphere* (the 6 or 7 shells corresponding to 1.5 billion lyr) and their distances:

CfA2 Great Wall (Northern Great Wall; the Wall): from 300 mlyr till 500 mlyr.
Sculptor Wall (Southern Great Wall): at a distance of 434 mlyr (z = 0.03).
Sloan Great Wall: at a distance of about 1 billion light years.

The earliest sexual reproduction which started 1.2 billion years ago in the *Proterozoic Eon* doesn't appear have had a big impact on the 5th shell (1.25 billion light years away) to which it corresponds, but at least the Sloan Great Wall lies roughly in that distance. The Northern Great Wall and the Southern Great Wall lie in distances that correspond to the heights of biological evolution, well after the Cambrian explosion. The movement of life in form of *invertebrates* (spineless animals) onto land began during the *Silurian Period* roughly 420 million years ago. That corresponds to the Southern Great Wall (434 mlyr).

By the *Carboniferous period* (360 myra), early *vertebrates* (animals with a spine), our four-legged ancestors, had in turn left the water and were feeding on the invertebrates. This corresponds to the Northern Great Wall (300 mlyr till 500 mlyr).

For comparison: the dinosaurs lived from 231 myra or 243 myra till 66 myra.

So we have a periodic concentric shell structure that marks the beginning of multicellular life (6 or 7 shells), a megastructure that marks the beginning of sexual reproduction and two close-by megastructures which mark the first heights of evolution after the Cambrian explosion.

How does Hartnett himself explain these structures? He explains the concentric nature (all galaxies move away from us) with a model in which the universe is a white hole. The periodicity in the pattern he explains with periodic acceleration and deceleration in the expansion of this white hole. That is very troublesome, because then one has to postulate a non-homogeneous acceleration of the universe, which just can't really be found in the data. One could only make adjustments within the relatively narrow bandwidth of measurement inaccuracy.

Another issue is that these concentric shells are not perfectly regular in their shape, so that they become hard to explain with periodic fluctuations of a homogeneous expansion. Therefore Hartnett has to assume that we are not in the middle of this structure, but about 135 million light years away from it!

A second possibility he considers is that these concentric shells are not a *real space effect*, and caused solely and merely by those fluctuations in the expansion speed. In that case we wouldn't be located in a special position in the universe. However, there doesn't seem to be any mechanism by which one could get such periodic fluctuations in an otherwise smooth accelerated expansion.

Also with this approach, generally referred to as *Carmelian cosmology*, it becomes impossible to account for the north-south asymmetry which can be found both in the cosmic microwave background and the galaxy distribution. Therefore Hartnett is forced to ignore it or blame it on unbalanced surveys in the northern and southern hemisphere respectively (2016; personal correspondence).

This shows in how much trouble one gets when approaching cosmology in a classical non-quantum fashion.

What about the cold spots in the CMB, can we explain their exact position?

Yes, we can: The plane associated with the rotation axis of the earth is called the equinox plane. It corresponds to the dipole structure in the CMB.

The plane associated with the motion of earth around the sun is called the ecliptic plane. It corresponds to the quadrupole structure in the CMB.

It is natural to assume that the measurement frequency is lowest at the poles of a certain rotation or orbit axis (very few conscious observers can survive at the poles).

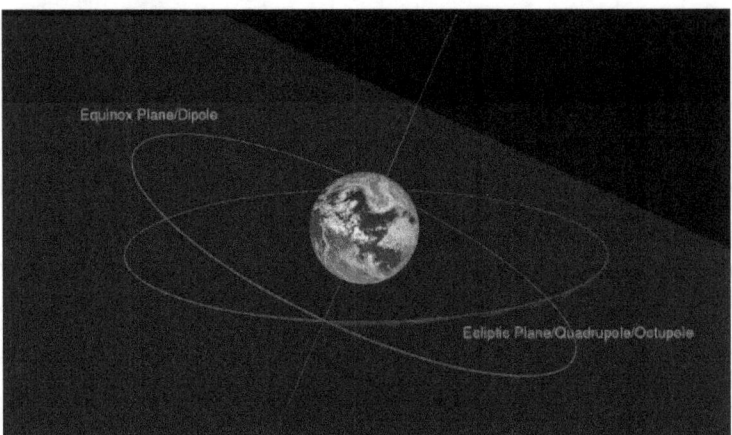

Fig. 39. The equinox plane corresponding to the dipole structure, and the ecliptic plane corresponding to the quadrupole structure.

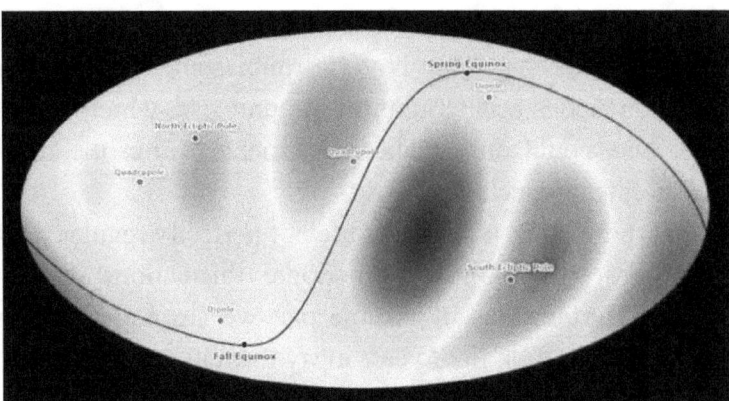

Fig. 40. The alignment of the CMB with the ecliptic poles and the equinox.

It is also natural that the south ecliptic pole has a larger cold spot with a bigger red shift (lower energy).

Future investigations will show if the galactic orbit axis of the sun can also be linked to a certain structure in the CMB, which if so, will most likely be the octupole structure (not depicted here).

Some scientists who try to explain away the CMB anomalies take it for granted that at least the dipole can be attributed to a local motion of the Milky Way galaxy relative to the CMB background. One can actually derive a speed for this motion using the CMB, which is 369 km/sec.[34] However, as the Indian physicist Ashok K. Sengal showed, these isotropies and alignments exists for other types of radiation as well, and trying to explain them by a motion of the local group leads to totally different speeds and directions for the Milky Way galaxy. For the radio sky it leads to a four times higher velocity, and for quasars it leads to a speed at 3% of light speed and an opposite direction for the motion (!). [35, 36] Thus it could well be that the Milky Way is instead not moving at all.

The likelihood for these CMB alignments to emerge out of mere chance is 1,500 and together with the concentric shell structure discovered by Hartnett this leaves only a 6 : 300,000,000 chance for both phenomena to be 'coincidences'. Adding in Filipovic's findings we arrive at

$$\frac{9}{25} \times \frac{6}{100,000} \times \frac{1}{3,500} = 6.171428571 \times 10^{-9}$$

And that is a certainty of 99.99999938%.

So far we have been using quantum mechanics in its orthodox interpretation only, without involving much of space particle dualism.
The CMB was released more or less simultaneous all over the universe, so it should be unaffected by the particularities of the expansion scheme.
That is not true for the galaxy distribution we see in the Sloan Digital Sky Survey.
As we will learn in chapter 4.12, the expansion scheme of space particle dualism has a huge impact on our assessment of distances in the universe.

Intergalactic distances according to space particle dualism

What light year distance do Hartnett's periodic redshifts ($\Delta z = 0.0246$) correspond to according to space particle dualism? That can be calculated using a scale factor we will derive in chapter 4.12. The value of this scale factor is:

$$\Delta x_z = 39,625,004,647 \text{ yr}$$

Accordingly we find that:

$z_1 = 0.0308$ is 1,220,450,143 light years.
$z_2 = 0.0554$ is 2,195,225,257 light years.
$z_3 = 0.0800$ is 3,170,000,372 light years.
$z_4 = 0.1046$ is 4,144,775,486 light years.
$z_5 = 0.1292$ is 5,119,550,600 light years.
$z_6 = 0.1538$ is 6,094,325,715 light years.

The distance between these redshifts is $974,775,114$ light years – roughly a billion light years. This is much more than the 250 million light year spacing that we have in standard cosmology. It destroys the correspondence with the time the solar system takes to travel from one corner of the dense bulk region to the other.

We could now employ a trick, and try to save our original explanation scheme by saying that we didn't account for the position of the sun relative to the galactic plane. For example, if the sun has to be both outside the bulk region and above the galactic plane in order to have maximal measurement frequency, then the periodicity of ideal observation would not be 250 million years, but 1 billion years.

What is the periodicity of oscillations of the sun over the galactic plane? Estimates from the 80's find a period of 26 to 37 million years.[37] Lisa Randall and Matthew Reece recently assumed a period of 35 million years when proposing possible correlations with short extinction cycles.[38] That is the time between each time the sun dives into the galactic disk. Twice this time is the time between each time the sun reaches the highest point over the galactic disk. Aside spending time in the bulk regions, the sun is located inside one of the various spiral arms about 60% of the time. Therefore, within 250 million years, the number of times the sun reaches the highest point while being outside the bulk regions is:

$$f_{peak} = \frac{250}{2\,(2 \times 35)} \times \frac{3}{5} = 1$$

This shows that it is not true that maximal exposure is reached only every fourth time the sun is outside the bulk regions.

Even if the estimates are wrong, and the oscillations are much slower, we are facing two other even more serious problem:

(1) If there is a measurement maximum every 1 billion years, then the expected pattern should disappear already after $1-3$ concentric shells, dependent on when we believe that consciousness emerged. We could now try and say that maybe only the first three shells are statistically significant, but that is not convincing, because when looking at a high resolution map of the galaxy distribution the pattern is clearly visible and recognizable by the naked eye, far beyond the third shell.
(2) If we only care about the exposure on the northern hemisphere, the pattern should appear on the northern hemisphere only, and if the effect is strong for both sides, the shells on the northern hemisphere and the southern hemisphere would have to be out of alignment.

How can it be that the *future observation density principle* we employed here works perfectly fine for the cosmic microwave background, but not for the distribution of galaxies?

Well, did it explain everything we see in the cosmic microwave background? It certainly explained the north south asymmetry and the position of the four cold spots, but it did not explain the small density fluctuations.

Similarly we do re-discover the north south asymmetry in the distribution of galaxies, and we do find cosmic voids approximately in the same directions in the sky as we found cold spots in the cosmic microwave background, both of which can be explained by the *density of future observation*, but the density fluctuations in the galaxy distribution seem to lie outside of this explanation scheme.

Short range periodicities

Hartnett's survey was searching for a periodic pattern and if there is any concentric structure within the galaxy distribution, then even if the claimed pattern spacing is strong for only two shells, it will still yield an impressive sigma value for the whole surveyed redshift collection.

It could be promising to try and find maxima of galaxy distribution density regardless of periodicity. If at a certain distance there are more galaxies to be found regardless of the direction in the sky, then that is remarkable, even if there is no periodicity. Expecting periodicity on the contrary makes us want to continue the pattern as far as possible.

Limiting the survey to smaller radii could reveal more significant periodicities on smaller scales. For example one could limit the surveyed region to 3.5 billion light years, which corresponds to the time since which multicellular life exists on earth, and look for periodic minima of galaxy population density every 250 million light years. Alternatively one could also be looking for patterns that correlate to the passage through ordinary spiral arms, which happens much more frequently.

In any way, if the whole of Hartnett's pattern is statistically significant, then such correlations can explain only parts of what we observe.

We will soon look at what could explain patterns that extend to infinity, but first we want to explore explanations for local maxima in galaxy population density within a limited range of 3.5 billion light years.

Something else we should notice here is that $\Delta z = 0.0246$ is not the only peak redshift periodicity that Hartnett found: others are $\Delta z = 0.0102$, $\Delta z = 0.0170$, $\Delta z = 0.0448$.

These all have their own statistical significance and could all correspond to different periodicities on earth.

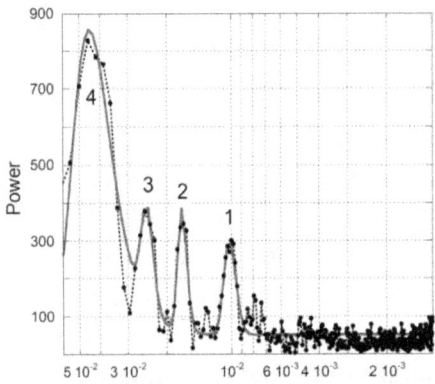

Fig. 41. The statistical significance of the various periodicities in the redshift
abundance. [fig. taken from Hartnett (2008)]

So, what distances do Hartnett's various maxima of periodic redshift abundance
correspond to according to space particle dualism?

Using the above mentioned equation yields:

$$\Delta z_1 = 0.0102 \text{ is } \Delta s_1 = 404{,}175{,}047 \text{ light years}$$
$$\Delta z_2 = 0.0170 \text{ is } \Delta s_2 = 673{,}625{,}079 \text{ light years}$$
$$\Delta z_3 = 0.0246 \text{ is } \Delta s_3 = 974{,}775{,}114 \text{ light years}$$
$$\Delta z_4 = 0.0448 \text{ is } \Delta s_4 = 1{,}775{,}200{,}208 \text{ light years}$$

Do they correspond to any periodicity on earth?
A passage through a spiral arm happens about every 142 million years. Entering the
dense bulk region happens every 250 million years. These are periodic events. The
above distances are those we obtain when looking for periodicity.

What if we don't look for periodicity, but for distinct events instead?
The highest galaxy population density we have at a redshift of $z = 0.0800$, which
according to SPD corresponds to 3,170,000,372 light years. The first evidence for
multicellularity is from cyanobacteria-like organisms that lived 3 – 3.5 billion years
ago, which would match that redshift value.[39] However, first direct evidence for brains
is from 521 million years ago.[40]

What about baryonic acoustic oscillations?

At this point we should consider that the concentric shells structure is not a genuinely
non-Copernican structure. An alternative explanation that would turn it into a more
mundane phenomenon is *baryonic acoustic oscillations* (BAO).
After Hartnett discovered these patterns other cosmologists analysed them further and
tried to come up with an explanation. The most common explanation now is that they
represent scaled up shock waves from the early universe. [41]
Such shockwaves would originate from places of higher density. According to SPD that

are primordial black holes, while in the mainstream they are believed to originate from quantum fluctuations in the mass density of the early universe.

According to space particle dualism theory irregularities in the matter distribution of the early universe can only originate from primordial black holes as well as irregularities in the measurements of future observers.

As we will see in chapter 4.15, the major periodicity of $\Delta z_3 = 0.0246$, corresponding to a distance of $\Delta s_3 = 974,775,114$ light years, can indeed be explained with baryonic acoustic oscillations, when using the full age of the universe that we derive there.

In the mainstream there is no hope of finding an explanation for the size of these oscillations, because there their size ultimately depends on how much microscopic vacuum fluctuations were amplified by the arbitrarily adjustable parameters of inflation.

Accepting baryonic acoustic oscillations as the proper explanation of the concentric shells pattern doesn't mean that *abundance of future observation* becomes irrelevant when it comes to the distribution density of galaxies.
Both the north south symmetry in the galaxy distribution, as well as large structures like the great wall indicate that there is more going on than just acoustic oscillations.

Short range periodicities and oscillations over the galactic plane

The sun is located about $26,490$ light years away from the center of our galaxy. The bulk region extends only to about $10,000$ light years. So when we talk about the sun diving into the bulk region, what we really mean is entering one of the two major spiral arms, which both originate at where the bulk region extends the farthest.
So, at the end, it might turn out that we really have to think more in terms of individual spiral arms, and not in terms of simple in-bulk-out-of-bulk model.

Within one spiral arm transit, the solar system doesn't change its position relative to the galactic plane much. We can therefore estimate the chance that it is both outside of a spiral arm and above the galactic plane as $1/4$ (above $\rightarrow$ in $\rightarrow$ below $\rightarrow$ in $\rightarrow$ above).

We know that this works only for spiral arm transits, but for simplifying the analysis, we will temporarily assume that it works for near-bulk-transits as well.

When doing so, we get a maximal measurement frequency due to out-of-bulk residence not every 250 million years, but every 4×250 million $= 1$ billion year.
So, although the main 1 billion years periodicity can be explained in terms of baryonic acoustic oscillations, within a radius of 3.5 billion light years this could be further enforced by the effect the *abundance of future observation* has.

The idea that measurement exposure is maximized when the solar system is above the galactic plane doesn't apply only to when it is located half way from one bulk region to the other, but also when it in between two spiral arms.

Let's divide the distances between the periodic redshift abundances by 4, in order to accounting for the times when the solar system is hidden beneath the galactic plane, and see if they match with periodicities we observe in the Milky Way:

$$\Delta s_1 / 4 = 101,043,761 \text{ ly}$$
$$\Delta s_2 / 4 = 168,406,270 \text{ ly}$$
$$\Delta s_3 / 4 = 243,693,779 \text{ ly}$$
$$\Delta s_4 / 4 = 443,800,052 \text{ ly}$$

$\Delta s_1 / 4$ and $\Delta s_2 / 4$ are both somewhat close to the 142 million years it takes for the solar system to go from one spiral arm to the other and when taking the average between the two, we arrive at a value that is really close to this time.

$$\frac{(\Delta s_1 / 4) + (\Delta s_2 / 4)}{2} = 134,725,016 \text{ lyr} \sim 142,000,000 \text{ yr}$$

As one would expect, these two are the weakest periodicities and they are not easily visible. Δs_3 is the redshift periodicities that was highlighted in fig. 31 and it corresponds precisely to the time it takes for the solar system to travel from one low-density side of the galaxy to the other and to at the same time be located above the galactic plane.

$$\Delta s_3 / 4 = 243,693,779 \text{ lyr} \approx 250,000,000 \text{ yr}$$

Δs_4 is the strongest periodicity and it would be strange if it wasn't easily visible in fig. 31 and fig. 32. The reason why fig. 31 & 32 seem to depict a single periodicity and not two is that $\Delta s_4 \sim 2 \times \Delta s_3$.

$$\Delta s_4 / 8 = 221,900,026 \text{ lyr} \sim 250,000,000 \text{ yr}$$

It is likely that this split into Δs_1 & Δs_2 and Δs_3 & Δs_4 is due to imprecision in the alignment between the oscillation over the galactic plane and the time the low-density regions of the galaxy are reached; that is lower density between spiral arms in the case of Δs_1 & Δs_2, and low density between the two bulk regions in the case of Δs_3 & Δs_4.

At this point it is unclear if the weaker of these periodicities exists even when cutting out the largest data sample with a radius of 3.5 billion light years.

Hartnett was analyzing redshift abundance within a redshift radius of $z < 0.1784$. Our own radius of interest here is $z < 0.088$. That is roughly 1/2 of it, yet as we will see in chapter 4.15, half of all supermassive black holes discovered within a range of $z < 10$ are actually found within redshifts of below $z = 0.088$. This shows that

$z < 0.088$ is the most densely observed volume in the sky.

If some of the weaker redshift periodicities, say Δz_1 & Δz_2 yield higher sigma values when restricting the survey to $z < 0.088$, then that is evidence that they do not represent baryonic oscillations, but the influence of the abundance of future observation instead.

However, Δz_3 and possibly Δz_4 are certain to be unrestricted in range, and in chapter 4.16 we will be exploring how they can be used to prove that the full age of the universe we will be deriving there is indeed correct.

Origins of the abundance of future observation principle

So far we have explained the cold spots in the cosmic microwave background only qualitative by assuming having a larger abundance of observers for certain directions in the sky leads to the wavefunction of the universe collapsing in a way that materializes more stars, galaxies and microwave background photons in those directions.

We have not so far touched upon the question of how this principle could be expressed mathematically or on if it can be derived from quantum mechanics, or possibly from the similar worlds interpretation.

In short, the abundance of future observation principle as we have used it so far is saying that more stuff appears where there are more measurements. It would mean that there is a bias in nature against zero-measurements.
This may sound good, but without precise math to quantify it, it is not worth much.

If this principle is to be applied universally, then it should be possible to demonstrate it in a laboratory. For example we could let a photon pass a beam splitter and then enter two boxes simultaneously, in superposition. One box shall have many detectors, and the other only one. If our principle is correct, then more photons should be detected in the box with the many detectors.
Fewer detectors increases the likelihood that the photon simply gets absorbed by the walls of the box, heating it up ever so slightly. That would correspond to a zero-measurement. However, we know that even when the experimenter is not in the room, both measurements and zero-measurements can occur. Therefore we have to seriously ask ourselves, what reason there can be for measurements to be favored over zero-measurements by nature.

Many things come to mind when pondering upon this question. Morphic resonance, which is the tendency of patterns to repeat, is one of them. It would favor quantum states which lead to chains of events that are more 'content rich', and have therefore more resonance with other eons (see chapter 4.18). However, it is hard to see how a detector detecting a few more particles can 'enrich' anything. Also, if this was a natural law, the effect could be much too strong.

If we simply want to compare measurements and zero-measurements, which is absorption by a wall or simply not showing up in the detector, without any strong resemblance with our cosmological problem at hand, then we can also imagine that we let photons pass a beam splitter that has a photon detector only on one side, let's say it is the reflection side, and those photons that pass through the beam splitter are simply getting absorbed by a wall.

It would be pretty surprising to find that more than 50% of the photons make it to the detector. For not having been discovered yet, the effect should be very small, lets say less than 1%; but if it is so small, then it would be very hard to distinguish from ordinary micro-psychokinesis. And if that is so, then why don't we use that as our working hypothesis instead?

Cosmokinesis

Could it be that the abundance of future observation principle in cosmology is simply long term micro-psychokinesis on a cosmological scale?

Could the CMB-anomalies be really a subtle 'psi effect' that has built up over hundreds of millions of years.

From half a century of human experiments with random number generators, we know that a quantum random event that originally has a 50% chance of happening, reaches 51% when intention is present.[42] When the single bits are replaced by pictures, this chance is in fact higher (see chapter 2.3 & 2.4), but we are concerned with single photons that reach us from the sky, and that is a situation which is much more similar to random binary bits, and therefore we have to stick to the 51% which was found for those.

50% correspond to random chance, and so 51% correspond to 0.5% deviation from random chance.

If we simplify the situation by dividing the earth into two regions, one with observers, and one without observers, then we can imagine that every time an observer is moving outside in the night, he has a strong intention of seeing things better. In an ideal setting this would lead to photons arriving at the observer inhabited part of earth with a 0.5% greater likelihood.

What this means for the cosmic microwave background is that there should be 0.5% more photons in the warmer regions than in the cold ones. We can now set out to check if this is the case.

The density of photons in the early universe and in the cosmic microwave background is given by:

$$\rho_\gamma = \frac{16\,\pi\,k^3\,\zeta(3)\,T^3}{c^3\,h^3} = 20{,}286{,}824.875 \times T^3$$

The precise temperature of the cosmic microwave background is 2.72548 ± 0.00057 K. The temperature fluctuations are in the range of $\pm\,0.00003$ K. We can now calculate how large the difference in photon number density is between cold and hot regions of the cosmic microwave background.

The number of photons is $410{,}703{,}815$ per cubic meter in cold spots and $410{,}730{,}940$ per cubic meter in hot spots.

The ratio between the two equals:

$$\frac{(2.72548 - 0.00003)^3}{(2.72548 + 0.00003)^3}$$

Which means that there are 0.0066% more photons in the hot spots than in the cold spots.

We do not expect observers to wish for more light on daytime, and therefore observer intention should only be active at night. This should lead to a decline in the overall effect to about the half, because there is sun light about half the time, corresponding to an effect size of 0.25%.

Most prehistoric animals are believed to be diurnal, which means that they were active during the day and sleeping during the night. If we take human sleep as an example, we sleep about 8 hours of the 12 hours that are dark on average. That leaves about 5 hours that we were spending in darkness, before torches or light bulbs were invented. If we were spending all that time outside, where the night sky is visible, that would give an expected effect size of 0.04%.
The discrepancy of this with the 0.0066% found in the CMB is only a factor of 6.

It is pretty clear that not during all of the sleepless time in darkness do we wish for better sight, and that probably explains the missing factor of 6.
On the other hand there were also night active animals in prehistoric times, but those were, most likely, a minority, same as in today's animal kingdom.

This quick approximate calculation shows that the expected effect size of psychokinesis and the actual temperature fluctuations in the cosmic microwave background are in the same order of magnitude. Arriving in the right order of magnitude is the best we can hope for with approximations like this.

Even when remaining skeptical about the above derivation of the expected psi effect size, there is one thing our calculation clearly shows, and that is that the temperature fluctuations in the CMB are well within the range of what is possible with micro-psychokinesis. If the fluctuations were larger, micro-psychokinesis would be ruled out right away.

Fluctuations that are smaller than the effect size of psychokinesis on random number generators on the other side are to be expected. Animals are not wandering outside in the dark all the time, and even when they do, they might not always wish to spot something. Sometimes they are just relaxing. Only when hunting or when investigating danger, is there a need for good sight. That is the only time when we can expect this intention to be present.

If we want to arrive at precisely the effect size we observe in the cosmic microwave background, then the time of focused intention should be 38 minutes per night.

Conclusion

This is just one example of how much impact the interpretation of quantum mechanics has on the analysis of observational data. Contrary to common believe, there are much more ways superpositions can be observed than just interference. Large superpositions can lead to large scale quantum tunneling around black holes (see chapter 4.3), they enable mind over matter interaction in and beyond conscious brains (see chapter 2.1), they help 5-fold symmetry quasi-crystals to grow (see end of chapter 2.1), and as we saw in the present chapter, they both allowed the formation of DNA and had a huge impact on the overall structure of the observable universe.

In this chapter we saw that life in the universe has to start from one point and end in one point – the α (Αλφα) and ω (Ωμεγα) if we want. I think that shows the oneness of all living beings. We will come back to that in chapter 4.17 and chapter 5.1, where we will discuss the nature of consciousness in detail.

Notes and references:
1. *Wanjek, C. (5 January 2006).* "Milky Way Churns Out Seven New Stars Per Year, Scientists Say". Goddard Space Flight Center. Retrieved 2008-05-08.
2. *Palmer, J. (11 January 2012).* "Exoplanets are around every star, study suggests". BBC. Retrieved 2012-01-12.
3. *Cassan, A.; et al. (11 January 2012). "One or more bound planets per Milky Way star from microlensing observations".Nature481(7380): 167–169.arXiv:1202.0903.Bibcode:2012Natur.481..167C.doi:10.1038/nature10684.P MID22237108.*
4. *Overbye, Dennis (4 November 2013).* "Far-Off Planets Like the Earth Dot the Galaxy".New York Times. Retrieved5 November2013.
5. Petigura, Eric A.; Howard, Andrew W.; Marcy, Geoffrey W. (31 October 2013). "Prevalence of Earth-size planets orbiting Sun-like stars". Proceedings of the National Academy of Sciences of the United States of

America.arXiv:1311.6806.Bibcode:2013PNAS..11019273P.doi:10.1073/pnas.131 9909110. Retrieved5 November 2013.

6. Life Itself: Its Origin and Nature (Simon & Schuster, 1981) ISBN 0-671-25562-2

7. Miyakawa, S., Joshi, P.C., Gaffey, M.J. et al. Orig Life Evol Biosph (2006) 36: 343. doi:10.1007/s11084-006-9009-6

8. *Sagan, Carl. "Cosmic Search Vol. 1 No. 2". Cosmic Search Magazine.* Retrieved 2015-07-21.

9. For a more thorough analysis on the Fermi-paradox see the last chapter of [22] ("the matter myth"), or watch this 23 episodes long documentary series: [https://www.youtube.com/watch?v=-1-J6zzHW-c].

10. The S here can stand for both slice and surface.

11. We shall here not bother to attach complex valued probabilities to the different states in this wavefunction. Entangled states with more than two particles become increasingly complicated with increasing numbers of particles. The + in between the different states indicates that they are entangled. Non-entangled states would be multiplied (which is sometimes denoted with the symbol "$\otimes$", representing a so called *tensor product*). However, being not entangled is usually just an approximation, since basically all things on earth are entangled with each other.

12. The I in I^- and I^+ stands for infinity. I^- is the *past null infinity*, while I^+ is the future null infinity. It is the border of the light cone, while *past* and *future light cone* is usually meant to include the interior.

13. "The energy of empty space that isn't zero", Outlet Edge.org 7506.

14. Fig 31 – 34, 41 & 42 are taken from the documentary "The Principle", which is available at [http://www.theprinciplemovie.com/] and discusses both the microwave background anomaly as well as the anomalous galaxy distribution found by John Harnett.

15. Hartnett, J.G.; K. Hirano (2008). "Galaxy redshift abundance periodicity from Fourier analysis of number counts N(z) using SDSS and 2dF GRS galaxy surveys".Astrophysics and Space Science. Cornell University Library.318: 13– 24.arXiv:0711.4885.Bibcode:2008Ap&SS.318...13H.doi:10.1007/s10509-008-9906-4.

16. This is a simplification: while disproving all non-consciousness based interpretations of quantum mechanics, these non-Copernican structures do not strictly disprove the many minds version of decoherence theory. It doesn't really make a difference if one says that the wavefunction collapsed or that one world split into two. What makes a difference is where the collapse or the split takes place. Both in the Von Neumann-Wigner interpretation and in the many minds decoherence theory it happens when signals are registered in the consciousness of an observer. So the two theories should be equal in their predictions. And yet, this is another simplification, because it would mean to pretend that decoherence theory is really a self-consistent theory, while we learned in chapter 2.1, that it would make quantum probabilities impossible, and that macroscopic interference would be inevitable on long term. Also the arguments in the present chapter are made using clear-cut criteria for what exactly is entangled. In standard quantum

mechanics these do not exist. There the most common view is that everything is entanglement, and that entanglement simply gets lost in the complexity of the intercorrelations of particles.

17. This theory was established in 2013 by M.D. Filipovic, J.Horner, E.J.Crawford, and N.F.H. Tothill and is available at [https://arxiv.org/pdf/1309.4838.pdf].

18. Of course there is no clear-cut edge between spiral arms and the area between them, but those concentric shells do also not have clear-cut edges.

19. It is unclear if being in a spiral arm could really notably obscure the sight, but even now we can see only a small part of our universe (less than half) because the disk representing our milky way is in our sight. We can only see what is above and below. When diving into a spiral arm it is plausible that even less would be visible.

20. It shall also be noted that tragic events like meteoroid impacts do not only diminish the number of observers, but they usually also result in a darkening of the sky and have therefore a similar obscuring effect as all the gas clouds and local star clusters found in the spiral arms of our galaxy.

21. Sparke, Linda S.; Gallagher, John S. (2007).*Galaxies in the Universe: An Introduction*. p.90.ISBN 9781139462389.

22. Benjamin, R. A. (2008). Beuther, H.; Linz, H.; Henning, T., eds. *The Spiral Structure of the Galaxy: Something Old, Something New...Massive Star Formation: Observations Confront Theory.* **387**. Astronomical Society of the Pacific Conference Series. p.375.Bibcode:2008ASPC..387..375B.
See also Bryner, Jeanna (June 3, 2008). "New Images: Milky Way Loses Two Arms".*Space.com*. Retrieved June 4, 2008.

23. Gerhard, O. "Pattern speeds in the Milky Way". arXiv:1003.2489v1.

24. "Massive stars mark out Milky Way's 'missing' arms", University of Leeds. December 17, 2013. Retrieved December 18, 2013.

25. Westerholm, Russell (December 18, 2013). "Milky Way Galaxy Has Four Arms, Reaffirming Old Data and Contradicting Recent Research". *University Herald.* Retrieved December 18, 2013.

26. Urquhart, J. S.; Figura, C. C.; Moore, T. J. T.; Hoare, M. G.; et al. (January 2014). "The RMS Survey: Galactic distribution of massive star formation". *Monthly Notices of the Royal Astronomical Society. 437 (2): 1791–1807.* arXiv:1310.4758 Bibcode:2014MNRAS.437.1791U. doi:10.1093/mnras/stt2006.

27. Source:[http://curious.astro.cornell.edu/physics/55-our-solar-system/the-sun/the-sun-in-the-milky-way/207-how-often-does-the-sun-pass-through-a-spiral-arm-in-the-milky-way-intermediate].

28. "The spiral structure of the Milky Way, cosmic rays, and ice age epochs on Earth", Nir J. Shaviv (2002); Link to the paper:
[http://www.phys.huji.ac.il/~shaviv/articles/long-ice.pdf].

29. A online article about this research together with related links and papers can be found here: [http://www.sciencebits.com/ice-ages].

30. Source: [http://web.hallym.ac.kr/~physics/course/a2u/cluster/wall.htm].

31. JOHN N. BAHCALL[*]&SAFI BAHCALL[†] (1985); "The Sun's motion perpendicular to the galactic plane"; Nature316, 706 - 708 (22 August 1985); doi:10.1038/316706a0.

32. Life on Earth: A - G.. 1. *ABC-CLIO. 2002. p.528.*ISBN9781576072868. Retrieved8 September2016.

33. Brain, C. K.; Prave, A. R.; Hoffmann, K. H.; Fallik, A. E.; Herd D. A.; Sturrock, C.; Young, I.; Condon, D. J.; Allison, S. G. (2012). "The first animals: ca. 760-million-year-old sponge-like fossils from Namibia". *S. Afr. J. Sci.* **108** (8): 1–8. *doi:10.4102/sajs.v108i1/2.658.*

34. However, accounting for the various contradicting speeds derived from other sources, this number has been estimated higher now, namely 631 km/sec. The source paper: Yehuda Hoffman, Daniel Pomarède, R. Brent Tully & Hélène M. Courtois (22 August 2016)."The dipole repeller". *Nature Astronomy.* doi:10.1038/s41550-016-0036.Archived from the original on March 3, 2017.

35. Ashok K. Singal (17 May 2013)."A large anisotropy in the sky distribution of 3CRR quasars and other radio galaxies"; Journal-ref:*Astrphys. Sp.* Sc., 357, 152 (2015); arXiv:1305.4134.

36. Ashok K. Singal (19 May 2014). "Extremely large peculiar motion of the solar system detected using redshift distribution of distant quasars"; arXiv:1405.4796.

37. Bahcall, J. N. and Bahcall, S. (1985) The Sun's motion perpendicular to the galactic plane. Nature, 316, 706–8.

38. Fan, J., Katz, A., Randall, L. & Reece, M. Phys. Dark Univ. 2 139–156 (2013).

39. Grosberg, RK; Strathmann, RR (2007). "The evolution of multicellularity: A minor major transition?" (PDF). Annu Rev Ecol Evol Syst. 38: 621–654. doi:10.1146/annurev.ecolsys.36.102403.114735.

40. Park TS, Kihm JH, Woo J, Park C, Lee WY, Smith MP, et al. (March 2018). "Brain and eyes of Kerygmachela reveal protocerebral ancestry of the panarthropod head". *Nature Communications.* 9 (1): 1019. Bibcode:2018NatCo...9.1019P.doi:10.1038/s41467-018-03464-w. PMC 5844904. PMID 29523785.

41. Anderson, L.; et al. (2012). "The clustering of galaxies in the SDSS-III Baryon Oscillation Spectroscopic Survey: Baryon acoustic oscillations in the Data Release 9 spectroscopic galaxy sample". Monthly Notices of the Royal Astronomical Society. 427 (4): 3435. arXiv:1203.6594. Bibcode:2012MNRAS.427.3435A. doi:10.1111/j.1365-2966.2012.22066.x.

42. "The conscious universe", Dean Radin (1997).

4.3 Black hole information paradox [17.09.2013 (horizon is a membrane); 14.09.2015 (more detailed analysis); 10.05.2016 (thermal tunneling & coherence horizon); 14.08.2016 (horizon is not a membrane); 16.09.2017 (radius of coherence horizon); 09.10.2017 (entropy of coherence horizon); 02.12.2017 (entropy of a photon); 07.12.2017 (black holes can't shrink); 15.06.2019 (cosmological event horizons are unobservable); 23.09.2019 (there is no Unruh radiation)]

Note: The original approach of solving the black hole information paradox was to assume that event horizons are like elementary spaces (2013). However after it was discovered that elementary spaces are much bigger than the critical circumference of particles, this had to be given up. Then later it was assumed that the information of the infalling matter is stored in measurement events and returned to the universe in Hawking radiation. As it turned out that Hawking radiation can't do that, quantum tunneling of the original matter was considered. First it was discovered that this tunneling can't be thermal, and then that it couldn't make the black hole shrink (gravity is a side effect of charge and photons have no charge). The consequence is that black holes don't evaporate and since the black hole information paradox arises from the concept of black hole evaporation, this solves the paradox.

However, it is still important to understand the full paradox nature of Hawking radiation and that is what this chapter deals with mainly. It also gives relevant information on black hole entropy.

There are numerous books exploring the inside of a black hole and what one could find there if one could get back to tell about it.[1] Seldomly it is noticed that this kind of analysis is very problematic from an epistemological point of view. There are no experiments one could do to verify if those analyses are correct; and if we analyse the equations for black holes correctly, we find that the inside of a black hole hardly exists (!). Yet, the general attitude of many researchers to not ask deep questions about conceptual problems (philosophical questions), the so called *positivist point of view*, prevented many from doubting what was presented to them. All kinds of fancy space-time topologies were analysed with less concern about their physical relevance.

In fact the detectable part of a black hole is all there is of a black hole. It ends or starts at the event horizon; and general relativity, correctly interpreted, was always hinting at this. In all first calculations of the space-time metric of a black hole, the black hole was ending at the horizon (this is referred to as the *apparent singularity*). Only a special trick allowed the relativists to continue the black holes structure further into its inside. From the perspective of any observer outside the black hole, the time dilatation on its surface is infinite, and the event of an astronaut passing through the event horizon lies in infinity (or *beyond the infinite future* of the outside observer). That is to say that for all other observers an astronaut falling through the horizon would need an endless amount of time to do so. From the perspective of an infalling astronaut his time is normal, while the time of the outside universe passes by with infinite speed. If we continue Einstein's equations into the inside of the black hole, space and time flip around. Space becomes time, and time becomes space. The centre of the black hole, the singularity, would not be in front of the astronaut, but in his future. From the perspective of observers outside the event horizon, the whole future of the infalling astronaut lies in a single present moment. From the perspective of space particle dualism theory that would mean the astronaut would stop to exist after passing the event horizon, because he would thereby leave the conscious now slice. He would have to have his own causally unrelated universe, which is regarded impossible in this theory (Leibnizian relationism doesn't allow things to be unrelated).

How is it possible for information directly at the horizon to get to the outside universe? Different from event horizons in general relativity, the event horizons of space particle dualism theory have many connections to the space around the black hole. It is just that most paths away from the horizon lead back to the horizon. That is because the density of elementary spaces is much higher at the horizon. Yet even small distances from the horizon would allow particles to eventually escape. On the other hand we have Hawking's model of virtual particles around the horizon, where a pair of virtual particles is torn apart by tidal forces, with one of them escaping from the black hole and becoming a real particle, and the other one falling back in. The information, and properties of the particles which have fallen into the black hole, must somehow be transferred to these escaping particles. Since escaping turns them into real particles, they should carry the information of the former particles which have fallen into the black hole, otherwise we would have new and random quantum information created by the black hole (the quantum world is random, but usually not to the degree that cause and effect are missing completely).

There is something very contradictive in the way the information paradox is often described in contemporary physics.
Since solving the paradox seems to require information to be stored at the horizon, there are many speculations on how this might happen. String theorists suggest there might be some M-theoretic brane formed at the horizon with string ends being attached to it. I regard this to be a highly artificial solution; especially when considering that string theory doesn't say anything about the structure of space-time (except of speculations on extra dimensions).
Interestingly all descriptions of the problem, and even all speculations on possible solutions, although suggesting that information might somehow be stored on the event horizon, claim that the particles would still fall through the horizon from their own perspective (the perspective of the particles). I find this rather inconsequent and highly contradictive.
The world consists of information only. If it is stored at the horizon, then it is there, and not somewhere else at the same time. There is no reason to believe that the horizon contains information about something happening inside the horizon. That would actually be a catastrophe! If there is an inside, and if the horizon contains information

about it, then it would have to contain the information about a stupendous mass energy increase within the black hole caused by an enormous redshift of the infalling light. That would make the black hole grow to infinity. So saying that information is stored at the horizon is as good as saying that the infalling matter is stuck at the horizon. If the information is there, then the matter is there, because matter is information.

There are many controversies when it comes to the information paradox. Penrose always favored the possibility that the infalling information is all lost. Hawking had the same opinion, but changed his mind in 2004. The main reason is that such an information loss would mean a violation of the unitary evolution **U**. For people who do not believe in the existence of the state reduction **R**, such a loss is a serious problem. For proponents of the state reduction **R** the unitary evolution is violated each time a measurement happens. Therefore there is nothing to fear from such a 'violation'.

Furthermore in space particle dualism both the unitary evolution **U** and the state reduction **R** represent the maximal gainable knowledge of an observer. **U** always stands for the observer's degree, as well nature's degree, of not knowing things. So **R** is always about obtaining information. It is important to understand that **U** is the opposite: the absence of information.

So although admitting the existence of **R**, we do not have to support a loss in information at black holes. There is nothing really random about the process **R**. Since it is caused by consciousness, it is non-computable but deterministic, just as consciousness itself.

Does avoiding the loss of information really crucially depend on the storage of information at the horizon? During the time a particle is not measured it becomes fuzzier, but that doesn't mean that all the information from the last measurement is lost. And it doesn't need to be stored anywhere, except of in the memory of collective consciousness (or the 'world of experience'; see chapter 5.1).

The information is in a large superposition when it arrives at the horizon. It should therefore be no problem for it to go over into the Hawking radiation. Even without basing space particle dualism theory, but just the orthodox Von Neumann-Wigner interpretation, the information must be preserved, because only the observer counts. Information is not something taken by particles from somewhere to somewhere else.

In his epic work "The Road to Reality" (2004) Penrose lists three standpoints for the information paradox:

LOSS: Information is lost when the hole evaporates away.
STORE: Information is stored in final nugget (black hole remnant).
RETURN: Information is all returned in final explosion.

Today's most popular view that information is stored at the horizon and returned by Hawking radiation is not among the alternatives. Although the holographic principle was discussed since decades, there was no proposal for a mechanism how this might happen. Recently there are (rather artificial) proposals from string theory and loop quantum gravity on that. So we have to add two alternatives more:

RETURN-B: <u>Stored at the horizon and returned in Hawking radiation.</u>
RETURN-C: <u>Stored in the knowledge of the last measurement and returned in Hawking radiation.</u>

From the perspective of space particle dualism theory we have to refuse all the three above mentioned alternatives, and stick to these two new patterns of information return.

In the contemporary literature the information paradox is mostly described as something rather subtle: something further restricting our ability to predict the future, or better the allegedly deterministic evolution of **U**, which is indeed deterministic when we deny the existence of **R**.[2] When we have two entangled particles we can use the results of measurements on the one particle to 'predict' properties of the other particle. When we measure a particle to have the spin 'up', we can 'predict' that the other particle must have the spin 'down'. The only thing which makes this seem to be a prediction is to look at the two particles as two separate things, while we should have learned from quantum mechanics that they are in fact *one thing* if they are entangled.

In fact, the information paradox is much more serious than that. In the core of this problem lies the *thermal nature* of the Hawking radiation. The no-hair theorem, which states that black holes can differ only in mass, angular momentum and electric charge, together with the laws of thermodynamics seem to suggest that Hawking radiation could not possibly contain any information about the matter that felt into the black hole. Even worse: since the radiation can only alter in its intensity depending on the mass of the black hole, black holes could become real <u>information shredders</u>.

What happens with the charge of a black hole which evaporates away? If the Hawking radiation is really <u>thermal</u>, and if it is dominated by photons for the most time of the evaporation process, could we be confronted with a tiny black hole that keeps a huge electric charge? The answer seems to be no. Processes in the ergosphere of a black hole tie the three properties together to some degree and thereby also prevent the formation of naked singularities.[3] The three properties of a black hole fulfill the following inequality

$$Q^2 + \left(\frac{J}{M}\right)^2 \leq M^2$$

Whereby Q is the charge, J the angular momentum and M the mass. It is for example not possible to destroy the event horizon by giving the black hole a very big angular momentum. Werner Israel showed that when a black hole reaches its maximal angular momentum, every attempt to let it rotate faster by throwing fast rotating matter into it would fail, because centrifugal forces would prevent that matter from falling in (Israel; 1986).

What about the ratio between matter and antimatter? If a black hole is completely thermal, then there should be no dark matter within the radiation. Space particle dualism theory predicts 60% primordial dark matter for the early universe (observations indicate 63% for the time the CMB was emitted). Now our universe has about 84% dark matter in from of both dark matter black holes and ordinary matter black holes, which is mainly due to the emergence of stars and their subsequent gravitational collapse. These two types of invisible matter cannot be clearly distinguished.

What if we wait long enough to see all black holes evaporating? If their radiation is absolutely thermal, this should result in a universe mainly consisting of photons.

The *photons per baryon number* could also be radically changed by those black holes. A universe which started with 60% dark matter could approach 100%, for then ending up with 0% dark matter after the evaporation process is completed.

Of course the evaporation takes very long time. For a big black hole it is

$$t_{ev} = \frac{5120\,\pi\,G^2\,M^3}{\hbar\,c^4}$$

For solar black holes with masses of about $10\,M_\odot$ that is in the magnitude of 10^{100} years! For large black holes most of the radiation would be in form of photons and neutrinos. According to calculations made by Don Page (1976) electrons and other rest mass particles can be ignored for black holes larger than 10^{17} gr.

Although we can be sure that infalling particles will never enter the horizon anytime before infinity, it is still difficult to get them back to the outer universe.

If the Hawking radiation is really thermal, then we have to expect severe violations in the conservation of

1) Lepton number
2) Baryon number
3) Entanglement information
4) Dark matter ratio (which can rise over 60% only through the formation of black holes)
5) Second law of thermodynamics

Violations of (1) and (2) are acceptable for many physicist, because the decay of all baryonic matter is still a possibility, and some theories, like most GUTs (grand unifying theories) and the CCC model of Penrose predict and strongly depend on the decay of seemingly stable baryons or leptons.[4] However, space particle dualism theory predicts that these apparently stable particles are indeed <u>stable for eternity</u>. We will see this in chapter 4.8, when we use the theory of entropic expansion to predict the maximal density at the border of the past.

(3) is not very serious from the perspective of a *collapse theory* where the nonlinear **R** is always regarded to be part of **U**.

Allowing the violation of (4) would introduce more degrees of freedom into the theory, which would reappear as fine tuning at the border of the past. If the universe could go under the default matter to dark matter ratio of 60%, then that would be equal to allowing such states to be possible initial states for the universe.

(5) could be avoided if (4) is avoided by having dark matter within the Hawking radiation. However, if black holes can really evaporate away into an ocean of light, we face serious problems with the second law of thermodynamics, because apparently black holes are the states with the highest entropy, and having black holes disappearing seems to reduce entropy. We could get some more entropy, because the number of particles (mainly photons) is highly increased. But that would be surely a smaller entropy than that of the black hole. The mainstream seems to resolve this issue in a different way. In his popular book "Black Holes and Time Warps" Kip Thorne explains that the Hawking radiation *distributes randomly in the outside universe*.[5] With other words, it radiates away in random directions and this corresponds to a higher entropy. This might seem to be a legitimate way of argumentation, but it is in fact very problematic.

When looking at a black hole we have no way of measuring any of its constituting particles. Its entropy is the number of possible ways it could have formed. However, for the radiation we can simply measure it all around the black hole, and determine the direction it came from. Measurements reduce entropy by collapsing wavefunctions. When particles are trapped in a black hole they are in superposition, and therefore have a huge entropy. When they are emitted outside the black hole, their entropy is still high, but gets reduced immediately when they are measured (directly or indirectly). The reason for the high entropy in and around black holes has to do with the high degree of quantum coherence in these areas. So the way we resolve the measurement problem, does not only gain importance when we examine the brain and consciousness, but also in a lot of other realms, such as black holes and quasi-crystals (see end of chapter 2.1). If a black hole could really evaporate into mainly just photons and other massless particles, it would locally reduce entropy and therefore cause a local reverse of the entropic expansion (see chapter 4.1).

Roger Penrose also argues that black hole evaporation implies a *drop in entropy*. For making that sound less alarming he says that the entropy is *newly defined* or *calibrated* after the black hole evaporated (see [5] "Cycles of Time" on that matter).[6]

If information is lost in a black hole, we should see the same amount of particles and antiparticles coming out of it in its final stages (when rest mass particles are within the radiation). If information is preserved, we should see almost only particles in the Hawking radiation, and no anti-particles.

Space particle dualism theory is based on the membrane paradigm and therefore the notion of a spacetime that bends beyond infinity having space and time flipped over doesn't make sense here. The black hole simply ends at the event horizon. That avoids the formation of a singularity where physical laws break down. But does standard physics really lead to the formation of singularities? If we recall that time dilation is already infinite at the event horizon we might wonder how a star is supposed to collapse all the way down to a singularity in finite time. I always imagined that an infalling observer would at some point meet the collapsing star that formed the black hole. I imagined that to happen some time after he enters the event horizon. But where? Surely no singularity can form, because the horizon would have long evaporated away, before the star can finish its collapse to a singularity.[7] Although it takes 10^{70} up to 10^{100} years for a black hole to evaporate, but with the infinite time dilation at the horizon and inside the black hole, that is short time when viewed from there.

Yet time dilation would start getting extreme even before the star goes below its critical circumference (which turns it into a black hole). Could a black hole fully form when the time dilation is already infinite at the event horizon? Some say no. Laura Mersini-Houghton from the University of North Carolina found that when we look at the collapse of a star into a black hole from the reference system of the star itself, we find that Hawking radiation evaporates the star away even before the local event horizon forms (L. M. Houghton; 2015).[8, 9] At first this sounds ridiculous: how could the extremely weak Hawking radiation evaporate a dying star even before it finishes its collapse? Usually we think of the Hawking process to be very slow and Hawking radiation to be very weak, but that is only how things look like when viewed from our reference system. When viewed from the reference system of the collapsing star, Hawking radiation is very intense and the Hawking process is very fast. This radiation looks only that weak to us because it is already so redshifted due to the gravitational pull of the black hole.

However, viewed from the outside these frozen stars still look exactly like black holes and there is still an event horizon for distant observers.

But a close observer would be seeing a collapsing star fastly destroyed by outgoing Hawking radiation. According to Houghton the Hawking radiation is so intense that it even reverses the collapse and pushes the matter outwards.

Although we can say that the event horizon exists only for the distant observers, that doesn't change too much: since a close approach to a black hole could not be survived, the distant observers are all what counts. The entropy and all other properties remain the same as previously known.

So although black holes having no singularity is a prediction of space particle dualism theory (and the similar worlds interpretation), standard physics can come close to this conclusion, as we can see from these recent results obtained by Laura Houghton.

And she is not the only one: long before her several scientists have pointed out that it doesn't make sense to speak of event horizons and singularities when time dilation becomes infinite even before an event horizon forms out. One of them is Dr. Abhas Mitra from the Homi Bhabha National Institute (HBNI) in Mumbai who wrote a paper titled "Mass of Schwarzschild black holes is indeed zero and black hole candidates are quasi black holes.".[10] Researchers like Mitra point out that the *apparent singularity* at the event horizon is in fact a true singularity and thus all black holes are just quasi black holes.

However, in order to solve the information paradox, we still have to find a way in which Hawking radiation could somehow carry the information of the matter which felt into the black hole. And for that to be achieved entanglement with the infalling matter is not enough. As I made clear above, the information paradox is not only about entanglement information being lost, but also about a violation of the lepton and baryon number conservation. If black holes evaporate mainly into light, then they could become real matter shredders on long term.

Houghton's calculations suggest that the Hawking radiation is so intense that it pushes matter out of the black hole.[11]

Of course, in a realistic situation a black hole always swallows much more matter than it can lose through Hawking radiation. But because the horizon exists only for the distant observer, it is always possible to have particles from inside leaked to the outside. Yet it would be quite troublesome to have the black hole losing mass energy both in form of leaking out particles AND random thermal Hawking radiation at the same time. That would not solve the information paradox.

From the perspective of an accelerated observer over the horizon, the Hawking radiation would originate from an ocean of real and not virtual particles. This is known as the Unruh effect, and is believed to be common to all accelerated observers in quantum vacua.[12] This radiation and the Hawking radiation are regarded to be the same just from different perspectives.[13]

At the end of the Hawking process, the black hole is able to produce massive particles. We might imagine that it could be possible to have more particles than anti-particles produced by the black hole. Since it is a matter of chance if we get a particle or an antiparticle, we may be tempted to suggest that for the sake of the baryon and lepton number conservation, we could have only particles in the radiation. But we forget that with every particle escaping the black hole, an antiparticle is falling into the black hole. The anti-particle would then annihilate with a particle from inside the black hole; again violating the baryon and lepton number conservation.

Just like Hawking radiation, the Unruh effect has never been proven by observations and is taken for granted as a consequence of the *equivalence principle*.

Quite contrary to this apparent consensus Douglas Singleton and Steve Wilburn from the physics department in Fresno, California made a detailed analysis of the Unruh effect and pointed out that it violates the equivalence principle of general relativity (Singleton & Wilburn; 2011).[14]

Laura Houghton is not the only one who realized that time dilation is important in respect to the information paradox. In a popular account physicist Fraser Cain noted that in a classical black hole the information of what felt into it is kept in the almost endlessly redshifted frozen images of those things.[15] I don't know if he also realized that this must include images of the original star itself as well.

Surely such images would be too redshifted to be measureable in praxis. And even in theory detection is close to impossible. So what happens with the wavefunction of all the matter that felt into the black hole? The extremely redshifted light reflected by the infalling matter is so weak, that it could never cause any visible macroscopic effects. Without macroscopic effects, there can be no difference for the conscious state of observers, and therefore no collapse of the related wavefunction. So there is another area before the apparent event horizon that might be regard as a gray area by many, but which is totally black in any practical way of sense. We can call this area the *coherence sphere*, and the border of it, the *coherence horizon*.[16]

So although it might appear to some of us that the information from those redshifted images is available in principle, it is in fact not, because they are too redshifted to cause any macroscopic changes which can constitute a measurement. Therefore they always remain in a superposition of having being emitted and not having been emitted.

What is the radius of this coherence horizon? Light beyond the visible spectrum may well cause macroscopic shifts in the matter distribution, but only with visible light being emitted the wavefunction can collapse, so that the event is settled. That doesn't mean radiation that is not visible can only be emitted when equipment is used to detect it. If the shifts in matter distribution caused by this invisible radiation are detectable through visible radiation, then the wavefunction will collapse. Yet in areas of space which can't emit any visible light, the wavefunction won't collapse, even if macroscopic shifts in the matter distribution of those regions are caused.[17]

Thus for knowing where the coherence horizon lies, we only need to calculate the radius of the area from which we can't expect *visible light*. If the redshift is bigger than the bandwidth of the whole visible spectrum, then there is no way we could see something there. This bandwidth is:

$$\lambda = 7 \times 10^{-7} \text{ m} - 3.9 \times 10^{-7} \text{ m} = 3.1 \times 10^{-7} \text{ m}$$
$$f = 7.7 \times 10^{14} \text{ Hz} - 4.3 \times 10^{14} \text{ Hz} = 3.4 \times 10^{14} \text{ Hz}$$

And that corresponds to a redshift of 1.79. The redshift depends on the ratio between the Schwarzschild radius R_h and the orbit radius R_O:

$$z_O = \sqrt{1 - R_O/R_h}$$

If we set this equal to the aforementioned critical redshift, we get:

$$z_O = 1.79 = \sqrt{1 - \frac{R_O}{R_h}}$$

$$1 - \frac{R_O}{R_h} = 3.2041 = z_O^2$$

$$-1 + \frac{R_O}{R_h} = -3.2041 = -z_O^2$$

$$R_O = |-2.2041 \times R_h| = (z^2 - 1) \times R_h$$

$$R_O = \left|-2.2041 \times \left(\frac{2\,G\,M_h}{c^2}\right)\right| = (z_O^2 - 1) \times \left(\frac{2\,G\,M_h}{c^2}\right)$$

Which means the coherence horizon lies at 2.2 of the Schwarzschild radius. That also means that the black area of black holes reaches much further than just the Schwarzschild radius. One could say they are more than twice as big as we imagine them to be.

From a true quantum perspective the really significant boundary of a black hole is not the Schwarzschild radius, but the coherence horizon. In relativity theory we can well ignore the question of weither or not endlessly redshifted photons are detectible, so we can take the Schwarzschild radius as the boundary of a black hole. In quantum mechanics on the other hand this question is of uttermost significance. If entropy is the logarithm of the number of indistinguishable states, then it must be proportional to the surface area of the coherence horizon, and not the event horizon (Schwarzschild radius). Thus the entropy of a black hole would be also 2.2 times higher, now given by:

$$S_{BH} = \frac{\pi}{2}\frac{c^2\,k_B}{G\,h} \times (z_O^2 - 1)\,A_{BH}$$

Yet the entropy of a black hole is linked to its temperature, and in the standard scheme that depends on the Schwarzschild radius.

As we saw, redshifted images can't carry any information out of the black hole and Hawking radiation can't either, because it is thermal and random. We also cannot modify it, because the Hawking process always involves both particles and anti-particles. Even if only particles had to come out, their original counterparts inside the black hole would be destroyed by the ingoing radiation. It also seems impossible to get massive particles long before the end of the evaporation process.

What we would need is the actual matter of the former star tunneling out of the *coherence horizon* and thereby maintaining the original lepton and baryon numbers as well as the dark matter ratio. If we want to maintain the analogy between black hole dynamics and thermodynamics, then this leaking must happen at a rate and with wavelengths that set the horizon surface in relation to a certain temperature.

Since the inside of a black hole is never measured, it becomes more and more superpositioned. We might be tempted to assume that this makes it easier for particles to tunnel out of the black hole. Yet a single particle has a position uncertainty related to its impulse and although it becomes totally uncertain where in the black hole it is located, we can be sure that it stays inside. Particles that were not measured for long time can't tunnel further than particles that are constantly measured. If they could, then gas from a gas container that wasn't observed for long time would for sure have to leak out completely.

That means tunneling can happen only for particles that have a wavelength as large as the diameter of the black hole.

When we conceive the existence of a new phenomenon that is derivable from known laws, we have to ask ourselves why this phenomenon hasn't been conceived by others. In this case the answer has to do with the way relativists think about the interior of a black hole. In a relativist fashion the matter of the former star is thought of as to have already collapsed to a singularity. In a singularity one can't anymore talk about position uncertainty of single particles but only of the singularity as a whole.

It is easy to have the Hawking process leading to a radiation with thermal characteristics. That is because there are almost endless many virtual particles in the quantum vacuum. Besides thermodynamics there is nothing else that could tell a black hole how many particles it should pull out of the quantum vacuum using its tidal power. That shows that Hawking radiation is a product of looking at a black hole as a thermodynamic system and the very reason Hawking started perceiving it as such was Bekenstein's claim that it has an entropy that is proportional to the horizon surface.

Is this true? Is the entropy of a black hole really measurable by its surface?

Entropy is usually a rather fuzzy concept. It can be interpreted as the logarithm of the number of microscopic arrangements which leave the macroscopic appearance and the thermodynamic properties of an object, such as temperature and pressure, unchanged. Yet this definition depends on how exactly we measure the state of an object. If we take a microscope there are far more states we can differentiate. And if we additionally try to keep track of a large number of particles within that object, the number of states that can be differentiated is even larger, lowering the entropy of the object even further.

In collapse theories we can measure entropy by the number of classical states the superposition consists of (number of indistinguishable worlds).

Using this definition for entropy we can give a very precise value for the entropy of single particles such as photons and electrons.

The position uncertainty of a photon is proportional to its wavelength. A wavelength

$$\lambda = \frac{c}{f}$$

can be divided into Planck lengths with each Planck length representing one possible position. Accounting only for the degrees of freedom in one dimension, the entropy of the photon could be defined as

$$S_\gamma = \log \frac{c}{f\sqrt{\frac{\hbar G}{c^3}}}$$

For a photon with a wavelength of one meter that would be a dimension-less entropy of roughly 35 (dimension-less because we didn't use the Boltzmann constant k). That is because the Planck length is roughly 10^{-35} m and the logarithm of 10^{-35} is 35. In reality the position uncertainty of a photon is extended radially in all three dimensions. The entropy of a photon is therefore more precisely given by

$$S_\gamma = \lg \left[\frac{4}{3}\pi \left(\frac{c}{2f\sqrt{\frac{\hbar G}{c^3}}} \right)^3 \right]$$

For our photon with a wavelength of one meter that is an entropy of roughly 105 (three times 35). A photon with a wavelength twice as large would have an entropy twice as high.

Yet if we look at the formula for the entropy of a black hole we notice that it is proportional to the surface area and therefore to the mass of the black hole. Since in general relativity gravity is caused by all forms of energy, including pure energy in form of photons, we arrive at the paradoxical conclusion that photons with low entropy falling into a black hole lead to a larger growth in the entropy of the black hole than high entropy photons. That already shows that the conventional Bekenstein-Hawking entropy of black holes is pretty much detached from ordinary entropy, or at least it is very different from the here presented concept of quantum entropy.

When viewed from the perspective of space particle dualism theory, the surface of a black hole seems to tell us even less about its entropy than before: here only rest mass contributes to the horizon surface area. All the infalling photons don't change the surface area of the black hole. We should also keep in mind that they become frozen in time, so that they don't really make it into the inside (except of those that were inside already when the original star collapsed).

Accounting for these, it seems quite obvious that we should no longer try to read too much into the analogy between the laws of black holes and the laws of thermodynamics. Black holes are central to the thermodynamics of the universe, but it is wrong to simplify them by mistaking their gravitational properties as their thermodynamic properties.

If the tunneling photons have wavelengths equal to the diameter of the black hole, then their energy is

$$E = \frac{h c^3}{4 G M_0}$$

It isn't sure if we have to use the aforementioned coherence factor here. A photon that tunnels only out of the event horizon might get endlessly redshifted, so that its wavefunction can't collapse. Yet many such photons together could cause macroscopic shifts inducing a wavefunction collapse. Using the coherence correction factor the energy of the tunneling photons is

$$E = \frac{h c^3}{8.8 G M_0}$$

Different from Hawking radiation there is a rather sharp peak at one wavelength, and not a thermal distribution.

Another difference is that with pre-existing particles tunneling, we need no energy to create them first (their rest mass), and we don't need two of them each time (particle and antiparticle, with one of them falling back into the black hole).
However, same as with Hawking radiation, only for very small black holes we can expect rest mass particles within the radiation.

Another interesting property of this evaporation mechanism is that although there is a standard minimal initial mass, that isn't the case for the final mass. While the minimal mass for the formation of a primordial black hole (dark matter black hole) depends on the size of elementary spaces (see chapter 4.10), the final minimal mass depends on the wavelengths of particles that formed the black hole or felt into it.
As long as there are particles with a smaller wavelength as the diameter of the black hole, the black hole continues to exist. That renders the evaporation from a run-away process into a gradual step by step destruction.

This would all be fine, but there are two problems, one serious and one fatal. The serious problem is that large black holes might not be able to radiate at all if the wavelengths they can use are so large that they can never be measured. The fatal problem is that, <u>if mass and gravity are side effects of charge, and if chargeless photons are the only particles that can escape out of macroscopic black holes, then having radiation cannot make black holes shrink.</u>

Yet this doesn't mean our considerations on quantum tunneling for matter in a black hole was completely in vain. In fact it means that one can't create a black hole that is smaller than the wavelength of particles one used to create it. We will use this at the end of chapter 4.10 to see what the minimal mass of non-dark matter primordial black holes is. In chapter 4.13 we will furthermore see that it is indeed meaningful to ascribe a temperature to a black hole, and that will help us to define the *temperature of the universe*.

Although it is well possible for a black hole to radiate, it doesn't need to. For example the star that is hidden within the black hole could theoretically be burned out completely already, so that the black hole ceases to have any radiation. The black hole's temperature would still be dependent on the black hole's size, but the total power that has been radiated away can be different for different black holes, violating the no-hair-theorem.

This makes black holes less paradoxical than thought: no longer are they objects that become hotter as they radiate away energy; and no longer do we have to directly link their temperature with their entropy. We can summarize:

(1) Temperature of a black hole is linked to its size.
(2) Radiation power is independent of its size, violating the no-hair theorem.
(3) Entropy is only roughly correlated to its size. Infalling photons have entropy but don't contribute to the size of the horizon.

There is no reason to restrict the usage of the term 'entropy' to thermodynamic systems only. As we already saw, within collapse theories, entropy can simply be defined as the number of constituent states within a superposition. This definition allows us to use the term entropy even with single photons, which obviously don't have a temperature.

Linking entropy to temperature one would expect low temperatures to be associated with low entropy. Yet if the temperature under consideration is the temperature associated with Hawking radiation, then the opposite is the case: the higher the entropy, the lower the temperature.
As laid out in this chapter, Hawking radiation would also make black holes violate the second law of thermodynamics. Both of these two properties make Hawking radiation highly non-thermodynamic.
Generally speaking a quantum mechanically defined entropy does not need to be dependent on temperature. For example an invisible gas has higher quantum entropy than any directly visible object, simply because it is measured with lower frequency (it has to push something visible or the skin of an observer in order to be measured).

It is a historically interesting fact, that the Hawking process was inspired by a model of J. B. Seldowitsch, where black holes were losing rotation energy through a similar effect, involving bypassing waves of vacuum fluctuations.[18] When Hawking proposed his model, Seldowitsch and his colleges were opposed to it. They believed that the black hole would stop radiating after it stopped rotating. They could not imagine the mass of the black hole being shredded to pure light, having all the previous information in it destroyed.

Do cosmological event horizons radiate?

Near a black hole event horizon, or a coherence horizon, redshift goes against infinity. Close to cosmological event horizons redshifts also reach higher and higher levels. Does the redshift go against infinity at cosmological event horizons?

That isn't the case: for all practical purpose the cosmological event horizon is defined by decoupling. We reach a maximal redshift of about $z = 1,100$, and then the universe becomes invisible due to the fact that before a certain time in its history it was in a plasma state in which light can't freely travel. Light is trapped between nuclei and free electrons.

This is the same type of state that exists inside the sun. Any region of space with a temperature of above 2,900 Kelvin is in this state, and if this would already qualify for an event horizon, then there would be an event horizon inside the sun too.

We do not expect Hawking-Unruh radiation from the sun, so we shouldn't expect that from the cosmological event horizon either.

In the mainstream the horizon defined by decoupling (last scattering) is called the 'optical horizon', while the horizon defined by neutrino decoupling is called 'neutrino horizon'. These are categorized as 'practical horizons'. However, if we acknowledge that these practical horizons always make the more theoretic horizons principally unobservable, then we have to also realize that theorizing about their entropy and temperature is meaningless.

Finally we have Rindler horizons that emerge for accelerated observers that temporarily avoid absorbing photons from certain regions of spacetime by moving away from them close to the speed of light. Obviously as soon as such observers stop accelerating those photons will catch up and reach them. Similar as with the optical horizon this is a rather arbitrarily drawn border. It isn't even a lasting border.

If we believe to be living in a sculpture-like space-time where past, present and future coexist and nothing changes, then it might make sense to talk about temporary event horizons, but when we realize that there is really just space and matter, with time simply denoting the rate at which the matter evolves in space, then we have to realize that theorizing about Rindler horizons is not fundamentally different from saying that running away from someone creates a barrier between oneself and that someone.

The radiation of Rindler horizons is called Unruh radiation. This radiation can be linked to the Rindler horizon, but it can also be thought of to arise from the apparent acceleration virtual particles seem to receive in accelerated frames of reference. According to radical relativism even properties such as real and virtual and the number of particles can be relative. Can this hold true in space particle dualism?

In space particle dualism with its similar world interpretation of quantum mechanics virtual particles are particles that do not exist in each and every of the similar worlds. Which worlds are deemed as similar is something that is not defined by a single observer alone, but rather the collective of all entangled observers together.

Notes and references:

1. A good popular account on the inside of black holes is [30] (In Search of the Edge of Time (1992); John Gribbin).

2. The viewpoint taken in this book, and which is discussed in greater detail in chapter 2.4, is that both **U** and **R** describe the knowledge of the observer.

3. Both the discovery of the ergosphere and the formulation of the cosmic censorship hypothesis are achievements that go back to Roger Penrose (1969). The ergosphere received its name mainly due to the fact that it is possible to gain energy out of it, utilizing something called the Penrose process. For a popular account see [35].

4. The CCC model is described in detail in Penrose's book "Cycles of Time" (2010).

5. This explanation can be found at the end of chapter 12 in [35].

6. Penrose needs the entropy to drop down to a minimum in order for a new eon to start, because he assumes the Big Bang of our eon to have been a low entropy state.

7. The reader may wonder how this obvious fact could have been ignored by the physical community for such a long time. Well, that has to do with the usage of space-time diagrams and the common practice of not bothering about issues related to the flow of time. Although 'beyond the endless future of the outside observer' may sound like a complicated way of saying 'never', but what it really means is 'around the corner in space-time', or beyond 'the edge of the outside light cones'. Inside the black hole space and time change their roles. What is space for the outside observers is time for the inside observer. Relativists have no problem with assuming that the infalling observer will NEVER pass the horizon from the perspective of the outside observer, but at the same time believing that for the infalling observer everything would be normal, and that he could pass the horizon without any complication. That is why no one bothers thinking about when the singularity is supposed to have formed out from the perspective of the outside universe. The answer would always be 'never' or 'beyond the endless future'.

8. This is true at least for the relative horizon (sometimes also referred to as the *apparent horizon*) in the reference system of the collapsing star. But there might still be a relative horizon for the remote observer, as well as an absolute horizon in 'space-time'. For more information on the historical relevance of the absolute horizon, read [35], info. box 12.2.

9. Her paper is available at [https://arxiv.org/pdf/1409.1837v2.pdf].

10. "Mass of Schwarzschild black holes is indeed zero and black hole candidates are quasi black holes"; Abhas Mitra; 2017; Cornell Electronic Preprint: arXiv:1708.07404, 2017.

11. One problem I see for the reversal of the collapse suggested by Houghton is that in the Hawking process one particle escapes while its antiparticle falls into the black hole. Therefore there should also be radiation inwards.

12. For possible experimental tests of Unruh radiation see: [http://www.phys.lsu.edu/mog/mog17/node8.html].

13. This duality is explained by having half of the wavefunction being lost in the

horizon. See also [35], chapter 12.

14. "Hawking radiation, Unruh radiation and the equivalence principle"; Douglas Singleton & Steve Wilburn; 2011 [http://arxiv.org/pdf/1102.5564v2.pdf].

15. This was part of a short popular account, available here: [https://www.youtube.com/watch?v=vzQT74nNGME].

16. These are new terms which I introduce here for the purpose of proper analysis. They are a result of applying the Von-Neumann interpretation, or, alternatively the similar-worlds interpretation, to the areas of astrophysics and cosmology. To my knowledge John A. Wheeler was the only one who made such applications by bringing quantum experiments to a cosmological level (*cosmological delayed choice experiment*).

17. From this perspective it seems that pointing radio telescopes at a black hole can shrink the coherence horizon slightly and affect how it radiates.

18. A popular account on the historic background of this can be found in chapter 9 of [32] (Kip Thorne; 1993).

4.4   A criticism of contemporary cosmology [14.09.2015]

Mainstream cosmology is based on the application of Einstein's field equations, which were originally designed for single object gravitational fields, on the universe as a whole. This means that every expansion of the universe must be one that is working against an universal gravitational pull. Only a universe with a very high initial expansion speed can fight such a gravitational pull, so that our universe must have reached the present low density rather quickly. It is the belief that gravity from different directions doesn't cancel out, that led the scientific community into believing that we live in a young universe.

The conventional Big Bang model is only concerned with what happened after the first second of the Big Bang. It does not try to explain what sets the initial expansion speed that in general relativity is required for a universe that exists long enough for giving rise to life and consciousness.

Usually it is imagined that the universe expanded due to some sort of 'explosive pressure'. The problem with this is that there is no pressure outwards coming from a singularity such as the one theoretical physicists suspect at the Big Bang. If singularities tended to cause expansion of space, then black holes would have to expand as well.

Stephen Hawking used Roger Penrose's black hole singularity theorem to prove that the Big Bang started off with a *space-like singularity*. Yet Roger Penrose's proof was going forwards in time, while Hawking had to apply it backwards in time in order to make it work. This way we could even prove, say, white holes existing near the Big Bang. If we ignore thermodynamics in our considerations, we could prove all sorts of things, including there to be no difference between past and future, but we know that's not true.

Hawking's proof was applying space-time geometry, or in other words, gravity, backwards in time, in order to show that the universe collapses to a singularity in the past, when we <u>look at it backwards in time</u>. But we could apply gravity forwards in time, and predict that the universe has to start collapsing 'right now'. The same way I could predict that I get born in the future and die in the past. All this is not telling us <u>why there is an expansion at all</u>.

That shows us that we cannot get valuable results basing time symmetric theories.

How are these issues treated in contemporary cosmology? In the early days of the Big Bang theory (1930) the expansion was treated as a basic property of our universe, one that wasn't subject to further attempts of explanation; it was seen as a boundary condition of the Big Bang. Later in the 1980's Alan Guth came up with *inflation theory*, which tries to explain why the universe is <u>flat</u>, has the <u>same temperature in all directions</u>, and why there are <u>no magnetic monopoles</u>. The flatness was explained with a huge inflationary <u>over light speed expansion of space</u> in the first few fractions of a second (from 10^{-36} sec. after the Big Bang to sometime between 10^{-33} sec. and 10^{-32} sec.) after the Big Bang. This expansion was thought to be led by an inflationary field, with *negative pressure*. The hypothetic particle to cause this inflation was called the inflaton (never found). In relativity, negative pressure has an anti-gravitational effect. This inflationary field depends on very special conditions, which have no justification. There is no proof that there was really an inflation. Roger Penrose pointed out that a universe which is flat because of inflation <u>would be even more rare and special than one which is flat from the beginning</u>. He also pointed out that a chaos expanding to infinity would still remain to be a chaos.[1] That is to say that fluctuations and irregularities in the curvature of space-time would not get ironed out by inflation.

Inflation theory tries to solve the so called horizon problem, the question why the universe has the same temperature in every direction we look, by saying that those opposite parts of it, was closer together before inflation.

Penrose (2004) again, criticizes that, by saying that it is not plausible to solve the low entropy initial condition problem (the horizon problem can be identified as such), by temperature compensation, while any temperature compensation must represent an increase in entropy!

The reader might wonder how inflation theory can be based on an expansion which is much faster than the speed of light. Relativity theory tells only particles not to move in space with a speed higher than the speed of light, but it doesn't say anything about how fast space itself can grow or about the growth of space in general. Inflation theory is using this gap in the laws of physics, to postulate an incredibly high speed for inflation (a growth to the factor of 10^{60} in less than 10^{-32} seconds).[2]

It should be clear that a final theory, which must be background independent in the sense of having space and particles tied together to a unity, could never predict an expansion of space with speeds much higher than the speed of light.

Hawking came to his singularity solution for the Big Bang by applying Penrose's singularity theorem of black holes backwards in time. That might be a feasible proof within the narrow scope of general relativity, but it doesn't explain the expansion and thereby falls short in describing the very phenomenon it tries to address. In his application of the singularity theorem he would just treat the expansion as a collapse. His ignorance for the arrow of time had caused him a lot of confusion, which at the end even led him to propose that a collapse of the universe would reverse time. We know that time goes forwards, and that there is nothing that could turn a contraction into an expansion.

So Einstein's universe, being a universe where the overall curvature is determined by the absolute mass energy density, could not stay stable, nor expand, but only collapse. Not knowing about the expansion of the universe, he had introduced a *cosmological constant* before, for keeping the universe stable, but had never clarified where this constant comes from. Choosing different values for it one can get an expanding universe as well, though usually the expansion of the universe is not described with this constant, but taken into the initial conditions of the Big Bang. Most theoretical cosmologists take it for granted that general relativity predicts the expansion of the universe, simply because the universe looks like a star collapsing to a black hole backwards in time. That is why Einstein called the introduction of a cosmological constant in its original purpose, keeping the universe stable, his 'biggest plunder'. Yet when Einstein introduced his cosmological constant, it was probably in desperation: he must have regarded it as a last resort, an expedient, to save his theory. After it was discovered that the universe is expanding, pre-existing solutions of Einstein's field equations found by Friedmann in 1922 were used to explain this. Einstein's theory was so successful in describing local gravitational fields that no one doubted that it would also be a proper description of the universe as a whole.[3]

The fact that an Einsteinian universe <u>could never expand, but only collapse</u>, was just wiped away, by applying gravity backwards in time – just as Hawking did with his Big Bang singularity theorem.

That is how a disadvantage of a theory was stylized into an advantage. Doubtless, relativity theory was a stroke of genius, but we would not be forced to describe nature using two completely different sets of rule (quantum rules and relativity rules), if relativity was a complete theory without problems of inner consistency. We face the incompleteness of relativity when we talk about singularities and the expansion of space; and we face the incompleteness of quantum theory when we talk about its interpretation – a logically complete theory, would include its interpretation.

Cosmologists seem to think about dark energy just since it was discovered that the universe's expansion is accelerating. Yet the expansion itself needs a driving force too, not only an accelerated expansion. A big pressure in a star causes the star to keep stable or even expand, but it doesn't cause space itself to expand; so why should it in the early universe?

Should we believe, as suggested by the mainstream, that vacuum energy is the driving force behind all this?

If this form of energy really had an effect on the expansion of the universe, then it would most probably have blown our universe apart long time ago. If we use general relativity for our cosmological models, it remains a mystery why vacuum energy didn't destroy the universe. This is a problem known as the *vacuum catastrophe*. Then the only solution within relativistic cosmology would be to have positive and negative energies in the quantum vacuum all levelled out by a hypothetic particle symmetry called *supersymmetry*, which is part of speculative unification theories.

Let's say supersymmetry really existed in nature; what tells us that supersymmetric partners would have a negative gravitational effect just when being part of vacuum fluctuations? And even if they have this effect and are capable of levelling out the huge energy in the vacuum, if the levelling out is exact, then there is no vacuum energy left to cause any expansion or collapse. Yet if there is a tiny little rest left over, then that would again be incredible fine tuning of highest precision: the smallest variation would destroy the universe immediately. The odds against chance for this fine tuning would be $1: 10^{120}$ (!). Yet for proponents of multiverse models, this isn't a problem: it is enough if this precision is reached in a few universes out of an infinite number of universes.

For this to be a legitimate position there has to be at least supersymmetry. Does supersymmetry exist in nature? Till this day, no single supersymmetric partner of any particle has been found. This is usually explained by assuming a symmetry break which led to a huge discrepancy between the masses of particles and their supersymmetric partners. Most supersymmetric partners of particles would be heavier than their ordinary counterparts, but that isn't true for the *spin-0 Higgs boson*. Yet it was discovered in 2013 without its supersymmetric partner being discovered prior to that. That already proves supersymmetry wrong, and along with that it proves relativistic cosmology, as well as superstring theory wrong.

Note:

1. See chapter 28 of [4] (*The Road to Reality - A Complete Guide to the Laws of the Universe* (2004); Roger Penrose).

2. The reader may take note of the fact this number is somewhat close to the chance for getting a flat universe, which is $1: 10^{62}$. Basically what this means is that inflation theorists artificially chose an expansion factor that is big enough to make the universe appear flat within a region as big as the visible universe. It doesn't take much to realize that there is nothing in the laws of physics that justifies such a choice. In fact the chance to get an inflation to the factor 10^{60} isn't really higher than to have a universe without inflation that is flat in a region 10^{60} times bigger than the initial universe (measured by the radius of its cosmological event horizon).

3. Einstein never seemed to have promoted his theory as a cosmological theory. He barely talked about the Big Bang or the expansion of space.

4.5 Cosmological thermodynamics [10.03.2015 (entropy equals number of worlds); 08.05.2015 (entropy depends on measurement problem)]

According to the *entropic expansion theorem* introduced in previous chapters, the entropy per volume on very large cosmic scales must be a constant. So how much is this constant, and can it be derived by a finite formula? By a formula as simple as the Bekenstein-Hawking formula for the entropy of a black hole? Such a formula would determine the expansion of any given universe. It would extend the holographic principle to the universe as a whole. However, it is important to notice that the word 'holographic' here is just metaphoric and doesn't in anyway imply information about the universe being stored in some border region. It is unclear on what properties or laws of our universe the holographic principle depends upon, but it may well be that it has to do with the two-dimensional nature of the constituents of space, the little surfaces called *elementary spaces*, which combine to form the 3-dimensional space we are so familiar with.

Interestingly 't Hooft and Susskind used the laws of black hole thermodynamics to argue for a general *holographic principle* of nature, which asserts that consistent theories of gravity and quantum mechanics must be *lower-dimensional*. Yet, space particle dualism theory seems to be the only theory with this property.[1]

Basing the similar worlds approach, we can give a new and more precise definition of entropy:

§ The absolute entropy S of the universe equals the logarithm of the number of coexisting presently indistinguishable realities.

This makes the notion of entropy less abstract and ties it neatly to that of a quantum state. This interpretation of entropy works only within the *similar-worlds interpretation*, which is an inseparable part of space particle dualism theory (see chapter 1.2). The usual formula for entropy, namely

$$S = \log V \, n \, k$$

With the Boltzmann constant k, the volume V of the considered area, and the number n of particles, is not directly referring to the number of microscopic states. k is a constant used to relate entropy to other measures in nature which are given in units. In particular an entropy increase is usually related to an energy flow from a hotter to a colder body. Therefore S usually has the unit $J\,K^{-1}$. The pure entropy, only related to the number of microscopic assemblies which do not change the macroscopic shape, is without k, and therefore dimensionless. We should also know the number of possible positions for every particle. As we saw when we considered the density of space, there are 10^{105} possible positions for every cubic meter. The logarithm is only for convenience here, to keep the numbers low. If we wanted to give an absolute number, it would be:

$$S = V \times n \times 10^{\,|\log V_{PL}|}$$

$$S = V \times n \times 10^{105}$$

Of course the absolute value for entropy does not really play an important role here. It only gains importance when we talk about the density of space. In general it is ok to just use the logarithm of the real entropy.

While in classical thermodynamics the notion of entropy depends on a rough differentiation between microscopic configurations and the macroscopic shapes and thermodynamic properties of an object, it can be a much more precise thing in quantum thermodynamics, where it ultimately depends on the size of superpositions. That makes it a very delicate issue, because superpositions are much larger without a collapse of the wave function. So strictly speaking, it depends on the interpretation we choose.

In space particle dualism the collapse of the wave function **R** and the unitary evolution **U** are both regarded as parts of the changing maximal attainable knowledge of the observer. This maximal knowledge is characterized by the number of worlds in our equivalence group of coexisting spaces (or space-times). We could say that entropy is characterizing the size of our superpositions.

In popular accounts entropy is often described as a measure of chaos, or, and which is quite the opposite, a measure for information. We might regard black holes, which have very high entropy, to be objects containing a lot of information. But the opposite is true: the high entropy of a black hole is due to the fact that it is very hard to tell something about its past. That is expressed in the *no-hair-theorem*, which says that any black hole of the same mass, rotation impulse and charge, would look exactly the same. We could never tell what it is made of. It could have been a collapsed star, but anything else as well.

Increased entropy is usually related to a loss of information about the system. This can be regarded as a minor information loss, if the information is still kept in the microcosm – in the positions and impulses of particles. If I for example delete files from my computer, there will be some information loss, but this is just a minor loss, because it is translated into heat information leaving my computer. Now what would be a serious loss of information? Two kinds of process in nature have been identified with serious loss of information: one is objects falling into a black hole; another one is the collapse of the wavefunction. According to space particle dualism theory the measurement is a gain in information, not a loss. How could a measurement anyway, be a loss of information? Well, in decoherence theory, or in the many-worlds approach, information is regarded as independent from observers. The wave function is not regarded as a combination of many states, but as a single state. The more free degrees this putative singular state has, the more 'information' it contains. If a human now measures this single state, his mind will be entangled with some certain parts of the superposition, so that he cannot see the rest of the 'whole' superposition. His mind is here also regarded to exist in a superposition, with each state of mind being represented by another consciousness. Therefore for proponents of decoherence theory a measurement is a loss of information. It ends the deterministic unitary evolution and brings randomness into the theory. But what is a unitary evolution without a measurement, without positions, without perceptions? It is doubtless something very abstract. Although counting to the realist interpretations of quantum theory, decoherence theory appears much more surreal, than the Copenhagen interpretation, which says that what we can observe is all there is. In my opinion that is much more realistic, and has less metaphysical ballast.

For proponents of decoherence theory, even a wave function of the whole universe, being in all possible states, would still be regarded as information. According to this line of thinking, the larger the superposition is, the greater is the amount of information contained. Here gaining information through measurements is regard as a loss of information (!). In my eyes that is quite absurd, but it is important to understand the argument. In decoherence theory this kind of 'information' is not lost completely, because interference between macroscopic bodies would still be possible in principle. In the many-worlds approach, which is supposed to have no collapse of the wave function, this quantum information is lost whenever a new world splits off from the existing worlds. There is no macroscopic interference even in principle in the many-worlds approach, because the different worlds are put on different space-times. From the viewpoint of the proponents of decoherence theory, there is an 'information' loss when the wave function collapses in the Copenhagen interpretation. Anyway, in every normal definition of information, information is something useful. It is not useful information to know that a particle is here and not here at the same time. Yes, it can be at many places at once, but only if we do not measure it. A measurement creates information, it does not destroy it. It does not create it out of nothing. It creates it out of possibilities, but still, it could never be regarded as a loss of information, not even in the strange world of quantum physics. Both the unitary evolution **U**, and the reduction of the wave function **R**, have to be regarded as parts of one single process – the changing accessibility of information.

In quantum mechanics any given superposition is growing bigger till the moment of measurement, getting smaller abruptly after it, for then slowly growing bigger again. The related quantum entropy does the same. If the measurement is something done by conscious observers, it is very natural to have it causing a temporal decrease in entropy. All human endeavour can be characterized as an effort to decrease entropy. A superposition is not always growing. It grows when a particle interacts with other particles, without being measured. Why is that? Imagine a billiard game in a superposition of many different initial states. Small differences in the initial conditions would cause small differences in the final conditions, if there is only one particle. But when there are many particles, small differences in the initial conditions, in the way the ball is hit at the beginning, would cause huge differences later in the game. That is why our superpositions are growing bigger and bigger.

The probability for a certain quantum state is high when this state is very common among the *similar indistinguishable worlds*. The number of worlds for a single particle is increasing when the particle is unmeasured. What about the total number of worlds for the whole universe? Is it increasing?

Ignoring the growth of the cosmologic event horizon just for a moment and applying the entropic expansion theorem to quantum mechanics, we see that the number of equivalent space-times should indeed be constant. Any increase of entropy is levelled out by a growth of space. That makes the average entropy per volume in the universe constant on large scales. What is the value of this *cosmological entropy constant*? In a paper entitled 'A larger estimate of the entropy of the universe' Chas A. Egan and Charles H. Lineweaver (2010)[2] found $S_{obs} = 3.1^{+3.0}_{-1.7} \times 10^{104}$ k for the entropy of the universe and a general entropy density of

$$\rho_{s_{obs}} \approx 8.4^{+8.2}_{-4.7} \times 10^{23} \, \text{k/m}^3$$

This average entropy must <u>always remain the same</u>. It can vary locally, but it must always remain to be the average. That is only possible because the expansion of space compensates every entropy increase. It is like there is some fixed information density for space-time. Or maybe the number of elements/worlds in the set of similar worlds has to remain the same, for yet unknown reasons. The entropic expansion theorem follows out of space particle dualism theory in a more or less direct way; yet the details on how exactly the universe manages to follow this law are not fully understood. The above value for the entropy constant did not account for dark matter. If we consider massive halo black holes with masses of about $10^5 \, M_\odot$, we might get 10^{25} k/m^3. If dark matter consists mainly of black holes, then even more (for details see Chas A. Egan and Charles H. Lineweaver, 2010). Space particle dualism <u>requires the existence of primordial black holes</u> for maintaining the expansion <u>before the formation of stars</u> and stellar black holes, and that in itself implies a pretty high entropy density. Also research on galaxy formation shows that galaxies formed so early that they are hard to explain without assuming <u>primordial black holes to have been the seeds for galaxy formation</u>. Entropic expansion needs primordial black holes for the expansion of space in the dark epoch of the universe (about 600 million years), and galaxy formation models needs them in order to explain the very early formation of galaxies, as well as the accurate correspondence between the size of supermassive black holes and the size of their harbouring galaxies.

The entropic expansion theorem is a new law of thermodynamics applied in cosmology only. Is it actually possible to do thermodynamics in a seemingly open system like the universe? If the second law of thermodynamics doesn't apply on a cosmic scale (because it isn't a closed system), then what about other laws of thermodynamics? What about the first law of thermodynamics, the conservation of energy?

Don't we have to develop a full new system of thermodynamic rules for large open systems? When we look at the universes as a whole, we find that not only the second law of thermodynamics is violated, but maybe even the first law as well (!). If we look at the microwave background, it is getting colder and colder because space itself stretches the waves of its photons. That means the microwave background is loosing energy. From a general relativity perspective we might suggest that this loss in energy is somehow accounted for by an increase in potential gravitational energy of the whole universe. In space particle dualism theory however this is not so clear, because here the universe can't collapse and therefore hasn't got any potential gravitational energy. Still the energy loss of the microwave background might be accounted for when considering that the expansion of space moves photons away from gravitational sources such as black holes.

In addition to all this, if we regard only the visible universe as real, we have to say that the universe gains energy as the cosmological horizon moves back. A precondition for such a finite universe is of course that we are alone in the universe, as suggested by the evidence presented in chapter 4.2.[3]
The reader might be worried that the energy loss caused by the cosmic redshift could distort the ratio between matter and dark matter. However, since black holes are indistinguishable from dark matter and since they grow bigger and bigger, transforming more and more mass from luminous matter into non-luminous matter, the ratio changes anyway.
Dark matter started at 60% at the beginning of physical time, grew to 63% until decoupling, for then arriving at todays value of 84%.[4]

The entropic expansion theorem states that every increase in entropy is levelled out by an expansion of space itself. That implies that the entropy per cosmic volume is always constant.

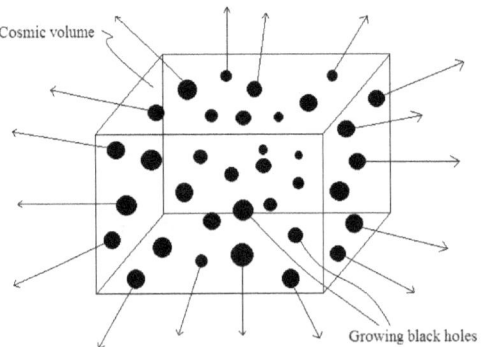

Fig. 42. The entropy in a sufficiently large (intergalactic) volume of space is always constant.

So what about the big miracle surrounding the low entropy condition at the beginning of time (the border of the past) frequently discussed by Penrose? There is no miracle: the entropy then was not higher, nor lower than the entropy of the universe at any time after.

For the layman it is hard to understand how the thermal equilibrium of the early universe can count as low entropy. It is usually regarded to be low, because it is supposed to not have contained any black holes. A state full of black holes has much higher entropy than a simple thermal equilibrium. Indeed there would be no time direction in such a thermal equilibrium. Time is introduced by black holes, and by the expansion of space itself. Yet since more than 300 years we are told that the second law of thermodynamics is bound to a low entropy initial condition. This believe which started in the early days of thermodynamics has been kept eversince and was carried into cosmology.

It made people believe that the beginning of physical time must represent a low entropy condition which was believed to represent a form of fine tuning of the universe; one that physicists were familiar with already before the time of cosmology.

Another reason for a majority of cosmologist to believe that the universe must have started with a low entropy is that black holes, which have a high entropy, usually would need huge irregularities in the initial matter distribution in order to form.

This however isn't the case for black holes that are made of dark matter: dark matter interacts only through gravity, and therefore there is nothing preventing it from condensing into an ocean of primordial black holes.[6]

Arriving here we already realize that the beginning of time did not at all represent a low entropy state. It was full of primordial black holes (in contrast to stellar black holes). It was Stephen Hawking who suggested the possible existence of primordial black holes, because that seemed to be the only way Hawking radiation could be observable.

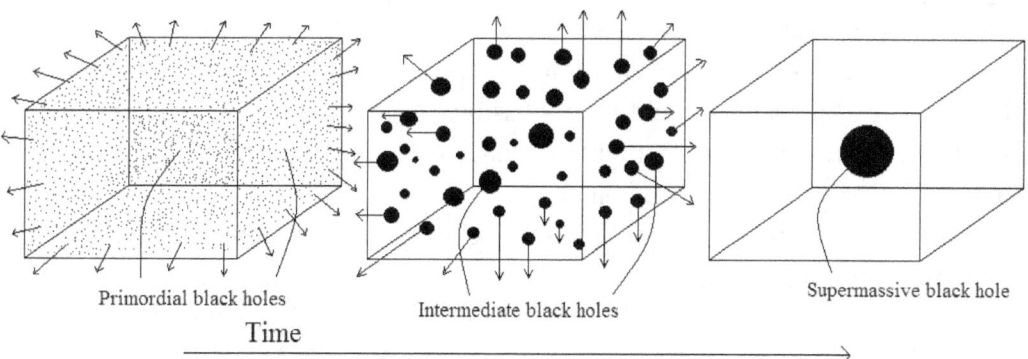

Fig. 43. Entropic expansion requires primordial black holes.

Arguments against the existence of primordial black holes have been the absence of radiation from evaporating or even exploding microscopic black holes. Firstly, microscopic black holes would <u>not explode</u> at the end of their evaporation process.

Without much unjustified assumptions, most researchers arrive in some way to the conclusion that black holes must leave back Planck-size remnants (see chapter 3.1 for an example). Secondly, they would most probably not manage to lose any weight, in the dense primordial soup of the early universe. Presumably each of them would become the seed for a galaxy. That is why we have a correspondence between the size of supermassive black holes and their harbouring galaxies (M-sigma relation).[5]

And thirdly, if mass and gravity are side effects of charge, then chargeless photons can't make a black hole shrink. That means Hawking radiation or any other form of radiation sourced from the rest mass (not rotation energy) of a black hole doesn't exist according to space particle dualism. Thus upper limits for the abundance of primordial black holes set by attempts to measure Hawking radiation are invalid within the framework of this theory.

People see it as a given fact that the world must have begun in a low entropy state, because the growth of entropy seems to be the only thing which could give the time a direction. But the real origin of the arrow of time is not a low entropy boundary condition but black holes and the entropic expansion of space. In any given universe, no matter which initial entropy it has, there would always be an arrow of time.

What would be the maximal entropy in a system without dark matter? It would be the thermal equilibrium. How can we get more entropy with dark matter? That can be achieved by adding many little dark matter black holes (60%) to this thermal equilibrium. Can we get more entropy? Yes, we can get more entropy by letting these black holes suck in the surrounding matter. Was there anything special about the entropy at the border of the past? No, and it was rather high. If it had no primordial black holes it wouldn't allow for the formation of galaxies, nor for an expansion of space, and thus there would be no chance for any further growth in entropy. Such a universe would always stay in its equilibrium state. It is furthermore important to notice that gravity alone, without dark matter, couldn't bring the universe out of such an equilibrium state, because without the black hole induced expansion of space there would be nothing that could stretch out and amplify small random fluctuations in the matter density. This way it would be impossible to get small islands of higher mass density that can provide even the slightest chance for a star to form out that could allow the universe to expand even a little bit.

This should make it clear that primordial black holes are an inseparable ingredient of the entropic expansion scheme.

It is certainly revealing that the universe did not start off in a minimal entropy state, but still one has to wonder what prevents the universe to start with a state in which all ordinary matter is already suck up by dark matter black holes.

Such a universe would be very cold, and although we could speculate about all the heat that could possibly be trapped inside black holes, there would be no way to measure it. It is not obvious what temperature to attest to such a universe. If it is indeed zero, then this raises the question how the universes temperature is correlated to other properties, such as matter density. We will explore this questions in chapter 4.13, where we will discover that the initial temperature of the universe can be derived from the uncertainty principle.

That shows that we don't have to resort to anthropic arguments in order to explain why the universe didn't start with black holes only. However, what is needed in every case, is primordial black holes. There must have been enough of them, to maintain a continuous entropy growth before the formation of stellar black holes. Same as for the present expansion (see chapter 4.1), we can ignore entropy increase not relate to the growth of black holes.

How does galaxy formation work in theories without primordial black holes? Usually quantum fluctuations in a thermal equilibrium are made responsible for little inhomogeneities in the distribution of matter. Those grow bigger in time, and finally lead to the formation of stars and later galaxies, a precondition being that the universe expands and that this expansion is independent from what happens inside the universe. There are two problems with this picture: firstly, it could neither explain why every galaxy has a central supermassive black hole, nor why galaxies formed so early. The second problem lies in the fluctuations themselves: as Penrose's pointed out in "The Road to Reality", without the state reduction **R,** all fluctuations remain in superposition. Hawking's wave function of the universe would, without a single state reduction since the beginning of time, already have developed into a state of everything and nothing. If one sticks to the many-worlds interpretation, as Hawking does, there can be no explanation for density fluctuations in the early universe. Most 'stories', as Hawking calls his different worlds, would not contain any fluctuations. Without entanglement with a conscious observer, we have no way to say that something really happened, or that there was a fluctuation.

If we didn't know about primordial dark matter, and that there was 60% of it at the beginning of time, we may say that the presence of stars and galaxies today led to the emergence of primordial black holes in the past. Since the initial state of the universe is not chosen before the emergence of life, there is nothing wrong with this form of teleological causalization.
Knowing about dark matter, we see that our dependence on this kind of teleological explanation is not as big as it may have seemed. And this is good, because such explanations couldn't account for the vastness of the universe in which the formation of galaxies all took a similar path.
In chapter 4.8 we will come back to the entropic expansion and examine to which maximal density for the border of the emergent past it leads.

Notes and references:
1. For an authoritative review, see Ofer Aharony; Steven S. Gubser; Juan Maldacena; Hirosi Ooguri; Yaron Oz (2000). "Large N field theories, string theory and gravity". *Physics Reports*. 323 (3–4): 183–386.

Bibcode:1999PhR...323..183A. arXiv:hep-th/9905111.doi:10.1016/S0370-1573(99)00083-6.

2. Egan, Chas A.; Lineweaver, Charles H. (2009). "A Larger Estimate of the Entropy of the Universe". arXiv:0909.3983 [astro-ph.CO].

3. If we instead regard the whole infinite universe as real, then it has an infinite amount of mass energy and it becomes even more difficult to talk about energy conservation.

4. Common diagrams on this give a quite misleading picture, because they include the non-existing dark energy, and therefore give another percentage for dark matter.

5. DISK ASSEMBLY AND THEMBH-σeRELATION OF SUPERMASSIVE BLACK HOLES; Victor P. Debattista,Stelios Kazantzidis, and Frank C. van den Bosch; 2013, February 13; The Astrophysical Journal, Volume 765, Number 1.

6. A precondition for this is dark matter being 'cold'. A theoretical justification for this will be provided in chapter 4.10.

4.6 No singularities [29.09.2015 (no Big Bang singularity); 12.10.2015 (causal event horizon)]

A singularity is a place of endless density and time dilation; a place were time starts (space-like singularity) or stops (time-like singularity), and where the laws of physics break down. From this it should be clear that they are not expected to exist in reality. Rather they are a hint that the theory reached its limits. Nevertheless most physicist assume that 'quasi-singularities' exist both at the Big Bang and the centre of black holes.

Contrary to that, as we saw in chapter 4.3 black holes are not singularities surrounded by event horizons, but simply *collapsing stars frozen in time*. Here the singularity is simply avoided by the infinite time dilation that is created when event horizons form out.

For the universe as a whole, there is no external reference system. Therefore time dilation can't set upper limits for the mass energy density here.

Similar as in other approaches to quantum gravity a maximal density is additionally set by the extended nature of the building blocks of reality. In all other quantum gravity theories particles or space quanta have a diameter equal to the Planck length (10^{-35} m). In space particle dualism things are different: here the size of the building blocks is proportional to the mass of the particles (see chapter 3.8).

Yet even if the universe were to reach an infinite density, according to space particle dualism theory that wouldn't lead to an infinite time dilation marking a beginning of time; it wouldn't lead to any time dilation whatsoever (!). That is simply because in this theory only energy differences matter. Time dilation same as gravity can only come about through differences in the density of elementary spaces.

If time dilation can't give us a beginning of time, does that mean that the past extends back into infinity? Wouldn't that mean there was a (physical) time before the dense state at what seems to be the end of the past?

According to the entropic expansion scheme described in previous chapters, the universe's expansion is tied to entropy increase. The expansion of space levels out the increase of entropy in the universe. This means the entropy density of the universe must be a constant. If this is the case, then there must be a maximal mass energy density for the universe. It is the density where black holes shrink to a size at which they disappear. We will come back to this in chapter 4.8.

If there was a universe before the maximal density state (called 'Big Bang' in the mainstream), it would have to be a collapsing universe from our perspective and that would mean that it had an opposite arrow of time. Having different directions of time doesn't fit in with the notion of a distinct present which is central to space particle dualism.

The initial density can not depend both on the entropy constant and the size of elementary spaces and surely the limit set by entropy is much higher than that set by the size of elementary spaces. The question which arises here is: what did the initial mass energy density depend on?
In standard cosmology the initial density is irrelevant, because it doesn't affect anything we can observe. In space particle dualism theory things are very different: here the initial mass energy density of the universe determines its overall entropy density at any other given time. In chapter 4.13 we will be exploring links between the entropy density and the initial energy density.

Aside from all this there seems to be at least one more way to get a limited past:
There are arguments which suggest that times where the universe exceeded a certain temperature can be simply ignored: as Steven Weinberg remarks in his famous book "The first three minutes" (1977), all the history of the universe before the end of the first second after the Big Bang, does not have any influence on what we can observe now.[1] As he explains that has to do with the fact that the universe was in a thermal equilibrium for the whole of the first second. In contrast to that inflation theory is concerned exclusively with things which are supposed to have happened in small fractions of the first second. It is therefore no wonder that it does not make any testable predictions, except of the putative reduction of fine tuning. We might in fact have a cosmologic event horizon placed one second after the Big Bang. That implies that there was no fixed quantum state whatsoever for the time before the first second.
At least we can argue this way within the framework of standard cosmology.
As we will see in chapter 4.8, the border of the past is reached at a much lower density according to space particle dualism, and that has indeed to do with the above mentioned entropy constant.

From all this it should be clear that time dilation and a space-like Big Bang singularity are not needed for have a beginning of time.

One should therefore be very skeptical about pre-Big Bang models that are solely based on avoiding the singularity. That includes models of one oscillating universe as well as multiverse models where vacuum fluctuations or black holes lead to the creation of new universes. The arguments brought forth in this chapter should make clear that such models are not to be taken lightly.

References:
1. See [7], end of chapter 7.

4.7 Anisotropic expansion [10.10.2014 (non-isotropic expansion)*; 14.11.2015 (Pioneer probes); 05.11.2017 (fluctuations in redshift)]

*In collaboration with Bernhard Umlauf (Germany).

Note: This chapter is more of historic value. A real up-to-date exploration of anisotropic expansion is to be found in chapter 4.15.

What would happen if we could move all galaxies into a certain region in space in order to create a non-isotropic universe? The entropic expansion would lead to a faster expansion of space for that area, and all the non-isotropy we tried to create would be ironed out. After all the overall entropy density must be a constant.

That means entropic expansion does not only explain the accelerated expansion of the universe, but solves also the *horizon problem* (isotropy of the universe). This is an additional and maybe better explanation for the isotropy of the universe than the afore-mentioned prediction of a high entropy for the beginning of time.

But isn't this requiring a *non-isotropic expansion*? Of course it would be impossible to move galaxies around like that, but as German physicist Bernhard Umlauf pointed out correctly, the entropic expansion of space particle dualism theory would lead to a anisotropy in the expansion of space at least sometimes, and this should be detectable in principle (2014; personal correspondence). He suggested that the extreme redshift of quasars could thus be explained by the extreme entropy increase caused in the central black holes of quasars. He further suggested that the expansion of space should be stronger near black holes.

Same as the usual cosmic expansion in standard cosmology, entropic expansion is not easy to detect within a galaxy. It is something mainly taking place on an intergalactic scale. It has even been argued that there should be no expansion at all below the scale of galaxy clusters (Misner et al. 1973).

How could a local expansion be measured? What we need is two distant bodies, which have no significant gravity between them and for which we know their distance with high precision. The Pioneer probes which were sent out in the 70's would be perfect for studying local expansion. And indeed there was an anomaly in the frequency of the signals send by the probe that led to speculations that it may be caused by cosmic expansion.[1] However, although the magnitude of the anomaly was close to the value of H_0 times the speed of light, it was not an acceleration, but a deceleration. Many theories evolved around the anomaly, even suggesting new physics, but at the end it turned out that the anomaly can be fully accounted for by *anisotropic radiation pressure*. That means although a local expansion could have been detected, there were no traces for such.

The usual explanation for the missing local expansion is that on the scale of the solar system and even the Milky Way, gravity is much stronger than the expansion of space.[2] Yet if it is really gravity that hides the effect, shouldn't we then be able to detect a deceleration of the Pioneer probes caused by gravity? If the gravity acting upon the Pioneer probes is not strong enough to cause a measurable deceleration, then how is gravity supposed to hide the effects of local expansion?

Also gravity within the Milky Way galaxy isn't strong enough to cause any redshift. Yet the expansion of space should definitely cause some measurable redshift even within the Milky Way galaxy.

Assuming a homogeneous expansion, it wouldn't make any sense if the expansion vanishes completely on small scales. However, when the expansion we are talking about is caused by entropy growth in black holes, then there is no reason to assume that space is expanding anywhere else than around black holes. In addition to that, space wouldn't always expand, but only when black holes swallow matter.

Along this line of reasoning it is natural that no expansion of space was measured between us and the Pioneer probes, because there are no black holes in between to cause such an expansion.

For proving the existence of non-homogeneous expansion it would be enough to detect local expansion at least somewhere. The missing expansion elsewhere would then already show that it isn't homogeneous.

If we could maintain contact with the Pioneer probes forever, and if we had enough time to wait till their signals exhibit the expansion of space, then they would be perfect for measuring it. Yet it is very unlikely that there are any active black holes near our solar system. Probes send out from earth are therefore probably not going to ever detect local expansion.

Another approach would be to look for redshifts in stars within the Milky Way. If the expansion of space is non-homogeneous, such redshifts would not occur uniformly for all stars in a certain area. Therefore it might be hard to distinguish such redshifts from redshifts caused by a local movement of the star.

Much more promising could be to search for fluctuations in the redshift of stars or galaxies. Such fluctuations would prove a non-homogeneous expansion regardless of the distance at which they occur.

The larger the distance is we are looking at, the higher is the chance to detect an expansion of space. Yet that doesn't mean the expansion of space must become more homogeneous on larger scales. If the expansion of space radiates away from black holes in a similar way as gravitational waves do, then it would be indeed plausible to expect a higher redshift for electromagnetic waves that come from active quasars. In order to test this, one could try to observe quasars and see if the redshift of light emitted from the centre is larger than the redshift emitted from the outer parts of the galaxy. Such a difference would mean that objects at the same intergalactic distance could have different redshifts and that would represent a definite proof for a non-homogeneous expansion.

We will re-examine the question of inhomogeneous expansion again in chapter 4.15 by conducting a thorough redshift survey on supermassive black holes.

Note:
1. Hans Jörg Fahr; Mark Siewert (2006): "Does PIONEER measure local space-time expansion?"; arXiv:gr-qc/0610034.
2. F. I. Cooerstock, V. Faraoni, D. N. Vollick (1998): "The influence of the cosmological expansion on local systems".

4.8 Maximal density at the border of the emergent past [29.11.2015 (critical entropy density); 10.12.2015 (no matter antimatter annihilation); 26.01.2016 (maximal density); 15.03.2019 (Egan & Lineweaver's estimate of the entropy density (2010) leads to the correct initial temperature; there is no 'critical entropy density'); 13.04.2019 (CMB & CNB have the same temperature)]

Note: Originally it was not clear how different entropic expansion is from the expansion in the standard model. Calculations in 2015 had shown that using general estimates on the entropy density of the universe, it must have begun at a temperature in the order of magnitude of 10^6 Kelvin. This was seen as a problem, because that is a temperature below the threshold for nucleosynthesis. In fact the extreme slowness of the expansion allows for spontaneous nucleosynthesis below threshold temperature to become relevant and to lead to an abundance of elements comparable to that predicted by the standard model.

However, the extended age of the universe we will get to know about in chapter 4.12 will put into question to what extend primordial nucleosynthesis is relevant anyway. In a very old universe stellar nucleosynthesis must have produced a large abundance of both light and heavy elements pretty much everywhere within interstellar space. Then in chapter 4.15 we will be conducting a thorough entropy density survey which will lead us to the conclusion that primordial nucleosynthesis never occured.

This chapter was originally full of speculations on how the entropy density of the universe could be higher than those independent entropy estimates suggest. Now that the estimates turned out to be correct, large parts of this chapter had to be deleted.

The entropic expansion of space particle dualism theory leads to a maximal density at the beginning of time. As mentioned before, the average entropy density of the universe was estimated (Chas A. Egan and Charles H. Lineweaver, 2010) to be

$$\rho_{S_{obs}} \approx 8.4^{+8.2}_{-4.7} \times 10^{21} \, k/cm^3$$

The number of photons in a volume in the early universe can be calculated as[1]

$$\rho_\gamma = \int_0^\infty \Delta N = 60.42198 \left(\frac{k\,T}{h\,c}\right)^3 = 20.28 \, [T(°K)]^3 \, photons/cm^3$$

$$\rho_\gamma = \frac{16\,\pi\,k^3\,\zeta(3)\,T^3}{c^3\,h^3} = 20,286,824.875 \, [T(°K)]^3 \, photons/m^3$$

The entropy per photon is about the magnitude of 1. Watching the universe backwards in time we would see all black holes shrinking. The maximal density can be reached when all black holes disappear. We can therefore set this equal to our value for the entropy density, and get

$$8.4 \times 10^{21} \, \gamma/cm^3 = 20.28 \, [T(°K)]^3 \, \gamma/cm^3$$

Now we solve this for T, and get

$$T_{max} = \sqrt[3]{\frac{\rho_S\,c^3\,h^3}{16\,\pi\,k^3\,\zeta(3)\,T^3}} = \sqrt[3]{\frac{8.4 \times 10^{21} \, \gamma/cm^3}{20.28}}$$

$$T_{max} = 7,454,246 \, K$$

This is below the threshold temperature for strong nucleosynthesis, which is 10^7 K, but still above the lowest temperature at which nucleosynthesis is happening inside the sun, namely 3 million Kelvin.
Using this initial temperature instead of the infinite temperature from which the universe in the standard model started seems to inevitably lead to less fusion products.

That would indeed be the case if the universe had started with a rapid expansion, which allows nucleosynthesis to take place for only a very short amount of time, but entropic expansion is different: it starts off very slowly and then speeds up continuously.

Occasional nucleosynthesis that happens because some nuclei are more energetic than others can usually be ignored, but it becomes significant when it happens over long periods of time. As we will see in chapter 4.12 the expansion was indeed extremely slow at the beginning.

This rather low maximal density and temperature also indicates that there was <u>no era of the universe where it consisted of light only</u>, as widely suggested. Having such an era would mean that when we look at the universe backwards in time stable protons and electrons must dissolve into pure light when high energies and densities are reached. That would mean they are unstable particles that decay after a certain time.
Since all attempts to detect the proton decay failed, driving its half-live up to a minimum of 8.2×10^{33} yr, I suggest that it is a stable particle. We shall therefore not aim to explain the matter dominance (as opposed to anti-matter) in the universe by a matter excess in a primordial matter-antimatter annihilation, but rather by assuming that electrons and protons are stable particles, which can survive from one eon to the other. In chapter 4.12 we will use the here derived initial temperature to derive the true age of the universe. Then, in chapter 4.14, we will be looking at causal relations between eons.

Is the cosmic neutrino background really colder than the CMB?

If the primordial matter-antimatter annihilation never took place, then the cosmic microwave background (CMB) cannot have been heated up by it, so that it can't be true that it is hotter than the cosmic neutrino background (CNB). Usually it is assumed that there are 40% less neutrinos than there are photons and that the CNB has a temperature of 1.95 K.

Despite from occasional single neutrinos, we can't really measure the CNB, so this assumed temperature of 1.95 K is pure conjecture.
If in fact matter didn't emerge out of a matter antimatter annihilation process, then the CNB should have the same temperature as the CMB.
Testing this experimentally will be challenging, but should be possible.

Note:
1. See [7].

4.9   Cosmic voids and dark matter galaxies [24.04.2016 (cosmic voids are filled with hypermassive black holes); 18.03.2018 (dark matter is 'cold'); 27.11.2018 (no observational constraints on the size of hypermassive black holes); 03.05.2019 (dead galaxies)]

Note: Same as the last chapter, this chapter was mainly speculating about how the entropy density of the universe could be higher than present estimates suggest. While it is clear now that an exceptionally high entropy density is not necessary, entropic expansion still requires cosmic voids to not be empty. This chapter is examining how cosmic voids form and what they contain.

The possibility that all dark matter is in black holes has often been put forward.
Back in 2015 it was assumed that it is absolutely necessary for the universe to begin from a very dense state, one that allows for primordial nucleosynthesis. It was regarded as a consistency requirement. That was coming along with a required very high entropy density. It was seen as an advantage of the theory that it does not only predict how much dark matter there should be, but also how much entropy it should have, implying certain mass distributions of black holes. These requirements seemed unrealistic, and are now dropped. As we will learn in chapter 4.12, entropic expansion implies very low expansion rates for the beginning of the universe, and thus it leads to a universe that is much older, changing much of the usual reasoning in these matters.

When we tried to calculate if the entropic expansion theorem holds true in our cosmic neighbourhood (chapter 4.1), we had to take the empty space around the Milky Way into account as well. That is because the empty space around pushes down the value for the entropy density. When we talk about a constant entropy density for the whole universe, we mean an average value that corresponds to the value we find when we go up to scales where the matter density per volume is also constant.
On the largest scales of the universe, on the scale of galaxy clusters, the overall structure of the universe becomes apparent. On this scale, the universe has a net-like structure, with huge empty regions, as we can see in Fig. 44.

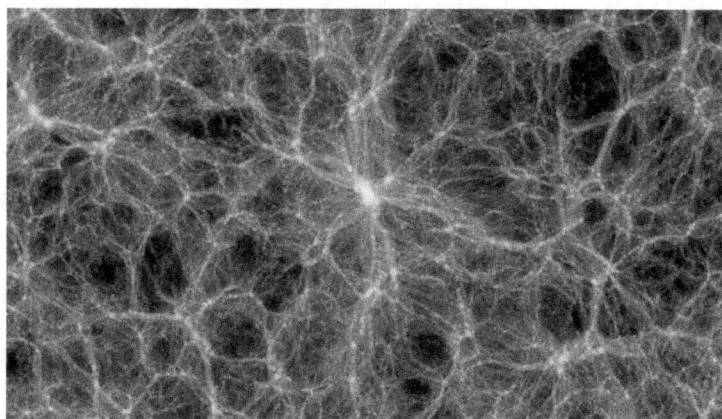

Fig. 44. The overall structure of the universe. Between the galaxy clusters there are huge empty regions called cosmic voids.

What created these huge empty regions of space? Following the entropic expansion scheme we would be more expecting galaxies clusters surrounded by a lot of empty space, but not a net-like structure.

Could it be that black holes much larger than the supermassive black holes known today are lurking inside those cosmic voids?

Hypermassive black holes

Such black holes would have almost no tidal powers. Things falling in would not be ripped apart even if they are quite large. There would be no accretion discs around them, and no jets at the poles. They would basically be 'silent eaters'.

Stars could disappear in them without any sign of 'struggle'; without being torn apart. Even if a star is torn apart slightly, without an accretion disc or jets it would probably go unnoticed. By surveying the sky one could hope to find stars that disappear in such *hypermassive black holes*. Yet one has to keep in mind that single stars are not easy to observe if they are located beyond our own galaxy.

However, a black hole that is so huge that it has a vanishing tidal force would swallow a lot of surround matter. That would cause a tremendous growth of its horizon, leading to a huge entropy increase as well as to the generation of gravitational waves.[1] Since 2015 we are finally able to detect gravitational waves. However, the larger the black hole is, the longer is the wavelength of the gravitational waves it sends out. The gravitational waves of black holes with cosmologic proportions might therefore be undetectable.

If cosmic voids harbour hypermassive black holes, then the expansion of space within them could be so enormous, that galaxies at their edges are pushed away and thereby prevented from entering them. Stars always form inside of galaxies, so it would be rare to see a star disappearing inside a hypermassive black hole located in a cosmic void.

Light rays on the other hand always traverse voids. It seems that if we assume black holes of cosmic dimensions to lurk inside these voids, then their presence should be obvious from the shadow they cast.

Do the considerations on the nature of black holes in chapter 4.3 change something in this picture? Indeed that seems to be the case: if black holes do not have absolute event horizons, then they can trap matter only, but not radiation. Radiation coming from the inside would become infinitely redshifted, but radiation from the outside could traverse such black holes. The only thing that could trap radiation from the outside would be being absorbed by the original star that formed the black hole and which is still there, hidden beyond the coherence horizon (see chapter 4.3).

If time dilation becomes infinite already before the critical circumference is reached, then the original star is filling up the whole volume of the black hole.

Therefore a black hole made from ordinary matter can't be traversed by light coming from the outside. Ordinary matter absorbs light and thus photons entering a black hole made from ordinary matter could never traverse it.

Dark matter on the other hand never absorbs light. It does not interact with ordinary matter except of through its gravity.

Since all dark matter must have condensed into black holes, it is natural to assume that the largest of them are residing in cosmic voids. If it is true that black holes don't have absolute event horizons, then cosmic voids could harbor dark matter black holes of cosmic proportions. The increased expansion rates in such regions would also give rise to larger galaxies with a lower gas density. Such galaxies would have shorter life-times, turning 'dark' earlier.

Without any solid body entering a void, it would be very hard to really find out the size of any black hole in it. Only the overall mass can be estimated then, and that doesn't allow for an estimate of the entropy in that region.

It could be that cosmic voids were formed by primordial dark matter black holes which were too big to be seeds for galaxies. They swallowed up all matter they encountered, without allowing matter to form stable orbits around them. Their growth was so enormous that the entropy increase in them led to such a tremendous expansion of space around them, that all the surrounding matter was either swallowed up or pushed away.

Stellar black holes have a limit for their growth. Once they reach a certain size, their accretation disks become so wide, that matter in it coalesces into stars that establish stabil orbits around it. According to recent findings, that happens with ultramassive black holes with masses above 50 billion solar masses.[2, 3, 4, 5]

However, that requires the black hole to be rotating. A black hole made of ordinary matter would always be rotating, but one that formed out of primordial dark matter would not, because dark matter particle's have no impulse (see chapter 4.10), and therefore also no angular momentum. It is thus possible for objects to fall into a dark matter black hole without being dragged around it.

Therefore growth limits don't apply to pure dark matter black holes. With this being said, there are no pure dark matter black holes. Primordial dark matter black holes always suck up ordinary matter and thereby become impurified. A primordial black hole would have to always consume more other primordial black holes than it consumes surrounding gas, only then it could maintain its 'exo-transparency' (see chapter 4.11) and its vanishing angular momentum.

Dead galaxies

It is sometimes suggested that altered laws of gravity might render dark matter unnecessary. Evidence against this sort of claim seems to be the discovery of dark matter galaxies, starting a decade ago.

Fig. 45. The galaxy VIRGOHI21 discovered by astronomers from Cardiff University is made entirely out of dark matter and ordinary hydrogen atoms. The hydrogen weights $10^8\ M_\odot$, while the invisible part is $1,000$ times heavier. So it is 99.9% dark.

Robert Minchin from Cardiff University and his team found a galaxy almost entirely consisting of dark matter. Unlike some other dark matter galaxies found later, this one was found mainly by the presence of circulating hydrogen gas which is gravitationally bound by invisible matter.

The search for dark matter galaxies was inspired by Neil Trentham from Cambridge University, who predicted their existence. The chances are good that there are many of these dark matter galaxies. And Michin himself commented the results with: "If they are the missing dark matter halos predicted by galaxy formation simulations but not found in optical surveys, then there could be more dark galaxies than ordinary ones".[6]

Many potential dark matter galaxies have been found before, but close observations always revealed some stars mixed in. This galaxy lacks stars because its density is not high enough for star birth.

Other candidates for dark matter galaxies are HE0450-2958 (a quasar with no visible host galaxy), HVC 127-41-330 (near Andromeda) and Smith's cloud (it passed through outer regions of our own galaxy).

As we will learn in chapter 4.13, all of the primordial dark matter that was around at the beginning of physical time (the border of the emergent past) must have been turned into well visible supermassive black holes. It is thus to conclude, that dark matter galaxies must be ordinary galaxies, which just happen to have used up all their gas necessary for star formation. Smaller galaxies use up their gas sooner than large galaxies. Considering that our own galaxy has already used up 90% of its initial interstellar gas,[7] it is not hard to imagine that there must have been other galaxies as well, which happen to use up their gas faster, and which turned dark entirely.

We can call such galaxies *dead galaxies*, analogous to dead stars.

It may be that cosmic voids are full of hypermassive dark matter black holes only, with no ordinary matter in them. However, considering the evolution of galaxies, it is much more likely that the gas density in those regions was lower, leading to galaxies with less gas, and shorter life-times, so that all galaxies in these voids have used up their gas and turned dark.

If 'dark galaxies' or 'dead galaxies' exist in ordinary regions of space, and if it is true that there are more dead galaxies than ordinary ones, then it would not be surprising if there are regions of space with dead galaxies only. Such regions would looks exactly like the cosmic voids we observe.

Gravitational lensing

Far more conclusive are surveys based on gravitational lensing. Strong gravitational lensing is even able to produce multiple images of the same galaxy. That happens when light is bended around a massive object under the influence of gravity. Thus, gravitational lensing can be used to conclude the mass in a certain area of space.
In many regions with gravitational lensing, it has been found that ordinary matter can account to only 10% of the effect; the remaining 90% are dark matter.
However, if most of this 90% is simply stellar black holes and black dwarfs (see chapter 4.13), then we don't need to expect such regions to contain unusually large black holes. Such regions did then simply used up their interstellar gas earlier than others. If galaxies that are observable with telescopes were anywhere near the border of the emergent past, then they would contain only supermassive black holes (made from actual dark matter) at their centers, and nothing 'black' or 'dark' anywhere else.

Fig. 46. Example for strong gravitational lensing.

The origin of cosmic voids

Why are cosmic voids so large and empty? If indeed galaxies running out of gas creates cosmic voids, then their existence seems natural. If instead we have to make extremely large quantum fluctuations at the beginning of time responsible for them, then we are basically resigning to viewing them as cosmic accidents or simply miracles.

Galaxies at a later stage of evolution, contain both larger stellar and central black holes. The entropy growth in black holes is exponential, thus regions of space with larger black holes expand faster.[8]

How cold is primordial dark matter?

What do other theories suspect behind dark matter?

Most physicists seem to suspect yet unknown weak interacting particles (WIP) behind it. Some decades ago, when the exact mass of neutrinos was unknown, some took neutrinos as dark matter candidates very serious. However, neutrinos travel always near the speed of light, and thus could hardly become gravitationally bound. The neutrino density throughout the universe should therefore be more or less constant. After the mass of neutrinos was found too small to account for dark matter, it had been given up as a candidate. Yet the same argument could be brought forward for any kind of WIP traveling near the speed of light.

What about WIPs which are heavier and slower? If they are slow, like baryonic matter, and they interact only and solely through gravity, on first sight there seems to be nothing to prevent them from forming black holes of all sizes.

But is that so? And if it is, how 'slow' would such dark matter particles have to be?

The problem with dark matter particles that move even just very slowly is that the do not repel upon each other and thus there is no way to stop such a movement, except of gravity, which has to be there in the first place.

Normal baryonic matter has temperature and pressure, but since dark matter can't press against anything, it has no pressure. Does temperature exist in absence of any pressure? If we can't measure the temperature of (primordial) dark matter through pressure, can we measure the impulse of individual dark matter particles?

If primordial dark matter really only interacts through gravity, then it is impossible to measure single dark matter particles. All we can measure are large accumulations of dark matter.

What does that say about the position uncertainty of dark matter particles?

It is surely close to infinity for all practical purposes.

We know that when particles are confined to a narrow space, they tend to exert what is called degeneracy pressure. The increased position certainty leads to an increased uncertainty in the particles impulse. That is dictated by Heisenberg's uncertainty principle. By analogy it is reasonable to assume that the high impulses of ordinary matter at the beginning of time were also caused by this confinement to very small volumes.

What can we expect to happen if the position uncertainty is almost infinite? If anything were to speed up or slow down dark matter particles, it certainly wouldn't be an expansion or contraction of space. To primordial dark matter the density of ordinary matter just doesn't 'matter'. If the uncertainty principle is truly what gave particles their initial impulses, then it is to conclude that dark matter particles did not have any impulse at all.

However, when they formed black holes, they reduced their position uncertainty, and that could have make them obtain a certain impulse. We can calculate that impulse as follows:

$$\Delta x \times \Delta p = \frac{\hbar}{2}$$

$$\frac{4\,G\,M_{BH}}{c^2} \times \Delta p = \frac{\hbar}{2}$$

$$\Delta p = \frac{\hbar\,c^2}{8\,G\,M_{BH}}$$

For a dark matter black hole of one solar mass we have a dark matter particle impulse of

$$\Delta p = 8.9 \times 10^{-39} \text{ Ns}$$

With one solar mass this corresponds to a speed of 4×10^{-69} m/s. This is far below the Planck length and thus it is safe to say that dark matter black hole can obtain impulse only through the in-fall of ordinary matter.

If dark matter is completely 'cold' as argued above, then it must all have condensed to black holes. The observational data seems to support this view: having almost empty regions with only 0.1% of the matter being visible, while in other regions visible matter dominates, indicates that dark matter isn't evenly dispersed but accumulates in certain regions. Hypermassive black holes could swallow intergalactic gas, stars and even other ordinary black holes without any visible effects. They could grow rather in secret.

In his popular book "bended space and curved time" Kip Thorne described how a black hole could be detected by the *lensing effect* it has on a star when passing in front of it. The book was from 1993, and Thorne stressed that it would be very unlikely to find any black hole by this method, because the universe would have to be "absolutely full of black holes" for such a coincident to ever occur. Yet, a decade later it really occurred: the star *MACHO-96-BL5* became brighter for a moment when an unseen object passed in front of it. The effect was in accordance with the lensing brightening effect.

In the mainstream small primordial black holes are ruled out by the absence of Hawking radiation, while hypermassive black holes are ruled out by the absence of large shadows in the night sky. In space particle dualism theory Hawking radiation doesn't exist and hypermassive black holes can be exo-transparent (see chapter 4.11) if they are foremost composed of dark matter. A way to spot them would be to search for star clusters or galaxies inside cosmic voids that slowly disappear without the presence of any visible black hole.

As mentioned earlier at the end of chapter 4.4 fluctuations in the density of the microwave background cannot be explained by quantum fluctuation. A quantum fluctuation needs a measurement in order to manifest, otherwise it will always remain in a superposition of having occurred and not having occurred.

Fig. 47. The Cosmic microwave background of the universe. Fluctuations in the temperature in the magnitude of 200 micro-Kelvin are made visible using different colors. According to the mainstream view these fluctuations represent quantum fluctuation in the density of the primordial particle soup at the Big Bang.
Roger Penrose pointed out that fluctuations can only become manifest when a measurement takes place (he used this as an argument for OR). According to space particle dualism theory these fluctuations can be explained by both primordial dark matter black holes and quantum cosmology as described in chapter 4.2.

In space particle dualism these fluctuation can readily be explained by the gravitational pull originating from primordial black holes, which immediately formed out of dark matter.

Additional observational evidence for primordial black holes comes from the recent discovery of gravitational waves originating from the border of the past. The border of the past is the time when dark matter melts together into endless many tiny black holes. Never again in the history of the universe did more black holes merge than at that time.[9]

However, inflation theorists see these gravitational waves as evidence for inflation theory. The inflationary expansion of space that is theorized to have occurred within fractions of the first second, would have created gravitational waves. Yet there is nothing that dictates how powerful inflation must be. Therefore no exact numerical prediction about the power of these waves can be made in inflation theory. In contrast to that, space particle dualism theory allows numerically exact predictions on the nature of these waves. This is yet another way to verify the theory.

Note:

1. Strong gravitational waves which are detectable with present technology are only produced when two black holes merge into one.

2. King, Andrew (February 2016). "How big can a black hole grow?". *Monthly Notices of the Royal Astronomical Society: Letters*. 456 (1): L109–L112. arXiv:1511.08502. Bibcode:2016MNRAS.456L.109K.

3. Trosper, Jaime (May 5, 2014). "Is There a Limit to How Large Black Holes Can Become?". *futurism.com*. Retrieved November 27, 2018.

4. Clery, Daniel (December 21, 2015). "Limit to how big black holes can grow is astonishing". *sciencemag.org*. Retrieved November 27, 2018.

5. Ap507. "Black holes could grow as large as 50 billion suns before their food crumbles into stars, research shows — University of Leicester". *www2.le.ac.uk*. Retrieved November 27, 2018.

6. [www.space.com/815-invisible-galaxy-discovered-cosmology-breakthrough.htm].

7. Interestingly mainstream cosmology also assumes that the Milky Way has already used up 90% of its initial gas. For 90% of the gas to be used in only a little more than one sun life-time, there would have to be very less available gas in the first place. From that we see that the universe in the standard model is not only very young but also extremely short lived. More on this in chapter 4.13.

8. It might even be the case that the expansion of space in cosmic voids absolutely dwarfs the expansion of space in our local group and that this is the true reason why it can only be observed at distances beyond the local group.

9. [https://www.scientificamerican.com/article/gravity-waves-cmb-b-mode-polarization/].

4.10 Initial masses of primordial black holes [15.06.2017 (minimal mass when using resting 'dark electrons', without accounting for tunneling); 07.12.2017 (black holes don't shrink); 09.12.2017 (accounting for tunneling); 12.12.2017 (using 'dark electrons' with high impulse); 18.03.2018 (primordial dark matter is 'cold'); 01.06.2018 (primordial dark matter doesn't tunnel); 09.05.2019 (minimal mass of ordinary matter primordial black holes using initial temperature from chapter 4.8)]

Note: This chapter was originally aiming to examine if there is a minimal mass for primordial black holes and if such a minimal mass can explain why we don't observe Hawking radiation. For achieving that this minimal mass would have to be above 4.5×10^{22} kg, which is close to the mass of the moon. At this initial mass primordial black holes are in thermal equilibrium with the cosmic microwave background and are thus unable to evaporate through the Hawking process.[1,2] After it was found that Hawking radiation cannot exist in a theoretical framework in which gravity is a side effect of charge (see end of chapter 4.3), the whole question became obsolete.

However, knowing the initial mass of primordial black holes is important for accurately describing entropic expansion, as we will see in chapter 4.12. It is also important when it comes to analyzing possible traces from gravitational waves in the microwave background.

Several theories predict gravitational waves at the Big Bang. Inflation theory predicts gravitational waves due to the inflationary expansion of space, but since nothing determines the exact strength of the inflationary expansion, any observed gravitational waves could be seen as 'evidence' for inflation theory.[3]
Penrose's CCC model predicts gravitational waves at the Big Bang which are leftovers from merging black holes in the last eon.
What about space particle dualism theory? As we saw in chapter 3.8, electrons have a diameter of 10^{-21} m, which is far above the Planck scale. That implies particles can never shrink below their Schwarzschild radius, and therefore a single particle can never form a black hole. If we want to know the strength of gravitational waves at the border of the past, we have to know the size of the first black holes. This size is also important for modeling the evolution from primordial black holes to the formation of galaxies.

How many electrons would we have to press together in order to get a black hole?
If we assume that the electron mass, or better, the naked electron mass, is a fundamental number, then this should also tell us how many 'dark electrons' we have to accumulate in order to create a black hole.
If the Schwarzschild radius of an electron is 10^{-57} m and its elementary space radius 10^{-21} m, then we need to accumulate 10^{36} of them in order to create a black hole.

The mass of an electron is $9.10938356 \times 10^{-31}$ kg and 10^{36} of them weight:

$$M_{DH_{min}} = 10^{36} \times m_e = 910{,}938 \text{ kg}$$

Replacing 10^{36} by the more precise term G_E/G we get:

$$M_{DH_{min}} = \frac{G_E \, m_e}{G} = 1{,}125{,}587.541 \text{ kg}$$

Such a black hole would be very small, having a radius of only 10^{-21} m. In order for a particle to be confined to a certain volume without immediately tunneling out of it again, its wavelength has to be smaller than that volume. When normal particles reach the impulse required to have a wavelength as short as 10^{-21} m, their elementary spaces have long grown over their wavelength, rendering the wavelength irrelevant for the localization of the particle.

However, here we deal with dark matter particles, and as argued in the last chapter, primordial dark matter is absolutely 'cold'. There is no impulse and therefore no meaningful wavefunction. All dark matter particles have is an elementary space.

If dark matter particles don't move at all, then their position uncertainty can be regard as infinite. Thus the minimal mass seems to be indeed as calculated above.

Minimal mass for primordial black holes made of ordinary matter

Differently from primordial dark matter, ordinary matter has temperature and pressure, making it harder for it to form black holes.

In today's universe the minimal mass for black holes is $2.17 \, M_\odot$, known as the Tolman-Oppenheimer-Volkoff limit (originally it was thought to be $2.5 \, M_\odot$), and that is because gravity needs to overcome the degeneracy pressure arising from the Pauli exclusion principle.

This is not the case for primordial black holes made of ordinary matter that are theorised to have been generated in the early universe. This is because such black holes would have to arise from irregularities in the matter density.

It is not very likely that sufficiently large irregularities existed in the early universe, so most likely all primordial black holes are dark matter black holes, or 'dark holes'. However, it is still interesting to see what minimal mass such ordinary matter primordial black holes would have and that is what we are going to look into now.

As we found in chapter 4.8, the initial temperature of the universe can be derived from the general entropy density. Our preliminary calculation there yielded $7{,}454{,}246$ K. The most likely impulse of electrons and other particles at a given temperature T is

$$p_p = k_B T$$

That corresponds to a wavelength of

$$\lambda_p = \frac{h}{k_B\,T}$$

$$\lambda_p = 0.0003577475 \text{ m}$$

Equating half of this with the Schwarzschild radius yields:

$$\frac{\lambda_p}{2} = \frac{2\,G\,M}{c^2}$$

$$M_{PBH_{min}} = \frac{\lambda_p\,c^2}{4\,G}$$

$$M_{PBH_{min}} = 1.205125256 \times 10^{23} \text{ kg}$$

The impulse p_p corresponds to an energy according to

$$E_p = \sqrt{p_p{}^2\,c^2 + m_0^2\,c^4}$$

And the diameter of the elementary space is thus

$$d_E = \frac{4\,G_E\,E_p}{c^4}$$

$$d_E = 3.343355713 \times 10^{-20} \text{ m}$$

This is much shorter than the particle's wavelength. We can therefore be sure that the above calculated minimal mass for ordinary matter primordial black holes is indeed correct.

If Hawking radiation doesn't exist, then small black holes can't be ruled out by surveys that try to detect this radiation. However, studies searching for primordial black holes using microlensing on stars, and the possible destruction of neutron stars through them seem to be ruling out even larger primordial black holes significantly contributing to dark matter. Yet if primordial black holes cluster in dense halos, then they couldn't be found by these methods.[4]
Because primordial black holes must have started clustering since the beginning of physical time, it is plausible to assume that almost all of them have been merged together into supermassive and *hypermassive black holes* (see chapter 4.9).
Supermassive black holes formed the seeds of galaxies, while hypermassive black holes would have had such enormous entropy growth rates that they pushed matter away from

them merely by the local expansion of space they caused, creating huge voids.
It might even be the case that no microscopic black holes at all can be left, if the initial masses were really that large (due to the mass minimum).
Computer simulations in the future using this initial mass may tell us if this is the case.

Future research will have to use this new minimal mass for black holes to predict the strength and nature of gravitational waves at the border of the past, as well as simulate the formation of galaxies.

Note:

1. Kumar, K. N. P.; Kiranagi, B. S.; Bagewadi, C. S. (2012). "Hawking Radiation – An Augmentation Attrition Model". *Adv. Nat. Sci.* **5** (2): 14–33. doi:10.3968/j.ans.1715787020120502.1817.
2. NASA - Dark Matter. Nasa.gov (28 August 2008). Retrieved on 16th November 2010.
3. The reader might have heard of recent claims from inflation theorists that gravitational waves from the Big Bang were detected. One may think that these were detected using the LIGO gravitational waves detector. In fact they were merely searching for patterns in the CMB that could indicate the presence of gravitational waves.
4. Clesse, S.; Garcia-Bellido, J. (2017). "The clustering of massive Primordial Black Holes as Dark Matter: Measuring their mass distribution with Advanced LIGO". *Physics of the Dark Universe* .10:142.Bibcode:2017PDU...15..142C. arXiv:1603.05234.doi:10.1016/j.dark.2016.10.002.

4.11 Black holes are eternally collapsing objects [02.11.2018 (how does gravity escape a black hole?); 27.11.2018 (dark matter black holes are exo-transparent); 28.11.2018 (magnetic fields are kept); 29.11.2018 (evidence for magnetic fields); 06.05.2019 (evidence for exo-transparency)*; 07.05.2019 (black hole mergers also partially exo-transparent)**]

*Evidence found by Brian Thomas Johnston and introduced to me by Wes Johnson and Savyasanchi Ghose (India).
**In collaboration with Thomas Southern (US).

This chapter is going to explore the differences between black holes in standard general relativity on the one hand and in space particle dualism on the other.

Black holes have magnetic fields

As we learned in chapter 4.3, black holes have only apparent event horizons, because time slows down infinitely shortly before a collapsing star can reach its critical circumference.

In general relativity a radical relativist standpoint is taken on this, arguing that although it may take an infinite amount of time for an object to enter an event horizon when viewed from the outside, from the perspective of the in-falling object everything is normal and the event horizon is trespassed in finite time. To take this relativist stand requires some rather radical assumptions about the behavior of space and time inside a black hole. It requires us to believe that space and time switch roles; space becoming time and time becoming space. Essentially falling into a black hole would mean to enter an orthogonal timeline.

As laid out in chapter 1.2, space particle dualism is refusing the spacetime block universe paradigm and holds that the passage of time is real, and not simply a psychological phenomenon arising from an unchanging sculpture-like spacetime continuum.

Recently there is an increasing number of physicists who point out that the usual view of black holes having absolute event horizons and containing singularities can't be right. While we relied on time dilation alone to argue against absolute event horizons, others have done this using additional arguments.

In 1998 Abhas Mitra argued that as a star collapses, the material of the star heats up more and more, approaching the so called *Eddington limit* at which outwards pushing radiation pressure brings the collapse to a halt. Mitra coined the term eternally collapsing object (ECO) for the resulting structure.[1, 2, 3]

As already mentioned in chapter 4.3, Laura Houghton quite similarly argued that Hawking radiation, which would be very intense when viewed from the frame of reference of the star itself, would create such a high outwards pressure, that it would stop and even reverse the collapse. Both assume that these processes are basically frozen in time when viewed from the outside, so that the observed structure is very similar to a black hole.

These additional arguments are important, because they lead the general relativist scheme ad absurdum.

Later on in 2003 Mitra's model was extended by Darryl Leiter and Stanley Robertson who pointed out that if no absolute event horizon forms out, then the collapsing star can maintain its original magnetic field. Mitra's ECOs became MECOs (magnetospheric eternally collapsing objects).[4]

In the standard model of black holes it is believed that black holes get rid of their magnetic field once they are formed, which had been shown mathematically by De la Cruz, Chase and Israel in 1970 and which was later further enforced by Price who showed that event horizons would radiate away all irregularities within the gravitational field; a result that further enforced Bekenstein's *no hair conjecture*.[5]

If a magnetic field is a direct way to see if black holes are in fact eternally collapsing objects, then is there observational evidence for that being the case?

It turns out there is such evidence and it was found by a team that includes the person who proposed the presence of magnetic fields in ECOs in the first place; Darryl Leiter. The magnetic fields were found to exist within a dense region inside a quasar which is believed to harbor a black hole.[6] Their abstract reads as follows:

"Recent brightness fluctuation and autocorrelation analysis of time series data and microlensing size scales, seen in Q0957+561A and B, have produced important information about the existence and characteristic physical dimensions of a new nonstandard magnetically dominated internal structure contained within this quasar. This new internal quasar structure, which we call the Schild-Vakulik structure, can be consistently explained in terms of a new class of gravitationally collapsing solutions to the Einstein field equations that describe highly redshifted Eddington-limited magnetospheric eternally collapsing objects that contain intrinsic magnetic moments. Since observations of the Schild-Vakulik structure within Q0957+561 imply that this quasar contains an observable intrinsic magnetic moment, this represents strong evidence that the quasar does not have an event horizon."

This is not the only way ECO are different from the black hole standard model. Magnetic fields are already a gross violation of the no-hair theorem, but as we examine ECO further, we find further differences and even grosser violations.

Classical black holes are very simple objects and from a classical perspective, they are elegant, but looking at them from a quantum perspective, various question arise which wouldn't come up in a strictly classical picture. One of them is:

How does gravity escape a black hole?

We know that photons emitted right above a classical event horizon can reach to the outside universe, they are just too red-shifted to ever be measured and therefore remain in a superposition of having and not having been emitted. This happens all the way up the border of the *coherence horizon* which lies at 2.2 of the Schwarzschild radius (see chapter 4.3).

Yet, since the event horizon, or a quasi-event horizon, exists only for outside observers, photons from any place within a black hole can escape to the outside.

It may seem that the infinite red-shift means that nothing really escaped, however, while for real particles we need wavefunctions to collapse in order for something to have effectively happened, that is not the case for virtual particles.

Gravity is based on a stream of virtual photons and gluons that is emitted by charged and hyper-charged particles. Those particles maintain their gravitational effect even if their prior charge effect is cancelled out. Their reach is infinite, because they have no mass and therefore there is no lower limit to their energy. That means that even near infinitely red-shifted virtual photons and gluons mediate gravity (as a side effect).

In general relativity it is not specified how gravity works on the quantum level. If we for example imagine space to be made of some kind of spin foam, as assumed by Loop Quantum Gravity, then the gravitational field would not be dependent on having virtual particles escaping from the black hole.

Virtual particles escaping black holes carrying their gravity to the outside universe is certainly interesting from a theoretical perspective, but it isn't something that leads to new predictions. What about real measurable particles, is there any case in which they can escape a ECO?
The answer to this question leads us to dark matter black holes.

How to distinguish exotic matter black holes from ordinary black holes

If black holes are in fact ECOs (eternally collapsing objects), that have only apparent event horizons, then light should be able to pass through them. Light they emit on their own is infinitely redshifted and light reflected by in-falling objects is also infinitely redshifted, but light rays that traverse the black hole would be subject to as much redshift when they exist as they were subject to blueshift when they entered.

If this is true, then one can see self-illuminating objects only from the opposite side of the black hole.

It would have mind boggling consequences for the size of black holes that are conceivable. If light can pass through them, then we can't rule out *ultra-massive black holes* lurking inside cosmic voids, because in that case such black holes would not be huge spherical dark spots.
They could swallow whole stars without forming an accretion disc and jets.

But is it really the case? If a black hole is a collapsing star frozen in time, wouldn't traversing photons get stuck inside the star bouncing back and forth inside it till they get lost? Yes, they would, if the black hole is composed of ordinary matter.

Dark matter doesn't interact with ordinary matter, so photons could traverse a dark matter black hole. If this is true, then dark matter black holes are truly remarkable objects being the only objects that are externally transparent, or *exo-transparent*; the inside being invisible, but self-illuminated objects on the other side shine through.

If we can distinguish between black holes made of ordinary matter and of dark matter, then it doesn't seem very right to speak of 84% dark matter, when in fact all that has increased since the beginning of time is ordinary matter black hole mass. Maybe we should rather speak of 60% dark matter and 84% black hole matter, with the latter including the former.

Most dark matter black holes have accumulated ordinary matter around them, forming galaxies. They have since grown much further, so that they already contain large amounts of ordinary matter, making them less exo-transparent. The only place to find pure dark matter black holes is potentially in cosmic voids.

Is there any evidence for this type of exo-transparency?
On the 10th April 2019 the first image of a black hole was released by the Event Horizon Telescope team lead by Katie Bouman. The release of the image was announced long time in advance. It shows the supermassive black hole *M87* (see fig. 48).

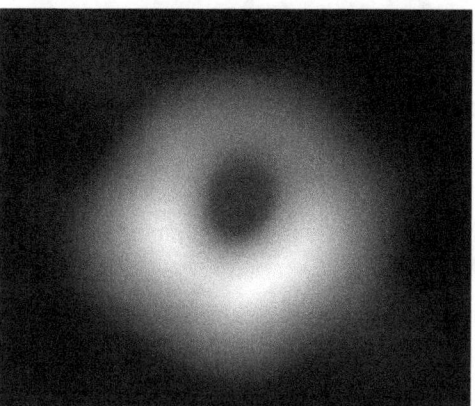

Fig. 48. At first sight the released image of M87 seems to show an ordinary black hole that fits the predictions of GR.

The image was created by combining the data from telescopes at six different locations on the globe, creating a virtual telescope with the size of the earth.
It is to be expected that such an image is a bit blurry. The bending of light around the black hole extends the dark area to about a factor of 5.3. This extended dark area we can call the 'shadow' of the black hole. The bending of light also means that things from behind the black hole can be seen from the front. However, they would always appear to be located besides the dark area. Is this what was observed?

The astronomer Brian Thomas Johnston analysed the original unedited images of *M87* and found that they do not agree with this ordinary view on how black holes should behave.[7] In his analysis he showed some of the original 'untrimmed' images (see fig. 49).

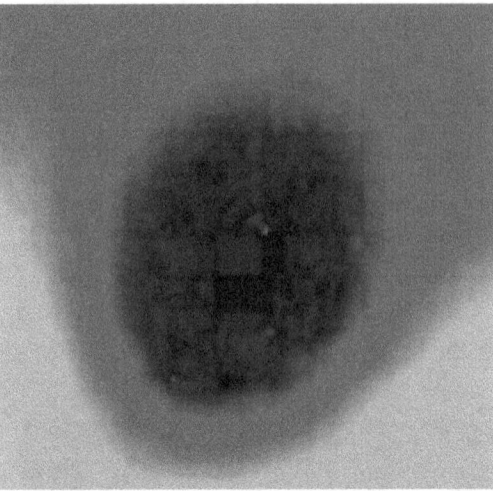

Fig. 49. Four clusters of objects apparently shining through the black hole. Is this direct evidence that M87 is a dark matter ECO, a DECO?

According to Johnston the black area in the released images was artificially made to look more black.

Apparently there are various objects to be seen within the dark region. As we zoom in, various objects become visible (fig. 50 & 51). The bottom left object shows very distinct spherical objects. Algorithms that were employed to create structure out of messy data was employed a lot in the event horizon project, however, that was always directed at the image as a whole, and not tiny portions of it. Therefore those points seem to be real objects, and not random background noise, which would have been filtered out anyway at a much earlier stage of processing.

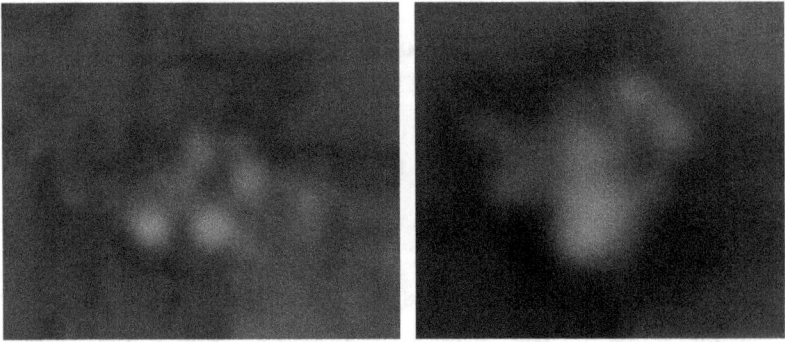

Fig. 50 & 51. This is what we see if we zoom in on the object below the center (left image) and the center (right image).

As Johnston points out, the object in the very middle of the dark area does not appear only on the original images of the event horizon telescope data, but also on photos from the orbit-based X-ray telescope *Chandra*.

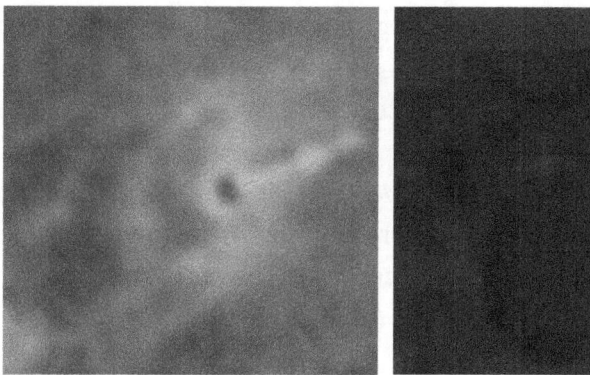

Fig. 52. & 53. On the left we see an image of the M87 taken by chandar. On the right side the dark region is magnified. The same central object as in the EHT reappears here.

As Johnston points out, the same object appears on images obtained from the orbit-based Chandar telescope as well, making it very implausible that it is a random fluctuation or a measurement error.

What procedure did the event horizon telescope team employ to produce the released picture? According to Katie Bouman it involved the following steps:

Step-1: Synthetic data tests
Various algorithms were tested on simulated objects, with the aim to find algorithms that could reproduce the image out of messy data without distorting it.

Step-2: Blind imaging of M87
Different groups of people used different algorithms to reconstruct the image of M87. Some of the used algorithms would blur out the whole picture, and those were sorted out.

Step-3: Objectively chosing parameters
This step was mainly about comparing imaging pipelines with different algorithms. That would for example involve training on disks, trying to avoid algorithms that turn a disk into a ring. Then judgements were made on what are believable images (Bouman used the term 'fiducial images').
In order to make the different images more similar to each other, they were all blurred out to a certain extend.

Step-4: Validation
A radius for the photon sphere was derived by averaging over all images. This was then used to derive the black holes mass. The derived mass fitted the constraints on the mass obtained by other methods, confirming what we know of the gravitational dynamics around black holes.

The blurring out was probably the most problematic step in this imaging process. In Bouman's own words: "We blurred them (the images) to a level such as that they were all consistent to some normalized cross-correlations.".

To what percentage does the black hole *M87* consist of primordial dark matter? If 60% of the matter at the border of the emergent past was primordial dark matter, then supermassive black holes should indeed consist of dark matter mainly, with only a little bit of ordinary matter mixed in.

Do the originals of the pictures rendered from the various telescopes all over the globe, show stars or galaxies shining through the black hole?

We may be inclined to think that this is an artefact of overlapping pictures or the data processing, but then the central object should not have appeared in the Chandar picture as well, because that was not created by a sophisticated algorithm.

Obviously the objects seen inside the dark area, cannot be in front, otherwise they would shine bright, and wouldn't appear dimmed out. It could be stars that shine through from the other side. This would confirm the prediction of space particle dualism, which was made already before the release of the EHT images, that supermassive black holes are partially exo-transparent.

Another idea, put forward by Thomas Southern, is that *M87* is not one single black hole, but several black holes prevented from totally merging through time dilation.
This is very much in line with the argument put forward here for black holes being ECOs. It would mean that light could not only penetrate dark matter black holes, such as those at the centers of galaxies, but black hole mergers as well.

The images of M87 are taken from the top, looking at the accretion disk from above. This means if those objects are stars, then they are not stars that orbit M87, and thus they should not be moving quickly.
Once we are able to observe Sagitarius-A in a similar way, we might be able to see stars that disappear behind the black hole reappear within the black hole shadow.
Currently an object is temporarily obscuring our sight to sagitarius-A. Once the sight is cleared, we will know more and with more certainty.

Shape of the coherence horizon and the black hole shadow

Kerr-black holes (rotating black holes) have a spheroidal horizon that is called the 'ergo-sphere'. This is the area around a rotating black hole where the frame dragging effect is so strong, that even light is dragged around the black hole. Thereby it is operationally very similar to an event horizon. The frame dragging effect exists in space particle dualism as well, so there is no difference here to expect.

What about other deviations from a spherical shape? What is the shape black holes in space particle dualism theory? Since black holes do only have apparent horizons with their only strict border being the coherence horizon (see the chapters 4.3 and 4.11), they do not radiate away irregularities like magnetic fields or bumps in the shape of their horizon.

Does that mean we have to expect highly irregularly formed event horizons? We can approach this question by looking at neutron stars.

A neutron star is the end stadium of a star that was not quite heavy enough to become a black hole. Neutron stars can rotate up to 766 times per second, yet the highest 'mountain' on their surface does not exceed 1 cm.

Through constraints from gravitational wave data we know that even the fastest neutron star cannot be deformed by more than $10 - 100$ cm.[8]

The formation of an absolute event horizon in the standard scheme eradicates all irregularities in the shape of a black hole. Accordingly a spheroidal object that collapses to a black hole would have to become perfectly spherical.

In space particle dualism a single collapsing star would lead to the formation of a very spherically symmetric black hole, however, two black holes merging into one would freeze in time well before merging completely. Therefore such black hole mergers could indeed have irregularly formed horizons. Since horizons are always surrounded by large shadows cast by bent light, it is unlikely to find evidence for such an irregularly formed join horizon. Much more likely it is to find evidence for it through light rays that pass in between the two merging black holes.

Note:

1. Mitra, A. (1998). "Final state of spherical gravitational collapse and likely sources of Gamma Ray bursts". arXiv:astro-ph/9803014.
2. Mitra, A. (2000). "Non-occurrence of trapped surfaces and black holes in spherical gravitational collapse: An abridged version". *Foundations of Physics Letters*. 13 (6): 543. arXiv:astro-ph/9910408. doi:10.1023/A:1007810414531.
3. A. Mitra, *Foundations of Physics Letters*, Volume 15, pp 439–471 (2002) (Springer, Germany). "On the final state of spherical gravitational collapse".
4. Leiter, D.; Robertson, S. (2003). "Does the principle of equivalence prevent trapped surfaces from being formed in the general relativistic collapse process?". *Foundations of Physics Letters*. 16 (2): 143. arXiv:astro-ph/0111421. doi:10.1023/A:1024170711427.
5. De la Cruz, Chase & Israel (1970); Price (1972).
6. Rudolph E. Schild, Darryl J. Leiter, and Stanley L. Robertson (2006). "Observations Supporting the Existence of an Intrinsic Magnetic Moment inside the Central Compact Object within the Quasar Q0957+561". *The American Astronomical Society*.

7. Brian Thomas Johnston (2019); "A Complete Analysis and Enhancement of the Image of the Core of M87: The Event Horizon Image.".

8. Nathan K. Johnson-McDaniel (2018). "Gravitational wave constraints on the shape of neutron stars". *Theoretisch-Physikalisches Institut, Friedrich-Schiller-Universität.*

4.12 The age of the universe [15.02.2019 (Hubble evolution); 18.02.2019 (age of the universe basing remote galaxy redshift); 05.03.2019 (age of the universe using CMB redshift); 06.03.2019 (irregular H in first second of standard model; H_{min} value derived from a Planck-temperature derived t_{min}); 10.03.2019 (distances of galaxies according to SPD; S_{min}; H_{min} value derived from M_{min}); 11.03.2019 (A_{min}; t_{min}; refined Hubble evolution and time-redshift equation); 12.03.2019 (corrected age of the universe; equation for maximal density); 13.03.2019 (corrected age of the universe using the maximal density); 14.03.2019 (extremely slow primordial nucleosynthesis); 16.04.2019 (we observe only 1% of cosmic history); 06.05.2019 (age of the universe based on the initial temperature from chapter 4.8); 28.06.2019 (more precise T_d); 18.03.2021 (corrected full precision time-redshift equation); 26.03.2021 (using more prec. figure for seconds in year)]

Note: This chapter is about deriving both the age of the cosmic microwave background (CMB) and the full age of the universe. However, the latter depends on the entropy density of the universe, which in turn depends on the SPD-specific intergalactic distances. In this chapter we don't account for these changes, so that from the two figures we derive here, only the first one can be trusted, namely the age of the CMB. The correct figure for the full age of the universe is derived in chapter 4.15.

In chapter 4.1 we learned that the universe expands according to a second square function. This means we only need to know the proportionality factor in order to calculate the value of H at any given time t. This can be done using the ratio between the present value of H, namely H_0, and the age of the universe t_0:

$$H = \frac{H_0}{t_0^2} \times t^2$$

When we enter $380,000\,\mathrm{yr}$ for t, which according to the standard model of cosmology is the age of the universe when it became transparent, we get:

$$5 \times 10^{-8}\, \frac{\mathrm{km}}{\mathrm{s}}/\mathrm{mpc}$$

More close to the border of the past this number is even smaller. In mainstream cosmology things are very different: there the universe starts off with an H-value that is even bigger than the present value. It is then assumed that the expansion of the universe slowed down due to the influence of gravity and that dark energy gained the upper hand somewhat 8 billion years after the Big Bang, accelerating the expansion. It is very convenient hereby that the supernova data that shows this acceleration goes back only 4.8 billion years, so that one can assume a slowing down phase for times we have no data for.

According to relativistic cosmology the universe started off with an H-value that is about 17 times higher than the present value.[1] This shows how radically different entropic expansion is from the type of expansion we know from the Friedmann models. In space particle dualism theory the universe starts off with an H-value close to zero and expands continuously, always with the same rate of acceleration.

In fact in this model the universe expands so slowly near the 'Big Bang' that we can hardly continue to call it 'Big Bang'; it is more appropriate to call it *slow drift, big slowness*, border of the (emergent) past or beginning of (physical) time.

Nothing special happens at this moment, except that black holes reach a minimal size while density and temperature reach a maximum ('reach' in a future to past sense).

With such a slow initial expansion, wouldn't it take longer for the universe to expand to its present size?

We can figure out how long it took the universe to expand from some early state to its present state by taking the redshift value of radiation that was sent out at that early state and set it equal to an age-redshift equation. We can obtain this equation by taking the above formula for H and form the sum of all the different expansion rates that act upon the wavelength of the photons in this radiation at different times, divided by the speed of light. We can then set this equal to our redshift of interest z_{t_0}:

$$\sum_{t=0}^{t=t_0} \frac{\left[\frac{H_0}{t_0^2} \times t^2\right]}{c} = z_{t_0}$$

H is describing the apparent speed of galaxies when viewed from earth. It tells us how much faster they appear the further away they are. It is usually given in km/s per mpc (megaparasec). It would be acceptable to use different values for H only for every subsequent year in the history of the universe, however, if we want to describe events close to the beginning of this eon, the 'Big Bang' in the usual terminology, we will have to use seconds instead of years.

According to latest measurements[2] the present value of H is

$$H_0 = 74.03 \, \frac{km}{s} / mpc$$

One mega-parsec equals 3,261,633.44 light years. Dividing through this tells us the apparent (redshift) velocity after one year. It is 0.02 m/s. A year has 31,557,600 sec.

Dividing through this tells us the redshift velocity after one second, which is:

$$H_0 = \frac{74.03\frac{km}{s}}{mpc} = 7.19231386 \times \frac{10^{-10}\frac{m}{s}}{ls}$$

Now that we have the right units, we can use this value in our above sum, which we can solve for t_0:

$$\sum_{t=0}^{t=t_0} \frac{\left[\frac{H_0}{t_0^2} \times t^2\right]}{c} = z_{t_0}$$

$$\sum_{t=0}^{t=t_0} \frac{H_0 t^2}{t_0^2 c} = z_{t_0}$$

$$\frac{H_0 t_0 (t_0+1)(2t_0+1)}{6 t_0^2 c z_{t_0}} = 1$$

$$\frac{H_0 (2t_0^2 + 3t_0 + 1)}{6 t_0 c z_{t_0}} = 1$$

$$2t_0^2 + 3t_0 + 1 = \frac{6 t_0 c z_{t_0}}{H_0}$$

$$2t_0^2 + \left(3 - \frac{6c z_{t_0}}{H_0}\right) t_0 + 1 = 0$$

We can simplify the term in the middle by the following replacement:

$$k = 3 - \frac{6c z_{t_0}}{H_0}$$

Using this k our above equation becomes:

$$2t_0^2 + k t_0 + 1 = 0$$

$$t_0 = \frac{-k \pm \sqrt{k^2 - 8}}{4}$$

$$t_0 \approx \frac{-k \pm k}{4}$$

$$t_0 \approx \frac{-k}{2} = \frac{3\,c\,z_{t_0}}{H_0} - \frac{3}{2}$$

The highest ever measured redshift from a celestial object is $z = 11.09$ and belongs to a galaxy by the code name *GN-z11*. Using this z-value we obtain:

$$\Delta t_{GN-z11} = 1.3867713 \dots \times 10^{19} \text{ sec}$$

$$\Delta t_{GN-z11} = 4.39441335121 \times 10^{11} \text{ yr}$$

This is even longer than the full age of the universe in the standard model, which is only 1.38×10^{10} yr, yet it merely represents the time that passed by since the emergence of the oldest galaxy we have discovered yet.
We can expect the full age of the universe to be even longer than that.

In order to rewrite the history of the universe according to space particle dualism, we have to examine how much the different phases of the universe are extended by the slowness of this new type of expansion.

How much time passed since the universe became transparent? That is when the cosmic microwave background (CMB) was created. At the moment of its creation, it had a temperature of $2,900$ K.[3] Today its temperature is 2.725 K. We can know the z-value of the photons in the CMB by dividing the initial temperature by the temperature it has today and subtract 1. It yields $z = 1,063$.[4] When inserting this into our time-redshift equation from above, we get:

$$\Delta t_{CMB} = \frac{3\,c\,z_{CMB}}{H_0} - \frac{3}{2} \cong \frac{3\,c\,z_{CMB}}{H_0} = 4.2130107931545 \times 10^{13} \text{ yr}$$

What is the full age of the universe? We can know that by using the maximal temperature of the universe from chapter 4.8:

$$z_{Tmax} = \frac{7,454,246 \text{ K}}{2.725 \text{ K}} - 1$$

$$\Delta t_{Tmax} = \frac{3\,c\,z_{Tmax}}{H_0} - \frac{3}{2} = 1.08394292472439538 \times 10^{17} \text{ yr}$$

However, as we will see in chapter 4.15 the true entropy density of the universe is much lower than the one assumed in chapter 4.8, so the true initial temperature must be lower, and therefore the true full age of the universe is in fact much shorter than the figure above. It is in fact in the same order of magnitude as the age of the CMB.

For the sake of argument we will however go along with the above age and initial temperature for a while.

Primordial nucleosynthesis mainly takes place at temperatures of 10^9 K until 10^7 K. This doesn't mean that below that no primordial nucleosynthesis (PNS) at all takes place. It still takes place below 10^7 K, but on much smaller scales. In the sun the minimal temperature for nuclear fusion is 3 million Kelvin. Our maximal temperature is above that, but not much. Below these temperatures there is only occasional nucleosynthesis, due to particles that are at the rare ends of the energy distribution. This form of occasional below-threshold nucleosynthesis can usually be ignored, but with this enormous age for the universe, it might become relevant.

Another possibility is that it is not relevant and that would mean that there was no significant primordial nucleosynthesis at all. In this case the only relevant thing that happened before decoupling was the growth of primordial black holes.

Computer simulations will have to show if primordial below threshold nucleosynthesis can be ignored.

One may think that abolishing primordial nucleosynthesis or changing its duration would significantly change the expected abundance of elements in the universe, but is that so?

In the seventies it was assumed that primordial nucleosynthesis lasted for about 3 minutes. Nowadays mainstream cosmology assumes it to have lasted for 20 minutes.[5] From this we can already see that observations do not give us any strict constraints for the duration of primordial nucleosynthesis.

In a young universe, primordial nucleosynthesis would have to have taken place; in an old universe however, there must have been so many star generations already, that the universe must be full of both light and heavy elements pretty much regardless of primordial nucleosynthesis.

Even if below threshold nucleosynthesis is relevant in our old universe model, it can't lead to the synthesis of heavy elements, because in an environment with a lot of highly energetic photons and a not so high density, heavier nuclei get destroyed quickly. That is why below threshold primordial nucleosynthesis can't lead to the synthesis of heavy elements even when it goes on for much longer than the primordial nucleosynthesis in the conventional Big Bang model. We don't have to expect it to produce more than hydrogen-1, helium-4, deuterium, helium-3 and trace amounts of lithium. In stars helium can be converted into carbon, but that takes tens of thousands of years.

In stars a rather high temperature and pressure is maintained during all of their life-time. When they can't keep up their temperature through nuclear fusion, their core collapses and the outer layers explode outwards, getting ejected and dispersed over the surrounding space. During primordial nucleosynthesis the temperature was not constant but instead very slowly dropping. Nucleosynthesis is not something that starts and stops abruptly. In order for the synthesized new elements to not be destroyed by radiation, nucleosynthesis must be faster than the destruction through radiation. Only when a certain 'safe' temperature is reached, can the final abundance of the different elements be assessed.

Given the extremely slow expansion rate in this model, it is to expect that significant amounts of Helium and Deuterium are produced.
The model could explain why less lithium is found in the universe than predicted by the standard model, a problem that is known as the *cosmological lithium problem*.[6] Synthesis up to lithium would certainly need regular fusion temperatures.

Further research needs to be conducted to determine the abundance of elements this type of vastly extended primordial below threshold nucleosynthesis leads to. Most likely this has to be done using computer simulations.

The fact that the most precise survey of the universe up to date, the *Sloan Digital Sky Survey*, didn't find any curvature within the observable universe means that there is no evidence whatsoever that the Friedmann models have anything at all to do with the real world. We therefore have to take radically different models very serious.

Entropy growth in black holes and the initial expansion rate

Our equation for the time dependence of the Hubble parameter H did allow H to become zero when $t = 0$. The expansion rate and the density are reciprocally proportional and so if one goes to zero, the other must go to infinity. Higher densities correspond to higher temperatures and as we found in chapter 4.8, there is a maximal temperature of $7,454,246$ K for the universe.

If the universe expands to compensate the entropy increase inside supermassive black holes, then the equation that describes the entropy increase over time inside a supermassive black hole must be analogous to the increase of the expansion rate over time, so we can write:

$$S = \frac{S_0}{t_0^2} \times t^2$$

The supermassive black hole we are looking at might be harbored by a very distant galaxy, so that the corresponding age of the universe t_0 is different, depending on the z-value of the galaxy, so we may write:

$$S = \frac{S_z}{t_z^2} \times t^2$$

We recall from chapter 4.10 that the mass minimum for primordial black holes, which are dark matter black holes, is 1,125,587.541 kg. Calculating the entropy of such a minimal mass black hole gives us the minimal entropy of a black hole:

$$S_{min} = \frac{c^3\, k\, A}{4\, \hbar\, G} = \frac{k\, 4\, \pi\, G\, M_{min}^2}{\hbar\ c} = 463,746.37455 \text{ J/K}$$

In order to prevent S to reach zero at $t = 0$, we can add this minimal entropy into its equation:

$$S = \frac{S_z - S_{min}}{t_z^2} \times t^2 + S_{min}$$

At $t = 0$ the left term disappears and leaves only S_{min}, as it should.

What is the value of S_z/t_z^2 ? Certainly it will be different for different galaxies.
To make an extreme example, the galaxy *S50014+81* has a central black hole with a mass of 40 billion solar masses and is located at a z-value of 3.366.
This z-value corresponds to a time 133,377,780,934 years before the present, seemingly a rather large difference in time, but compared to the full age of the universe it is short, so we don't need to bother subtracting this from it. When dealing with such an old universe, all galaxies we see around us, even the most distant ones, can be regarded to exist in roughly the same era.

The entropy of this supermassive black hole is:

$$S_{S50014+81} = 1.157844 \times 10^{75} \text{ J/K}$$

The average entropy growth is:

$$\Delta S_{S50014+81} = \frac{S_{S50014+81}}{t_0} = 1.038197882 \times 10^{58}\ \frac{\text{J}}{\text{K}}/\text{yr}$$

The increase in entropy growth over time is:

$$a_{S_{S50014+81}} = \frac{S_{S50014+81}}{t_0^2} = 9.309154282 \times 10^{40}\ \frac{\text{J}}{\text{K}}/\text{yr}^2$$

In seconds that is:

$$a_{S_{S50014+81}} = 2.951913458 \times 10^{33}\ \frac{\text{J}}{\text{K}}/\text{s}^2$$

In order to find the average entropy increase rate we have to find the average mass of supermassive/primordial black holes.

Using a sample of 129 supermassive black holes[7] gives an average mass of 5,757,999,302 sun masses.[8]

If every increase in entropy inside black holes is levelled out by a corresponding expansion of space, then it must be true that:

$$\frac{H_{min}}{H_0} = \frac{S_{min}}{S_0}$$

And therefore:

$$H_{min} = \frac{S_{min} \, H_0}{S_0}$$

Using entropy is a bit cumbersome. In fact the entropy depends on the square of the mass only, so we can simplify the above equation as follows:

$$\frac{H_{min}}{H_0} = \frac{M_{min}^2}{M_{DH\mu}^2}$$

$$H_{min} = \frac{M_{min}^2 \, H_0}{M_{DH\mu}^2}$$

Using the above average mass of supermassive black holes for M_0 this yields:

$$H_{min} = 2.748425505 \times 10^{-17} \, \frac{\text{m}}{\text{s}} / \text{ls}$$

This initial H_{min} essentially means that our $H(t)$-graph moves to the left, so that all H-values below H_{min} represent negative t-values. The time required for H to climb up from the value it has at $H = H_0/t_0^2 \times 1$ to the value H_{min} is now missing. In order to account for that we can simply subtract H_{min} from H_0:

$$H = \frac{H_0 - H_{min}}{t_0^2} \times t^2$$

In order to avoid $H = 0$ when $t = 0$ we add H_{min} to the equation, because that is the initial value of H:

$$H = \frac{H_0 - H_{min}}{t_0^2} \times t^2 + H_{min}$$

The equation we used for the age of the universe was derived from the above equation, but without the correction terms. We want to now re-derive our time-redshift equation using our corrected Hubble evolution equation:

$$\sum_{t=0}^{t=t_0} \frac{\left[\frac{H_0 - H_{min}}{t_0^2} \times t^2 + H_{min}\right]}{c} = z_{t_0}$$

$$\sum_{t=0}^{t=t_0} \left[\frac{(H_0 - H_{min})\, t^2}{t_0^2\, c} + \frac{H_{min}}{c}\right] = z_{t_0}$$

$$\frac{(H_0 - H_{min})\, t_0\, (t_0 + 1)\, (2\, t_0 + 1)}{6\, t_0^2\, c} + \frac{t_0\, H_{min}}{c} = z_{t_0}$$

$$\frac{(H_0 - H_{min})\, (2\, t_0^2 + 3\, t_0 + 1)}{6\, t_0\, c} = z_{t_0} - \frac{t_0\, H_{min}}{c}$$

$$2\, t_0^2 + 3\, t_0 + 1 = \frac{6\, t_0\, c}{H_0 - H_{min}} \left(z_{t_0} - \frac{t_0\, H_{min}}{c}\right)$$

$$2\, t_0^2 + 3\, t_0 + 1 = \frac{6\, c\, z_{t_0}}{H_0 - H_{min}}\, t_0 - \frac{6\, H_{min}}{H_0 - H_{min}}\, t_0^2$$

$$\left(2 + \frac{6\, H_{min}}{H_0 - H_{min}}\right) t_0^2 + \left(3 - \frac{6\, c\, z_{t_0}}{H_0 - H_{min}}\right) t_0 + 1 = 0$$

$$t_0 = \frac{-\left(3 - \frac{6\, c\, z_{t_0}}{H_0 - H_{min}}\right) + \sqrt{\left(3 - \frac{6\, c\, z_{t_0}}{H_0 - H_{min}}\right)^2 - \left(2 + \frac{6\, H_{min}}{H_0 - H_{min}}\right)}}{2\left(2 + \frac{6\, H_{min}}{H_0 - H_{min}}\right)}$$

Under the root the right side term is dwarfed by the left side term, and so our equation simplifies to:

$$t_0 = \left(\frac{6\, c\, z_{t_0}}{H_0 - H_{min}} - 3\right) \Big/ \left(2 + \frac{6\, H_{min}}{H_0 - H_{min}}\right)$$

The value of the denominator is 2.00000022928. If we round that down to 2, then we are left with:

$$t_0 \approx \frac{3\,c\,z_{t_0}}{H_0 - H_{min}} - \frac{3}{2}$$

Which is close to our original approximation, namely:

$$t_0 \approx \frac{3\,c\,z_{t_0}}{H_0} - \frac{3}{2}$$

The value of the different terms in the precise version of the time-redshift equation is:

$$\frac{(z \times 2.4992286 \times 10^{18}\ \text{sec}) - 3\ \text{sec}}{2.00000022928}$$

If we remove the insignificant '− 3 sec', we can turn the whole equation into just z with a time scale factor:

$$t_0 = z \times 39{,}625{,}004{,}647\ \text{yr}$$

Entering the redshift value z_{CMB} of the cosmic microwave background yields:

$$\Delta t_{CMB} = 4.2130104711679 \times 10^{13}\ \text{years}$$

That agrees with the value we got from the approximate equation in the first 7 digits. Our non-zero value for H_{min} makes the cosmic microwave background merely 3,219,866 years younger than it would otherwise be.

Our original approximation gives us an absolute maximum for the time scale factor, which is:

$$t_{0_{max}} = z \times 39{,}625{,}007{,}675\ \text{yr}$$

That agrees with the precise scale factor in the first 7 digits.

How much of the universe have we seen so far?

If the CMB is in fact 42 trillion years old, and the most distant ever observed galaxy *GN-z11* is 439,149,903,204 light years away, then the history of the transparent universe we have observed so far is merely 1.0430577% of its entire history.

Meanwhile we have been led to believe that what we are seeing is 97.104123157% of the universe's past.

If the universe was really only 13.8 billion years old, we would have to truly wonder why galaxies that supposedly are close to the beginning of time are so similar to the galaxies we observe in our direct neighbourhood.

True distance of various galaxies

In order to get a feeling for what all this means for the distances in the universe, we will make a list of important astronomical objects and structures together with their previously assumed distance and the distance given by the entropic expansion scheme:

Note: There are two values given for the actual distance, the first one is calculated using the approximation, which is the maximal distance (corresponding to $H_{min} = 0$), the second one is calculated using the precise time-redshift equation.

Name	Type	Redshift	Assumed distance in billion light years	Actual distance in light years
Hercules-Corona Borealis Great Wall	Supercluster	1.6 to 2.1	9.612 to 10.538	63,400,012,280 63,400,007,435 to 83,212,516,118 83,212,509,759
Hyperion proto-supercluster	Supercluster	2.45.	11.	97,081,268,804 97,081,261,385
GN-z11	Most distant galaxy	11.09.	13.4.	439,441,335,116 439,441,301,535
3C 273	Closest quasar	0.158.	2.443.	6,260,751,213 6,260,750,734
ULAS J1342+0928	Most distant quasar	7.54.	13.1.	298,772,557,870 298,772,535,038
Centaurus A	Closest radio galaxy	0.00182.	0.010 – 0.016.	72,117,514 72,117,508
TN J0924-2201	Most distant radio galaxy	5.19.	12.523.	205,653,789,833 205,653,774,118
Circinus Galaxy	Closest Seyfert galaxy	0.00142.	0.013.	56,267,511 56,267,507
Markarian 421	Closest blazar	0.030021.	0.397.	1,189,582,355 1,189,582,265
Q0906+6930	Most distant blazar	5.47.	12.3.	216,748,791,982 216,748,775,419

As can be seen in the above table, the discrepancy increases with the distance, reaching up to a factor of 39.625.

The bizarre expansion scheme in the first second of the standard model

If we imagine our exponential expansion to not be entropic for a moment, we can lay out how it would look like to have it starting from a singularity, just like the expansion in the standard model. How long would it in that case take the universe to expand from the so called singularity up to the actual maximal density?
A true 'singularity' would have the Planck-temperature, which is 1.417×10^{32} K. Dividing this through the current temperature of the CMB, gives us an imaginary redshift, which we can use to calculate our imaginary expansion time, yielding:

$$\Delta t_{im} = \left(\frac{3\,c\,z_{TPL}}{H_0} - \frac{3}{2}\right) - \left(\frac{3\,c\,z_{CNB}}{H_0} - \frac{3}{2}\right) = 2.0605004 \times 10^{42}\ \text{yr}$$

In the standard model this is supposed to have taken just 1.08 sec. We are led to believe that the temperature of the universe dropped from a staggering 10^{32} K to 10^{10} K in a single second (!). H is not supposed to change dramatically from one second to the next, so using the same rate of expansion and therefore temperature drop, the universe would have to have arrived at $T = 0$ K already in the second second.

What the standard model of cosmology is really doing here is quite bizarre from a scientific point of view. Instead of using any definite expansion rate H, particle physicists disguised as cosmologists have simply matched the different time scales to different energies in particle physics, using the famous time-energy uncertainty equation:

$$\hbar = \Delta E \times \Delta t$$

Within this line of thinking the temperature after a Planck-second must be the Planck-temperature, because radiation with this temperature would have its maximum at the Planck energy.
From this we can see that the whole history of the universe within the first second as narrated according to the standard model is simply particle physics sold as cosmology. Then after the first second passed, suddenly an ordinary expansion is assumed, supposedly because the universe somehow leaves the quantum realm by leaving the high energy sector.

Notes and references:
1. Using the Friedmann equations one arrives at an initial Hubble value that is 17 times higher than the present value (Bernhard Umlauf; personal correspondence; 2019).

2. Riess, Adam G.; Casertano, Stefano; Yuan, Wenlong; Macri, Lucas M.; Scolnic, Dan (18 March 2019), Large Magellanic Cloud Cepheid Standards Provide a 1% Foundation for the Determination of the Hubble Constant and Stronger Evidence for Physics Beyond LambdaCDM, arXiv:1903.07603, doi:10.3847/1538-4357/ab1422.

3. The figure one usually encounters is 3,000 K, but that looks like a rounded up number, and it indeed is: according to "Quark-Gluon Plasma: Theoretical Foundations: An Annotated Reprint Collection", the decoupling temperature T_d is 150 MeV, which is about 2,900 K (1 degree Kelvin is 0.0862 MeV). The precise value remains uncertain (according to this source), but the Graussian distribution around the mean is fairly narrow (David Kahana; 2017; informal correspondence).

4. Here again the usually given figure is somewhat imprecise, namely 1,100 K. It is a result of dividing 3,000 K by 2.725 K and subtracting 1. Using the more precise figure 2,900 K instead of 3,000 K leads to $z = 1,063$ instead.

5. Coc, Alain; Vangioni, Elisabeth (2017). "Primordial nucleosynthesis". International Journal of Modern Physics E. 26 (8): 1741002. arXiv:1707.01004. Bibcode:2017IJMPE..2641002C. doi:10.1142/S0218301317410026. ISSN 0218-3013.

6. R. H. Cyburt, B. D. Fields & K. A. Olive (2008). "A Bitter Pill: The Primordial Lithium Problem Worsens". Journal of Cosmology and Astroparticle Physics. 2008 (11): 012. arXiv:0808.2818. Bibcode:2008JCAP...11..012C. doi: 10.1088/1475-7516/2008/11/012.

7. The used list can be accessed here: [https://en.wikipedia.org/wiki/List_of_most_massive_black_holes].

8. There could be hypermassive black holes within cosmic voids, which would push the average mass of dark matter black holes further up, however, the H_0-value we measure is not the H_0-value one would measure from within a cosmic void, so we don't need to include them in our survey.

4.13 The status of energy conservation [26.11.2018 (temperature of the universe); 18.03.2019 (initial energies were sourced from the uncertainty principle); 06.04.2019 (average photon energy); 08.04.2019 (contribution from other particles; matching orders of magnitude; temperature of black holes)]

The relativity of energy in GR and the absolute frame of reference in QM

According to the theory of relativity, while every observer sees a different total energy within any given closed system, each frame of reference is regard as equally valid, so that a violation of energy conservation is avoided.[1]

In space particle dualism theory things are different: here there are several relations that define a preferred frame of reference:

1. Dark matter is theorized to be absolutely cold, so that it has an impulse of zero. As such it can be used as a reference system: if something is not moving relative to a pure dark matter black hole, then it is truly at rest.

2. The quantum vacuum provides the elementary spaces that form the general background space on which everything is moving. In principle it should be possible to measure redshifts and blueshifts within the quantum vacuum.

3. The wavefunction collapses simultaneous all over the universe. Here the question arises in what frame of reference it is simultaneous. It seems that if one observer measured a particle to be at the position x, it cannot be found at any other position. If we used the plane of simultaneity of a fast moving observer, we would have to believe that for resting remote observers the wavefunction of a particle he measures collapsed already thousands of years in advance.

4. As we saw in chapter 4.2 all wavefunctions in the universe are collapsed by conscious observers here on earth. Considering all the dynamics the earth has with other celestial bodies within the solar system, it seems counter-intuitive to regard it to be a fundamentally preferred frame of reference, but it is not entirely impossible. After all it was shown that the north-south asymmetry that is observed in the cosmic microwave background (CMB) can't be a Doppler effect caused by a local movement of the earth or the Milky Way relative to the CMB (see chapter 4.2).

5. In space particle dualism a superposition is defined an assembly of all classical states that are presently indistinguishable. This definition relies on an absolute frame of reference in order to define 'presently'.

6. Collective consciousness implies instantaneous information exchange over unlimited distances. This in itself defines yet another absolute frame of reference.

Having an absolute frame of reference has serious consequences for energy conservation:

1. In the frame of reference of any conscious observer the photons of the cosmic microwave background are losing energy. Photons leaving the gravitational field of a planet become redshifted too, but they do gain potential energy within this gravitational field. The photons of the CMB on the other hand lose energy regardless of where they are in respect to gravitational sources. At least we can say that their redshift is not at all related to their distance from gravitational sources.

2. Inside black holes (or ECOs) time dilation is almost infinite, which means that everything on the inside has lost its kinetic energy according to observers on the outside.

The universe started off with a high temperature and will end with black holes and cold emptiness.

For quantum mechanics to work, there must be a preferred frame of reference. In this frame of reference we have a violation of the conservation of energy for open systems at a cosmological scale.

This violation of the conservation of energy is a direct consequence of the expansion of the universe and the condensation of matter into black holes. We can characterize it by looking at what we may call the 'temperature of the universe'.

The temperature of the universe

Particles inside a black hole are never measured and their wavefunctions cease to collapse. Can we assign a temperature to them?

We know that particles have a greater impulse when they are confined to a smaller region of space. Particles inside a black hole are confined, but they are never measured again after they enter the black hole. We can therefore say that their position uncertainty equals the diameter of the black hole; or to be more precise, the diameter of the coherence horizon. We recall from chapter 4.3 that the coherence horizon is at 2.2 of the Schwarzschild radius. Knowing that gravity is merely a side effect of the strong force, we replace the mass-dependent gravitational constant G in our equation for the coherence radius by G_S, which is used in combination with the absolute hypercharge (it will be formally introduced in chapter 6.6), which is essentially the number of quarks (n_q). Accordingly the wavelength of particles inside a black hole can be given as:

$$\lambda_{BH} = \frac{4\,(z_0^2 - 1)\,G_S\,n_q}{c^2} \approx \frac{8.8\,G_S\,n_q}{c^2}$$

And the energy and temperature as:

$$E = \frac{h\,c^3}{4\,(z_0^2 - 1)\,G_S\,n_q} \approx \frac{h\,c^3}{8.8\,G_S\,n_q}$$

$$T_{BH} = \frac{h\,c^3}{4\,(z_0^2 - 1)\,G_S\,n_q\,k_B} \approx \frac{h\,c^3}{8.8\,G_S\,n_q\,k_B}$$

We can now compare this position uncertainty deduced temperature with the Bekenstein-Hawking temperature:[2]

$$E = \frac{h\,c^3}{8\,\pi\,G\,M}$$

$$T_{BH} = \frac{h\,c^3}{8\,\pi\,G\,M\,k_B}$$

Of course, an observer that had fallen all the way along with the collapsing star that formed the black hole, would see something very different. For such an observer the black hole would be a collapsing star, an object that is not cold at all. Yet we know that such an observer would be frozen in time forever if he were to exist somewhere inside the black hole. From the outside perspective things that are frozen in time forever certainly have a temperature close to zero.

Things with a temperature usually have thermal radiation. What about black holes?

In chapter 4.3 we reached the conclusion that black holes don't emit Hawking radiation, because photons have no charge and if gravity is a side effect of charge, then photons cannot reduce the gravity of a black hole.

However, we also learned that black holes are in fact eternally collapsing objects (ECO), and therefore they do not have absolute event horizons and do in fact emit thermal radiation. After all there is still a star inside every stellar black hole; a star which is still fairly above its critical circumference. We can therefore expect it to emit thermal radiation. Above we calculated the energy of photons within the black hole.

Those can be photons that are trapped inside the original star, bouncing back and forth between atoms. Once they are emit to the outside universe, their wavelength will equal the diameter of the coherence horizon.[3]

This means that black holes indeed emit a type of radiation that is very similar to Hawking radiation, but it simply consists of photons that can get in and out of it, and they don't make the black hole evaporate.

The fact that black holes don't evaporate in space particle dualism theory also tells us that there can't be any situation in which this radiation becomes intense.

What makes it similar to Hawking radiation though, is that despite originating from the frozen star that is trapped inside the black hole, it is still perfectly thermal black body radiation. This is why in effect, we have a loss in information somewhat similar to that in Hawking's improvised quantum gravity, and that must be so, otherwise black holes wouldn't have such huge entropies.

On the other hand, in chapter 4.11, we learned that black holes do have magnetic fields, and that means less information is lost here, than in the standard model.

The above equation for the temperature of a black hole can serve us as a measure for the temperature of both stellar and dark matter black holes. For literally all black holes in the universe it is close to zero. Therefore, when assigning temperatures to different forms of matter in the universe, we can regard the matter inside black holes to be perfectly cold, as cold as dark matter.[4]

The above method of assigning a temperature to a black hole is the easiest, but not necessarily the only one. A more direct but harder to access approach would be to look at how time dilation would impact the perception of impulses within a black hole, if they could be observed directly.[5] However, there doesn't seem to be any straight forward way to obtain a value for the overall time dilation within a black hole, as it is different at different radii. Furthermore the fact that particles within a black hole don't have definite positions anymore also means that there is no definite value for the time dilation acting upon a single particle. It is therefore more practical to use an impulse that is derived from the uncertainty principle, than one that is derived from gravitational time dilation.

The universe began with 60% dark matter, and is now at 84%; that is 84% matter with almost no impulse at all.

While degrees of freedom in impulse increasingly get destroyed during the evolution of the universe, degrees of freedom in space are created as the universe expands.[6]

Using the matter to dark matter ratio we can calculate the average temperature of the universe as:

$$T_{U_1} = 0.16 \times 2.725 \text{ K} = 0.436 \text{ K}$$

From chapter 4.12 we know that the universe began with a temperature of:

$$T_{max} = 7,454,246 \text{ K}$$

Accounting for the 60% (cold) dark matter that was present already back then, we can calculate the 'temperature of the universe' back then as:

$$T_{U_0} = 0.4 \times T_{max} = 2,981,698.4 \text{ K}$$

We can thus establish that the universe constantly loses energy as it expands.

Is this energy turned into potential energy of gravitational fields?
When we elevate an object, we can say that we give it potential energy which will be turned into kinetic energy as soon as we release it and let it drop. However, in case of a black hole this kinetic energy is lost as soon as the object enters the coherence horizon.[7] The formation of black holes turns objects with both inertia and gravity into objects that have gravity only. All the energy associated with their inertia is lost forever.

If energy is not preserved, then we may wonder even more where it came from in the first place. The energy of virtual particles is sourced from the uncertainty principle. Could this very same principle be also in some more subtle way be responsible for the energy of permanent particles?

The uncertainty principle and the initial impulses of particles

What if not only the energy of particles in the quantum vacuum comes from the uncertainty principle, but all energy, including the energy of all particles we see around us?

We could assign a certain volume Δx^3 to every single particle which is determined by the particles position uncertainty according to the uncertainty principle:

$$\hbar = \Delta x \times \Delta p$$

Quantum mechanics tells us that when we confine a particle to a space with the volume Δx^3, it will have an impulse Δp according to the uncertainty equation. We tend to think that this applies only to confined particles.

What if every particle aside from carrying its own quantum of space, generates a volume Δx^3 of space around itself?

If we take a cubic meter of space and divide it through the number of particles in it, then we can know the corresponding impulse uncertainty according to:

$$\hbar = \sqrt[3]{\rho^{-1}} \times \Delta p$$

$$\Delta p = \frac{\hbar}{\sqrt[3]{\rho^{-1}}}$$

The density of photons in the early universe is given by:

$$\rho_\gamma = \frac{16 \pi k^3 \zeta(3) T^3}{c^3 h^3} = 20{,}286{,}824.875 \times T^3$$

Combining the two and multiplying by the speed of light gives us the average photon energy, which is:

$$\Delta E_\gamma = \frac{\hbar c}{\sqrt[3]{(20{,}286{,}824.875 \times T^3)^{-1}}}$$

Using the new temperature estimate from chapter 4.12 yields:

$$\Delta E_\gamma = 1.685553433 \times 10^{-17} \, \text{J}$$

This is the average amount of energy per photon we get for free, from the uncertainty principle.

What is the average energy of photons in a thermal bath with the initial temperature T_{CNB_max}? According to Wien's displacement law[8], the intensity maximum for our photon bath must be given by:

$$f_{max} = T_{CNB_{max}} \times \frac{\alpha k}{h} = T_{CNB_{max}} \times 5.879 \times 10^{10} \, \frac{\text{Hz}}{\text{K}}$$

Multiplying this by the Planck constant yields:

$$E_{\gamma_\mu} = 7.614951312 \times 10^{-17} \, \text{J}$$

We note that $\Delta E_\gamma \approx E_{\gamma\_\mu}$, which shows that indeed all energy of particles at the border of the emergent past could be sourced from the uncertainty principle.

We did not account for the presence of other particles. How many of them are to expect? As a matter of fact there is only one proton, neutron or electron for each 10^9 photons, so we can fairly ignore those particles in our calculation.
The only particle which should have been present in a similar abundance as the photon is the neutrino. We can account for its presence by simply multiplying the photon density with a factor of 2. This yields:

$$\Delta E_\gamma = 2.123664288 \times 10^{-17} \, \text{J}$$

And that represents 27.88808754% of the entire energy that was present. Accounting for the minimal mass primordial black holes that were around will probably account for the rest of this energy.

Is the relation $\Delta E \approx E_{\gamma\_\mu}$ dependent on the temperature T? It turns out no, because

$$\Delta E_\gamma = \frac{\hbar c}{\sqrt[3]{\left(\frac{16\pi k^3 \zeta(3) T^3}{c^3 h^3} \times 2\right)^{-1}}}$$

Can be simplified into:

$$\Delta E_\gamma = \frac{kT}{2\pi} \sqrt[3]{32\pi \zeta(3)}$$

Which grows in exactly the same way as:

$$E_{\gamma_\mu} = T \times \alpha k$$

The cosmic microwave background radiation is a black body radiation. Some have raised the question of how black body radiation can emerge in something that is not a cavity. As demonstrated above, high density can restrict particles in a similar away walls do. Stars don't have physical walls too, but their density makes their radiation closely approach black body radiation. Black holes then are so perfect cavities that smaller wavelengths disappear, so that the prevalent wavelength depends on the size of the cavity alone. It is unclear if non-gravitational cavities can have a similar effect.[9]

In principle we could try to generate energy by enclosing particles in order to increase their position certainty and thus give them more impulse. This however must fail, because particles tunnel through cavities which are smaller than their wavelength.

In a star, degeneracy pressure generates additional kinetic energy, but that came from the gravitational potential energy.

Any type of energy that is not lost forever can be thought of as being preserved, as what is called 'potential energy'. When a space ship accelerates, time slows down inside it, so that when viewed from the outside, the initial individual kinetic energies of the particles seem to be lost, but as soon as the space ship slows down, time runs normal again and all the kinetic energies go back to their original values.

The same can't happen with black holes, and it also can't happen with the cosmic microwave background. The energy lost there can't be restored, and therefore we can't say that it is transformed into potential energy.

This is precisely why cosmologists prefer closed universes that collapse, astrophysicists black holes that evaporate and particle physicist particles that decay:

1. In a closed universe the positive mass-matter energy and the negative gravitational potential energy level each other out perfectly. All the energy that photons lose from being stretched out during the expansion, they get back when the universe collapses.
2. All the energy that a star loses through the infinite time dilation that takes place when it becomes a black hole, is restored when the black hole evaporates.
3. If all particles decay into photons at some point in time, it becomes easy to imagine that they all have been created out of photons in the past.

Universes that don't collapse, don't get back the energy they lost during the expansion. Black holes that don't evaporate, freeze everything in time forever without ever restoring any of the energy they swallowed.

We also can't say the energy a black hole swallowed is turned into energy of the gravitational field, because according to space particle dualism, energy is not contributing to gravity.

We are left with a universe that doesn't look too different from the one we knew before, but we realize that it is a universe where the distinction between past and future is not at all an illusion, where reversibility is simply a fancy concept that comes as a side effect of too much reductionism.

It seems that even now that it has been discovered that the universe is accelerating its expansion, many cosmologists still favor the model of a closed universe.

No evidence for global curvature has ever been found, and currently for the universe to be a closed one, the visible universe would have to be only 0.25% of the entire universe.

One would think that closed universe don't accelerate their expansion, but today, with all sorts of artificial ingredients such as 'inflation' and a 'variable cosmological constant', everything becomes possible.

Behind all this is not only the felt need to protect the conservation of energy, but as mentioned earlier in this book, the feeling that a story that has a beginning but no end, is 'unnatural' and 'philosophically unsatisfying'. In chapter 4.19 we will be looking at how the end of this story is connected to its beginning. This will hopefully take away a lot of the unease people feel when confronted with a world that is in constant decay.

Failing attempts to understand black hole entropy in a classical framework

The physicist Carlo Rovelli was once asked how a black hole can represent a high entropy state if the degrees of freedom in position are obviously reduced.[10]
He responded that gravitational collapse while causing a loss of degrees of freedom in position space, for the object in consideration, it leads to an increase of degrees of freedom in impulse space. Accordingly it is this increase in degrees of freedom in impulse that justifies the high entropy of black holes.
Rovelli is here using an interpretation of black hole entropy that is not very common and which aims to approach the problem in a more classical way.

The common way of interpreting black hole entropy is that it represents the logarithm of the number of ways the black hole could have been formed. This corresponds to the multitude of possible states on the inside of the black hole.
In chapter 5 we will see that everything that has ever been measured is stored in what can be called the 'world of experience' or the 'nonlocal realm'. It might be that the wavefunction is based on all of the potentially accessible knowledge within this 'nonlocal realm', and in that case it makes more sense to identify black hole entropy only with the degree of superpositioning within the black hole, and not with the multitude of possible ways it could have formed.

Carlo Rovelli's interpretation might work for stars, but if we are looking at black holes, then it becomes very problematic: if black holes are in fact ECOs (eternally collapsing objects), then they are frozen in time, and if we believe that (1) there is a preferred frame of reference, and (2) black holes don't evaporate, then we have to conclude that the thermal energy of the original star is practically lost and also not going to be returned.
If on the other hand we believed that black holes have event horizons and singularities, it even more seems unclear where the degrees of freedom in impulse space went to; a singularity can hardly have a temperature. The black hole as a whole can have a temperature, but it is by every means lower than that of the original star.

No classical explanation exists for the entropy of black holes, thus breaking with the belief that macroscopic objects can be reasonably approximated in a classical way.

Note:

1. In the cosmology of general relativity there is the theoretical possibility of a spatially closed universe in which positive mass-matter energy and negative gravitational potential energy level each other out, making any further discussion of energy conservation unnecessary. That is why closed universes always were so popular among cosmologists. However, by definition closed universes must collapse and since the discovery of the accelerated expansion of the universe, we know that we do not live in a closed universe.

2. It is a rather interesting coincidence that the additional dividend 8π brings the average energy per photon of the Hawking radiation closer to the value of our 'coherence radiation'.

3. Photons bouncing back and forth between atoms seems to require a wavelength much shorter than the one we calculated here. We have to recall that for observers inside the black hole, time flows normally, so to such an observer particles do not have a temperature close to zero.

4. As we saw in chapter 4.9, dark matter starts off with an impulse of absolute zero, for then receiving a tiny impulse that results, via the uncertainty principle, from being trapped inside a black hole. This tiny impulse is of course again so close to zero, that it can be ignored for all practical purposes.

5. It is an interesting question if it is possible to observe regions of space with extreme time dilation. In the case of gravitational time dilation, redshift hides such regions from us. In the case of speed induced time dilation, it is both blueshift and redshift that hides them from us. A video transmission from a space ship moving away from earth at say 80% the speed of light could only be received with gigantic arrays of radio antennas, spanning an area of many kilometers.

6. It might well be that entropic expansion is resulting from a transfer in degrees of freedom in impulse space to degrees of freedom in position space. That is however pure speculation at this moment.

7. One can shrink the coherence horizon by using radio telescopes or passing a black hole as close as possible. In that way some of the energy that was lost in the black hole can be released. Energy that can't be restored in any case is energy that lies within the *potential event horizon*. It can be defined best as a *time-like anti-horizon*, a border of no entry that is a border only to matter (time-like), but not to radiation (light-like).

8. This is its frequency dependent formulation.

9. If a cavity could be so reflective that it spreads the wavefunction of incoming particles so perfectly over space that they have the same position probability everywhere on the inside, they could effectively lose much of their energy. However, at the end even the most advanced super-mirror works only for certain wavelengths, so that the photon would get absorbed eventually.

10. This was at the *Quantum Gravity Physics & Philosophy conference*; October 24 – 27; 2017 at the *Institut des hautes études scientifiques* (IHES).

4.14 Consequences of living in an old universe [10.04.2019 (estimate using star birth rate; refinement using average masses); 19.04.2019 (corrected number of stars in the Milky Way); 20.04.2019 (accounting for black dwarfs); 30.04.2019 (age confirmed by chemical analysis); 01.07.2019 (expected star birth rate; correct dark matter prediction)]

It is currently believed that there 7 new stars in the Milky Way every year ($R_* = 7$). One out of thousand becomes a black hole ($q_{BH} = 1/1,000$). If it was true that the universe is only 13.8 billion years old ($t_0 = 1.38 \times 10^{10}$ yr), then we would have to expect

$$n_{SBH} = t_0 \, R_* \, q_{BH} = 96,600,000$$

Stellar black holes within the Milky Way.

If it is in fact 84 trillion years old (see chapter 4.13) and $\Delta t_d = 4.2130109541636 \times 10^{13}$ yr years passed by since decoupling, as we found in chapter 4.12, then we have to expect $2.94910766791 \times 10^{11}$ stellar black holes. The average mass M_{BH_μ} of black holes we observe within the Milky Way is 8.4 solar masses.[1] The average star mass M_{S_μ} is 0.36 solar masses and there are roughly a trillion stars in the Milky Way.[2, 3, 4, 5] Using the old-universe estimate, this yields:

$$10^{12} \text{ stars} \times 0.36 \, M_\odot = 3.6 \times 10^{11} \, M_\odot$$

$$2.94910766791 \times 10^{11} \text{ BHs} \times 8.4 \, M_\odot = 2.477250441048 \times 10^{12} \, M_\odot$$

$$q_{dark} = 1 - \frac{n_* \, M_{*\mu}}{\Delta t_d \, R_* \, q_{BH} \, M_{BH_\mu} + n_* \, M_{*\mu}}$$

$$q_{dark} = 85.46775917\%$$

That would be 85% dark matter in form of stellar black holes. Almost the right amount, but not quite. The Milky Way consists of 90% invisible matter.[6]

A star does not need to turn into a black hole in order to leave behind something invisible. Most stars turn into white dwarfs at the end of their life-time, so they are still visible, but even white dwarfs have to burn out one day and turn into black dwarfs.

Standard cosmology doesn't consider black dwarfs, because it is based on a young universe model and there is just not enough time for white dwarfs to cool out and become black dwarfs in such a model.

Black dwarfs are essentially burned out cores of stars and we can account for them, by assuming that the left-over core of a star with the average mass of $M_\mu = 0.36 \, M_\odot$ has a mass of $M_{f\_\mu} = 0.2 \, M_\odot$.[7] Adding that to all the stellar black holes yields:

$$q_{dark} = 1 - \frac{n_* \, M_{*\mu}}{\Delta t_d \, R_* \, q_{BH} \, M_{BH_\mu} + \Delta t_d \, R_* \, M_{f_\mu} + n_* \, M_{*\mu}}$$

$$q_{dark} = 99.41765857\%$$

Now this is too much. Is any of the involved factors uncertain?

$n_* \, M_{*\_\mu}$ is a rather imprecise term for the mass of the visible matter. The visible matter is comprised of both stars and gas, and the gas is equal to about $10\% - 15\%$ of the mass in form of stars. According to latest estimates the total mass of the visible stars is $4.6 \times 10^{10} \, M_\odot$. Multiplying with 1.15 gives us the visible mass M_{vis}.

The star birth rate R_* was originally thought to be roughly at about 1. Now it is believed that there are 3 sunmasses in new born stars every year. Is this star birth rate observed or assumed? If we take the latest estimate on the total star mass M_{stars} and divide it by the average star mass of $0.36 \, M_\odot$, we arrive at a total number of $127,777,777,778$ stars. Using the mainstream age of the universe this would imply that $R_* \approx 9$. This value being very close to the actually assumed value $R_* = 7$ shows that the star birth rate is not something that is measured directly through the observation of actual star birth, but something that is obtained simply by accepting the age of the universe general relativity predicts and then looking at the total number of stars in a system.

In space particle dualism theory obtaining the star birth rate is not that simple. Here the visible stars are regarded as representing only one generation among hundreds of star generations, with all the stars from past generations being invisible by now.

$$q_{dark} = 1 - \frac{M_{vis}}{R_* \left(\Delta t_d \, q_{BH} \, M_{BH_\mu} + \Delta t_d \, M_{f_\mu} \right) + M_{vis}}$$

$$R_* = \frac{\dfrac{M_{vis}}{1 - q_{dark}} - M_{vis}}{\Delta t_d \, q_{BH} \, M_{BH_\mu} + \Delta t_d \, M_{f_\mu}}$$

$$R_* = \frac{M_{vis} \, q_{dark}}{\Delta t_d \, (1 - q_{dark}) \left(q_{BH} \, M_{BH_\mu} + M_{f_\mu} \right)}$$

$$R_* = 0.0669457266$$

This is very close to the star generation rate we obtain when dividing the full mass of the Milky Way, which is estimated to be $0.8 - 1.5 \times 10^{12} \ M_\odot$,[8, 9, 10, 11] by the average star mass of $0.36 \ M_\odot$ and the age of the universe. This yields:

$$R_* \left(t_d \right) = \frac{M_{tot}}{M_{*_\mu} \Delta t_d} = 0.0527466519 - 0.0988999723$$

The $R_* = 0.09$ figure is what we get if we assume $M_{tot} = 1.5 \times 10^{12} \ M_\odot$ and $R_* = 0.05$ is what we get when we assume $M_{tot} = 0.8 \times 10^{12} \ M_\odot$.

Using this figure for R_* we can recalculate the percentage of dark matter, and arrive at:

$$q_{dark} = 1 - \frac{M_{vis}}{R_* \left(\Delta t_d \ q_{BH} \ M_{BH_\mu} + \Delta t_d \ M_{f_\mu} \right) + M_{vis}}$$

$$q_{dark} = 89.74828277 - 94.25769801\% \approx 90 - 94\%$$

Above we mentioned that the gas inside the Milky Way comprises about $10\% - 15\%$ the mass there is in form of stars. Above we took that to be 15%. Using 10% instead yields:

$$q_{dark} = 90.15010598 - 94.49361579\% \approx 90 - 94\%$$

The lower end gives the more correct result, as 90% is the observed percentage of invisible matter. We can take this to mean that the total mass of the Milky Way must indeed be $0.8 \times 10^{12} \ M_\odot$.

The above suggests a big paradigm shift: the actual dark matter that space particle dualism predicts is thus not the dark matter that is missing in surveys on the gravitational dynamics of galaxies, but rather only the dark matter that gave rise to supermassive black holes in the cores of galaxies and possibly 'exo-transparent' black holes in cosmic voids (see chapter 4.9 and 4.11). At the same time, what we usually mean by 'dark matter' is entirely comprised by ordinary mass black holes and black dwarfs (see fig. 54 & 55).

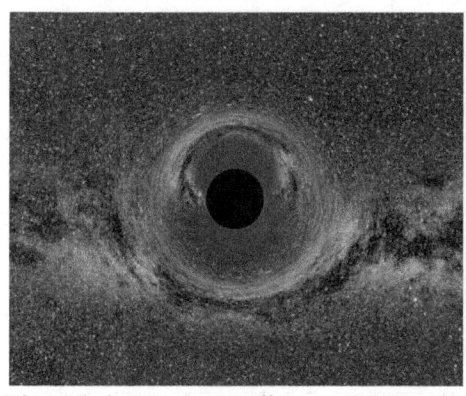 

Fig. 54 & 55. According to SPD primordial dark matter is only responsible for the formation of supermassive black holes and does not contribute to the missing matter in star surveys. That can be fully accounted for by loner stellar black holes (left) and black dwarfs (right).
[Illustration taken from: Ute Kraus/Wikipedia; CC BY-SA]
[Illustration taken from: "*MCG+01-02-015 Interstellar Fight*"]

Using the young universe model of mainstream cosmology there can be only 1% invisible matter, while using the old universe model of space particle dualism, there must be 90% invisible matter. What both models agree on is that 90% of the gas the galaxy initially had is already turned into stars.[12] Apparently the standard model is assuming that galaxies started off with only enough gas for one or two generations of sun-like stars.

The mistake that is made here is assuming that the central black hole formed out of ordinary matter and the 90% invisible matter we see is in fact primordial dark matter. That is why the mainstream assumes that the galaxy was comprised of so few gas, that it was already used up in only 13.8 billion years.

The average mass for stellar black holes of $8.4\,M_\odot$ used above is based on a list of black holes that are part of black hole-star systems. Those can have formed from double star systems through gravitational collapse of one of the stars, or it could have been a loner black hole that captured a star. There is no reason to assume that the resulting mass is not representative. However, it is interesting to also look at loner black holes that were discovered through gravitational waves.

If stellar black holes have masses of $8.4\,M_\odot$ and primordial dark matter black holes all turned into supermassive black holes with masses of $10^6 - 10^8\,M_\odot$, then the only explanation left for *intermediate mass black holes* is stellar black hole mergers in an old-universe model. Let us look at the masses of intermediate mass black holes that were discovered through gravitational waves:

Name	Detection time	Redshift	Mass
GW150914	2015-09-14	0.093	$35.6\,M_\odot$ $+\ 30.6\,M_\odot$
GW151012	2015-10-12	?	$23.3\,M_\odot$ $+\ 13.6\,M_\odot$
GW151226	2016-06-15	0.09	$13.7\,M_\odot$ $+\ 7.7\,M_\odot$
GW170104	2017-01-04	0.18	$31\,M_\odot$ $+\ 20.1\,M_\odot$
GW170608	2017-06-08	0.07	$10.9\,M_\odot$ $+\ 7.6\,M_\odot$
GW170729	2017-07-29	?	$50.6\,M_\odot$ $+\ 34.3\,M_\odot$
GW170809	2017-08-09	?	$35.2\,M_\odot$ $+\ 23.8\,M_\odot$
GW170814	2017-08-14	0.11	$30.7\,M_\odot$ $+\ 25.3\,M_\odot$
GW170818	2017-08-18	?	$35.5\,M_\odot$ $+\ 26.8\,M_\odot$
GW170823	2017-08-23	?	$39.6\,M_\odot$ $+\ 29.4\,M_\odot$

Adding all these up results in an average mass of

$$M_{GW_\mu} = 29.065\,M_\odot$$

Assuming that the average mass of stellar black holes is $8.4\,M_\odot$ and that black hole mergers are rather unlikely for average mass black holes, we would need each of these black holes to have merged with other black holes about 4 times already. The fact that all of them are roughly in the same mass range shows that:

1. Merging is extremely unlikely for black holes that haven't merged before.
2. Merging events peak at $\approx 30\,M_\odot$.
3. The mass gap between these larger stellar black holes and supermassive black holes, show that they cannot be both based on the same formation principle.

Black holes with masses of roughly $3\,M_\odot$ can be explained through simple gravitational collapse of a star.
Black holes with the average mass of $8.4\,M_\odot$ can only be explained by star-black hole encounters in a universe that is much older than currently believed.
Black holes with masses of $30\,M_\odot$ can only be explained with much more rare black hole-black hole encounters. If only one out of a thousand stars becomes a black hole, these events must be roughly a thousand times more unlikely.

Supermassive black holes with masses of $10^6 - 10^8 \, M_\odot$ then are only explainable with the condensation of primordial dark matter in an old universe.

The fact that supermassive black holes with such masses are to be found even at the very border of how far we can see, shows that what we are seeing can't be the beginning of time, but instead it must represent very recent events in the history of the universe.

The abundance of elements

When the sun formed, the heavy elements moved to the outside, forming the planets. Arguably, the 0.1% metals in the sun have all been generated by the sun itself.

The sun exists roughly since 5 billion years and it will continue to exist another 5 billion years.

To go from the 0.00022% iron to the 34.6% the earth consists of, would require

$$t_{Fe} = \frac{q_{Fe_\oplus}}{q_{Fe_\odot}} \times \frac{1}{2} t_\odot$$

$$t_{Fe} = 7.8636364 \times 10^{14} \, yr$$

Of course there are stars much larger than the sun, and it is those which produce the largest amounts of iron.

However, the average star has only 0.36 solar masses, and thus it certainly doesn't produce more iron than the sun. The very large stars that do, do that only for rather short time, as large stars have lower life-expectancies, being in the range of million years, instead of billion.

The age of the CMB according to SPD is $4.2130109541636 \times 10^{13}$ years. That is only 5.3% of the above number (it is 18 times smaller), but it is certainly closer and more plausible than the 13.8 billion years we get from the mainstream.

In figure 56 and 57 we can see and can compare the chemical composition of the sun and the earth respectively. Following the same scheme for some of the other elements we obtain:

Sulfur (S 硫): 8.6363636×10^{12} yr
Silicon (Si 硅/矽): 3.1666667×10^{13} yr
Magnesium (Mg 鎂): 2.4423077×10^{13} yr
Nickel (N 鎳): 1.875×10^{12} yr

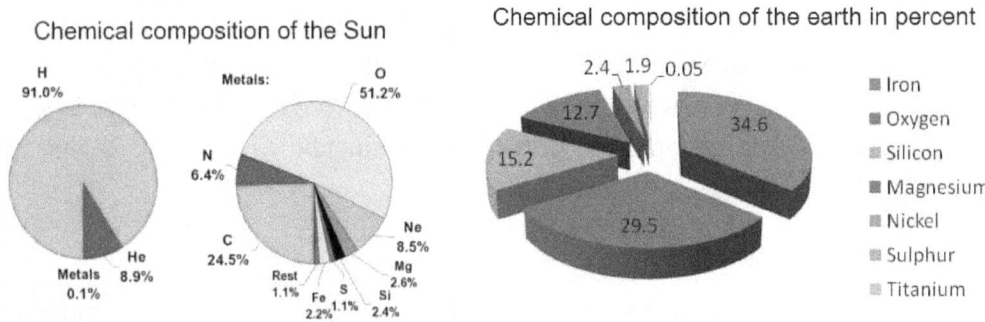

Fig. 56 & 57. Both on the right side of the left figure and in the right figure, hydrogen and helium are taken out. If the sun were to explode right now, the ratio of the metals that would be left over, seems very different from what the earth apparently was provided from previous star generations. The discrepancy can only be accounted for when assuming that the earth does consist of the star dust of thousands of star generations.

Iron is produced to larger extends by more massive stars. Accounting for those stars would change the time necessary for reaching the observed iron abundance. We may therefore leave out iron for the moment, and look at the other elements only. Without including the figure for iron, the average time for reaching the observed abundance is:

$$1.6650277 \times 10^{13} \text{ yr}$$

This is 40% of the age of the CMB (it is only 2.7 times shorter).

The above estimates were all obtained by looking at a single example, our earth-sun system. It did not account for lighter and heavier star, which produce elements at different rates and with slightly different abundance ratios. Considering the approximative nature of the approach, arriving at the right order of magnitude can be regarded as positive evidence.

How does mainstream cosmology makes sense out of this?
In the young universe model of the mainstream, it is believed that the heavy elements, from oxygen up through iron, were produced in stars that are at least ten times more massive than the Sun. It is assumed that the very first generation of stars was supermassive. This seems to be the only way the observed abundance of elements can be made plausible in the standard model.

Note:
1. This average mass is calculated based on a list of black hole candidates that can be found accessed on the Wikipedia page on 'Stellar black holes'.
2. Ledrew, Glenn (February 2001). "The Real Starry Sky". Journal of the Royal

Astronomical Society of Canada. 95: 32. Bibcode:2001JRASC..95...32L.

3. The average mass of $0.36\ M_{\odot}$ is obtained by assuming an exponential distribution for each type of star (approach proposed by statistician Aaron Brown (Quora; 2017)).

4. "The initial mass function 50 years later"; Edwin E Salpeter; Edvige Corbelli; F Palla; Hans Zinnecker; Dordrecht, the Netherlands; New York: Springer, ©2005. [This book mentions two averages, the second of which is $0.35\ M_{\odot}$]

5. "Counting the Stars in the Milky Way"; Dr. Sten Odenwald; NASA Heliophysics Education Consortium; *Huffington post*; 03/17/2014; ET Updated Dec 06, 2017.

6. McMillan, P. J. (July 2011). "Mass models of the Milky Way". Monthly Notices of the Royal Astronomical Society. 414 (3): 2446–2457. arXiv:1102.4340. Bibcode:2011MNRAS.414.2446M. doi:10.1111/j.1365-2966.2011.18564.x.

7. V. Weidemann (2000). "Revision of the initial-to-final mass relation". Astron. Astrophysics. 363, 647-656 (2000).

8. McMillan, P. J. (July 2011). "Mass models of the Milky Way". Monthly Notices of the Royal Astronomical Society. 414 (3): 2446–2457. arXiv:1102.4340. Bibcode:2011MNRAS.414.2446M. doi:10.1111/j.1365-2966.2011.18564.x.

9. McMillan, Paul J. (February 11, 2017). "The mass distribution and gravitational potential of the Milky Way". Monthly Notices of the Royal Astronomical Society. 465 (1): 76–94. arXiv:1608.00971. Bibcode:2017MNRAS.465...76M. doi:10.1093/mnras/stw2759.

10. Kafle, P.R.; Sharma, S.; Lewis, G.F.; Bland-Hawthorn, J. (2012). "Kinematics of the Stellar Halo and the Mass Distribution of the Milky Way Using Blue Horizontal Branch Stars". The Astrophysical Journal. 761 (2): 17. arXiv:1210.7527. Bibcode:2012ApJ...761...98K. doi:10.1088/0004-637X/761/2/98.

11. Kafle, P.R.; Sharma, S.; Lewis, G.F.; Bland-Hawthorn, J. (2014). "On the Shoulders of Giants: Properties of the Stellar Halo and the Milky Way Mass Distribution". The Astrophysical Journal. 794 (1): 17. arXiv:1408.1787. Bibcode:2014ApJ...794...59K. doi:10.1088/0004-637X/794/1/59.

12. [https://www.nasa.gov/centers/goddard/news/topstory/2006/milkyway_seven.html] (Paragraph 6: "About ten billion years into its life, the Milky Way galaxy has now converted about 90 percent of its initial gas content into stars.").

4.15 Entropy and redshift survey to test anisotropic expansion and refine the age of the universe [15.07.2019 (average redshift is proportional to mass; phenomenal evidence for entropic expansion); 16.07.2019 (entropy of the mapped universe); 17.07.2019 (entropy density according to SPD; full age of the universe)]

Using the time-redshift equation from chapter 4.12, and basing the latest Hubble measurement of $74.03\ \text{km s}^{-1}\ \text{mpc}^{-1}$ we can formulate the time-redshift equation in form of a single distance constant Δx_z, which is given by:

$$\Delta x_z \; = \; 39{,}625{,}004{,}647 \; \text{lyr}$$

And so the time-redshift equation simplifies to:

$$\Delta t_\gamma \; = \; \Delta x_z \; \times \; z$$

Similarly we can express the entropy of black holes by using a single black hole entropy constant ΔS_M:

$$\Delta S_M \; = \; 3.660344593 \; \times \; 10^{-7} \; \text{m}^2 \; \text{s}^2 \; \text{K}^{-1} \; \text{kg}^{-1}$$

Or for the use in combination with sun masses:

$$\Delta S_M \; = \; 1.447960126 \; \times 10^{54} \; \text{m}^2 \; \text{s}^2 \; \text{K}^{-1}$$

And thus we can give the entropy of a black hole simply as:

$$S_{BH} \; \cong \; \Delta S_M \; \times \; M_{BH}^2$$

We now want to use these two constants to efficiently create a large chart of all known supermassive black holes together with their SPD-distances and their entropies:

SMBH	Mass in sun-masses	Redshift	Distance in light years according to SPD	Entropy in J/K
TON 618	6.6×10^{10}	2.219	87,927,895,393	$6.307314311 \times 10^{75}$
IC 1101	$(4-10) \times 10^{10}$	0.0777	3,078,863,214.1	$7.095004617 \times 10^{75}$
S5 0014+81	4×10^{10}	3.366	133,377,780,933	$2.316736201 \times 10^{75}$
SDSS J102325.31+514251.0	$(3.31 \pm 0.61) \times 10^{10}$	6.3	249,637,557,897	$1.586399593 \times 10^{75}$
H1821+643	3×10^{10}	0.2970	11,768,627,729	$1.303164113 \times 10^{75}$
NGC 6166	3×10^{10}	−0.030354	490,000,000	$1.303164113 \times 10^{75}$
APM 08279+5255	$1.0^{+0.17}_{-0.13} \times 10^{10}$	3.911	154,973,410,942	$1.447960126 \times 10^{74}$
NGC 4889	$(2.1 \pm 1.6) \times 10^{10}$	0.021665	858,475,824.1	$6.385504155 \times 10^{74}$
Central BH of Phoenix Cluster	2×10^{10}	0.597	23,656,130,486	$5.791840504 \times 10^{74}$
SDSS J074521.78+734336.1	$(1.95 \pm 0.05) \times 10^{10}$	3.220	127,592,529,592	$5.505868379 \times 10^{74}$
OJ 287 primary	1.8×10^{10}	0.306000	12,125,252,812	$4.691390808 \times 10^{74}$
NGC 1600	$(1.7 \pm 0.15) \times 10^{10}$	0.015614	618,704,893.49	$4.184604764 \times 10^{74}$
SDSS J08019.69+373047.3	$(1.51 \pm 0.31) \times 10^{10}$	3.480	137,895,031,981	$3.301493883 \times 10^{74}$
SDSS J115954.33+201921.1	$(1.41 \pm 0.10) \times 10^{10}$	3.426	135,755,281,485	$2.878689526 \times 10^{74}$
SDSS J075303.34+423130.8	$(1.38 \pm 0.03) \times 10^{10}$	3.590	142,253,782,992	$2.757495263 \times 10^{74}$
SDSS J080430.56+542041.1	$(1.35 \pm 0.22) \times 10^{10}$	3.759	148,950,409,545	$2.638907329 \times 10^{74}$
Abell 1201 BCG	$(1.3 \pm 0.6) \times 10^{10}$	0.169	6696626553.1	$2.447052612 \times 10^{74}$
SDSS J0100+2802	$(1.24 \pm 0.19) \times 10^{10}$	6.30	249,637,557,897	$2.226383489 \times 10^{74}$
SDSS J081855.77+095848.0	$(1.20 \pm 0.06) \times 10^{10}$	3.700	146,612,534,003	$2.085062581 \times 10^{74}$
NGC 1270	1.2×10^{10}	0.016561	6,562,297,772	$2.085062581 \times 10^{74}$
SDSS J082535.19+512706.3	$(1.12 \pm 0.20) \times 10^{10}$	3.512	139,163,032,275	$1.816321182 \times 10^{74}$

SDSS J013127.34-032100.1	$(1.1 \pm 0.2) \times 10^{10}$	3.512	139,163,032,275	1.752031752 $\times 10^{74}$
PSO J334.2028+01.4075	1×10^{10}	2.060	81,627,518,931	1.447960126 $\times 10^{74}$
RX J1532.9+3021	1×10^{10}	0.3613	14,316,515,820	1.447960126 $\times 10^{74}$
QSO B2126-158	1×10^{10}	3.268	129,494,530,033	1.447960126 $\times 10^{74}$
Holmberg 15A	1×10^{10}	0.055672	2,206,003,511.6	1.447960126 $\times 10^{74}$
NGC 1281	1×10^{10}	0.014343	568,341,506.81	1.447960126 $\times 10^{74}$
SDSS J015741.57-010629.6	$(9.8 \pm 1.4) \times 10^{9}$	3.572	141,540,532,827	1.390620905 $\times 10^{74}$
NGC 3842	$9.7^{+3.0}_{-2.5} \times 10^{9}$	0.021068	834,819,693.61	1.362385682 $\times 10^{74}$
SDSS J230301.45-093930.7	$(9.12 \pm 0.88) \times 10^{9}$	3.492	138,370,532,091	1.204332147 $\times 10^{74}$
SDSS J075819.70+202300.9	$(7.8 \pm 3.9) \times 10^{9}$	3.761	149,029,659,564	8.809389406 $\times 10^{73}$
CID-947	$6.9^{+0.8}_{-1.2} \times 10^{9}$	3.328	131,872,030,584	6.893738159 $\times 10^{73}$
SDSS J080956.02+502000.9	$(6.46 \pm 0.45) \times 10^{9}$	3.281	130,009,655,152	6.042569279 $\times 10^{73}$
SDSS J014214.75+002324.2	$(6.31 \pm 1.16) \times 10^{9}$	3.379	133,892,906,053	5.765212517 $\times 10^{73}$
Messier 87 "Powehi"[35]	6.3×10^{9}	0.00428	169,595,039.33	5.74695374 $\times 10^{73}$
NGC 5419	$7.2^{+2.7}_{-1.9} \times 10^{9}$	0.014593	578,247,759.11	7.506225293 $\times 10^{73}$
SDSS J025905.63+001121.9	$(5.25 \pm 0.73) \times 10^{9}$	3.373	133,655,155,998	3.990940097 $\times 10^{73}$
SDSS J094202.04+042244.5	$(5.13 \pm 0.71) \times 10^{9}$	3.276	129,811,530,106	3.810582183 $\times 10^{73}$
QSO B0746+254	5×10^{9}	2.979	118,042,902,377	3.619900315 $\times 10^{73}$
QSO B2149-306	5×10^{9}	2.35	93,118,771,597	3.619900315 $\times 10^{73}$
SDSS J090033.50+421547.0	$(4.7 \pm 0.2) \times 10^{9}$	3.290	130,366,280,235	3.198543918 $\times 10^{73}$
Messier 60	$(4.5 \pm 1.0) \times 10^{9}$	0.003726	147,642,784.24	2.932119255 $\times 10^{73}$
SDSS J011521.20+152453.3	$(4.1 \pm 2.4) \times 10^{9}$	3.443	136,428,906,641	2.434020971 $\times 10^{73}$
QSO B0222+185	4×10^{9}	2.69	106,591,274,721	2.316736201 $\times 10^{73}$
Hercules A (3C 348)	4×10^{9}	0.155	6,111,876,424.4	2.316736201 $\times 10^{73}$
Abell 1836-BCG	$3.61^{+0.41}_{-0.50} \times 10^{9}$	0.0363	1,438,387,833.6	1.886996115 $\times 10^{73}$
SDSS J213023.61+122252.0	$(3.5 \pm 0.2) \times 10^{9}$	3.272	133,615,530,989	1.773751154 $\times 10^{73}$
SDSS J173352.23+540030.4	$(3.4 \pm 0.4) \times 10^{9}$	3.432	135,993,031,540	1.673841905 $\times 10^{73}$
SDSS J025021.76-075749.9	$(3.1 \pm 0.6) \times 10^{9}$	3.337	132,228,655,667	1.391489681 $\times 10^{73}$
NGC 1271	$3.0^{+1.0}_{-1.1} \times 10^{9}$	0.019183	760,126,551.29	1.303164113 $\times 10^{73}$
SDSS J030341.04-002321.9	$(3.0 \pm 0.4) \times 10^{9}$	3.233	128107654711	1.303164113 $\times 10^{73}$
QSO B0836+710	3×10^{9}	2.17	85,986,269,942	1.303164113 $\times 10^{73}$
SDSS J224956.08+000218.0	$(2.63 \pm 1.21) \times 10^{9}$	3.311	131,198,405,428	1.001539539 $\times 10^{73}$
SDSS J030449.85-000813.4	$(2.4 \pm 0.50) \times 10^{9}$	3.287	130,247,405,208	8.340250325 $\times 10^{72}$
SDSS J234625.66-001600.4	$(2.24 \pm 0.15) \times 10^{9}$	3.507	138,964,907,229	7.265284728 $\times 10^{72}$
ULAS J1120+0641	2×10^{9}	7.085	280,743,190,111	5.791840504 $\times 10^{72}$
QSO 0537-286	2×10^{9}	3.10	122,837,528,489	5.791840504 $\times 10^{72}$
NGC 3115	2×10^{9}	0.00221	87,571,270.31	5.791840504 $\times 10^{72}$
Q0906+6930	2×10^{9}	5.47	216,748,800,269	5.791840504 $\times 10^{72}$
QSO B0805+614	1.5×10^{9}	3.033	120,182,652,873	3.257910283 $\times 10^{72}$
Messier 84	1.5×10^{9}	0.00327	129,573,780.05	3.257910283 $\times 10^{72}$
Abell 3565-BCG	$1.34^{+0.21}_{-0.19} \times 10^{9}$	0.013	515,125,119.47	2.599957202 $\times 10^{72}$
NGC 7768	$1.3^{+0.5}_{-0.4} \times 10^{9}$	0.02612	1,035,005,240	2.447052612 $\times 10^{72}$
NGC 1277	1.2×10^{9}	0.016898	669,583,405.29	2.085062581 $\times 10^{72}$

MS 0735.6+7421	1×10^9	0.216	8,559,001,985	$1.447960126 \times 10^{72}$
QSO B225155+2217	1×10^9	3.668	145,344,533,709	$1.447960126 \times 10^{72}$
QSO B1210+330	1×10^9	2.502	99,141,772,993	$1.447960126 \times 10^{72}$
NGC 6166	1×10^9	−0.030354	490,000,000	$1.447960126 \times 10^{72}$
Cygnus A	1×10^9	0.056075	2,221,972,390.3	$1.447960126 \times 10^{72}$
Sombrero Galaxy	1×10^9	0.003416	135,359,031.39	$1.447960126 \times 10^{72}$
Markarian 501	1.2×10^9	0.033640	1,332,985,309.2	$2.085062581 \times 10^{72}$
PG 1426+015	$(1.298 \pm 0.385) \times 10^9$	0.086570	3,430,337,045.6	$2.439529012 \times 10^{72}$
3C 273	5.5×10^8	0.158339	6,274,184,330.1	$4.380079381 \times 10^{71}$
ULAS J1342+0928	8×10^8	7.54	298,772,569,293	$9.266944806 \times 10^{71}$
Messier 49	5.6×10^8	9.4	372,475,086,386	$4.540802955 \times 10^{71}$
NGC 1399	5×10^8	9.9	392,287,590,981	$3.619900315 \times 10^{71}$
PG 0804+761	$(6.93 \pm 0.83) \times 10^8$	0.00136855	54,228,806.327	$6.953814025 \times 10^{71}$
PG 1617+175	$(5.94 \pm 1.38) \times 10^8$	0.1120	4,438,001,029.3	$5.10892459 \times 10^{71}$
PG 1700+518	$7.81^{+1.82}_{-1.65} \times 10^8$	0.292	11,570,502,683	$8.831992064 \times 10^{71}$
NGC 4261	4×10^8	0.007465	295,800,693.6	$2.316736201 \times 10^{71}$
PG 1307+085	$(4.4 \pm 1.23) \times 10^8$	0.155	6,141,876,424.4	$2.803250803 \times 10^{71}$
SAGE0536AGN	$(3.5 \pm 0.8) \times 10^8$	0.1428	5,658,451,312.3	$1.773751154 \times 10^{71}$
NGC 1275	3.4×10^8	0.017559	695,775,536.37	$1.673841905 \times 10^{71}$
3C 390.3	$(2.87 \pm 0.64) \times 10^8$	0.056159	2,225,300,891.1	$1.192670276 \times 10^{71}$
II Zwicky 136	$(4.57 \pm 0.55) \times 10^8$	0.0630780	2,499,466,329.7	$3.024050243 \times 10^{71}$
PG 0052+251	$(3.69 \pm 0.76) \times 10^8$	0.155	6,141,876,424.4	$1.971556987 \times 10^{71}$
Messier 59	2.7×10^8	0.00136855	54,228,806.327	$1.055562931 \times 10^{71}$
PG 1411+442	$(4.43 \pm 1.46) \times 10^8$	0.089	3,526,625,817.9	$2.841607267 \times 10^{71}$
Markarian 876	$(2.79 \pm 1.29) \times 10^8$	0.138512	5,488,539,272.9	$1.127106641 \times 10^{71}$
Andromeda Galaxy	2.3×10^8	−0.001001	2,540,000	$7.659709066 \times 10^{70}$
PG 0953+414	$(2.76 \pm 0.59) \times 10^8$	0.239	9,470,377,196.4	$1.102998105 \times 10^{71}$
PG 0026+129	$(3.93 \pm 0.96) \times 10^8$	0.142	5,626,751,305	$2.236359935 \times 10^{71}$
Fairall 9	$(2.55 \pm 0.56) \times 10^8$	0.048175	1,908,934,817.7	$9.415360719 \times 10^{70}$
Markarian 1095	$(1.5 \pm 0.19) \times 10^8$	0.03230	1,279,887,796.8	$3.257910283 \times 10^{70}$
Messier 105	1.7×10^8	0.00307	121,648,778.21	$4.184604764 \times 10^{70}$
Markarian 509	$(1.43 \pm 0.12) \times 10^8$	0.034076	1,350,261,813.2	$2.960933661 \times 10^{70}$
OJ 287 secondary	1×10^8	0.306000	12,125,252,812	$1.447960126 \times 10^{70}$
RX J124236.9-111935	1×10^8	0.05	1,981,250,459.5	$1.447960126 \times 10^{70}$
Messier 85	1×10^8	0.00243465	96,473,028.624	$1.447960126 \times 10^{70}$
NGC 5548	1.23×10^8	0.01651	654,208,901.73	$2.190618874 \times 10^{70}$
PG 1211+143	$(1.46 \pm 0.44) \times 10^8$	0.0809	3,205,663,243.5	$3.086471804 \times 10^{70}$
Messier 88	8×10^7	0.007609	301,506,694.93	$9.266944806 \times 10^{69}$
Messier 81 (Bode's Galaxy)	7×10^7	−0.000113	12,000,000	$7.095004617 \times 10^{69}$
Markarian 771	7.586×10^7	0.06373	2,525,301,835.7	$8.332633476 \times 10^{69}$
Messier 58	7×10^7	0.00506	200,502,546.5	$7.095004617 \times 10^{69}$
PG 0844+349	2.138×10^7	0.064	2,536,000,588.2	$6.618689446 \times 10^{68}$
Centaurus A	5.5×10^7	0.001826	72,355,266.781	$4.380079381 \times 10^{69}$

Markarian 79	5.25×10^7	0.022296	883,479,204.9	$3.990940097 \times 10^{69}$
Messier 96	4.8×10^7	0.00299656	118,738,717.54	$3.33610013 \times 10^{69}$
Markarian 817	4.365×10^7	0.031455	1,246,404,664.1	$2.758831007 \times 10^{69}$
NGC 3227	3.89×10^7	0.00386681	153,222,381.79	$2.191067742 \times 10^{69}$
NGC 4151 primary	4×10^7	0.003262	129,256,779.98	$2.316736201 \times 10^{69}$
3C 120	2.29×10^7	0.035	1,386,875,321.7	$7.593247696 \times 10^{68}$
Markarian 279	4.17×10^7	0.030601	1,212,564,906.2	$2.517843383 \times 10^{69}$
NGC 3516	2.3×10^7	0.008816	349,334,081.02	$7.659709066 \times 10^{68}$
NGC 863	1.77×10^7	0.02609	1,033,816,489.8	$4.536314278 \times 10^{68}$
Messier 82 (Cigar Galaxy)	3×10^7	0.000677365	26,840,594.35	$1.303164113 \times 10^{69}$
Messier 108	2.4×10^7	0.00233434	92,498,243.953	$8.340250325 \times 10^{68}$
M60-UCD1	2×10^7	0.00431227	170,873,738.38	$5.791840504 \times 10^{68}$
NGC 3783	$(2.98 \pm 0.54) \times 10^7$	0.008506	337,050,328.17	$1.28584651 \times 10^{69}$
Markarian 110	$(2.51 \pm 0.61) \times 10^7$	0.03552	1,407,480,326.4	$9.122293589 \times 10^{68}$
Markarian 335	$(1.42 \pm 0.37) \times 10^7$	0.0261257	1,035,231,102.6	$2.919666798 \times 10^{68}$
NGC 4151 secondary	1×10^7	0.003	118,875,027.57	$1.447960126 \times 10^{68}$
NGC 7469	6.46×10^6	0.016317	646,561,274.95	$6.042569279 \times 10^{67}$
IC 4329 A	5.01×10^6	0.016	634,000,147.04	$3.634394395 \times 10^{67}$
NGC 4593	8.13×10^6	0.008344	330,631,076.68	$9.570567565 \times 10^{67}$
Messier 61	5×10^6	0.005224	207,001,048.01	$3.619900315 \times 10^{67}$
Messier 32	3.25×10^6	−0.000667351	2,490,000	$1.529407883 \times 10^{67}$
Sagittarius A*	4.3×10^6	0	26,000	$2.677278272 \times 10^{67}$

Space particle dualism theory predicts that the larger a supermassive black hole (SBH) is, the more expansion of space it causes, and therefore the higher should be the host galaxy's redshift on average. Is that true?

We can use the above chart to find out if there is evidence for this type of anisotropic expansion. For doing so we split up the masses of the various supermassive black holes into classes according to the order of magnitude of their masses and then calculate their average redshifts:

Average redshift for black holes with $M \sim 10^{10} \, M_\odot$:

$$z_{\mu_{M \sim 10^{10} \, M_\odot}} = 2.1316612963$$

Average redshift for black holes with $M \sim 10^9 \, M_\odot$:

$$z_{\mu_{M \sim 10^9 \, M_\odot}} = 2.0740522$$

Average redshift for black holes with $M \sim 10^8 \, M_\odot$:

$$z_{\mu_{M \sim 10^8 \, M_\odot}} = 1.0063487845$$

Average redshift for black holes with $M \sim 10^7 \, M_\odot$:

$$z_{\mu_{M \sim 10^7 \, M_\odot}} = 0.0175947293$$

Average redshift for black holes with $M \sim 10^6 \, M_\odot$:

$$z_{\mu_{M \sim 10^6 \, M_\odot}} = 0.0076475$$

We find that:

$$z_\mu \propto M_{SBH}$$

A striking proportionality between the masses of supermassive black holes and their average redshift.

The differences in these averages are furthermore highest at the middle of the mass distribution: 1.0277761072; 2.0609675611; 57.196036798; 2.3007164825.

In ordinary young universe cosmology, the opposite would be expected. There it is assumed that the black holes with the highest redshifts are near the beginning of time, so that they should be much smaller.
Or, if one assumes that their growth was taking place mainly through the accretion of dark matter before decoupling, then one would assume the average redshift of all mass classes to be the same.
The above results make thus only sense in space particle dualism theory.

After this phenomenal conformation of anisotropic expansion, let's now divert our attention to the entropy of these black holes. Adding up all the above entropies we find that the entropy of the already mapped part of our universe is:

$$S_{U_{obs}} = 2.71101 \times 10^{76} \, \text{J/K}$$

We can now analyze the entropy density by looking at different redshift volumes:

Entropy density for $z < 1$:

$$\rho_S = \frac{S_{z \leq 1}}{\frac{4}{3} \pi \Delta x_z^2} = \frac{1.30792 \times 10^{76} \text{ J/K}}{2.202311095 \times 10^{80} \text{ m}^3}$$

$$\rho_S = 0.0000593885 \, \frac{\text{J}}{\text{K}}/\text{m}^3$$

This entropy is a dimensional entropy and can therefore not be subjected to the assembly interpretation of entropy, according to which entropy is the logarithm of the number of indistinguishable configurations that don't change the macroscopic properties of a physical body.

In order to transform the above entropy into what we may call configuration entropy, we have to divide it by the Boltzmann constant:

$$\rho_S = \frac{S_{z \leq 1}}{\frac{4}{3} \pi \Delta x_z^2} = \frac{9.47322929 \times 10^{98} \text{ k}}{2.202311095 \times 10^{80} \text{ m}^3}$$

$$\rho_S = 4.30149460370689 \times 10^{18} \text{ k/m}^3$$

Entropy density for $z < 2$:

No supermassive black holes surveyed in this redshift range!

We could now go on to $z < 3$ which is mapped rather well, and then cut out the section in between, but it wouldn't really improve the accuracy of our entropy density value.

Inserting our above value for ρ_S into our temperature-density equation from chapter 4.8 yields:

$$T_{max} = \sqrt[3]{\frac{\rho_S c^3 h^3}{16 \pi k^3 \zeta(3) T^3}} = \sqrt[3]{\frac{\rho_S}{20,286,824.875}}$$

$$T_{max} = 5,963.0497688 \text{ K}$$

This is surprisingly close to the decoupling temperature of 2,900 K, which indicates that the full age of the universe can't be much larger than the age of the CMB.

Using the above initial temperature we arrive at a new full age of the universe of:

$$t_0 = \Delta x_z \left(\frac{T_{max}}{T_0} - 1 \right)$$

$$t_0 = 86{,}670{,}797{,}702{,}863 \text{ yr}$$

In the next chapter we will be using this new full age to see if the redshift periodicity in the Sloan Digital Sky Survey (SDSS) can be explained with Baryonic Acoustic Oscillations (BAO) within the framework of space particle dualism.

4.16 Baryonic acoustic oscillations and the full age of the universe [18.07.2019 (speed of sound in plasma)*; 18.07.2019 (dominant size of BAO)]

*In collaboration with David Chester (UCLA & MIT; US).

In chapter 4.2 we made our first encounter with the fractal structure in the galaxy distribution around us. We want to now see whether or not this structure can be explained in terms of sound waves originating from density differences induced by primordial dark matter black holes in the early universe.

The speed of sound in a plasma is given by:

$$c_s = \sqrt{\frac{\gamma}{m_i} Z k_B T_e}$$

Whereby γ is the adiabatic index, which approaches $4/3$ at relativistic speeds (3 is related to the 3 dimensions of space), Z is the charge state, in this case 1, because nucleosynthesis hasn't taken place yet, T_e is the electron temperature, m_i is the ion mass, which is in this case simply the mass of the proton, and k_B is the Boltzmann constant as usual. Inserting the average T_μ between the decoupling temperature T_d and the maximal temperature T_{max} yields:

$$c_{s_{t<t\_b}} = \sqrt{\frac{4}{3} \frac{1}{m_p} k_B T_\mu}$$

$$c_{s_{t<t\_b}} = 6{,}983.7505 \text{ m/s}$$

We can use this sound speed to find out the radius of baryonic acoustic oscillations in the early universe.

$$r_{BAO} = (t_0 - \Delta t_d) \times \frac{c_{St<t\_b}}{c}$$

$$r_{BAO} = 1,037,587,987.7 \text{ lyr}$$

This corresponds pretty neatly to Hartnett's famous redshift spacing:

$$\Delta z_3 = 0.0246$$

$$\Delta s_3 = 1,002,923,528.9 \text{ lyr}$$

In standard cosmology this is a distance of about 250 million light years.
There are four periodicities $\Delta z_{1,2,3,4}$, but Δz_3 is the only one which appears to the naked eye, without any computer analysis.[1] Δz_4 has the highest statistical significance, but it is pretty much just twice Δz_3, so it is obvious why it doesn't appear to the naked eye.

Now we want to project the baryonic acoustic oscillations in the present universe onto the density fluctuation we observe in the cosmic microwave background. Using simple trigonometry we should be able to find the angle the smallest of these fluctuations span in the sky:

$$2\,r_{BAO} = \frac{\theta}{360°}\,2\,\pi\,\Delta t_d$$

$$\theta = \frac{2\,r_{BAO}\,360°}{2\,\pi\,\Delta t_d}$$

$$\theta = 0.0028221817$$

Using the SPD-specific distances of Hartnett's periodic spacings Δz_3 and Δz_4 instead of our above r_{BAO} leads to:

$$\theta(\Delta s_3) = 0.0027278963$$

$$\theta(\Delta s_4) = 0.0049678762$$

An approximation that can be taken for relating the angular scale θ to the multipole moment l of the spherical harmonics that are used to analyze the cosmic microwave background is:

$$\theta = \frac{180°}{l}$$

An angle of $\theta = 0.002$ would then correspond to:

$$l = \frac{180°}{0.002} = 90{,}000$$

This is beyond the presently accessible multipole range, which is at $l = 17{,}283$, corresponding to an angular scale of $\theta = 0.0104148$.[2]

According to the above we can predict that the cosmic microwave background should be free of fluctuations at a multipole value of $l = 90{,}000$. Beyond that range we can expect to see inhomogeneities only if we were to reach a resolution that is high enough to resolve the imprint of single primordial black holes, which is rather unlikely.

However, it should be noted, that this is not quite how the CMB data is interpreted usually. Because the CMB has a peak at $\theta = 1°$, it is believed that this should be the identified with the dominant baryonic acoustic oscillations.

Yet when we look into a single $\theta = 1°$ area we can see up to 100 hot and cold spots. If the CMB was really merely 13.8 billion light years away, we could be convinced that maxima in angular scale of $\theta = 1°$ could correspond to inhomogeneities caused by sound waves, but when it is 42 trillion light years away, things are very different. We then have to regard the smallest visible irregularities as the dominant fractal structure, while maxima in angular scale are more likely to be correlated to larger clusters.

Note:
1. Hartnett, J.G.; K. Hirano (2008). "Galaxy redshift abundance periodicity from Fourier analysis of number counts N(z) using SDSS and 2dF GRS galaxy surveys". *Astrophysics and Space Science*. Cornell University Library.318: 13–24.arXiv:0711.4885.Bibcode:2008Ap&SS.318...13H.doi:10.1007/s10509-008-9906-4.
2. Planck Collaboration et al. 2013a.

4.17 Probing the SPD-specific distances using supernova data [27.11.2019 (SPD-specific luminosity of most distant supernovae; most distant supernova have unnaturally low GR-specific luminosity); 14.12.2019 (surveying all Ia-supernovae ever observed)*; 15.12.2019 (average GR- and SPD-magnitude; 46% of GR-luminosities violate stellar physics); 01.01.2020 (comparison with expectation curve for average GR- and SPD-magnitude); 04.01.2020 (overlaying the graphs)**; 07.01.2020 (four models)]

*In collaboration with Hải My Thị Nguyễn (Vietnam).
**In collaboration with Việt Khôi Nguyễn (Vietnam).

As we saw in chapter 4.12, intergalactic distances are much larger than assumed in general relativity based cosmology. This also implies that the supernovae we observe have higher luminosities than currently assumed.

A certain type of supernovae, so called Ia-supernovae, are often used as so called 'standard candles', because their typical luminosity is known and can be used as a measure for their distance.

Instead of luminosity in cosmology often another measure of brightness is used, namely magnitude. Absolute/apparent magnitude is probably the most counter-intuitive concept/measure in astronomy: high values correspond to low luminosity, while low values correspond to high luminosity.

The absolute magnitude M is related to the luminosity L according to:

$$M = -2.5 \times \log_{10}(L/L_0)$$

Where L_0 is the zero-point luminosity, equal to 3.0128×10^{28} W, while the apparent magnitude m is related to the absolute magnitude M according to:

$$m = M - 5 + 5 \times \log_{10}(D)$$

With D being the distance. At a distance of 10 mpc the apparent and the absolute magnitude are equal.

Theoretically, all Type-Ia supernovae have a standard blue and visual magnitude of

$$M_B \approx M_V \approx -19.3 \pm 0.3$$

However, when we go through supernovae catalogues, we find many supernovae which are categorized as type-Ia, but which have lower or sometimes higher absolute magnitudes. What absolute magnitude we ascribe to a certain supernova depends on how we translate redshifts into distances. We know that distances are larger in space particle dualism theory than in the standard model, therefore supernovae must accordingly also be more powerful.

How much more powerful at least are supernovae according to space particle dualism theory?

As we learned in chapter 4.12, SPD-specific distances are given by:

$$t = d = \frac{3\,c\,z}{H_0 - H_{min}} - \frac{3}{2} + \frac{H_{min}}{c}$$

For short distances, which means $z \ll 1$, we can simplify that to:

$$d = \frac{3\,c\,z}{H_0}$$

Which is very similar to the mainstream formula for $z \ll 1$, namely:

$$d = \frac{c\,z}{H_0}$$

That means the discrepancy between standard model luminosities and SPD-specific luminosities is at least of a factor of 3. However, the further away we go, the larger the discrepancy becomes. That is because the precise formula for the look-back time (light travel distance) in the mainstream depends on complicated differential equations, which simplify to the above term only at low redshifts.

What are the magnitudes of some of the Ia-supernovae in our own galaxy?
There are only two records of such supernovae in our own galaxy, one is from the year 185 and one from the year 1006. Both observations were made by Chinese astronomers and without instruments. Although archives list their absolute magnitudes respective as -16 and -12.49, that doesn't seem to be data we can rely on.[1]

Magnitude is a logarithmic measure, and so a factor of 3 translates here roughly to $+1$. We have to keep that in mind when looking at the catalogue of all supernovae. When doing so, we however face a very serious issue, and that is the fact that supernovae rarely have the theoretically expected magnitude.

In recent years it has often be pointed out that Ia-supernovae might after all not be very good 'standard candles' and especially that in the earlier universe the typical magnitude of supernovae might have been different.[2] As we learned in chapter 4.12 what the mainstream deems as the remote past is just recent history according to the cosmology of space particle dualism theory. What indeed is very different at the redshifts standard cosmology deems as representing the remote past is the distance assessments of standard cosmology and space particle dualism theory respectively.

When the distance assessments become more and more incorrect, that influences the assessments of absolute magnitudes of supernovae too.

If our distance assessments are correct, we have to expect observing that the most distant observed supernovae have record luminosities. If the luminosities for the most remote supernovae the mainstream finds turn out to be rather low, or even below average, then that shows that there is something terribly wrong with the distance assessments uses. Let's now look at some of the most distant supernovae and see what magnitudes they have.

As mentioned above, the smaller the magnitudes the higher the luminosity. Absolute luminosities are normally negative numbers, so a big negative number implies a big luminosity as well. So while those are technically very small numbers, they look big, so it is easier for us to associate them with large luminosities.

What magnitude should we deem as anomaly low for a most distant supernova?
Since many supernovae which are relatively close to the Milky Way have magnitudes of below -30, we should deem everything above -25 as at least surprisingly low (reminder: high magnitude means low luminosity). And if a most remote supernova doesn't even reach the magnitude which is considered 'standard', namely -19.3, then we can surely consider it as highly anomalous.
If none of them reaches this standard, then the mainstream is in deep trouble once again.

How do we calculate SPD-specific magnitudes?

Step (1): We take the redshift z and multiply it with $\Delta x_z = 39,625,009,190$ lyr to get the SPD-specific distance.

Step (2): We enter the redshift z into a mainstream cosmology calculator like for example [www.kempner.net/cosmic.php], and choose $H_0 = 70.4$ km s^{-1} mpc^{-1}; it is the most frequently used Hubble value in the mainstream, because it lies right between the misguided CMB-based value of 67.04 km s^{-1} mpc^{-1} and the latest supernova based value of $H_0 = 74$ km s^{-1} mpc^{-1}.
The CMB-based value is derived from the CMB indirectly and is entirely relying on general relativity. However, since we are comparing mainstream magnitudes to those of space particle dualism, we have to use the most mainstream value for H_0 when we compute GR-specific supernova properties.

Step (3): We then calculate the ratio between the square of the GR-specific distance and the SPD-specific distance in order to find the difference in luminosity:

$$\frac{L_1}{L_2} = \frac{d_1^2}{d_2^2}$$

Step (4): We translate the GR-specific magnitude into a luminosity, multiply that luminosity with the factor from step (3), and then translate that back into a magnitude.[3]

Let's now look at the top-5 of the most distant supernovae and compare the magnitude and luminosity they have according to the standard model with the magnitude and luminosity they have according to the cosmology of space particle dualism theory:

1. SN1000+0216 (Date: 2006/11/22)
   m_{max} [band]: 24.37
   M_{max} [band]: −21.6622 (SPD-specific: −27.3)
   Redshift: $z = 3.8993 \pm 0.0074$
   GR-distance: 11,660,900,000 light years.
   SPD-distance: 154,509,798,335 light years.
   SPD-brightness: 6,387,431,900,554.343 $L_\odot$.
   According to SPD it is 175.569029864 times brighter than currently believed.

2. UDS10Wil (Date: 2010/12/30)
   m_{max} [band]: 27.13
   M_{max} [band]: −17.604 (SPD-specific: −22)
   Redshift: $z = 1.914 \pm 0.001$
   GR-distance: 9,916,250,000 light years.
   SPD-distance: 75,842,267,589.7 light years.
   SPD-brightness: 50,667,101,177.3 $L_\odot$.
   According to SPD it is 58.496203988 times brighter than currently believed.

3. SCP-0401 (Date: 2004/04/03)
   m_{max} [band]: 25.2
   M_{max} [band]: −19.316 (SPD-specific: −23.57)
   Redshift: $z = 1.713$
   GR-distance: 9,559,710,000 light years.
   SPD-distance: 67,877,640,742.5 light years.
   SPD-brightness: 211,330,614,393 $L_\odot$.
   According to SPD it is 50.4154957499 times brighter than currently believed.

4. SN1997ff (Date: 1997/12/23)
   m_{max} [band]: 26.8
   M_{max} [band]: −17.764 (SPD-specific: −22.06)
   Redshift: $z = 1.755$
   GR-distance: 9,639,310,000 light years.
   SPD-distance: 69,541,891,128.4 light years.

SPD-brightness: $52{,}239{,}555{,}416.5\ L_\odot$.
According to SPD it is 52.0476467906 times brighter than currently believed.

5. SN Primo (Date: 2010/10/10)
m_{max} [band]: 24
M_{max} [band]: -20.31 (SPD-specific: -24.4)
Redshift: $z = 1.545 \pm 0.001$
GR-distance: $9{,}210{,}200{,}000$ light years.
SPD-distance: $61{,}220{,}639{,}198.5$ light years.
SPD-brightness: $462{,}653{,}626{,}737\ L_\odot$.
According to SPD it is 44.1832438073 brighter than currently believed.

Looking at GR-specific luminosities:
From these only (1) and (5) even get above the 'standard' luminosity, while (3) is right at it. (2) and (4) are below standard luminosity, despite being found at record redshifts.

Looking at SPD-specific luminosities:
All of these supernovae are exceptionally bright, with all of them being at or above the magnitude $M \approx -22$.

According to the mainstream the farthest ever observed supernovae *UDS10Wil* and *SN1000+0216* have absolute magnitudes of merely -17.604 and -21.6622 (reminder: high magnitude means low luminosity), while randomly going through the catalogue of nearby supernovae reveals many with magnitudes of -34.99 (code name: SN1999gp) or -33.7 (code name: SN1999dk).
These exceptionally large supernovae lie at redshifts of merely 0.02675 and 0.01496.

Wouldn't we expect the farthest ever detected supernovae to at the same time be the brightest?
From the viewpoint of someone that firmly believes in the standard model, the fact that the furthest ever observed supernovae had rather ordinary luminosities could be seen as a mere coincidence.

It is easy to explain Ia-supernovae with more than the standard luminosity, while supernovae with less than the standard luminosity are very problematic. That is because the standard candle luminosity is calculated assuming a supernova that goes off because a white dwarf exceeds the Chandrasekhar-limit of $1.4\ M_\odot$ (once believed to be $1.3\ M_\odot$) through accreting mass from a companion star, while in reality many supernovae go off at masses that can be at $2.4 - 2.8\ M_\odot$. Such types of ultra-luminus Ia-supernovae are theorized to originate from the merging of two white dwarfs, which means that the Chandrasekhar-limit is exceeded only for a short moment, namely at the moment when the two stars merge. We can call them *Super-Chandrasekhar mass supernovae*. Examples for this subset of supernovae are *SN 2006gz, SN 2007if* and *SN 2009dc*.[4]

If single-white dwarf supernovae already have magnitudes $- 19.3 \pm 0.3$, and if double-white dwarf supernovae are fairly common, then a plausible average magnitude should surely by far exceed $- 19.3 \pm 0.3$. Not only that, but we should not be seeing any or almost any supernovae above $- 19$ (which is 'below' when we translate it into luminosity).

If SPD makes Ia-supernovae more natural, more like they are expected to be according to theoretical models, then that is further evidence for the intergalactic distances predicted by SPD.

In order to more thoroughly demonstrate that the supernova data support the SPD-specific distances over the GR-specific distances (GR for general relativity) we have to look at the average magnitudes of supernovae at all redshifts and see which theory does a better job at not exceeding the magnitude $- 19$ (meaning 'dropping below' in terms of luminosity). We can find a complete list of supernovae here: [https://sne.space/].[5]

The magnitudes listed in the catalog are based on a mainstream value for the Hubble constant, and thus we should use that value for computing the mainstream GR-distances. When looking at a very large number of supernovae, we can no longer use the cosmology calculator to compute the GR-distances. We can use windows excel to simultaneously calculate the GR- and SPD-specific properties of all supernovae, but for the GR-distances using the actual GR-equations would exceed the limitations of excel. We will therefore split the catalogue into Ia-supernovae with $z < 0.2$ and supernovae with $z \geq 0.2$.

Using the aforementioned equation for GR-light travel times at short intergalactic distances yields a GR-redshift-distance factor for $z < 0.2$ (on the basis of $H_0 = 70.4$ km/s/mpc) of:

$$13{,}889{,}390{,}656 \; lyr$$

For larger redshifts ($z \geq 0.2$) we have to instead use this approximation:

$$T = \frac{0.589 + 13.87 \times z^{1.25}}{0.852 + z^{1.25}}$$

As we know already from the last chapter, the SPD-redshift-distance factor (on the basis of $H_0 = 74.03 \; \frac{\text{km}}{\text{s}}$/mpc) is:

$$39{,}625{,}009{,}190 \; lyr$$

The last thing we need for our chart is to turning what we learned before into excel commands:

Magnitude to luminosity: $= 3.0128 * POWER(10, (70 - M)/2.5)$
SPD-luminosity: $= POWER((SPDdis./GRdis.), 2) * GRlum.$
SPD-magnitude: $= (-2.5) * LOG(SPDlum./(3.0128 * POWER(10,28)))$
Average: $= AVERAGE(A:B)$

Using this we can generate a list of all Ia-supernovae with their respective GR- and SPD specifics:

Ia-Supernova	Redshift	GR Distance	SPD distance	GR magnitude	GR luminosity	SPD luminosity	SPD magnitude
SN2011fe	0.000804	11167070.1	31858507.4	-18.28	6.17974E+35	5.02971E+36	-20.55643
SN2013dy	0.003889	54015840.3	154101661	-18.381	6.7822E+35	5.52005E+36	-20.65743
SN2012fr	0.0054	75002709.5	213975050	-20.2	3.62218E+36	2.9481E+37	-22.47643
SN2012dn	0.010187	141491223	403659969	-19.606	2.0959E+36	1.70586E+37	-21.88243
SN1998bu	0.002992	41557056.8	118558027	-18.5	7.56781E+35	6.15946E+36	-20.77643
SN2015F	0.00541	75141603.4	214371300	-18.8	9.97632E+35	8.11975E+36	-21.07643
SN2016dxj	0.094	1305602722	3724750864	-22.4	2.74771E+37	2.23636E+38	-24.67643
SN2018bac	0.0372	516685332	1474050342	-19.06	1.26757E+36	1.03167E+37	-21.33643
SN2013gh	0.0088	122226638	348700081	-19.2	1.44202E+36	1.17366E+37	-21.47643
SN2012cg	0.001458	20250731.6	57773263.4	-17.35	2.62404E+35	2.13571E+36	-19.62643
SN2012ht	0.003559	49432341.3	141025408	-18.079	5.13535E+35	4.17968E+36	-20.35543
SN2003du	0.006381	88628201.8	252847184	-19.5	1.90095E+36	1.54719E+37	-21.77643
ASASSN-14lp	0.005101	70849781.7	202127172	-19.841	2.60238E+36	2.11808E+37	-22.11743
LSQ12gdj	0.03	416681720	1188750276	-21	7.56781E+36	6.15946E+37	-23.27643
SN2013aa	0.003999	55543673.2	158460412	-19.962	2.90918E+36	2.36779E+37	-22.23843
SN2019kg	0.079906	1109845650	3166275984	-20.9151	6.99859E+36	5.69617E+37	-23.19153
SN2017ckc	0.054	750027095	2139750496	-18.6	8.29794E+35	6.75371E+36	-20.87643
SN2019yb	0.055	763916486	2179375505	-19.1	1.31513E+36	1.07039E+37	-21.37643
SN2003cg	0.00413	57363183.4	163651288	-18.3	6.29463E+35	5.12322E+36	-20.57643
SN2005hc	0.044983	624786460	1782451788	-19.23	1.48242E+36	1.20654E+37	-21.50643
SN1972E	0.001358	18861792.5	53810762.5	-19.4	1.73369E+36	1.41105E+37	-21.67643
SN2016dxv	0.02	277787813	792500184	-19	1.19942E+36	9.76209E+36	-21.27643
PS15ahs	0.031	430571110	1228375285	-20.6	5.23565E+36	4.26131E+37	-22.87643
SN2005ki	0.019207	266773526	761077552	-19	1.19942E+36	9.76209E+36	-21.27643
SN2013gy	0.014023	194770925	555661504	-19.3321	1.62858E+36	1.32551E+37	-21.60853
iPTF13dge	0.015854	220202399	628214896	-19.1496	1.37661E+36	1.12042E+37	-21.42603
SN2017erp	0.006174	85753097.9	244644807	-18.766	9.66875E+35	7.86942E+36	-21.04243
SN2017fgc	0.007722	107253875	305984321	-19.173	1.4066E+36	1.14483E+37	-21.44943
SN2007sr	0.005417	75238829.2	214648675	-18.2	5.74077E+35	4.67243E+36	-20.47643
SN2017deu	0.05	694469533	1981250460	-19	1.19942E+36	9.76209E+36	-21.27643

SN2016hvl	0.013	180562079	515125119	-18.6	8.29794E+35	6.75371E+36	-20.87643
ASASSN-15fa	0.027408	380680419	1086042252	-19.5747	2.03634E+36	1.65738E+37	-21.85113
SN2012hr	0.008	111115125	317000074	-19	1.19942E+36	9.76209E+36	-21.27643
SN2011by	0.002843	39487537.6	112653901	-18.12	5.33298E+35	4.34053E+36	-20.39643
SN2008hs	0.017349	240967038	687454284	-18.4461	7.20129E+35	5.86115E+36	-20.72253
SN2002er	0.008569	119018189	339546704	-18.4	6.90193E+35	5.61749E+36	-20.67643
SN2017ito	0.043	597243798	1703875395	-18.9	1.09388E+36	8.90313E+36	-21.17643
SN1989B	0.002425	33681772.3	96090647.3	-19	1.19942E+36	9.76209E+36	-21.27643
SN2006ax	0.016725	232300059	662728279	-19.4	1.73369E+36	1.41105E+37	-21.67643
SN2016fbo	0.034	472239282	1347250312	-20.7	5.74077E+36	4.67243E+37	-22.97643
SN2018gv	0.005274	73252646.3	208982298	-19.014	1.21498E+36	9.88878E+36	-21.29043
SN2017drh	0.005554	77141675.7	220077301	-17.476	2.94693E+35	2.39852E+36	-19.75243
SN2016coj	0.004483	62266138.3	177638916	-18.61	8.37472E+35	6.81621E+36	-20.88643
SN2016zc	0.033749	468753045	1337304435	-19.6901	2.2647E+36	1.84324E+37	-21.96653
SN1994ae	0.004266	59252140.5	169040289	-20.11	3.33404E+36	2.71358E+37	-22.38643
SN2009an	0.009228	128171297	365659585	-18.661	8.77749E+35	7.14402E+36	-20.93743
SN2006le	0.017432	242119858	690743160	-18.4	6.90193E+35	5.61749E+36	-20.67643
SN1980N	0.005871	81544612.5	232638429	-18.22	5.8475E+35	4.7593E+36	-20.49643
PS15bom	0.02	277787813	792500184	-19	1.19942E+36	9.76209E+36	-21.27643
SN2017hbi	0.04	555575626	1585000368	-20	3.0128E+36	2.45213E+37	-22.27643
SN1981B	0.006031	83766915	238978430	-18.72	9.26766E+35	7.54297E+36	-20.99643
SN2005ag	0.079402	1102845397	3146304980	-19.5711	2.0296E+36	1.6519E+37	-21.84753
SN2007fb	0.018026	250370156	714280416	-18	4.77497E+35	3.88636E+36	-20.27643
SN2018kyi	0.0214	297232960	847975197	-18.27	6.12309E+35	4.98359E+36	-20.54643
SN2013cs	0.009	125004516	356625083	-19	1.19942E+36	9.76209E+36	-21.27643
SN2018bhc	0.056	777805877	2219000515	-21.2	9.09851E+36	7.4053E+37	-23.47643
SN1937D	0.00518	71947043.6	205257548	-19	1.19942E+36	9.76209E+36	-21.27643
SN2017glx	0.011294	156866778	447524854	-19.1006	1.31586E+36	1.07098E+37	-21.37703
SN2005gj	0.0616	855586464	2440900566	-21.9	1.73369E+37	1.41105E+38	-24.17643
SN2005hj	0.056948	790973019	2256565023	-19.3382	1.63776E+36	1.33298E+37	-21.61463
SN2003fa	0.006004	83391901.5	237908555	-19.69	2.26449E+36	1.84307E+37	-21.96643
SN2015N	0.019	263898422	752875175	-19	1.19942E+36	9.76209E+36	-21.27643
SN2002dp	0.011639	161658618	461195482	-19	1.19942E+36	9.76209E+36	-21.27643
SN1995D	0.006561	91128292.1	259979685	-20	3.0128E+36	2.45213E+37	-22.27643
SN2017ckx	0.027675	384388886	1096622129	-21.5759	1.28626E+37	1.04689E+38	-23.85233
SN2018aqe	0.044557	618869579	1765571534	-22.5791	3.24049E+37	2.63744E+38	-24.85553
SN2007co	0.026962	374485751	1068369498	-19.19	1.4288E+36	1.1629E+37	-21.46643
SN2006lf	0.013189	183187173	522614246	-18	4.77497E+35	3.88636E+36	-20.27643
SN2011B	0.00467	64863454.4	185048793	-19.19	1.4288E+36	1.1629E+37	-21.46643
SN2001ep	0.013012	180728751	515600620	-18.9	1.09388E+36	8.90313E+36	-21.17643
SN2007ca	0.014066	195368169	557365379	-18.9	1.09388E+36	8.90313E+36	-21.17643
SN2009Y	0.009316	129393563	369146586	-17.8	3.97164E+35	3.23253E+36	-20.07643
SN2007fs	0.01719	238758625	681153908	-18.4	6.90193E+35	5.61749E+36	-20.67643

SN2006mq	0.0032	44446050.1	126800029	-18.1	5.23565E+35	4.26131E+36	-20.37643
SN2001bf	0.015501	215299445	614227267	-19.8	2.50594E+36	2.03959E+37	-22.07643
SN2015bw	0.03	416681720	1188750276	-18	4.77497E+35	3.88636E+36	-20.27643
LSQ13ry	0.03	416681720	1188750276	-20	3.0128E+36	2.45213E+37	-22.27643
AT2017dzs	0.017269	239855887	684284284	-18.0261	4.89114E+35	3.98091E+36	-20.30253
SN1999gp	0.02675	371541200	1059968996	-34.99	2.98518E+42	2.42964E+43	-37.26643
SN2018aoz	0.005801	80572355.2	229864678	-19.801	2.50825E+36	2.04147E+37	-22.07743
SN2001bt	0.01464	203340679	580110135	-18.6	8.29794E+35	6.75371E+36	-20.87643
SN2003W	0.020071	278773960	795313559	-18.9	1.09388E+36	8.90313E+36	-21.17643
LSQ11ot	0.027	375013548	1069875248	-17.6	3.30347E+35	2.6887E+36	-19.87643
SN2007kk	0.041045	570090039	1626408502	-18.8989	1.09277E+36	8.89411E+36	-21.17533
SN2014ek	0.023	319455985	911375211	-18.1	5.23565E+35	4.26131E+36	-20.37643
SN2013dr	0.016835	233827892	667087030	-19.6106	2.1048E+36	1.7131E+37	-21.88703
LSQ11bk	0.037	513907454	1466125340	-18.9	1.09388E+36	8.90313E+36	-21.17643
SN2001en	0.015871	220438519	628888521	-19.8	2.50594E+36	2.03959E+37	-22.07643
SN2002ha	0.014046	195090381	556572879	-19.4	1.73369E+36	1.41105E+37	-21.67643
SN2003kf	0.007388	102614818	292749568	-18.437	7.14119E+35	5.81223E+36	-20.71343
SN2018kmu	0.018	250009032	713250165	-17.8	3.97164E+35	3.23253E+36	-20.07643
SN2011ao	0.010694	148533144	423749848	-19.4817	1.86918E+36	1.52133E+37	-21.75813
SN2000bh	0.02281	316817001	903846460	-18.8	9.97632E+35	8.11975E+36	-21.07643
SN2008ar	0.026147	363165897	1036075115	-18.9917	1.19028E+36	9.68774E+36	-21.26813
SN2017ets	0.032	444460501	1268000294	-20.1	3.30347E+36	2.6887E+37	-22.37643
SN1999cl	0.007609	105684374	301506695	-17.6	3.30347E+35	2.6887E+36	-19.87643
SN2001ba	0.02942	408625873	1165767770	-19.54	1.97229E+36	1.60525E+37	-21.81643
SN2017him	0.05	694469533	1981250460	-18	4.77497E+35	3.88636E+36	-20.27643
SN2017glq	0.011755	163269787	465791983	-19.1677	1.39975E+36	1.13926E+37	-21.44413
SN2002cs	0.01577	219035691	624886395	-19	1.19942E+36	9.76209E+36	-21.27643
...	...	...	...	...	...	...	...

The full list would exceed the format of this book and therefore only a small sample of 100 is printed here. The full list has 8,830 entries and was obtained by filtering out all supernovae which (1) don't belong to the type 'Ia', (2) have multiple different redshifts listed, (3) have negative redshifts (two of them had) and (4) have missing information.

Filter (2) was necessary for being able to analyze the data using excel; filter (3) was necessary for obtaining meaningful distances.

What are the average magnitudes for general relativity and space particle dualism respectively?

Average GR-specific magnitude: -18.851
Average SPD-specific magnitude: -21.2627

If even the average is below what should be physically possible, then general relativity is in serious trouble. How bad it really is we can visualize better when looking at the number of supernovae with very low absolute luminosities using general relativity and space particle dualism theory respectively:

Theory	< -19	< -18	< -17	< -16	< -15	< -14
GR	4,099	1,062	287	93	23	13
SPD	213	64	22	13	4	2

Theory	< -13	< -12	< -11	< -10	< -9	< -8
GR	1	1	1	0	0	0
SPD	7	2	1	1	1	0

Translating these into percentages yields:

Theory	< -19	< -18	< -17	< -16	< -15	< -14
GR	2.41223%	0.7248%	0.24915%	0.14723%	0.0453%	0.02265%
SPD	46.42129%	12.02718%	3.25028%	1.05323%	0.26048%	0.14723%

Ia-supernovae are standard candles only when they emerge from the explosion of a single white dwarf. The collision of two white dwarfs will always generate a supernovae that exceeds the standard luminosity.

When using general relativity we find that 4,099 supernovae don't even reach the luminosity that is expected for single-white dwarf supernovae, while when using space particle dualism theory, we find that only 213 supernovae don't reach it.

That means when using general relativity to do our calculations we find that 46% of supernovae have luminosities so low that it violates the laws of physics. When using space particle dualism that figure is down to 2%.

We can summarize that almost half of supernovae violate the laws of physics when we use general relativity to estimate their luminosity and magnitude.

Apparent magnitudes

The apparent luminosity of an object drops with the square of the distance, and so the magnitude, which is an inverse logarithmic way of expressing luminosity, should increase exponentially with the distance. In order to test in which theory this law is followed more strictly, or followed at all, we will now look at how the apparent magnitudes drop with increasing distance:

Name	m_{max}	GR luminosity	M_{max}	z	GR distance	SPD distance
SN2011fe	9.48	4.86374E+24	-18.28	0.000804	11167070.1	31858507
SN2013dy	12.8	2.28544E+23	-18.381	0.003889	54015840.3	1.54E+08
SN2012fr	11.74	6.06695E+23	-20.2	0.0054	75002709.5	2.14E+08
SN2012dn	13.67	1.02558E+23	-19.606	0.010187	141491223	4.04E+08

SN1998bu	11.44	7.9978E+23	-18.5	0.002992	41557056.8	1.19E+08
SN2015F	13.1	1.73369E+23	-18.8	0.00541	75141603.4	2.14E+08
SN2016dxj	15.77	1.48242E+22	-22.4	0.094	1305602722	3.72E+09
SN2018bac	17.04	4.60225E+21	-19.06	0.0372	516685332	1.47E+09
SN2013gh	13.75	9.52731E+22	-19.2	0.0088	122226638	3.49E+08
...	...	...	...	...	...	...

This time our data contains 8,846 supernovae, 16 more than in our absolute magnitude survey. Now lets plot the data onto a distance to magnitude graph. This is the data for space particle dualism theory:

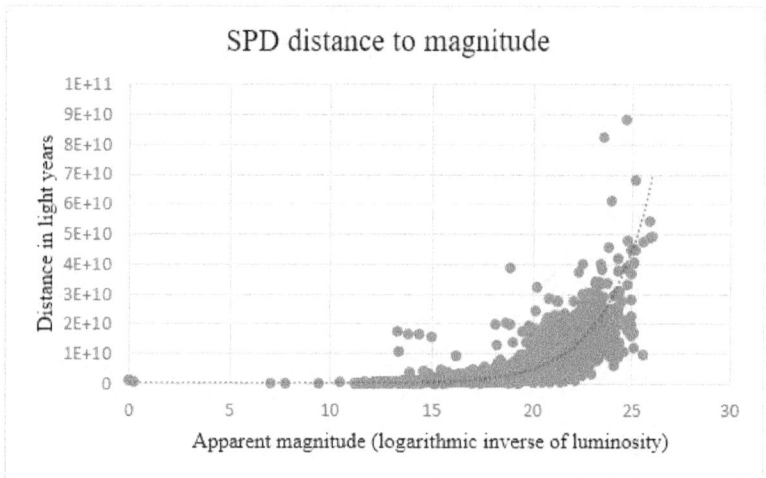

Fig. 58. Using space particle dualism theory to map different apparent magnitudes to distances in light years results in this distribution. The red line is the exponential trendline.

In order to check if the above exponential trendline in red fulfils the inverse square law for luminosities, we will create another plot, this time using the SPD-average supernova magnitude of $\bar{x}_M = -21.12740383$. The apparent magnitudes m can be calculated using this formula:

$$m = M - 5 + 5 \times log_{10}(D)$$

Where D is the distance in parasec and M is the absolute magnitude.
Doing so gives us this graph:

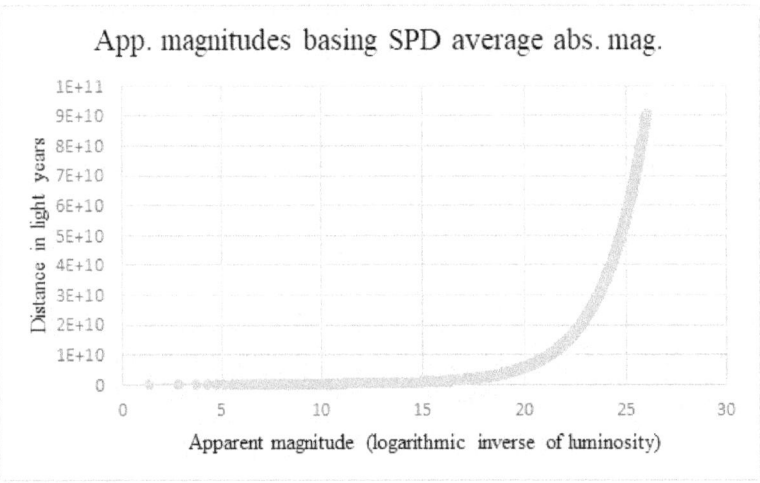

Fig. 59. Taking an SPD-average star and sending it from here far out into the universe results in this distance to apparent magnitude graph.

Now we overlay the two graphs, in order to compare them (see fig. 60):

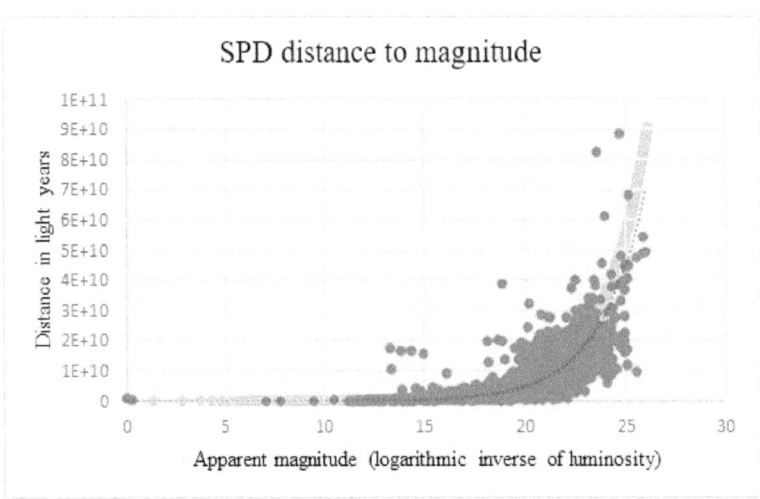

Fig. 60. The exponential trendline of the actual data in red tightly follows the inverse square law expectation line in yellow.

The result shows that the intergalactic distances space particle dualism theory provides us with fulfil the inverse square law with rather high precision.

Now we repeat the same procedure for general relativity. The result is shown in figure 61:

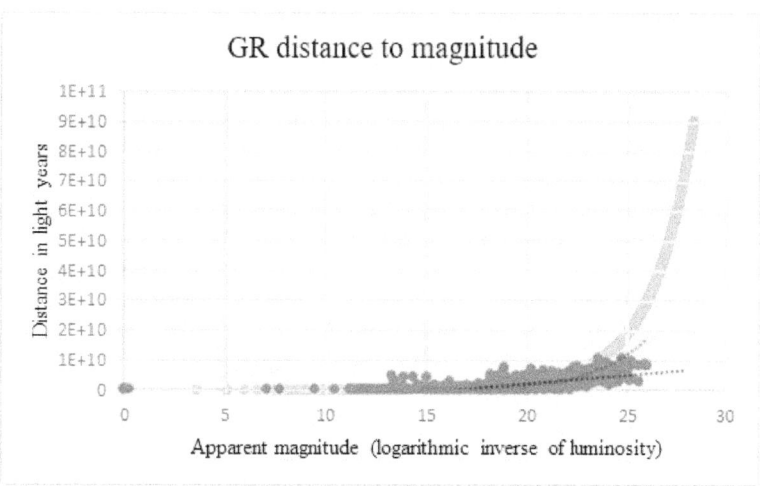

Fig. 61. The way apparent magnitudes correspond to various distances in general relativity (red line) is deviating quite strongly from what would be expected basing the inverse square law of luminosities (yellow line). At this scale the data almost seems to follow a linear trendline (black line).

The red trendline crosses the $m = 25$ mark at 12 billion light years, while the yellow expectation line crosses it at 19 billion light years. That is a discreptancy of 7 billion light years for GR.

Due to the vastly different distance assessments the GR table is almost empty. In order to give GR a second change, we examine things at shorter ranges, by changing the final value of the y-axis. For a start we set it to 20 billion light years. The results at this scale for SPD and GR respectively can be seen in fig. 62 & 63:

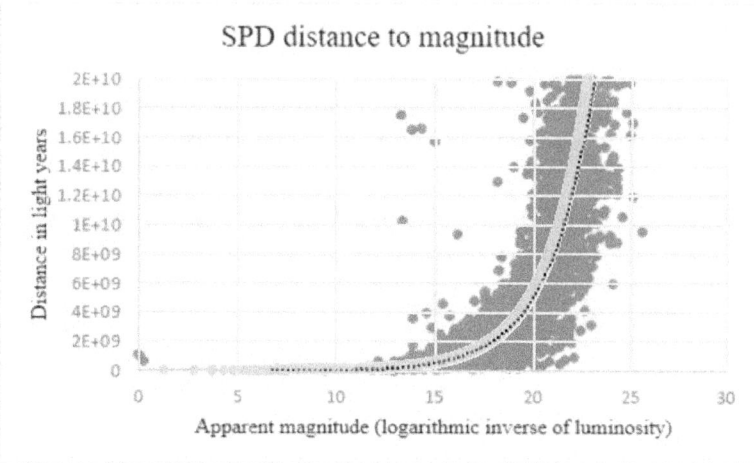

Fig. 62. Again the SPD-specific distances fulfil the inverse square law for luminosities & magnitudes.

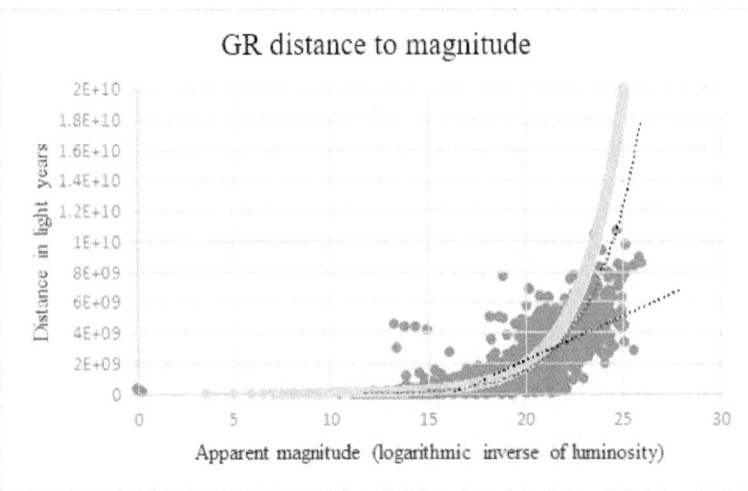

Fig. 63. Looking at it at smaller scales doesn't make things better for GR.

Again we get this discrepancy of 7 billion light years for GR.

The GR data ends pretty much at 10 billion light years. Maybe things look better for GR if we go down to this range (see fig. 64 & 65):

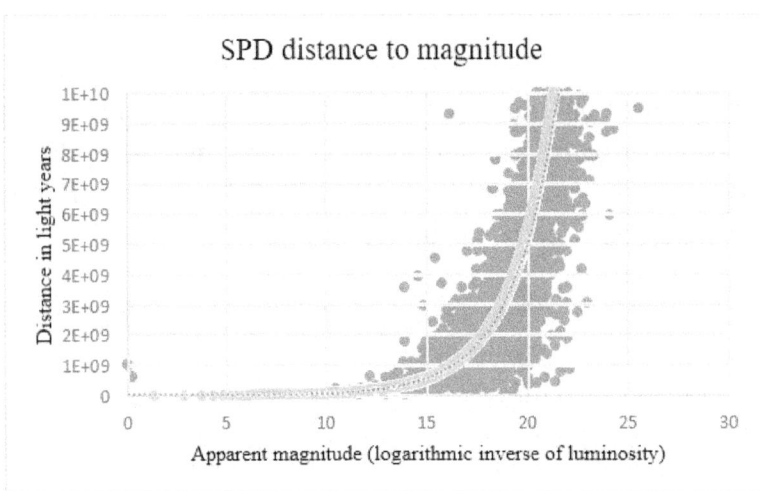

Fig. 64. As expected, the SPD-specific distances fulfil the inverse square law for luminosities & magnitudes.

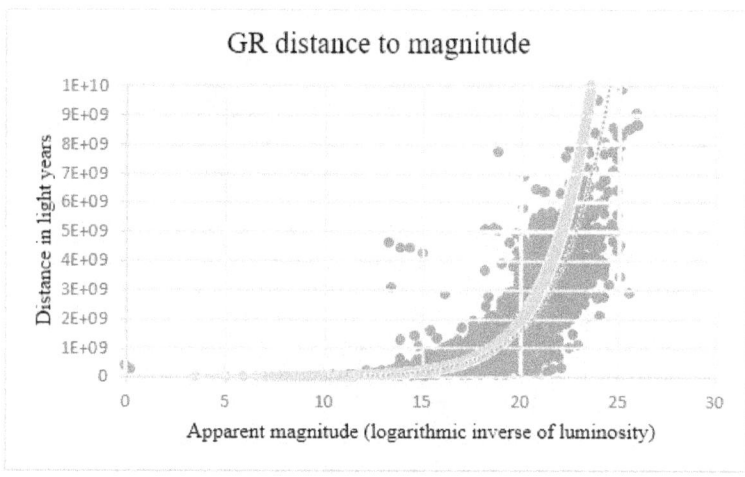

Fig. 65. The red trendline still crosses the 25 mark at 12 billion light years, just as before.

The discrepancy remains.

z > 0.2 and z < 0.2 are using different GR approximation equations. If GR is wrong, then this split will create chaos, and that is why beyond z = 0.2 things get messy for GR. z = 0.2 corresponds to 2.4 or 2.5 billion light years, depending on the chosen Hubble value.

Because the split is missing below 2.5 billion light years, it is to be expected that SPD and GR perform equally well in regards of the inverse square law of luminosities & magnitudes. In fig. 66 & 67 we can see that this is indeed the case:

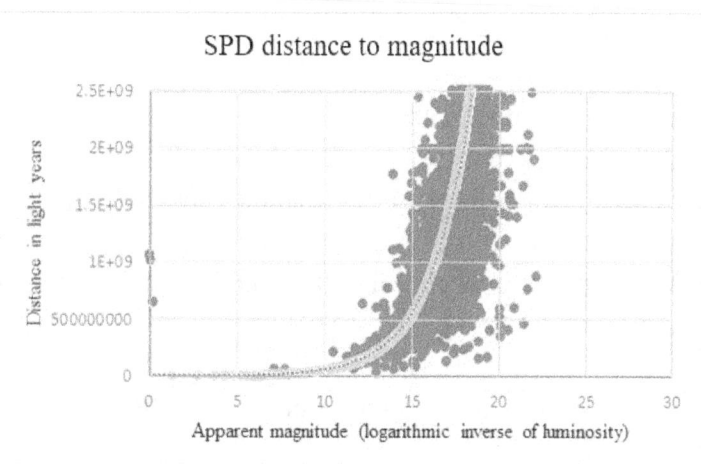

Fig. 66. Again a perfect match for SPD.

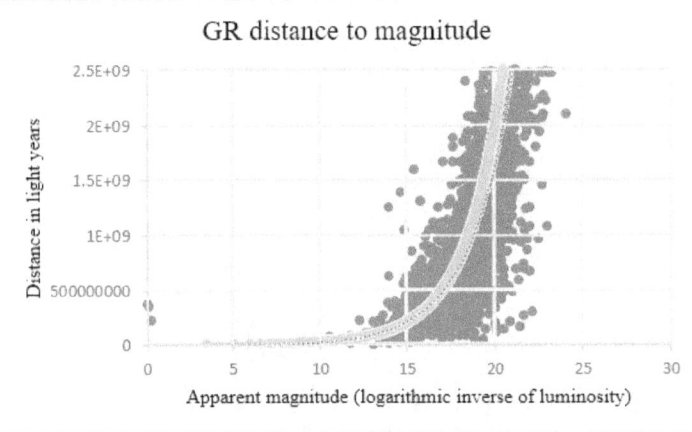

Fig. 67. At distances of below 2.5 or 2.4 billion light years the discrepancy vanishes and one doesn't notice anymore that the GR-specific distance assessments violate the inverse square law.

While the SPD-trendlines all stay inside the thick yellow line, they are not always in the middle of it. Slight inaccuracies here can be stemming from variations in the Hubble values that were used in the source data and on which the corrected SPD-specific data are based on. In the process of generating the SPD-specific data it was assumed that all the GR-source data was using the same Hubble value of $H_0 = 70.4$ km/s/mpc.

Implications

The analysis of supernova data by Riess, Perlmutter and Schmidt in the late 90's which has led to the consensus of an accelerated expansion was in many aspects similar to our above analysis.[6, 7] It ruled out

(1) a universe that decelerates due to gravity.

Their data sample was too small for them to be able to conclude more than this. Using just a few hundred supernovae can't reveal the even more serious discrepancies with GR-expectations we found here, especially when most of these supernovae are short range supernovae.

The reader may now wonder how much exactly the alignment with the yellow expectation line from above implies.
If we are concerned with only this aspect of the supernova data, then it proves at the very least that gravity is not influencing the expansion of the universe. It rules out:

(2) a universe that decelerates due to gravity and then accelerates again due to some mysterious dark energy.

This leaves two possibilities:

(3) a linearly expanding universe with light travel distances given by cz/H.

(4) an exponentially expanding universe with light travel distances given roughly by $3cz/H_0$.

Possibility (3) would have led to the same alignment with the yellow expectation curve, but it would have also led to ridiculously low average luminosities, pretty much like (1) and (2).

An average absolute magnitudes of only about -18.5 is highly implausible, and that is why (4) is the only possible conclusion.

Mainstream reaction to overly faint supernovae

The mainstream doesn't seem to have noticed that the average luminosity of supernovae is below the theoretically expected value. What it has noticed is that high redshift supernovae are too faint.

One would think that this would make researchers doubt the equations that are used to derive distances and thereby absolute luminosities, but that is not the case.

Instead the general approach now is to claim that the theoretically expected brightness of supernovae is only what is to be expected now, and that in the early universe supernovae must have had lower absolute luminosities.

Prof. Young-Wook Lee (李英旭) from Yonsei (延世) University in Seoul, Korea, and his team, even went so far to propose that there is a continuous 'luminosity evolution' which mimics dark energy.[8]
Using this luminosity evolution, Lee is not only explaining away the apparent problem with GR-specific distances, but at the same time he is using that to deny the acceleration of the expansion of the universe all together.

The hypothesis that supernovae were fainter in the early universe is in gross contrast to another mainstream assumption, namely that the first generation of stars was dominated by supermassive stars with masses in the range of $100\ M_{\odot}$.
Prior to noticing that high redshift supernovae are too faint, nobody had ever proposed or expected supernova luminosities to evolve over time. There was no theoretical reason for such an assumption.

Instead of adding more and more unnatural assumptions to save the existing models, we have to accept what the data is telling us, namely that there is no connection between these models and reality.

It should be clear by now that general relativity is not a useful theory in the realm of cosmology. Things like this proposed luminosity evolution, supermassive first stars, dark energy and inflation are just some of the many tricks employed to hide the fact that the Friedmann equations do not describe our universe.

Note:

1. It might be worth noting that this gives an average of 14.2, while the average magnitude of all supernovae ever observed within the Milky Way is 14.6.

2. Linden, S.; Virey, J.-M.; Tilquin, A. (2009). "Cosmological parameter extraction and biases from type Ia supernova magnitude evolution". Astronomy and Astrophysics. 506 (3): 1095–1105. arXiv:0907.4495. Bibcode:2009A&A...506.1095L. doi:10.1051/0004-6361/200912811.

3. A very useful magnitude-luminosity calculator can be found here: [www.omnicalculator.com/physics/luminosity].

4. Hachisu, Izumi; Kato, M.; et al. (2012). "A single degenerate progenitor model for type Ia supernovae highly exceeding the Chandrasekhar mass limit". The Astrophysical Journal. 744 (1): 76–79 (Article ID 69). arXiv:1106.3510. Bibcode:2012ApJ...744...69H. doi:10.1088/0004-637X/744/1/69.

5. The Astrophysical Journal, Volume 835, Issue 1, article id. 64, 15 pp. (2017). DOI: 10.3847/1538-4357/835/1/64. arXiv:1605.01054. Bibcode: 2017ApJ...835...64G. Link: [https://sne.space/?fbclid=IwAR1_OUBJGr2pLTpnupvdiiUzqmqnPkIS8sfjGQo2BsKe4X6Fm_-KgGHT_II].

6. Riess, Adam G.; Filippenko; Challis; Clocchiatti; Diercks; Garnavich; Gilliland; Hogan; Jha; Kirshner; Leibundgut; Phillips; Reiss; Schmidt; Schommer; Smith; Spyromilio; Stubbs; Suntzeff; Tonry (1998). "Observational evidence from supernovae for an accelerating universe and a cosmological constant". Astronomical Journal. 116 (3): 1009–38. arXiv:astro-ph/9805201. Bibcode:1998AJ....116.1009R. doi:10.1086/300499.

7. Perlmutter, S.; Aldering; Goldhaber; Knop; Nugent; Castro; Deustua; Fabbro; Goobar; Groom; Hook; Kim; Kim; Lee; Nunes; Pain; Pennypacker; Quimby; Lidman; Ellis; Irwin; McMahon; Ruiz-Lapuente; Walton; Schaefer; Boyle; Filippenko; Matheson; Fruchter; et al. (1999). "Measurements of Omega and Lambda from 42 high redshift supernovae". Astrophysical Journal. 517 (2): 565–86. arXiv:astro-ph/9812133. Bibcode:1999ApJ...517..565P. doi:10.1086/307221.

8. Early-Type Host Galaxies of Type Ia Supernovae. II. Evidence for Luminosity Evolution in Supernova Cosmology, Astrophysical Journal arxiv.org/abs/1912.04903.

4.18 Using white dwarfs to measure the age of the universe [19.08.2020 (cooling age); 21.08.2020 (*PSR J2222-0137 B* is a black dwarf; all WD with temp. below 4000 K are older than GR-age of universe); 24.08.2020 (all other astronomical objects can be ruled out); 28.08.2020 (99.99% of coolest WD fit SPD-age of universe); 29.08.2020 (refined pre-white dwarf age; all coolest WD fit age of universe); 02.09.2020 (from all WD known, 99.3% fit SPD-age of universe); 03.09.2020 (no lower limit for temp. of *PSR J2222-0137 B*)]

Note: it is a common misconception that brown dwarfs are something in between black dwarfs and white dwarfs. In fact they are not, as black dwarfs evolved from white dwarfs, and white dwarfs from active stars. Brown dwarfs on the other hand are something in between a planet and a star, with only rudimentary deuterium fusion to justify being called a 'star'.

A universe that is very old is characterized by containing 'black dwarfs'. Those are white dwarfs that have burned out completely and are thereby rendered invisible.
It takes very long time for a white dwarf star to burn out to the point where it becomes invisible. In the young universe of the mainstream, 'black dwarfs' are not supposed to exist yet.

The age of a universe that contains black dwarfs must be at least in the realm of trillions of years; therefore, finding one would be very strong additional evidence for the old universe model of space particle dualism theory.

There are many other astronomical objects with low luminosities, and astronomers that assume we live in a very young universe might mistake black dwarfs for them. Those could be:

1. Exceptionally cold white dwarfs (?).
2. Exceptionally large brown dwarfs (?).
3. Exceptionally small black holes (?).
4. Exceptionally heavy exo-planets (?).

Black dwarfs have the same mass range as white dwarfs, because the cooling process leaves the mass unchanged. The lowest mass for a white dwarf encountered so far is $0.17\,M_\odot$.[1]
The most massive brown dwarf found so far is *EPIC 212036875 b*, which has a mass of $50\,M_{al}$ (50 jupiter masses).[2] That is 9.5×10^{28} kg, and corresponds to $0.08\,M_\odot$; less than any white dwarf. None of the black dwarfs can therefore have been mistaken for brown dwarfs.

The lightest black hole discovered so far is *XTE J1650-500*. Its mass was originally estimated to be $3.8\,M_\odot$, but that was later retracted and replaced by a new estimate of

$5 - 10 \, M_\odot$.[3, 4, 5] That is well beyond the Chandrasekhar limit, which is the maximal mass of a star that can become a white dwarf. Historically this limit, which is the maximal mass at which degeneracy pressure coming from electrons can stop the collapse of a star, was believed to be $1.3 \, M_\odot$, and it is now refined to $1.4 \, M_\odot$. That means no black dwarf can have been mistaken for a black hole.

The heaviest exo-planet discovered so far is TrES-4. It has a mass of $1.7 \, M_{2\!|}$. That is well below the minimum for a white dwarf, which rules exo-planets out as potential black dwarfs.

That leaves white dwarfs as the only candidates for black dwarfs. The definition for a black dwarf that I will be using here is:

A black dwarf is a white dwarf which is so cold that its derived cooling time plus the time spend in the pre-white dwarf state exceed the GR-specific age of the universe. Ignoring the 'trouble with Hubble', we can take that to be 13.8 billion years.

Now lets go through each one of them, and see if we can find black dwarf candidates among them:

End stage-black dwarf

In 2014, a double star system consisting of a pulsar and an invisible companion was discovered. The mass of the companion, named as *PSR J2222-0137 B*, was calculated to be $1.05 \, M_\odot$. This exceeds the mass of a brown dwarf, and therefore it must be a burned out white dwarf. Due to it being invisible, the upper limit for its effective temperature is $3,000$ Kelvin. In fact the researchers who found it assume that it is not much cooler than that, because that would create a conflict with the age of the universe assumed by the mainstream.[6] The abstract of the related paper ends with this statement:

"For the implied age to be consistent with the age of the Milky Way requires the white dwarf to have already crystallized and entered the faster Debye-cooling regime."

This means that in fact the temperature could be anywhere between $2.7 \, K$ and $3,000 \, K$, and it is only the young universe model which dictates that we believe the upper limit is the actual temperature.

The age of the coolest white dwarfs

A way of empirically finding out how old the universe is, is to survey the coldest white dwarfs and calculate their approximate age. The average effective temperature of a white dwarf is about $100,000$ Kelvin.

The energy of a star can be approximated by:

$$E = N k T$$

Where N is the number of atoms inside the star, k is the Boltzmann constant and T is the effective temperature.

The age can then be approximated by:

$$t = \frac{E}{L}$$

Where L is the luminosity in power output (Watt).

We will simplify the calculation slightly by pretending that stars consist of 73% hydrogen and 27% helium, while in fact it is more something like 73% hydrogen, 25% helium and 2% other heavier elements; at least that is the composition of the sun.

The number of atoms N is then simply:

$$N = \frac{0.73\ M}{1.6735575 \times 10^{-27}\ [kg]} + \frac{0.27\ M}{6.6464731 \times 10^{-27}\ [kg]}$$

This is then multiplied with the drop in temperature, which is given by:

$$\Delta T = 100,000\ [K] - T_{eff}$$

And multiplied with the Boltzmann constant k, as stated above.

The last step is to divide by the Luminosity L, which can be derived from the absolute magnitude M (not to be confused with the M in the equation for N, which represents mass), according to:

$$L = 3.0128 \times \log_{10}\left(\frac{70 - M}{2.5}\right)$$

In Windows Excel this can be entered as '$= 3.0128 * POWER(10, (70 - M)/2.5)$'.

The time spent as an active star can be calculated according to:

$$t = \left(\frac{M_\odot}{M}\right)^3 \times t_\odot$$

Whereby $t_\odot$ is the life expectancy of the sun, which is approximately 10 billion years.

Doing so with all known white dwarfs that have an effective temperature equal or below 4,000 Kelvin, yields the following age estimates:

Name	Effective temperature in K	Mass in $M_\odot$	Absolute magnitude	Cooling age in years	Together with time spent in pre-white dwarf state
2MASS J10145164+4541479	4,000	0.301	19.92244	3.71×10^{10}	4.04×10^{11}
2MASS J13411316+0100266	3,901	1.103	19.30623	7.73×10^{10}	8.47×10^{10}
2QZ J114947.9-012045	3,098	0.907	19.20103	5.82×10^{10}	7.16×10^{10}
EGGR 536	3,835	0.334	17.35061	3.87×10^{9}	2.72×10^{11}
EGGR 539	3,890	0.57	18.43087	1.78×10^{10}	7.18×10^{10}
Gaia DR2 10122712212682003136	3,750	0.395	18.7701	1.69×10^{10}	1.79×10^{11}
Gaia DR2 1039078998380506880	2,774	0.091	20.45888	1.87×10^{10}	1.33×10^{13}
Gaia DR2 105240786245136256	3,951	0.376	18.37905	1.12×10^{10}	1.99×10^{11}
Gaia DR2 1067421801098857728	3,818	0.194	18.75916	8.20×10^{9}	1.38×10^{12}
Gaia DR2 1447925976193082368	3,845	0.413	18.80856	1.83×10^{10}	1.60×10^{11}
Gaia DR2 1626531606635150008	3,777	0.307	18.67885	1.20×10^{10}	3.58×10^{11}
Gaia DR2 2395984030477271680	3,841	0.395	18.68643	1.57×10^{10}	1.78×10^{11}
Gaia DR2 2469566617077493248	3,736	0.344	18.80817	1.52×10^{10}	2.61×10^{11}
Gaia DR2 2482539166362050304	3,918	0.126	19.00796	6.69×10^{9}	5.01×10^{12}
Gaia DR2 256087115787263616	3,124	0.084	18.82662	3.82×10^{9}	1.69×10^{13}
Gaia DR2 2761401084970307712	3,735	0.145	18.32116	4.13×10^{9}	3.28×10^{12}
Gaia DR2 3346603611845335424	3,941	0.443	18.36308	1.30×10^{10}	1.28×10^{11}
Gaia DR2 3979751266665795456	3,113	0.239	18.85554	1.11×10^{10}	7.44×10^{11}
Gaia DR2 4484289866726156160	3,970	0.439	18.5543	1.54×10^{10}	1.34×10^{11}
Gaia DR2 5181816233750415488	3,982	0.449	18.11749	1.05×10^{10}	1.21×10^{11}
Gaia DR2 5722788525986605568	3,989	0.346	19.54854	3.02×10^{10}	2.72×10^{11}
Gaia DR2 6044265144466741888	3,934	0.425	18.51832	1.44×10^{10}	1.45×10^{11}
Gaia DR2 6250213984568447872	3,885	0.422	18.03543	9.17×10^{9}	1.42×10^{11}
Gaia DR2 675615677264385536	3,990	0.449	18.34234	1.29×10^{10}	1.23×10^{11}
LEHPM 921	3,703	0.169	19.60712	1.57×10^{10}	2.09×10^{12}
LHS 2068	3,490	0.5	17.95028	1.01×10^{10}	9.01×10^{10}
SDSS J003105.86+010600.1	3,222	0.622	19.37727	4.69×10^{10}	8.84×10^{10}
SDSS J012339.94+405241.9	3,789	1.011	20.59736	2.33×10^{11}	2.42×10^{11}
SDSS J013441.30-092212.7	3,710	0.077	20.48513	1.60×10^{10}	2.19×10^{13}
SDSS J034424.72-011201.1	1,583	1.531	19.35085	1.15×10^{11}	1.17×10^{11}
SDSS J090027.55+204559.5	3,766	0.765	19.46061	6.19×10^{10}	8.42×10^{10}
SDSS J091145.12+353135.6	3,813	1.069	17.85935	1.98×10^{10}	2.80×10^{10}
SDSS J110217.48+411315.4	3,485	0.462	18.77232	1.99×10^{10}	1.21×10^{11}
SDSS J111828.49+331158.6	3,811	0.998	19.32307	7.11×10^{10}	8.12×10^{10}
SDSS J114202.49+121805.5	3,891	0.228	19.19452	1.44×10^{10}	8.58×10^{11}

SDSS J114829.00+482731.2	3,961	1.319	18.35895	3.84×10^{10}	4.27×10^{10}
SDSS J141418.88-012215.1	3,428	0.956	19.14167	5.78×10^{10}	6.93×10^{10}
SDSS J160839.60+450358.8	3,134	0.305	19.65751	2.98×10^{10}	3.82×10^{11}
SDSS J162714.07+263101.5	3,314	0.476	20.99997	1.60×10^{11}	2.52×10^{11}
SDSS J172748.71+080819.6	3,684	0.215	19.77202	2.32×10^{10}	1.03×10^{12}
WD 0222-291	3,240	0.57	18.24608	1.51×10^{10}	6.91×10^{10}
WD 0343+247	2,970	0.39	18.59727	1.43×10^{10}	1.83×10^{11}
WD 0351-566	3,950	0.58	20.0422	7.99×10^{10}	1.31×10^{11}
WD 1253+385	3,252	1.088	20.9294	3.42×10^{11}	3.50×10^{11}
WD 2115-078	3,834	0.378	19.53707	3.28×10^{10}	2.18×10^{11}
WD 2143+108	3,524	0.544	20.36728	1.02×10^{11}	1.64×10^{11}
WD 2143+108.1	3,961	0.574	20.0037	7.65×10^{10}	1.29×10^{11}
WD 2202-005	3,787	0.395	20.71262	1.01×10^{11}	2.64×10^{11}
WD 2253+812	3,940	0.129	17.67148	2.01×10^{9}	4.66×10^{12}
WD J0205-053	3,920	0.419	18.13858	1.00×10^{10}	1.46×10^{11}
WD J1403+4533	2,670	0.57	19.06409	3.24×10^{10}	8.64×10^{10}
WISEP J074509.38+262659.1	3,620	0.58	18.7483	2.44×10^{10}	7.56×10^{10}
[SLH2007] J080230.00+072858.2	3,735	0.665	19.76348	7.11×10^{10}	1.05×10^{11}

It turns out that all 53 white dwarfs with an effective temperature of below 4,000 Kelvin are much older than the mainstream age of the universe. At the same time none of them exceed the SPD-specific age of the universe, which is 42 trillion years (see chapter 4.12).

This can be regard as a good test for space particle dualism, because if the age of the universe provided by the theory was not correct, then the age of some white dwarfs could exceed this age.

From the above list of coolest white dwarfs, the following three came somewhat close to this age:

1. *Gaia DR2 1039078998380506880*: 13.3 trillion years.
2. *Gaia DR2 256087115787263616*: 16.9 trillion years.
3. *SDSS J013441.30-092212.7*: 21.9 trillion years.

Does the SPD-age of the universe give us a minimal temperature for the invisible white dwarf *PSR J2222-0137 B*?

According to space particle dualism, there are even much colder white dwarfs, or better, black dwarfs out there, just waiting to be discovered. The aforementioned pulsar companion *PSR J2222-0137 B*[7] has a mass of 1.05 $M_\odot$, and should therefore have had an active time of 8,638,375,985.31 years.

Its minimal temperature can be calculated as:

$$T_{min} = 100,000 \, [\text{K}] - \frac{(t_d - t_{active}) \, L}{N \, k}$$

Whereby t_d is the time since decoupling, also called the 'pragmatic age of the universe'; t_d the active time of the star that gave rise to the white dwarf, L the luminsotity, N the particle number, and k the Boltzmann constant.

We do not know the luminosity L of *PSR J2222-0137 B*, but we can make an estimate, by simply taking the average of all the white dwarfs with a mass of $1.05 \, M_\odot$. That average is 8.9233333×10^{19} W. Doing so yields:

$$T_{min} = -8,527,083.82095 \text{ K}$$

There are no negative temperatures, and so we know that if *PSR J2222-0137 B* is a remnant of a star that was born right after decoupling (At the time $-t_d$), then it has cooled down completely, so that it is now at equilibrium with the 2.7 Kelvin microwave background.

The age of the least massive white dwarfs

What about other types of stars? Are there other types of stars that are even older than the above white dwarfs? When trying to answer this question we obviously have to look for stars that have masses much smaller than the sun.

70% of the stars in the universe are categorized as 'red dwarfs'. Those are stars that are too small to fuse helium to heavier elements. The nuclear fusion threshold for helium is $0.5 \, M_\odot$. The minimal mass for fusing hydrogen to helium is $0.08 \, M_\odot$. Stars that are below this mass are regarded as something in between a star and a gas giant, called brown dwarfs. All they generate is deuterium. Stars that are able to fuse helium, such as our sun, are called 'yellow dwarfs'.

The following chart shows the different star categories together with their definition:

Classif-ication	Fusion	Fusion By-product	Minimum Core Temperature	Minimum Core Density	Minimum Stellar Mass
Red Dwarf	Hydrogen	He	1.3×10^7 K	100 gm/cc	$0.08 \, M_\odot - 0.5 \, M_\odot$
Yellow Dwarf	Helium	C, O	10^8 K	100,000 gm/cc	$0.5 \, M_\odot - 4 \, M_\odot$
Subgiant	Carbon	O, Ne, Mg, Na	5×10^8 K	200,000 gm/cc	$4 \, M_\odot - 8 \, M_\odot$
Giant	Neon	O, Mg	1.2×10^9 K	4×10^6 gm/cc	$\leq 8 \, M_\odot$
Giant	Oxygen	Mg, Si, S, P	1.5×10^9 K	1×10^7 gm/cc	$\leq 8 \, M_\odot$

Giant	Silicon	Si, S, Ar, Ca, Ti, Cr, Fe, Ni	3×10^9 K	3×10^7 gm/cc	$\leq 8\,M_\odot$

Red dwarfs with a mass of less than $0.35\,M_\odot$ are fully convective, which means that hydrogen is circulating throughout the whole star, being constantly refueled into the core.[8] This means that all hydrogen is used, not just the hydrogen inside the core, as it is in more massive stars. That is why red dwarfs have much larger life spans.

The sun, which is a yellow dwarf, has a life span of 10 billion years. It is believed that red dwarfs exceed this life span by the third or fourth power of the ratio of the solar mass to their masses.

$$t = \left(\frac{M_\odot}{M}\right)^3 \times t_\odot$$

Red dwarfs with masses lower than $0.25\,M_\odot$ do not evolve into red giants at the end of their lifetime, but rather turn into so called 'blue dwarfs'. When the blue dwarf has fused all of the remaining material, it turns into a Helium-white dwarf.[9]

At this point we should search the Montreal White Dwarf catalogue not for the coolest stars, but for the stars with the smallest mass.

If we ask Google what the white dwarf with the lowest mass is, it will tell us that it is a white dwarf by the name of *SDSS J0917+46*, and a mass of 0.17 sun masses. That information comes from 2006. It is the same information that is provided on Wikipedia. If we instead go to the Montreal White Dwarf database and search for White Dwarfs with masses below $0.17\,M_\odot$, we get 3,060 results!

Now, we should keep in mind that none of these should exist in a young universe, because stars with such low masses take almost forever to become white dwarfs. Their existence falls right into the face of the standard model.

The situation is apparently so embarrassing, that the scientific community avoids talking about it altogether.

According to the mainstream, all white dwarfs should have formed from yellow dwarfs, similar to our sun. In fact, if the active period is not supposed to exceed the GR-specific age of the universe, which is believed to be 13.8 billion years, then we should be seeing a sharp drop of white dwarfs at a mass of:

$$M = \sqrt[3]{\frac{M_\odot^3\, t_\odot}{t}}$$

Or, with M expressed in sun masses:

$$\frac{M}{M_\odot} = \sqrt[3]{\frac{t_\odot}{t}}$$

That yields:

$$M_{WD_{min}} = 0.89820121431\ [M_\odot]$$

Only 36% of the White Dwarfs we know are above this minimum mass. That means, that going by the standards of the standard model, 64% of all white dwarfs observed so far, are not supposed to exist.

When doing the same calculation using the SPD-specific age of the universe, which is 42,130,109,541,636 years (see chapter 4.12), we arrive at a minimal mass of:

$$M_{WD_{min}} = 0.06191593997\ [M_\odot]$$

The White Dwarf database contains 56,708 white dwarfs, and there are only 7 white dwarfs below that mass. That means in space particle dualism 99.99% of the observed white dwarfs are within the expected mass range.

The 7 white dwarfs that step out of the line are:

Name	Mass in $[M_\odot]$
[CSC85] 2233.9+1380	0.057999999
[SLH2007] J045248.17-003934.3	0.05
[SLH2007] J154656.40+214201.4	0.04
WD J1257+5428	0.032
[BGK2006] J151718.75+072609.2	0.025
WD 1559+006	0.008
[SLH2007] J173314.15+604735.6	0.005

None of them seem to be mentioned anywhere outside of the database. It is very likely that their masses have been miscalculated.

It is interesting to note that the universe age derived minimal mass for white dwarfs, which is $0.06\ M_\odot$, is very close to the minimal mass of red dwarfs, namely $0.08\ M_\odot$. This means the universe is so old now, that not a single red dwarf of the first generation can have survived till now.

So far we have only looked at the least massive white dwarfs among the coolest white dwarfs.

We shall now calculate the age of all white dwarfs. The Montreal White Dwarf database has 56,712 entries, yet, when filtering out entries with missing mass estimate or missing absolute magnitudes, then there are only 34,130.

34,086 of these white dwarfs exceed the GR-specific age of the universe, while only 215 exceed the SPD-specific age of the universe. That means that 99.37% of the known white dwarfs are within the SPD-specific age of the universe, while only 0.13% of them are within the GR-specific age of the universe.

In other words, white dwarfs confirm space particle dualism theory 99.37% of the times, while at the same time proving genral relativity wrong 99.87% of the times.

The missing first generation

When reading articles on cosmology we often encounter the aspiration that future telescopes might be able to look far enough into the past in order to sight the first stars in the universe. Stars that consist of only hydrogen and helium and nothing more.

It is rather strange to expect such stars only near the border of the visible universe. In fact, if the universe was really only a bit older than the life expectancy of a sun-like star, then any star of the first generation with a mass of about 0.8 sun masses should have turned into a Helium-white dwarf by now. Those with masses lower than that would still be shining and be categorized as Helium-yellow dwarfs and Helium-red dwarfs.

In reality, none of the stars we find in our galaxy are pure enough to be first generation stars. They all contain elements heavier than hydrogen and helium. This contradicts the Big Bang model, according to which we should be seeing low metallicity red dwarfs and white dwarfs that contain only hydrogen, helium and trace amounts of lithium.

If the universe was only 13.8 billion years old, then any red dwarf of the first generation would have to still exist today, considering their enormous life spans. The fact that we have not found a single one of these first generation low metallicity red dwarfs is hard to accommodate with the believe in a violent and very recent Big Bang.

This problem is somewhat toned down by the fact that the first generation of stars is expected to be heavier than the stars we are seeing today. That is because the first stars have formed in an environment that contains no metals. Metals can function as cooling material for gas clouds, allowing them to collapse easier.

There are two problems with this explanation.
Firstly, it is hard to see how the 2% metals that exist only at the outside boundaries of

the sun can make such a huge difference. This doubt is even more justified when we realize that these are 'metals' only in the context of astronomy, namely simply elements heavier than hydrogen and helium. The actual metals make up only about 0.1% of the sun.

Secondly, while the lack of 'metals' makes low-mass stars more scarce, it does not prevent their formation, even if we were to fully accept the significance of metals in cooling gas clouds.

Far from rendering low mass stars impossible to form, we rather merely have a shift to another kind of mass distribution in the early universe. The present day mass distribution is called the 'Salpeter mass distribution', while the first stars are believed to have followed what is referred to as the 'top-heavy initial mass function'.

Therefore, even according to the most daring Big Bang models, it is not anticipated to find no first generation red dwarfs or white dwarfs at all. Yet, this is the situation astronomers find themselves in.

For long time, the difficulty of detecting objects as dim as red dwarfs was thought to account for this discrepancy, but improved detection methods have only confirmed it.[10]

The situation the mainstream has arrived in is quite desperate.
On the one hand, preventing the formation of all low-mass stars in the first generation would solve the problem that we do not find any first generation stars. High mass stars burn out too quickly and form black holes, which hide their chemical composition.

On the other hand, having no low mass stars in the first generation would mean that all the low mass star remnants we see, would have to be from second generation stars, which makes their alleged short age even more implausible than it was to begin with. Instead of white dwarfs having 13.8 billion years to form, they would, in order to be second generation stars, have only about 10 billion years to form.

From all this it should be clear that general relativity does not have anything useful, or even non-nonsensical to say about cosmology. It is merely a theory that works for local gravitational fields, and as we will see in chapter 6, even there it is only an approximation.

Notes and references:
1. Kilic, M.; Allende Prieto, C.; Brown, Warren R.; Koester, D. (2007). "The Lowest Mass White Dwarf". The Astrophysical Journal. 660 (2): 1451–1461. arXiv:astro-ph/0611498. Bibcode:2007ApJ...660.1451K. doi:10.1086/514327.
2. Carina M. Persson, Szilárd Csizmadia & 56 more. "Greening of the Brown Dwarf Desert. EPIC 212036875 b -- a 51 MJ object in a 5 day orbit around an F7 V star".

Earth and Planetary Astrophysics (astro-ph.EP). A&A 628, A64 (2019). DOI: 10.1051/0004-6361/201935505. arXiv:1906.05048v2 [astro-ph.EP].

3. Smallest, lightest black hole identified Technology & science - Space - Space.com By Andrea Thompson, updated 4/1/2008 4:32:08 PM ET.

4. Andrea Thompson (1 April 2008). "Smallest Black Hole Found". Space.com.

5. Determination of Black Hole Masses in Galactic Black Hole Binaries Using Scaling of Spectral and Variability Characteristics Shaposhnikov, Nickolai; Titarchuk, Lev; The Astrophysical Journal, Volume 699, Issue 1, pp. 453-468 (2009) doi:10.1088/0004-637X/699/1/453 Pdf.

6. Kaplan, David L.; Boyles, Jason; Dunlap, Bart H.; Tendulkar, Shriharsh P.; Deller, Adam T.; Ransom, Scott M.; McLaughlin, Maura A.; Lorimer, Duncan R.; Stairs, Ingrid H. (2014-07-01). "A 1.05 $M_\odot$ Companion to PSR J2222-0137: The Coolest Known White Dwarf?". The Astrophysical Journal. 789: 119. arXiv:1406.0488. Bibcode:2014ApJ...789..119K. doi:10.1088/0004-637X/789/2/119. ISSN 0004-637X.

7. 'J2222-0137' is the name of the pulsar, and the 'B' is added to refer to its invisible companion.

8. Reiners, A.; Basri, G. (March 2009). "On the magnetic topology of partially and fully convective stars". Astronomy and Astrophysics. 496 (3): 787–790. arXiv:0901.1659. Bibcode:2009A&A...496..787R. doi:10.1051/0004-6361:200811450.

9. Adams, Fred C.; Laughlin, Gregory; Graves, Genevieve J. M. (2004). "Red Dwarfs and the End of the Main Sequence" (PDF). Gravitational Collapse: From Massive Stars to Planets. Revista Mexicana de Astronomía y Astrofísica. pp. 46–49. Bibcode:2004RMxAC..22...46A.

10. Elisabeth Newton (Feb 15, 2012). "And now there's a problem with M dwarfs, too". Retrieved 2019-07-10.

4.19 The end and the beginning – The last conscious observer and biological evolution [22.07.2015 (general idea); 08.08.2015 (wrongly suggested the last observer would dematerialize); 30.08.2015 (early formation of life); 13.04.2016 (external superpositions are only restricted by high frequency of measurement); 29.11.2018 (last observer is connected to first observer through morphic resonance)]

Space particle dualism takes away all the fine-tuning we have in standard cosmology where the flatness and isotropy of the universe depends on a very sensitive balance between initial expansion speed, matter density and even dark energy density.[1] Using the concept of entropic expansion all these incredibly fine-tuned parameters disappear.

This takes away all the rare boundary conditions at the very early universe, but according to the emergent universe scheme introduced in chapter 4.2, the very early universe simply emerges out of a life-inhabited present, as a logical backwards-continuation of it, with the real 'beginning' being the emergence of conscious observers. This raises the question of how this initial state containing the first conscious observer was chosen after life in the previous eon went extinct.

If life gets extinct in this universe, the world will have to find a way back into a new consciousness-inhabited quantum state; but which? It would surely be a state long time after the beginning of biological evolution, because the early stages or evolution did not bring forth conscious beings. That would mean to get a great deal of complexity out of nothing. From all the abundance of possible states, which one would the universe choose as the initial state?
Was the last conscious observer in the previous universe in any way involved in the choice of the initial state of this universe?

What happens when only one observer is left?
One is tempted to assume that when there is only one observer, the difference between dream and reality vanishes, and the observer obtains full control over his environment; after all the difference between dream and reality seems to be the fact that dreams happen in one mind, while reality is the product of many minds.

However, the last observer would still measure his environment with the same high rate as before, keeping reality just as 'stable' as it is today.
It is also wrong to assume observers to be singular entities, because any form of separation in the brain can create more individual consciousnesses. We know at least that there is consciousness and subconsciousness, giving us a limit of at least two consciousnesses per individual.

If the mind of the last observer, or maybe collective consciousness[2], has any influence on the new initial quantum state of the universe, it might then be a blueprint of the images from the world of experience from the previous universes (as we will see in chapter 5, consciousness consists of relations within an entity I call the *world of experience*).

It would be easy to accept such an initial state to be set in a primordial epoch of biological evolution. But let's say the initial state was yesterday; that would mean that all our past is mere illusion or that it is something that happened in a previous eon/universe. That is a very disturbing thought.

Is there anything that makes initial states more likely to be set in an early period of biological evolution?
If we didn't know that the overall entropy density is always constant, we might suggest that high entropy states were more likely to be chosen as initial states, simply because

a higher entropy seems to imply a higher degree of randomness. That way the initial state would always happen to be near the end, which is obviously not the case – hopefully not.

As we saw in the previous chapters there is no general entropy increase on a cosmological scale. Yet the phase space volume is still increasing, because the visible universe grows as the cosmic event horizon moves backwards with the speed of light; so if states with a big volume in phase space (corresponding to high overall entropy) were to be chosen, then the story would always start-off right at the end, contradicting everything we see around us now.
Even if it was not right at the end, but near the end, it would still mean that all our memories were built into such a randomly chosen new state. We have relics of our ancestors which are results of genuine conscious creativity, which cannot possibly appear in a randomly chosen state.

On the other hand we also have to ask ourselves if a high entropy state is really the most 'random' state.
If the only requirement for the newly chosen state is to contain observers, then the most likely chosen state would always turn out to be a single *Boltzmann brain*, living for fractions of a second, only long enough to think the phrase 'I think, therefore I am'.

This already tells us that the new state can't possibly be randomly selected.

What if the new initial state is a state in which the universe had been in during the last eon/expansion? It would be like a repetition of history with different results. But what mechanism could choose such a state that previously existed?
Could the past of our civilization possibly be an illusion? False memories built into an arbitrarily chosen and revived past quantum state? That would mean that our ancestors with all the relics they left us don't belong to our present eon/world, but to a long forgotten previous eon; not even necessarily the last one, but just any one.

When we think about which was the first moment after the universe got reinitiated, it seems that the border of the past (corresponding to the 'Big Bang' of standard cosmology) is not a possible choice, because it is not a state inhabited by any conscious observer. We would think that the world would then have to continue existing in a superposition for all the time which is needed to have conscious life emerging on earth – a very long time. But we have to recall that only mental time is time in the original way of sense. A moment in mental time is one measuring event – one collapse of the wave function. The world before the emergence of consciousness was a superposition of all possible states: a superposition of having a primordial soup and not having one; a superposition of earth existing and not existing. Nothing really happens in such a superposition, and time is not actually flowing in it. So the first real *moment* would be the time when consciousness emerged. The wave function of the universe collapses into a certain state, when one part of it contains a conscious observer. It would be odd to have the universe reinitiated to a moment long after the emergence of consciousness.

This might explain why life on earth, or in the universe in general, formed out so early. If our universe started off with a superposition of all possible states, this superposition would collapse in the earliest possible moment which allows conscious observers.
The relevant criterion for this initial consciousness-inhabited state would not be to have low entropy, but to lead to conscious life in an early period of time. This seems to be a somewhat stronger version of Wheeler's *participatory anthropic principle*. It does not only require conscious observers, but also an early formation of life in the universe. This seems like a plausible hypothesis, but is it?

The above explanation assumes that the universe would not directly chose a new observer-inhabited state, but first transition into a primordial Omnium state, a mix of all possible states, setting off from a maximal density state. We tend to imagine such an Omnium state as being an assembly of universes with different initial conditions that evolve simultaneous and in superposition. We forget that there is really no time parameter in such a state, so a state which allows the wavefunction to collapse in the 'earliest possible moment' is not really a useful choice criterion if there are no moments.

From this it is clear, that the new initial state can't possibly be chosen randomly, because all probabilistic means of explaining how such a choice might take place seem to fail.

What about our initial speculation that a previous state remembered by the last observer, or by collective consciousness, is revived? The last observer could, without attempting to access the nonlocal realm, surely only remember states within his own limited lifetime. Collective consciousness has access to all that has been experienced (see chapter 5.1), but in its entirety it is certainly impersonal, and thus can't possibly make a choice.

Morphic resonance between the first and the last observer

What if we imagine the re-initialization of the universe after the death of the last observer to be some sort of reincarnation process? Reincarnation will be discussed in chapter 5.4. There we will see that it is governed by what is called *morphic resonance*. It is a nonlocal type of natural law that is not bound to local interactions but instead governs nonlocal properties of the world, giving patterns the tendency to re-occur.

If the universe always choses the most self-similar state when confronted with a choice between vastly different alternatives of equal probability, then what is the state most self-similar to the state the universe is in before the dead of the last observer?

From the perspective of collective consciousness, the state the universe is in after the emergence of the first observer, is very similar to the state it is in before the death of the last observer. Both are states of maximal unity: all of collective consciousness is concentrated in one single brain.

Therefore, if morphic resonance is a natural law, then the universe will naturally transition from the last observer to the first observer. We may call them the Omega (Ω) and the Alpha (A), as already suggest at the end of chapter 4.2.

Morphic resonance was first suggested by Rupert Sheldrake in 1981 as an alternative to standard Darwinian evolution.[3] Since then it has successfully passed many scientific tests in both biology and cognition. One will find descriptions of these tests in most of his publications and in other publications that reference him.

Applications of morphic resonance in astronomy

Linking the last and the first observer is not all morphic resonance can do for us in cosmology. After all containing only one conscious observer is not much of a restriction on the new initial condition of the universe. What else do we get applying morphic resonance to the universe as a whole?

For maximizing the morphic resonance between the new eon and previous eons, one obviously needs to have a maximum of relational content and patterns in the new eon. Including our first-last-observer linkage from above, one arrives at the following set of rules for the selection of the new eon:

1. Allows to maintain life for a maximal amount of time, because such states are most rich in content, being interrelated with the largest amount of previous eons.
2. Allows the development to highly intelligent life, for the same reason as (1).
3. Starts off most similar to where the last state ended, namely with only one conscious observer.
4. Starts off as early as possible, because that further maximizes the amount of time life is maintained. See (1).

If morphic resonance is a fundamental law of nature, we should expect it to apply in every realm, not only biology, but to the universe as a whole. If (1) and (2) are correct, then we can conclude that the earth and everything surrounding it is fine-tuned for long sustained life by morphic resonance between eons. If this is true, it makes the *rare earth hypothesis* appear in new light.

It could mean that the earth-moon system is infinitely fine-tuned for life. We can test this prediction by looking at how sensitive this system is to change.

According to the here presented scheme it doesn't take long for life to emerge, because the formation of RNA and DNA structures is always something that belongs to the emergent past, a past that emerges out of a consciousness-inhabited present. It therefore always happens automatically and at the earliest possible geological stage; see (4).

If DNA formed out by mere chance, the opposite would be expected: a very late formation of life. If DNA would have formed out in a non-superpositioned past, simply by random collisions of molecules, then, as already mentioned in chapter 4.2, according to Crick, the chance against odds for that to happen would be $1: 10^{260}$, and thus life would be most likely to emerge in a dying universe, short before its very end.

Morphic resonance between eons and the perfection of the earth-moon system

If (4) is true and the formation of life doesn't take any time, then *climate stability* is not a precondition for life, thus a climate stabilizing moon would not be a precondition for life in general, but only for long sustained life and highly intelligent life, turning evidence for the perfection of the earth-moon system into direct evidence for (1) and (2).

So, how perfect is the earth-moon system?
First of all we have to note that while all other moons are regarded to be simply captured asteroids, our moon is so large that a planet the size of earth could not have captured it and that it can therefore only have formed as a result of a rare collision between two planet-size objects.[4] In most circumstances such collisions don't result in the formation of a single large satellite; rather the debris of the collision partially escape the planet and partially fall back to the surface. Sometimes a Saturn-like belt remains. According to computer simulations by Alan P. Jackson and Mark C. Wyatt only $8 - 10\%$ of collisions between planets result in a single moon.[5]
However, even when a moon is formed, it is not very likely for it to have the very right size and distance for having a stabilizing effect on the earth's rotation and therefore on earth's climate.

A very striking property of the moon is that it has exactly the right size to make total solar eclipses possible. That is because the sun is about 400 times further away than the moon, but also 400 times larger.

For the size of sun and moon we can note two simple facts:

   I.   When a planet, like Earth, is in the habitable zone around a star, that star will appear to be a certain size in its sky.
   II. A habitable planet like Earth needs to have a stabilizing Moon of a certain size in its sky.

What if these two apparent sizes in the sky need to be equal in order for the optimal constellation for long sustained life to be reached?

If this is indeed the optimum, then, how close are we to this optimum? For seeing this, we have to look closer at the numbers. The ratio between the radii of the sun (日) and the moon (月) is:

$$\frac{R_日}{R_月} = \frac{695,700 \text{ km}}{1738.1 \text{ km}} = 400.26465681 \approx 400$$

The ratio between the distances to earth is:

$$\frac{s_日}{s_月} = \frac{384,402 \text{ km}}{1738.1 \text{ km}} = 389.17037554 \approx 389$$

That is a difference to the factor of 11, and and that gives us a chance against odds of

$$\frac{11}{400} = 0.0275 = 0.000275\%$$

If (4) is true, then life always emerges as early as possible, and for simply allowing any form of conscious life, the requirements on the stability of the climate are not very strict. Multicellular (conscious) life in the oceans started long before land life set off. If the universe requires only that for its existence, and not human-level intelligence, then a rare stabilizing moon can't be explained with the anthropic principle or the biocentric principle, but only with morphic resonance between eons.

Here it is important to note that there are different ways to use the anthropic principle. Mostly it is taken literal, namely to mean that if humans (*anthropi*) were not around, there would be nobody there to talk about the universe in the first place. Yet, from a quantum cosmological perspective, we know that our consciousness is not different in any fundamental way from that of other previously existing and more primitive life forms. We therefore might want to replace anthropic here by *biocentric* or *sentiocentric*.[6]

We don't recall other eons, at least not directly, so we don't know if our eon is 'special', but if we are not fundamentally different from other animals, then the mediocrity principle might suffice for arguing that the perfect conditions we find ourselves in are (most likely) not a coincidence.

If we self-identify with only ourselves, then our existence will always appear to us as representing a very unlikely coincidence. If 'we' are only the humans of the 21th century, then the rareness of the moon is what we expect, but if we identify with collective consciousness; with all of sentience; then we need to assume that we live in an ordinary time in an ordinary eon and if that is the case, then the extraordinary fine-tuning for long sustained life we observe around us needs an explanation.

However, not only the moon is special: the earth is very special on its own in regards of the eccentricity of its orbit around the sun. The orbital eccentricity is a measure for how ecliptic an orbit is. A value of 0 represents a perfectly circular orbit, values between 0 and 1 represent ecliptic orbits and 1 as well as values above 1 mean the object is escaping.[7]

While a stabilizing moon and orbits with low eccentricity are important for the development to highly intelligent life, it doesn't really contribute to the survival of life in face of threads such as large asteroids on a collision course with earth. Therefore these factors only contribute to (2), not much to (1).

The main factors contributing to (1), namely long sustained life, is the presence of a magnetic field that protects a planet from cosmic radiation, as well as a nearby gas giant like Jupiter that captures asteroids and even moon size objects, that could otherwise collide with the planet, putting an end to the evolution of life.

In a very definite sense Leibnitz was right when he theorized in 1710 that we live in the *best of all possible worlds*.[8] Morphic resonance between eons explains why.

Morphic resonance in biological evolution

The emergent universe scheme can explain the formation of DNA, which is regarded to happen retrospectively here. Morphic resonance between eons can explain the life-enhancing conditions on earth. However, morphic resonance was first suggested as an explanation for the incredible efficiency of biological evolution. Why was this necessary? Can't biological evolution be explained by random mutations alone?

In his book "The Emperor's New Mind" Roger Penrose addressed this question in a sub-chapter with the title "Natural selection of algorithms?". There he pointed out that most Turing machines (a Turing machine is the abstract notion of a computer working according to a certain algorithm) from an arbitrary assembly of all possible Turing machines are completely useless, giving out only wrong answers. He then raised the question of how natural selection could have ever been effective enough to produce correct algorithms for the brain, at least for the computer-like parts of it, such as the entirety of the cerebellum for instance.

As mentioned in previous chapters consciousness has the non-algorithmic ability of creating new algorithms and judging upon their correctness. This access to truth is unique to consciousness. Robots don't have such an access and depend entirely on the programming skills of their creators.

We certainly expect non-algorithmic selection of algorithms in the brain, but what we are concerned with here is the evolution of the genetic material, which doesn't reside in the brain and which has no consciousness of its own.

Is it somehow possible that the *world of collective experience* (see chapter 5.1) has influence on genetic mutations? How can it, if our mind apparently has no influence on genetic mutations?

Different from the non-algorithmic consciousness emphasized by Penrose, Sheldrake's morphic resonance is not restricted to the brain.

Penrose's collaborator Hameroff often describes the notion of the non-algorithmic consciousness as something which has access to *Platonic information* embedded in the structure of space (or space-time) itself. If it is embedded in the structure of space itself, it is easy to imagine that not only conscious brains, but the whole universe to have access to it.

If we replace 'Platonic information' by 'collective experience', then it would be like allowing the world to have a memory. On first sight it seems hard to go so far, and not ending up with panpsychism, but as we will see, it is not.

Consciousness is responsible for the collapse of the wave function in the entire universe and not only in the tiny fraction it is directly aware of. That includes a horrendous amount of complexity under the direct control of the consciousness that each one of us has. We share this control, and it seems to be mostly taking place on a subconscious level. That is why we don't feel like having any direct control on the external world at all.

How would the sub-conscious choose between quantum states which are not of any direct relevance to the observer? It may be that the subconscious is of a more impersonal nature; that it is more part of our collective consciousness and that it has a more direct access to the world of collective experience, as well as to the Platonic world, which is part of it (Platonic information results from summarizing all eternal information from the world of collective experience; see chapter 5.1 for details).

Morphic resonance as proposed by Rupert Sheldrake is a tendency of patterns to repeat, but in his model this patterns are always patterns that have shown up before in this eon, as he doesn't make any assumptions on the existence of a pre-Big Bang era (or pre-maximal density era, using SPD terminology) or a previous universe/eon. Assuming there to be a succession of universes, and most importantly an evolution of universes, is a big game changer, because it allows both past and 'future' patterns to influence biological evolution; the future patterns being patterns in the 'future' of past eons.

An evolution process that has access to all information in the past and which is guided by an assembly of possible futures is extremely effective. It does not coincidentally lead to the development of humanoid species, but is almost inevitably driven towards it.

There have been attempts to explain the efficiency of evolution with less 'radical' ideas. In "The Self and its Brain" Karl Popper says "Organisms search for new niches, even if they did not change their organic structure. Then they change through the external selection pressure, the section pressure of the niche they actively chose.".

It may seem that this active choice imports the non-algorithmic element into genetic evolution, but that doesn't change the blind mechanisms of genetic mutation itself and therefore can't significantly increase their overall efficiency.

Since 2009 there is a revival of Lamarckism, because more and more scientific studies show that acquired characteristics can be passed to the offspring. According to the neuroscientist Guy Barry we can understand the evolution of the brain only by going beyond Darwinian evolution.[9] In an opinion paper he published in 2013 he stated:

"Darwin's hypothesis pangenesis coupled with Lamarckian somatic cell-derived epigenetic modifications and de novo RNA and DNA mutations could explain the evolution of the human brain."

Taking account of Gödel's incompleteness theorem and general properties of Turing machines, natural selection cannot be all there is. Especially not for the brain.

Note:
1. This fine-tuning problem becomes even more serious in inflation theory, where we additionally have an inflation field with arbitrarily chosen strength.
2. When there is only one observer we have the interesting case where all of collective consciousness is running on one single brain. In some sense such an observer would come very close to what one might want to call a 'god'. It would be interesting to know if the last observer would be able to feel this 'oneness', or if he would simply feel lonely instead.
3. Sheldrake, R. (1981, sec. ed. 1985); *A New Science of Life: The Hypothesis of Formative Causation*, Blond & Briggs, London.
4. The moons of Mars and Pluto are also sometimes theorized to have formed from a collision, but this is speculative and most astronomers seem to regard them to be captured asteroids.
5. Alan P. Jackson and Mark C. Wyatt (2012). "Debris from terrestrial planet formation: the Moon-forming collision". *Institute of Astronomy, University of Cambridge*. Mon. Not. R. Astron. Soc. 425, 657–679 (2012) doi:10.1111/j.1365-2966.2012.21546.x.
6. The term *sentiocentrism* puts an emphasis on sentience, while biocentrism seems to suggest that having biological organisms is sufficient. *Biological pan-psychism* is rather popular among large groups of people, so it seems necessary to make this distinction here.
7. Chambers, J. E. (1998). "How Special is Earth's Orbit?". American Astronomical Society, DPS meeting #30, id.21.07; Bulletin of the American Astronomical Society, Vol. 30, p.1051. 1998DPS....30.2107C.
8. Leibnitz, Gottfried Wilhelm (1710). "Essays of Theodicy on the Goodness of God, the Freedom of Man and the Origin of Evil".
9. "Lamarckian evolution explains human brain evolution and psychiatric disorders". Guy Barry, Neuroscience Division, Garvan Institute of Medical Research, Darlinghurst, NSW, Australia, doi: 10.3389/fnins.2013.00224.

5. CONSCIOUSNESS IN DETAIL

5.1 Relationism theory – The metaphysical basis of consciousness [06.10.2004 (identity theory is wrong); 14.10.2004 (holistic nature of consciousness); 16.10.2004 (benefit of the subconscious); 22.10.2004 (temporally extended nature of consciousness & world of experience); 26.10.2004 (world of relations); 23.02.2005 (theory on qualia); 28.02.2006 (theory on positive and negative relations); 20.01.2006 (Fregian set theory); 09.04.2015 (shift from conscious relations to conscious sets)]

What is consciousness? Consciousness is what the word says: to be con- (with) sciencia (knowledge). One might be tempted to think a computer has knowledge too, but it does not, because the symbols it uses make only sense if they are interpreted by humans. At the end a computer only consists of ones and zeros. Consciousness mainly consists of two parts: one is *conscious knowledge*, one is *Qualia*. Qualia is the way colours, smells, and feelings appear as contents of our Consciousness. Colours do not exist in the external physical world. There are only frequencies out there.

So far so good, but what is consciousness actually made of? Let us start by asking what physical bodies are made of. The old Greeks imagined the world consisting of four different fundamental elements, namely fire, air, water and soil. The ancient Chinese differentiated five fundamental elements, namely gold, wood, water, fire and soil (金木水火土). The ancient Greeks were clever enough to come up with the idea of atoms; inseparable units of matter. So they imagined their four elements to always appear in little portions of these so called atoms. They imagined atoms to be indivisible. Some contemporary scholars criticized this concept by asking what would happen if we come across a big atom and try to cut it in pieces. But a continuous nature of matter caused logical problems too. If one was able to cut something into endless many pieces, one would eventually produce an endless number of new entities.

But the ancient Greeks did not think that the world does only consist of these four elements and nothing more. They furthermore believed in the existence of immaterial objects, in particular in the objective and eternal existence of mathematical notions. They realized the objectivity and universality of mathematical findings. This was an achievement of Plato, whom this notion is named after: the *Platonic world of mathematical entities*. It does not only consist of mathematical entities though. Plato imagined it including the notions of *beauty*, *rightfulness*, and even *god*. He seems to have find these objective and eternal to some extent as well. According to him, it is the ability to have access to this Platonic world which makes us conscious.

Although he stated that Platonic bodies do not exist in the physical world, but he influenced other scholars who saw a closer connection to the physical world. Some associated the four elements with the five Platonic solids (see fig. 68).

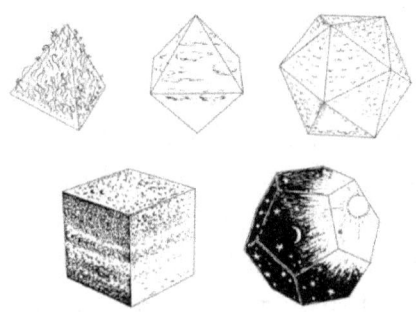

Fig. 68. Some ancient Greek scholars made associations between the four different elements and different Platonic solids. Here the dodecahedron represents the heavenly firmament in addition.

[Illustration from: Roger Penrose's "The Road to Reality", 2004]

If we compare this with our modern understanding of matter, there seems to be some truth to this. Different materials are nowadays viewed as atoms with different numbers of always same electrons, protons and neutrons. They are not made up of different materials, nor do they have substance. They are mere geometry. If we could switch off the magnetic charge of the electrons, we could walk through walls without being hindered at all, because any repulsion is only due to the repulsion of electric charge of electrons. The classical notion of 'touch' or 'hit' does not exist in quantum physics.

The ancient Greeks tried to come to an understanding of consciousness as well. They regarded the heart as the seat of the mind. The brain was thought to be only responsible for cooling the body. And some even claimed to know what consciousness consists of: 'fire' was a common answer!

As we see, their concept of consciousness was quite misled. By identifying consciousness with a certain substance we are making it equivalent to that substance. Some make this kind of wrong association even in modern times. One of the most renowned neural scientists, Sir John Eccles, for example, made so called 'psychons' responsible for mind brain interactions (see "the self and its brain", 1977). The well known philosopher and founder of the discipline of neuroscience, Rene Descartes, identified consciousness with the brain fluid. A one-to-one identification between a material substance and an immaterial entity as the mind is not helpful. It is like reducing the mental to the physical.

Only Platon himself had understood that consciousness is immaterial. This viewpoint gained popularity in the following centuries and nowadays most people correctly associate the mental with *immaterial existence*. What other things do we regard as being immaterial? The Platonic world is surely immaterial. But we saw that the physical world is not as material as we thought too. Earlier in this book we found that the world consists mainly out of information gained in measurement events. What do the physical and the mathematical world have in common? They both appear to be obeying some inner logic. What do they consist of? Mathematical notions are mostly about relations between numbers and shapes. But what are numbers? The definition of numbers is itself a tricky issue of formal logic. Cantor defined numbers as sets of empty sets with different cardinality. In this system the numbers one, two and three are defined as:

$$0 = \{\emptyset\}, \ 1 = \{0\} = \{\{\emptyset\}\}, \ 2 = \{0, 1\} = \{\{\emptyset\}, \{\{\emptyset\}\}\},$$

$$3 = \{0, 1, 2\} = \Big\{\{\emptyset\}, \{\{\emptyset\}\}, \big\{\{\emptyset\}, \{\{\emptyset\}\}\big\}\Big\}, \ldots$$

The German mathematician Gottlob Frege defined numbers in a less abstract way. He established a method to define things by natural sets. According to him the number three would be defined as the set of all sets with the property of threeness (the cardinality of three). That could be the set P_{R_x} of all people in a particular room, or the set A_{S_x} of all astronauts in a particular spaceship. Analogous other properties can be similarly defined in terms of set theory. 'Redness' would then be defined as the set of all red things. So '3' is a generalisation. It can stand for 3 cars, 3 kg, 3 sec, and so on. Some examples:

$$0 = \{\{\emptyset\}, \{\emptyset\}, \ldots\}, \quad 1 = \{\{Newton\}, \{QT\}, \ldots\}$$

$$2 = \{\{Sky, Sheila\}, \{Max, Moriz\}, \ldots\}$$

$$3 = \{\{Tick, Trick, Track\}, \{proton, neutron, electron\}, \ldots\}$$

$$3\ particles = \{\{electron, muon, tauon\}, \{proton, \ldots\}, \ldots\}$$

$$red = \{carrot, heart, apple, blood, meat, \ldots\}$$

According to Frege's definition of numbers, the Platonic world cannot be thought of as independent from the physical world. I prefer Frege's definition over the definition of Cantor, because it is more intuitive, and ties the physical and the Platonic world together. Frege's definition does not only help us to define numbers, but it also helps us understanding consciousness. As mentioned before in chapter 2.2, the perception of the colour red as a warm colour is tied to our experience with it. It tends to appear together with things which are hot or warm and this experience constitutes the colour. The feeling of seeing a colour is generated from all experiences one had with it (if *Qualia* is based on collective consciousness, then it is generated from the experience of all individuals). So when looking at the colour red, without knowing which object we are looking at, our consciousness is the set of all red things. This set has certain properties. One of them is the attribute 'warm'. We could find many more, such as 'violent', 'passionate', and so on. Our consciousness of the colour red is a cross relation between all events in which we (or others) encountered the colour red. We now said that consciousness is a *relation*. What are mathematical entities? They are relations as well. We are now coming closer to the real nature of consciousness.

One tends to regard only separate physical bodies as real. One does not see that they exist only in relation to each other. Although quantum entanglement is not mysterious anymore when viewed from the perspective of space particle dualism, but it shows an important point: nothing exists independently from anything else.

Regarding mathematical entities as 'real' requires quite a bit of abstraction. Many might ask what it helps us to do so. It doesn't seem to make a big difference. The reader might feel that mathematical entities are only pieces of logic, so that their existence is doomed to be regard as trivial as the existence of logic itself.

If we were not to discuss consciousness, then the Platonic view would not be of too much interest. Yet if we want to understand the metaphysical basis of consciousness it is absolutely essential to understand the idea of Platonic existence. We are used to think about things separately. We give them names, and view them as entities. These entities have relations among each other. We don't realize that this is an approximation; that entities themselves are relations. We talk about the colour red as an entity, without realizing that it is a relation between different experiences we had with it. We view a person as an entity, and ignore that people change all the time, and are never identical with themselves.

What kind of objects are most close to be real entities? Those are elementary particles. We can count them to world-1, the physical world. Relations between objects in the world-1, we can count to world-2, the world of relations. Some of these relations are of symbolic nature. Processes in the brain are related to things happening in the outside world. Although mere representation is not enough for consciousness, it is still a precondition for it. These kinds of relation belong to world-3, the mental world. A camera has a representation of the outside world within itself as well, but it has no personal experience, and therefore no Qualia. Without Qualia it cannot have any subjective sensations. So the mere representation of the external world does not create consciousness. What is it then? The basic units of physical objects are particles. What are the building blocks of the contents of consciousness? Qualia! But Qualia is based on experience. Experience can only be made actively, not passively. The behaviour of a camera or a computer is fully algorithmic and deterministic. Viewed from world-2, it is a relation, but doesn't contain any Qualia information. Any algorithm is a compact unit. Even if we look at learning algorithms, it is still possible to list all their deriving algorithms and sum them up to a new bigger algorithm. Everything an algorithm is capable of doing is potentially already existing in it. Therefore an algorithm cannot produce subjective experience or a flow of time. It is something static. For consciousness we need representations of the world, connected by some minimum of memory. We know that an algorithm is not capable of representing the outside world, because it is always limited. Any formal system must be incomplete and cannot be proven within the system (Gödel, 1930). Any judgement about the correctness of a formal system can only be conducted from outside the system. If a computer were to have any consciousness, it could be only a steady state; a consciousness of always the same algorithm. After all it does not make any difference if an algorithm is read out, printed or used to control some machine. It is always there, and always the same. It never changes. Without change there is no Qualia. Without Qualia, a relation is not any different from the relations of any random particles out there, or from the geometric relations between the walls of someone's house. We will soon see what is so special about relations in the brain that they can be self-aware relations.

Traditionally the world is thought of as to be consisting of three different realms. The old Greeks were trialists. They differed between the mathematical world, the physical world, and the mental world. The well-known philosopher Karl Popper added a fourth world into this picture: the world of culture. What I call the world of collective experience can be roughly identified with Popper's world of culture. Roger Penrose defines consciousness as something having access to the Platonic world of mathematical truth and beauty. Popper realized that if we want to understand consciousness as something connected to abstract immaterial entities, then these can not be only of mathematical nature, because most people do not think about mathematics only. From the viewpoint of world-1, a physical book is only the material it consists of. But from the viewpoint of world-4 it is the content it carries. It is to be assumed that before Popper most people counted all the culture content to world-3; the mental world. However, the mental world is not extended in time. We perceive every moment separately. Yet our consciousness at a given moment is a relation between experiences we made at different times. Even the understanding of a spoken sentence is extended in time. The main difference between Popper's world of culture and the here presented *world of collective experience* is that the latter contains all of our subjective experiences, and not only those which are recorded or remembered. In the here presented scheme, we was referring to the second world as the *world of relations*. Is that different from *Platon's world of mathematical truth*? It is. Platon's world contains only generalisations, not relations between real bodies. Yet if we view the generalisations as being sets consisting of more specific sets, then the world of relations is just a subset of Platon's world.

A world that is a *set universe* would at its highest level indeed look like the Platonic world. In set theory it is no problem to have a set containing itself. That is why in space particle dualism, space itself is a number set with positive, negative, imaginary and complex parts. As it seems, we indeed live in the Platonic world, and that is why our world is understandable through science and logic.

We know that any algorithm or mechanism is point-like. Everything it is and everything it can be is already there, in a potential sense, just like the past, present and future are all there at once for a completely computable universe. The flow of time, as well as consciousness comes with non-computable meaningful behaviour. Scholars who hold the view that consciousness is not computable, that we have a free will, as well as access to truth, always suggest some kind of *mind body interaction*. But here one can rightly ask what it is which is interacting here.

Consciousness is a special type of relation between representations of the external world projected into the brain at different times and remembered within the world of experience. What is special about them? They are special because they contain information about the system which are nonlocally obtained by having large interrelations between objects in the world of experience and culture represented back into the brain. If the brain was just a camera or a computer, then it could only merely project the outside world into itself without creating any additional information about it. And if consciousness were just a relation between the neurons of just the present state of mind, it could not possibly become any more than just those neurons. Even if it is a relation between the relations of neurons at different times, it is still not becoming more than just that.

For being conscious we need relations between all those representations to be represented back into the brain. How could an immaterial relation influence a physical body? How could nonlocally obtained facts about the world be represented back into the world?

Imagine a computable *toy universe* with many little points or pixels evolving in time according to certain algorithms suddenly having some of those points forming patterns encoding true information about the system as a whole. It would be like discovering an inscription in the *Mandelbrot set*, which says something like: "*I am the Mandelbrot set, I think, therefore I am (!).*" (the Mandelbrot set is a beautiful self-repeating pattern which evolves by representing a certain complex function ($z \rightarrow z^2 + c$) on the complex plane) – obviously impossible. Even the fact that we have language and correct algorithms would be unexplainable in a computable universe. It requires consciousness.

Fig. 69. The Mandelbrot set: It is an algorithmically created pattern of incredible beauty evolving on the complex plane repeating itself endlessly when we zoom in on it. Despite the presented complexity, would we ever expect to find it forming meaningful expressions in some kind of language?

Why should that be possible in a quantum universe? In chapter 2.2 we realized that all three worlds are depending upon each other. The material world is just a gathering of points, without scale or colour, and strictly speaking it can even not exist without being looked upon. Equally the mental world can not exist without having the physical world. Why do we have superpositions in physics? Because the physical world depends upon the mental world. If two states lead to the same mental state, they count as the same, and coexist in a superposition.

Let's now imagine our toy universe to be in a superposition. In this case it would be indeed possible for it to have nonlocal information about itself materializing in it. But in what language? Without information processing entities and without continuous subjective experience it would be impossible for it to have a language. Language is depending on experience as much as Qualia is. Just like the feeling of the colour red is an interrelation between all our experiences with things having this colour, as much is your understanding of a word an interrelation of all the situations we encountered it.

The main difference between Popper's world of culture and my world of collective experience is that the former does only contain information which is practically accessible to anyone, while the latter contains all subjective experiences of each of us. Basing Poppers world there is no way to understand how larger sentences can be understood before existing in written form and becoming a part of culture (he might have suggested that being expressed in a language is enough in order to belong to world 3). Another main difference is that Popper's world of culture seems to depend upon the ongoing existence of some physical representation. What implications does it have to accept the existence of the world of experience? First of all, in this world, all subjective experiences become part of one thing, without a differentiation between different individuals. There is no separate world of experience for every observer. Why? Because there are no different materials for the consciousness of different individuals. And sometimes one individual can even become two individuals, when certain areas of the brain are disconnected (see the description of different split brain patients described in "The Emperors New Mind", Penrose, 1989). But then, how comes that only relations between own thoughts are represented back into the brain? That is because such relations cannot require a too large changes in the state of the mind. They must be within the scope of the superposition the brain is in. If they exceed this superposition, they cannot be incorporated into the brain.

In this terminology we can call relations between physical bodies in the present time *type-1 relations*. Relations between the thoughts of a single person can be called *type-2 relations*. Those are the most common relations for thoughts. Then there are *type-3 relations* which are relations between the thoughts of two persons or even more. These are quite rare, and cannot easily be incorporated by the brain (or they remain subconscious, but that is speculative). Relations between the experiences of all individuals can be called *type-4 relations*, and they might be the base for Qualia. Interestingly mathematical relations can also be classified as type-4 relations, because they are generalisations of all possibilities, and therefore relations between all experiences made.

Not only Qualia, but all building blocks of the mental world might be based on type-4 relations. The consciousness regarding the meaning of a word is a relation between all the situations we might encounter it. To me it seems that we understand the meaning of new words much too faster for this understanding to be based solely on our own experience with those words. Many things which are more felt than thought can be based on type-4 relations. Consciousness consists of two parts: feelings and thoughts. Feelings come from the subconscious and they might all be based on type-4 relations. That means the subconscious has connections to other minds. These connections do not require the exchange of any signals.

Before we described Qualia in terms of sets; now we talk about consciousness in terms of 'relations'; which description is more accurate?
In the early days of the theory the emphasis was on relations and I used the name 'relationism' for it. However, what we called 'relations' can all be expressed in terms of sets as well. Although both descriptions are equally valid, it is more appropriate and generally applicable to talk about consciousness in terms of sets, because the notion of a set already includes the notion of a 'relation'. Therefore we can refer to this theory more accurately as 'conscious set theory'.[1]
The notion of a set can help us to understand the nonlocal nature of consciousness. A

set can contain another set, simply by definition and without information being transmitted from one place to another. Similar to entanglement information, this kind of information does need no time, nor any form of 'signals' in order to be obtained. This can readily explain the phenomenon of 'sinchronicity in thought', which is referred to as 'nonlocal consciousness' in the parapsychology literature and as 'telepathy' in common language. We will look at some cases of nonlocal consciousness in chapter 5.4.

For a certain set, to become part of a new consciousness, or a new conscious moment, the main criterion is its cardinality, and its grade of being related to the present consciousness. The property of being related can be described by saying that it is a subset of the set which is presently conceived by consciousness. Let's say a person is considering the set of natural numbers. It is then relatively easy for him to find the subset of prime numbers. A bit more difficult it is for him to find a set containing the natural numbers as a subset, like the set of whole numbers, or rational numbers, if he never learned about such a set. The cardinality of the set of natural numbers is $\aleph_0$ (read as Aleph-zero), which is the smallest infinite cardinality. The set of real numbers contains endless many real numbers between two natural numbers or whole numbers. Therefore its cardinality must be ∞^2 or $\aleph_1$ (Cantor called it 'C', because he could not prove within the system that $C = \aleph_1$). Mathematical information is however a kind of information which can be derived by logic, independently, without information exchange, at least in principle. More of interest here are sets of experiences or sets of culture based content. Let's say two people know a person called 'Frank'. Frank could be described as a set of all experiences anyone ever had with Frank. A subset $\mathbb{F}_A$ of this set would be the experiences person A had with Frank. Another subset $\mathbb{F}_B$ would be the experience person B had with Frank. Both sets are parts of the full set $\mathbb{F}$. His own experience with himself we can call $\mathbb{F}_{\mathbb{F}}$. When we refer to Frank, we always mean $\mathbb{F}$, except of when we say "the Frank I know". To be conscious of $\mathbb{F}$ does not mean to know all of its content but just to have a feeling for its properties. As we saw before, the feeling of seeing the colour red can be described as a set $\mathbb{r}_S$ (S for self) of all experiences one had with this colour, or as the set $\mathbb{r}$ of all experiences all conscious observers ever had encountering it. It is plausible if the consciousness of something develops from specific sets like $\mathbb{F}_A$ or $\mathbb{r}_I$, to more general sets like $\mathbb{F}$ or $\mathbb{r}$. Why? Because all these sets are standing for different kinds of Qualia content, and therefore for feelings: the feeling of seeing the colour red; the feeling of seeing Frank. These feelings have to be represented back to the brain and the body by reactions of the *amygdala nuclei* and corresponding hormone release into the body.

To move from a specific set to a more general set which contains it as a part can be called an *extrapolation*. We can, for example extrapolate the existence of the infinite set of natural numbers $\mathbb{N}$, by understanding some of its elements $1, 2, 3, 4, 5, 6, \ldots$. A computer is not able to do this kind of extrapolation, because it requires the non-local integration of information. The classical Platonic view is that we humans get in contact with the immaterial Platonic world of mathematical entities and truths when we do mathematics. The more correct and modern view should be that consciousness itself consists of this kind of Platonic bodies. 3 is as much a Platonic body as 'red' is, or as 'beauty' or even 'rightfulness' is. Those are all generalisations. We would not count them to the world of experience of any specific person. All these very general sets are type-4 relations and they all are Platonic entities. The notion of a certain person called Frank is not that Platonic. It is only accessible to a finite number of individuals living in a certain period of time and in a finite number of places. It can therefore only be a type-3 relation. Type-4 relations are rich in general Platonic information, which is useful in any situation. Type-2 relations, based on one's own experience, and type-3 relations, based on the experience of two or a few individuals are what we usually call information. It is specific. Information related to type-3 relations is *telepathy*.

When we talk about Qualia it is more useful to describe consciousness by the notion of sets. When talking about more specific contents of consciousness, it is more convenient to use the notion of relations. The notion of sets might give us the idea every person is a compact, independent and well defined entity, which is not true. The only thing that gives a person identity is his memory.

How can we have Qualia for colours or smells we see or smell the first time? This kind of Qualia could never be based on one's own experience. It can therefore only be a type-4 relation. I would suggest that all basic units of sensation are such kind of type-4 relations (relations between collective experiences).

I said that the cardinality of a set is important in order to tell if it can easily be incorporated by consciousness. Sets with high cardinality are very general, and can be incorporated easily. For example all Qualia sets are infinite. Limited sets are very specific and restricted. They represent what we usually regard as 'real information'. The feeling of seeing the colour red, or of knowing the meaning of 'sad' or '3' are usually not regarded as information. Yet they are the basic building blocks of our consciousness. Each of them is a set with a cardinality reaching at least $\aleph_0$. Here it is useful to measure the size of a set in a way different from the one Cantor suggests. We could again say that $\mathbb{N}$ has the smallest infinite cardinality, namely ∞, while $\mathbb{Z}$, which has a partner for every element in $\mathbb{N}$, has the cardinality $2 \times \infty$ (or $(2 \times \infty) + 1$ in order to include 0). '∞' can here be defined as the set of all sets with an infinite number of elements. Accordingly $\mathbb{R}$ would be $2 \times \infty^2$ (or $(2 \times \infty^2) + 1$), and $\mathbb{C}$ would be $2 \times \infty^4$. We may imagine the exponent here to be representing something like the dimensionality of the set, but it does that only indirectly, since the power of 4 here stands for a only 2-dimensional continuum. Some hold the view that the world is, on its most fundamental scale, based on whole numbers, not on real numbers. That was also one of the motivations behind Penrose's spin networks and the later twistor theory. The exact cardinality is not of too great importance for consciousness, but it is still interesting to see that all things we regard as entities have some inherent infinity in them.

How can we describe other notions of the mental world in terms of conscious set theory (or 'relationism')? For example the notions of 'good' and 'bad', 'happy' and 'sad', 'exciting' and 'boring', 'comfortable' and 'painful'? Let's start with sounds. When do we like sounds, and when do we dislike them? Usually they need to have some sort of inherent harmony. Having harmony means to be rich in <u>internal relations</u>. Internal relations are <u>arithmetic and geometric relations</u>. Another kind of relations are <u>relations to subjective experiences</u>. If we associate a certain melody with good/positive memories, we tend to like it. What are good memories then? Usually they include being accepted in a group and having many diverse experiences in different fields. Two things seem to be important here: harmony and rich experience. Both lead to rich relations. Why does pain feel bad? Pain is related to death. Death means a loss in interpersonal relations. Why is it so terrible to go to jail? Because isolation means the lack of excitement. Fewer things happen there, and those few things which happen, are usually bad things.

If consciousness likes meaningful truths, rich interrelations and harmony, which are all based on logic, then why does it like jokes and humour as well? For most jokes and humour is even more exciting than serious and meaningful discussions based on real valuable arguments. Well, paradoxes are an important part of logic, and they play an important role in science. 'Reductio ad absurdum' is one of the most powerful and often the only way to prove the non-existence of something, and thereby often the existence of an alternative. This is a very special type of relation. We could call it an 'anti-relation'. Poor relations can be called 'negative relations', and rich relations can be called 'positive relations'. What has to count to which depends on the level of consciousness of a person. Some are easy to excite, some easy to get bored.

Note:
1. Another example for a set theory of consciousness was developed by the mathematician and AI expert Ben Goertzel (2011). His theory however didn't aim to explain Qualia, but only 'self-awareness' (using hypersets that contain themselves). Reference: Goertzel, Ben. (2011). Hyperset Models of Self, Will and Reflective Consciousness. International Journal of Machine Consciousness. 03.10.1142/S1793843011000601.

5.2 Preconditions for consciousness [14.10.2004 (holistic nature of consciousness); 16.10.2004 (benefit of the subconscious)]

Let's see if this theory of consciousness, can lead to clear-cut preconditions for consciousness. Basing what is already said, we can name the following preconditions for consciousness:

1. Information processing (<u>excludes plants</u>).
2. Quantum coherence (<u>excludes computers</u>).
3. Large entangled quantum superpositions (holism) (<u>rules objective reduction theories out</u>).
4. Memory (<u>excludes certain types of fish</u>).

5. Non-algorithmic response pattern (<u>excludes insects and computers</u>).

6. Sensibility of the information processing to microscopic quantum states (the opposite of resilience; <u>excludes computers</u>).

7. Inventory of already existing useful algorithms obtained through a combination of Darwinistic and Lamarckian evolution (some guidance through collective consciousness based upon the evolution in former universes is also conceivable) (<u>excludes plants</u>).

8. Preconditioned algorithms for emotional reactions (related to Qualia) (<u>excludes insects</u>).

9. An information filter, which separates the conscious from the subconscious and allows their fruitful cooperation (<u>excludes computers</u>).

Let's now discuss the single points:

1) Information processing:

The brain is very different from a computer in the way it processes information. First of all, there is no certain area in the brain for memory. Memories related to different senses are saved in areas specialized for those senses. Visual information is kept in the visual cortex, language related information in the Broca's area, smells in the *olfactory bulb*, and so on.

For the conscious part of the brain, there are no fixed reactions to incoming signals. There is even not a part of the brain which integrates all information from different parts of the brain, to produce some reaction.

The whole working process of the brain depends crucially on the power of different synapses. On synapses which are so small that they are necessarily always in superpositions. Without consciousness, the brain would not be able to work at all. It is not like a computer with some interfaces for mind-brain interaction attached. Its whole construction is made for this kind of interaction. Most of its interconnections are built up long after birth. And they are not stable. The connections have to be reused for many times in order to be stable (at least 1 - 3 years!). Imagine a computer would need to run a game for many times in order to remember it. That is only one aspect where we see how different brains and computers are.

Brains with fixed path ways, and fixed input-output channels, could not develop consciousness. Both computers as well as unicellular organisms operate entirely different from the way a biologic brain operates.

2) Quantum coherence:

The operating pieces of a system which is supposed to generate consciousness ought not to be so large that they are regarded as macroscopic by other observers. They ought to be able to operate without affecting the macroscopic appearance of the system after each operation. A macroscopic model/duplicate of the human brain where each operation is visible to the naked eye could not operate, because quantum coherence would be destroyed in it. Every signal has to remain in superposition for as long as possible.

For reaching an at least subconscious level, it has to pass the *amygdala nucleus* in order to get emotionally evaluated. Only then the incoming signals can be represented by a certain Qualia. Although Qualia is something immaterial, more like a Platonic body, all the emotions it implies have to be accurately presented in the brain.

3) Large entangled quantum superpositions (holism):

It has often been pointed out that quantum coherence, with its holistic properties is the only way to explain the holistic nature of consciousness (Penrose; Eccles). Each neuron for itself represents only a little part of our whole consciousness; a small part of our sight; a face; a line; never a whole picture. But we seem to be aware of the whole picture at once. At least we feel that way. Why is that? It is because the wave function collapses at the same time in a big area of the brain. Why? Because the different neurons are entangled with each other. All we need for having entanglement covering the whole brain, or at least the whole cerebrum, is a superposition which is maintained long enough to reach all parts of the brain. That is no problem if signals are kept in superposition until they reach the conscious (or subconscious) parts of the brain. Although each signal remains in superposition for a half second (and sometimes even a bit more), the overall coherent state collapses with a frequency as high as 20 times a second.

Microtubules may well, as suggested by Penrose and Hameroff, be supportive for maintaining quantum coherence over the brain, but that does in no way prove Penrose's OR collapse theory. Quantum coherence in microtubule is possible in all interpretations of quantum mechanics. It just appears to be the only kind of quantum coherence in the brain which is not ruled out by OR, because OR doesn't allow superpositions of firing and not firing neurons. So OR needs microtubules, while microtubules don't need OR. Do microtubules improve or enlarge quantum coherence in the brain basing similar-worlds theory? Yes, they might. Microtubules are nets of molecule transporting pipes which remain in superposition for a respectable amount of time. Thereby they can adjust the power potentials on the different synapses of the brain without having an impact to the brains macroscopic state. That doesn't mean there are no neurons firing during that, but the wave function inside the microtubule doesn't collapse before it significantly alters the potentials in synapses located in the conscious parts of the brain. If we take OR, then the collapse happens all the time, starting even outside the brain, occurring again in the human eye, and again many times after. If we were to believe this, we would have to associate every such objective reduction as a proto-conscious event (in the words of Hameroff). Hameroff's microtubule theory is good, but it does in no way require OR as a basis.

4) Memory:

Memory is important in two aspects: one is qualia; another is the development of new algorithms. For the first aspect, qualia, memory might not be of too great importance. At least the very basic various qualia, as for example the vision of colours, could be represented by some evolutionary developed hormone reactions. For the second aspect, the development of new algorithms, memory is of crucial importance.

Not all contents of consciousness are represented in the brain. Lots of information in the brain exist non-locally and are, although being part of consciousness, not represented by any particular neuron. These contents still cause reactions and the activation of certain neurons. After many such activations the brain gets used to these reactions and memorizes them by establishing fixed reaction pathways. Those pathways (neural connections) still need to be used frequently, either in dreams or in real life, in order to not get lost.

5) Non-algorithmic response pattern:

Many animals respond in a very predictable way. Many see it as a sign of intelligence that bees have a very rigid and organized life. But being predictable is actually a proof of not being intelligent or of being unconscious.

I remember a bee being stuck in my dad's red van. The side door was widely opened, but the bee still tried to get out from the back door window pane. I was watching it for long time, to see if it would find the way out. After a while I gave up and left for about one hour. When I came back, the bee was still trying to get out from the back door window pane. It reminded me strongly of a computer having a hangover or a robot doing the same all over again, because its program doesn't allow it to do differently.

I also tested a worm by letting it walk over the same little stick over and over, till it would stop and turn into another direction. It repeated the action for 9 times, till it stopped.

Behaviours which are automatic do not require consciousness. Even if some types of behaviours appear sophisticated at times, like for example the organized behaviour of bees, they still can be modelled algorithmically, with algorithms that can be conceived to have developed in biologic evolution.

6) Sensibity of the information processing to microscopic quantum states:

The firing of a synapse on a neuron depends on its potential, and that depends on molecule movements among microtubules, which are generally always in superposition. If the successful transmission of signals over circuits in a computer would depend on instable potentials like this, it would only work very inefficiently. Yet in the brain it allows for the emergence of consciousness. Frequent activation makes the potentials stronger, but they never stop being in superposition of different potentials. Having all these synapses depending on microscopic quantum states makes the brain strongly dependent on consciousness. For a computer there are only single reaction pathways for every input. For the brain there are many possible reaction pathways for each input. Probably most of them are rather nonsensical and unlikely to be chosen. But what chooses them? It is consciousness which chooses. Without it the brain would become completely useless. Most pathways would become dead end roads or lead to nonsensical reactions.

7) **Inventory of already existing useful algorithms obtained through a combination of Darwinistic and Lamarckian evolution (some guidance through the collective consciousness based upon the evolution in former universes is also conceivable):**

An information processing system could hardly begin from zero. It would need useful algorithms to start with. Simple algorithms could easily be incorporated from the world of collective experience. The most important seem to be to have such Platonic information projected onto quantum states, which can be magnified and brought into the macrocosm. So in this picture, the brain is more like a magnifying glass for *Platonic information* at the quantum level – something entirely different from a computer, which is just a simple Turing machine. An evolution without Lamarckian elements couldn't bring forth such algorithms. Superpositions of different genes wouldn't of any use, because genes are just the basic building plans for the body and the brain, they do not process information themselves. Also they do not usually change during a lifetime.

8) **Preconditioned algorithms for emotional reactions (related to Qualia):**

Although emotional reactions are very easy to simulate, we tend to regard them as real when performed by animals. When computers simulate emotions, we know that those emotions are not real, and not felt by any subject. We know that computers have programs for the simulation of such emotions. It is quite misleading when the android *Data* in the famous TV series *Star Trek* always understanding everything, but being unable to simulate any emotion. Since Data is an android, any understanding would be merely simulated. But as we know emotions are much easier to simulate than understanding. It would be no hard thing to construct a robot which cries when being hurt, or smiles, when having a full battery. We could regard these emotional reactions, which can, to a certain extent, come by automatically, as mere symbols or representations of certain Qualia or more primarily of certain somatic conditions. Only in a fully conscious organism these symbols gain their actual meaning.

9) **An information filter, which separates the conscious from the subconscious and allows their fruitful cooperation:**

The lack of such a filter does not make one unconscious. It would even have some few advantages, but all in all it makes useful responses harder. If there was no separation between the middle brain and the cerebral cortex at all, then the brain activity would probably end up in chaos. Some hierarchic structure might be essential for the way brains work.

According to this scheme insects and plants have no consciousness, because they do not have a brain. For animals whose behaviour is not simulable on a computer, we know for sure that they must have consciousness. But even for others, we can assume that they have consciousness, if they share more or less the same brain structure. If an animal's behaviour is based on more than just basic instincts, then it feels and knows. Of course, in a biological sense, plants are alive; so are insects; but they do not think and feel. A plant is not in any essential way different from a rock, and an insect is not in any essential way different from a robot.

How could we test if an animal has consciousness? We could watch out for the typical 1/2 second reaction time for conscious reactions. Unconscious reactions are always as fast as calculations on a computer or basic reflexes like blinking with the eyes or the Achilles reflex.

Another means would be to check if certain animals are able to cause a reduction of the wave function. As described in chapter 2.5 this can be done using the *quantum eraser experiment*. At first sight one would assume that all one has to do is to look if a measurement done by a certain animal could be undone by the quantum eraser. Yet, same as for humans, signals in conscious animals would need some time to reach the conscious level. That means we would have to see if the presence of the animal changes the time window within which the quantum eraser can operate. If the consciousness of the animal is slower than that of a human, it would be hard to test for it. In that case one would have to isolate the animal in a faraday cage.

The quantum eraser experiment shows us that the wave function does not collapse before it reaches consciousness. However, some others take it as a proof for decoherence theory (the collapse of the wave function to be an illusion).

5.3 Psychokinesis and thermodynamics [16.11.2016 (binary psychokinesis); 19.11.2016 (entropy decrease in target might be compensated by entropy increase in brain); 02.12.2016 (metal bending due to high entropy of metals); 12.01.2017 (quantum randomness in chaotic systems); 31.01.2017 (maximal shifting speed for PK); 10.03.2017 (binary microkinesis weak because small time window); 02.12.2017 (calculating the chance for MMI on RNGs); 23.12.2017 (shift to the frequency of subconsciousness); 01.05.2018 (entropy drop in brain & no compensation); 08.07.2018 (realistic shifting speed); 11.01.2020 (excluding four non-genuine telekinets)*; 26.01.2020 (Girard's levitation); 12.02.2020 (Maurer's levitation)*]

*In collaboration with Benedikt Maurer (Germany).

The main reason why most physicists tend to ignore the measurement problem is their assumption that quantum probabilities are always obeyed and that consciousness doesn't have any influence on them. In classical physics it would indeed be a violation of the physical laws if observers could influence physical behaviour with their minds. That is because in classical physics the world follows strict rules with an unambiguous deterministic evolution. There is no place for free will in this picture. In quantum mechanics things are quite different: here there are only probabilities for this and that to happen. However, physicists who inherited the old deterministic thinking patterns would insist that these quantum probabilities have to be followed strictly by nature. Yet even they would have to admit that a violation of probabilities is much less serious than a violation of laws. After all, no law is violated when something unlike but possible occurs.

In classical physics unlikely things happen very rarely and correspond to a huge drop in entropy. If, for instance, a cup of tea falls down from a table and hits the floor, the gravitational potential energy it has from being positioned on the table is turned into kinetic energy and then, when hitting the floor, transformed into heat of the floor's molecules. It would be a possibility for the molecules of the floor to all move in the right way for the cup to be reassembled and thrown back onto the table. Since all movements of particles are reversible, this represents a true possibility, but it is so unlikely that it is so unlikely that it never happens.

In quantum mechanics that is not the case: here, everything that is possible always happens and it happens all at the same; in quantum superposition. It is then, at the end, the conscious observer which chooses one state among all the possible states included in the superposition. If we assume that the observer has no influence on the choice of states, then this doesn't change much: events which which were unlikely to occur in classical physics would then still be unlikely to occur in quantum physics.

Yet as we saw in previous chapters, the observer must have an influence on the choice of states. That will not make cups which scattered on floors reassemble and jump back on tables, but it makes a whole series of things possible which would have to be regard as 'magic' in classical physics.

Macro-psychokinesis: shifting objects

Coming back to the cup of tea, we could imagine that at a particular moment in time the microscopic heat movement of the cup's constituting particles happened to be such that all particles move into the same direction, just by mere chance. That is very unlikely to occur from a classical perspective, but we have to keep in mind that these particles obey the laws of quantum mechanics; and that means when they move, they move in superposition. If we for instance sit in front of a cup of tea, this cup has a certain degree of fuzziness to it. It is measured about 40 times a second and that might seem like a lot, but for the fastly moving constituting particles a 40th second is enough time for a lot of collisions and therefore changes in impulses which bring particles on new courses, rearranging all their impulses, again all in superposition, so that after every 40th second we are confronted with a new choice regarding all those endless many impulses. As if this wasn't chaos and superpositioning enough, we also realize that observing a cup of tea does only determine its overall impulse, still allowing a long list of impulse assemblies, expecially for those particles below the surface of the object. Consequently the observer should have a huge variety of states he can choose among.

And yet, we never see a cup moving on its own. Similar as with the scattered cup reassembling and jumping back on the table, this corresponds to a huge drop in entropy. Using the quantum mechanical approach to entropy introduced in previous chapters we can say that the drop in entropy is evident from the fact that only very few and special states within the vast superposition of the warm tea cup on the table have it moving across the table pushed only and solely by the thermal energy of its molecules. If we take the case of an entropy reduction only in impulse space and only concerning the impulse direction, maintaining the thermal character of the considered object, we can try to find out what impulse, or speed, we get when all particles happen to move into the same direction in space.

The most probable speed of particles in an ideal gas is given by the Maxwell-Boltzmann distribution, namely

$$v_p = \sqrt{\frac{2\,k_B\,T}{m}} = \sqrt{\frac{2\,R\,T}{M}}$$

Where R is the gas constant, m the molecule mass and $M = N_A\,m$ the molar mass.[1]. The *mean speed*, which is the expected average value after many single measurements is

$$\langle v \rangle = \frac{2}{\sqrt{\pi}}\,v_p$$

In liquids particles travel at about 8 times lower speed, and in solid objects they are just jiggling around. That is because they are constantly hit by surrounding particles, due to the higher density. That seems like we would need another formula for calculating their speed, but since the many hits distribute the kinetic energy only among the same particles, nothing really slows down. The apparent slowing down is only due to zigzagging. That means the same formula applies to molecules in solid bodies as well.

For a nitrogen molecule we have

$$\langle v \rangle = 470.5 \text{ m/s}$$

For iron this speed is

$$\langle v \rangle = 337.4 \text{ m/s} = 1,214.64 \text{ km/h}$$

Of course this is simply a maximal speed. In a realistic situation only a part of the particles could move into the intended direction, so that the overall speed is much lower.

Is there any way to get estimates for how one could reasonably move objects by the means of telekinesis?

A macroscopic object of one kilogram is composed of somewhat 10^{25} atoms. If we are interested only in two directions, say left and right, then it is possible to look at the thermal movement of particles as representing some form of random number generator (RNG). If we then take Dean Radin's 51% average chance for influencing such a RNG, then we can derive a certain speed from that.[2]

If this chance was 50%, then the object would not move, if it was 100%, then it would move at the above calculated maximal speed. So a chance of 50% corresponds to 0%

of the maximal speed, while a chance of 100% corresponds to 100% of it. Therefore a 51% chance corresponds to 2% of the maximal speed, and that is:

$$\langle v \rangle \times 0.02 = 6.748 \, \text{m/s} = 24.2928 \, \text{km/h}$$

Shifting speeds observed so far are well below this.

It is important to note here, that although this speed corresponds to the average success rate for influencing single quantum events, the entropy drop associated with doing this on so many particles at once is enormous. If 0.02 of 10^{25} atoms are affected, then the drop in entropy is:

$$\Delta S = \log [0.02 \times 10^{25}] = 23.30103 \, k$$

While the odds against chance are one part in 2×10^{23}.

This is a tremendous entropy drop, and if this was to be compensated by an entropy increase in energy consumption of the brain, then the after-effects would be devastating. In fact the brain doesn't need to directly coordinate 2×10^{23} atoms; instead it only has to coordinate the corresponding perceptional neurons; their number is what counts. A typical object we focus our attention on is displayed in our brain with a resolution of about $4^{+3}_{-3} \times 10^6$ pixels (one to seven megapixels), depending on size and distance. Each pixel corresponds to one neuron, and thus we can now calculate the required entropy drop based on coordinated neurons alone, which yields:

$$\Delta S = \log [0.02 \times 4^{+3}_{-3} \times 10^6] \approx 5 \, k$$

If this approach to entropy in psychokinesis is correct, then it should be easier to do psychokinesis thought a screen or from greater distance.

Real examples of shifting by means of psychokinesis

How do real (non-theoretic) cases of mind matter interactions (MMI) or psychokinesis (PK) look like? We can find that out by look at the performance of some of the telekinets that have performed under scientific testing. After studying all of the available material one thing becomes very apparent: in most cases shifting of objects by means of PK happens in single pushes or pulls that are rather light and if the object is exposed to friction, that can't make it move more than just a few centimeters each time.

The most renown example for someone who was able to shift objects by the means of psychokinesis is *Nina Kulagina* from Russia (1926 - 1990). She has been studied and tested very carefully by hundreds of scientists, both domestic and foreign over a period of two decades. These tests were going on until before she died. In the years before her death her health had deterioted so much that she had already lost her ability to shift objects.

When PK is performed on objects with low friction, the movement is smooth. That can be inducing a rotation or making a wheeled object move.[3] But when it is performed on inflexible objects on a flat surface the movement can only happen in small steps of which each takes only a small fraction of a second.[4] The fact that all seriously claimed cases for PK have this inherent discontinuity, having all target objects move only in small steps is another indicator that we deal with a real phenomenon. Any real phenomenon has real restrictions and rules.

Kulagina often suffered from headache and even pain in her spine. After performing she often looked exhausted. This may suggest that the entropy reduction involved in those acts of psychokinesis is fully compensated by an entropy increase in the brain & body of the practitioner.

Of course, how long the object keeps moving can't tell us anything about the frequency of wavefunction collapse; all we can infer from it is that the object is moved in distinct single pushes.

The more particles an object is made of, the higher is the entropy reduction involved in moving it by means of PK. Yet for certain objects the entropy from which it starts off is important too and also if it is a chaotic system or not.

Binary macro-psychokinesis (no entropy reduction)

One example for a chaotic system with high entropy is the atmosphere that surrounds us. That is due to it's low measurement rate. It is measured only when a current within it; a blow of wind; hits another object or the skin of the observer. That means superpositions of different gas states can be sustained for much longer than the 1/40 seconds one has for clearly visible objects. However, the measurement happens via a target object which has to be pushed by the air, and the measurement frequency for that object is again 40 Hertz. Yet a flexible object like a wind wheel measures only a small fraction of the air around it. It can not give any information about air currents far away from it. That means, most of the air in a room remains unmeasured and in superposition. Thus it should be much easier to have air currents influenced by ones intention than any other physical objects.

This is not only because of the high quantum entropy of air, but also because no entropy reduction is involved when objects are simply moved by air currents.

Let's take a thin aluminium structure that is balanced on a needle, a so called psi wheel, and placed in a room with no regular air currents, just chaotic thermal fluctuations. Generally there is a 50% chance for it to rotate clockwise and a 50% chance to rotate counter-clockwise. From a classical physics perspective we would have to think that the decision about the spin direction is depending upon some microscopic initial conditions. Taking into account quantum mechanics we however have to see that all the air around the freely rotatable object is in superposition. That means the 50% chance is a real quantum probability, real quantum randomness, instead of classical chaotic behaviour. And if we can make something which has a 50% chance of happening happen at each trial, we have violated probabilities, without lowering entropy, because both states correspond to the same coarse-graining volume in phase space, meaning that both events have the same likelihood.

So, is it really that easy for an observer to influence such a system simply using his conscious (or subconscious) intention? I have been experimenting a lot with psi wheels, which are very sensitive to tiniest currents of air. Usually it is only considered a demonstration of psychic abilities if one is able to make the psi wheel rotate while it is totally isolated from any air current; typically placed inside a glass hub or a glass cylinder. However, if we choose a certain preferred rotation direction for it, and find it choosing this direction in 90% or even 100% of the trials over a long period of time; say a half hour; then there is a strong case for mind matter interaction, even without the use of a glass hub. This is basically what turned out to be the case in most of my sessions with the psi wheel, some of which I have recorded on video and which are available on the internet.

Binary micro-psychokinesis (no entropy reduction)

In chapter 2.1 I mentioned that the smallness of the effect the observer has on random number generators (RNG) can be explained with the high timing precision required. The observed chance to successfully influence a RNG is 51%. Can we derive this chance from something?

In a first attempt we may assume that a person needs one second time to maximize his 'willpower' and that he can release it only into one of the 20 or 40 frames he sees per second (40 Hertz is the frequency of subconsciousness (f_{SC})). We know that a wilful action takes about 1 second at least and therefore this assumption seems slightly justified.

Using this approach we could calculate the chance of influencing a random number generator as 1/40. That would be 2.5% and with half of that being part of the first 50% which one gets for free (mere chance), that is

$$z_{RNG} = \frac{1}{2} + \frac{1}{2 \, f_{SC}}$$

$$z_{RNG} = 0.5125$$

51.25% above mere chance. So 51% is what we observe and this model predicts 51.25%. Some drop down is to be expected due to bad performance/condition of some participants.

If we choose consciousness and not subconsciousness as the thing that causes psi, then we get 52.5%. According to this model influencing random number generators is hard mainly due to the required timing. One has to focus ones intention into one singular event. That is very different from shifting objects where one has unlimited time to do so.

Dice tossing

In the early years of parapsychology it was more common to use a pair of dice or a single die to test for mind over matter interactions, than to use a RNG. According to Dean Radin the success rate at such experiments is 51.2%, which is slightly above the success rate for RNGs. A die is macroscopic and one may think that only people able to perform large scale psychokinesis are able to influence it. However, if the die rotates fast enough it can complete a full rotation within a 20th or 40th second and in that case we have distinct superpositioned states which occupy the same volume in phase space, corresponding to the same entropy. In simple words: different outcomes have the same likelihood. The measurement frequency can even be lowered by the use of a non-transparent dice shaker held by a robot arm. When the shaking is done by a robot, measurement through the skin is avoided.

We can now differentiate between several forms of PK, namely:

Level-0 PK: Influencing *binary chaotic systems* or chaotic systems where there are multiple macroscopically distinct states which share roughly the same likelihood of occuring.
Description: to make (macroscopic) events which have a $10 - 50\%$ chance of occurring take place every time (50% is the binary case).
Entropy drop: None.
Timing: System needs to be influenced before random movement in the unfavoured direction starts.
Expected success rate (with training): $\sim 90\%$.
Note: Although this is the easiest form of telekinesis, it seems to never have been studied in detail.
Practitioners: Sky Darmos (binary macro-PK), Micheal Grubb (aerokinesis).

Level-1 PK: Influencing singular quantum decisions at the moment of measurement (binary micro-PK).[5]
Description: To make (quantum) events which have a 50% chance of occurring take place every time at precisely distinct moments.
Entropy drop: None.
Timing: The system must be influenced at the exact moment of measurement.
Success rate: 51% (found experimentally with odds against chance of $1 : 10^{12}$ (one to one trillion)).[6]
Note: In some years the success rate was as high as 53% (1968; 1972).
Practitioners: Uri Geller (influencing Geiger counters), but mostly random subjects.

Level-2 PK: Influencing chaotic systems with a small number of distinguished states (usually dice-tossing) in the moment they reach a final state.[7]
Description: To make macroscopic events which have a $1 : 6$ (one die) or $1 : 36$ (two dice) chance of occurring take place at precisely distinct moments.

Entropy drop: Reduced by a factor of 6 or 36.

Timing: The system must be influenced shortly before reaching the final stationary state.

Success rate: 51.2% (found experimentally with odds against chance of $1:10^9$ (one to one billion)).[7]

Practitioners: Usually random subjects.

Level-3 PK: Shifting[8], bending[9, 10] or scattering[11] by inherent thermal energy (high level macro PK).

Description: To make macroscopic events which have a vanishing chance of occurring (almost zero) occur in a short period of time.

Entropy drop: Enormous.

Timing: System doesn't need to be influenced at a particular moment in time.

Success rate: About 90%, given enough time is provided (15 sec. up to 2 hours).

Level-4 PK: Levitation (almost impossible PK).

Description: To continuously force other objects or oneself into a certain very unlikely state.

Entropy drop: Same as for Level 3, but prevented from rising again for significant amounts of time.

Timing: System doesn't need to be influenced at a particular moment in time.

Success rate: Same as for level 3.

Level 3 and 4 are pretty spectacular; so spectacular that only very few people in the world are able to perform them. Some of these people have been scientifically tested and found genuine, such as the aforementioned Nina Kulagina; others have not been tested yet. Mostly it is not that these people don't want to be tested, but they are simply not approached by scientists. Many psychics believe the scientific establishment is so settled in its views that no experiment in the world could cause a paradigm shift. And indeed, paradigm shifts are usually caused by observation backed theories, not observation alone.

A skill possessed by very few, which is so strange that it seems to defy logic itself is always doomed to be regard as fraud or trickery, no matter for how long the practitioner performed it without any evidence for the use of a trick to ever emerge. In many cases the rejection of the paranormal is based on nothing but unsustainable rumors of fraud spread by a few influential individuals. Those are a group of people one could best describe with the term 'militant materialists'.

Uri Geller

Probably the most widely known psychic of the 20th century is Uri Geller (full name: Uri Geller Freud). Most people know that he was tested by Russell Targ and Harold Puthoff at the Stanford Research Institute (SRI) and that he worked for various secret agencies. Very few know that he was not only tested by the SRI and the CIA, but also by many other institutions all over the world and by a number of very renowned world class physicists like David Bohm, John Taylor and Richard Feynman.[12, 13]

Fig. 70. Russell Targ with Geller who is successfully making a one gram weight on an electronic scale lighter several times. This was one of the few PK tests at SRI. Most of the testing at SRI was concentrating on extrasensory perception (ESP) instead. PK was tested more thoroughly at Birkbeck college, London and other universities or institutions.

It is interesting to note that when David Bohm was going about testing Geller, his long term collaborator Basil Hiley warned him that this might undermine the scientific credibility of their work. Today it is pretty much the same: most physicists would risk their carrier if they really went about to test psychics. Even convinced materialist scientists admit that experiments are not encouraged and unwelcomed within the community (Paris Weir; personal correspondence; 2017). Even Geller himself told me that testing psychics is harmful for ones carrier and that only well established scientists should even consider to take the risk (Uri Geller; personal correspondence; 2017).
For exactly this reason most of the parapsychology research in the early years was done by secret government agencies rather than open research institutes. The government wanted to enable scientists to research these phenomena without being distracted by the controversies surrounding them. Some researchers, such as Stephan A. Schwartz, refused to work for secret agencies because of ethical concerns (Stephan A. Schwartz; personal correspondence; 2017). The main purpose of such agencies is not truth seeking but practical military and espionage applications such as remote viewing of enemy facilities, mental erasing of floppy disks with secret data and remote mind control. A large body of these projects is now declassified and freely available.

David Bohm, who was not working in secrecy, tested Geller's psychokinetic abilities at Birkbeck college in England. The SRI experiments had concentrated on his telepathic and remote viewing abilities because they are of most use to secret agencies, but Bohm was foremost a quantum physicist, so what interested him most in Geller was his ability to influence how the wavefunction collapses – the so called observer effect.

Fig. 71 & 72. Left: Bohm with Geller at Birkbeck college. Right: Bohm with Geller who attempts to mentally bend a thick metal cylinder. After that he successfully influenced the Geiger counter on the left.

One of the people involved with the experiments judged the outcome as follows:

"My personal judgment as a PhD physicist is that Geller demonstrated genuine psychoenergetic ability at Birkbeck which is beyond the doubt of any reasonable man, under relatively well-controlled and repeated experimental conditions."

Jack Sarfatti PhD. Physicist

Geller also went to France to the Foch hospital in Suresnes, to be tested by renowned physicist John G. Taylor.

Taylor was fairly convinced by Geller and wrote a book that explored how electromagnetism might explain extrasensory perception, psychokinesis and other paranormal phenomena.[14] He failed to link these phenomena to the observer effect and went for a materialistic explanation using forces. When he experimentally and theoretically ruled out both electromagnetism as well as a new fifth force for the witnessed phenomena, he gave up on it and resigned to doubting the evidence for the paranormal altogether. His doubt, or disbelieve, was not based on evidence for fraud, because there was never such evidence, instead it was solely based on his inability to find an explanation.

Taylor is a classical example for a physicist that was trapped in a mechanical world view that seeks to explain everything in the world in terms of forces and signals.

Most other physicists who witnessed the 'Geller effect' first hand instead found that it reinforces the orthodox interpretation of quantum mechanics in which the wavefunction collapse should, at least in principle, allow for the existence of PK. It is what is to be expected from a world that is based on quantum mechanics as opposed to the world of classical physics.

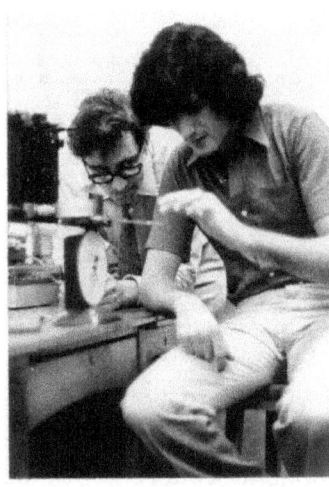 

Fig. 73 & 74. Left: John G. Taylor observing Geller while he mentally bends a piece of metal. The metal is attached to a scale that is supposed to measure if any force is applied to it by Geller. Right: Taylor also measured Geller's brain waves during many of the experiments.

All in all there are about 24 scientific papers from scientists (foremost physicists) that have tested Geller and found him to be genuine. In addition to that the *Central Intelligence Agency* of the United States had found him to be genuine as well and had employed him as a professional psychic working undercover for many years.

Going into the accounts of all the scientists who witnessed Geller's skill would go beyond the scope of this book. If the reader is interested in these accounts, I can recommend his website, which is www.urigeller.com. It contains endorsing statements from at least 60 scientists, 32 trick-magicians and 13 celebrities.

Scattering by means of psychokinesis

Level-3 PK includes shifting, bending and scattering. Geller does bending mainly; and shifting only on some very few occasions. There is no example of him scattering glass by psychic means.

An early example for *psychic scattering* is the Belgian Erwin Wiesner. Different from Geller, Erwin Wiesner never tried to reach out to scientists and although he is very skilled, he knows conjuring tricks as well. This blurs the boundary between fact and fiction, but as he says himself, this blurring is not intended (Personal correspondence; 2018).

In fig. 75 & 76 we see him demonstrating psychic scattering on German television:

Fig. 75 & 76. Erwin Wiesner using PK to scatter a wine glass under a glass hub
[Taken from an episode of the German show "Na siehste!", 1989].

In 2020 he reinforced the claim that this was genuine psychokinetic scattering (personal correspondence; 2020). In his words (translated from German):

"The requsites were from the ZDF - the table & the glass - I had no chance to bring anything with me. I got the tripod table and the beautiful plastic salad bowl after the show from the ZDF. After that I got about 400 m away from the screen and did the same thing again. A light bulb bursted. There is no explanation for it, except for Chi. Read the book about Chi-powers and they have the best explanation. First you have to believe in the thing - without that nothing works. Regards, Hardy."

A German psychic that has also mastered the art of psychic scattering is Benedict Maurer. Benedict Maurer has ten years of experience with PK and is very consistent. Different form many other subjects, his body seems to show high degrees of electromagntic activity. This made him believe that psychokinesis is an exclusively electromagnetic phenomenon, and not based on quantum mechanics and wavefunction collapse (Benedikt Maurer; personal correspondence; 2017 & 2018).
Recently however we have agreed that it can be a mix of both, and that in many, if not most, cases it is quantum mechanics alone which provides the explanation (Benedikt Maurer, personal correspondence; 2020).

Benedikt Maurer has been tested by various institutions, including the BND and the skeptics group GWUP. Since the establishment of GWUP 15 years ago he was the only one to pass one of their tests. Interestingly that was about detecting magnets.
Maurer had succeed 9 out of 13 times in detecting in which of 10 boxes a magnet was hidden. The odds against chance for this are roughly 1 in 10^9. 7 hits would have already been enough to pass the test.[15]

Fig. 77 & 78. The GWUP tested Bendikt Maurer's magnetic sense, and he passed the test 9 out of 13 times. The chance against odds for this is already 1 in 10^9, but the GWUP wants to do the same test again scanning for magnets inside his body, before giving him his prize money. It is unclear when the second test will take place.

While this test was not examining his skill of psychokinesis, it shows that his body has unusual electromagnetic activity. This has an influence on how he moves objects. Sometimes it is through psychokinesis, and sometimes through somatically induced electromagnetism. It is important to keep this in mind when examining what he does.

Levitation

The levels-0 to 3 can all be described by the observer influencing how particles distribute in impulse space. Level-4 is different: if levitation was achieved by constantly shifting an object upwards, then it would be very unstable. The impulse upwards would have to exactly equal the acceleration downwards. That would require almost infinite precision and on that ground it seems unlikely as an explanation.
Could it then be a manipulation of position space? All particles have both impulse and position uncertainty and it shouldn't be harder to influence the position than to influence the impulse. After all, if perception creates reality, then all what would be required for levitation would be some form of consciously induced perception loop, where neurological signals somehow repeat themselves in such a way as to create the sensation of levitation and thereby making it a real phenomenon.

At the beginning of this chapter we noted that psychokinesis means inducing an entropy drop within a target object. In order to save the second law of thermodynamics into the human realm, we then argued that this drop in entropy is probably compensated by an entropy increase in the brain.
But is this really the case? If we measure the difficulty of a certain psychokinetic skill by the entropy drop caused within the target object, then we need to count all the particles it is made of, while only a small fraction of them is actually measured by the observer, namely those on one side of the objects surface.
If we count in all particles the object is made of, then the entropy drop is tremendous and if that entropy drop was supposed to be compensated by some form of entropy increase inside the brain, then it would certainly be a devastating increase.
If the brain is a reflection of the outside world, then the entropy drop should occur within the brain foremost.

If we only look into the neural activity occupied with the perception of the target object, then the entropy drop that is required for macroscopic psychokinesis is actually smaller the further away the practitioner is from the target object.

Are there examples for people doing this in real?

Indeed there are: the French telekinet Jean Pierre Girard, who has demonstrated both telekinetic shifting and bending of objects in the early 70's, has in addition demonstrated levitation on several occasions under laboratory conditions and under the scrutiny of the physicist professor Zbigniew William Wolskowski, the nuclear physicist Dr. M. Troublé, the Chemist Raymond Villetange, the biophysicist C. Bogdanski, and the quantum physicist R.D. Mattuck.[16, 17, 18]

Girard reproduced this type of phenomenon in various laboratories, in particular in Utrecht in 1997 (see fig. 79).

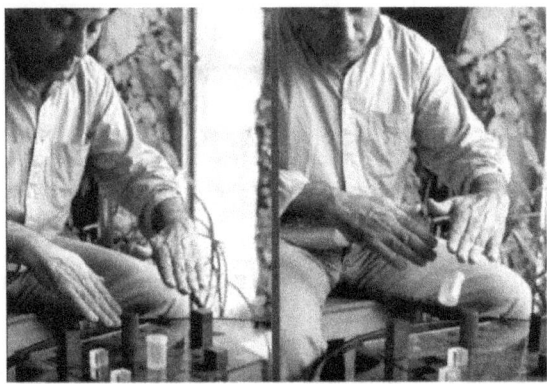

Fig. 79. Jean Pierre Girard is regarded to be the French "Uri Geller". His skill of shifting objects over a table was however beyond what Geller did and more close to Nina Kulagina. A very rare phenomenon was the brief levitation of a small object.

We can see that the object in fig. 77 reaches a hight of approximately 0.1 m. This corresponds to an initial speed of:

$$v^2 = v_0^2 + 2\,g\,h$$

$$v_0^2 = -\,2\,g\,h$$

$$v_0 = \sqrt{-\,2\,(-\,g)\,h} = \sqrt{2\,g\,h}$$

$$v_0 = 1.4\ \text{m/s}$$

That is rather close to the maximal speed $v_{max} = 6.748\ \text{m/s}$ for telekinesis which we calculated earlier; it is 20.7% of it. For something to be called 'levitation' it needs to go beyond a little hop, and that will require considerable fractions of the maximal PK-speed. That is why it is so rare. However, only if an object is really suspended in mid-air without rising or dropping can we speak of true levitation.

Using this definition we have to say that this feat has not yet been demonstrated under laboratory conditions. Furthermore it would be hard to explain using the here presented scheme, if observed one day.

Levitation events are extremely rare. Benedikt Maurer claims to also have acchieved this 5 times in his life. Only two times it was filmed. One was just a tiny jump (see fig. 80), the second one was very high (see fig. 81), but he doesn't have yet disclosed the full video of it. According to his personal estimate the hight was approximately 0.6 meters (personal correspondance; 2020). That would be 3.4 m/s, which is 50.8% of our theorized maximal speed for psychokinesis.

From many experiments it is evident that Maurer is not always doing psychokinesis, but that at least sometimes he moves objects through electromagnetic activity in his body. That becomes obvious when the objects he moves are made of metal and when they follow the exact movement of his hands. When the target objects are not made of metal, the movements are more intermittent and take more effort. Therefore the little jump from fig. 80 is most likely not due to psychokinesis, but due to electromagnetic activity. This activity can be mentally induced, but because controlling one's own body is not an example for a nonlocal phenomenon, so this cannot be categorized as psycho- or telekinesis.[15]

Fig. 80. This little levitation event was not telekinetically induced, but rather it was an example for electromagnetism in the human body and the control of it. It is strictly speaking not in the realm of parapsychology.

Fig. 81. This is the only example of psychokinetic levitation from Benedikt Maurer. His hands are far away and the tennis ball is made from non-conductive material.

Benedikt Maurer seems to be the only person in the world, so far, who can move objects both through the observer effect, as well as through somatically induced electromagnetism. In his first book he was focusing on the electomagnetic aspect only, while in his recent work he also talks about quantum effects.

List of telekinets

Name	Tested by	Feats
Benedikt Maurer (Germany)	BND (2 ×); Professor Walter von Lucadou; GWUP (Society for the Scientific Study of Para Sciences)	Shifting; scattering; bending; rotating; electromagnetic activity in the body.
Silvio Meier (Germany)	B. Waelti & the University of Freiburg.	Bending; shifting.
Claus Rahn (Germany)	Physics institute of the University Munich.	Rotating; bending; shifting.
Nina Kulagina (Russia)	KGB (over twenty years).	Shifting.
Uri Geller (Israel)	Mossad; CIA; SRI; MI5; MI6.	Bending.
Jean Pierre Girard (France)	C. M. Crussard; J. Bouvaist; J. B. Hasted; D. Robertson.[20, 21]	Bending of very thick metal; shifting of objects over a table; brief levitation (only once).
Michael Grubb (England)	Only his students.	Aerokinesis only.
Nicholas Yanni (Australia)	ASIS.	Shifting.

Unclear cases: **Thomaz Green Morton** (could be genuine in his skill, but was convicted of murder in 2010); **Erwin Wiesner** (could be genuine, but is a magician, and therefore difficult to take serious).

List of YouTube channels from telekinetic people by category, including many that haven't been tested:

Moving wheeled or suspended objects: **Tribor Seven**.
Aerokinesis: **Micheal Grubb**.
Psi-wheel: Telekinesis Mind Possible (**Sean Mc Namara**)*; **Darryl Sloan***; **Ashura Gaeden**.
Shifting objects: **Nicholas Yanni**; **Seven Rise**; **Mad Cat Sphere**; **Toghrul Gasanov**; **Sarah Grant**.
Shifting and bending objects: Psicoquinesis (**Diego L Medrano**); **Benedikt Maurer**.

*Have done it blind as well.

List of frauds, magicians or people who do not make it clear enough that they are using tricks:

1. Guy Bavli (Israel): Has once been 'tested' by the television show "Superhumans". He shared this episode on his own Youtube channel, but added a disclaimer into the description, saying "Guy Bavli never claimed to have super natural powers in his words, yet the show made all voice over to make it look like he claim it. He never claimed to have real powers, he uses lots of techniques he learned and developed over many years to create what others call telekinesis.".
2. Leon Benjamin (England): Always claimed to be genuine, but in his fifth telekinesis video one can see how he put two videos together, and how in the process of doing so, the window frame in the video was distored. This was pointed out by Benedikt Maurer (Benedikt Maurer; 2019; personal correspondence).
3. Danny Wolverton (US): Does levitation on a stick he carries with him. He claims this is an example for real magic, but at the same time he says it is a type of balance. The fact that there is always either sand, soil, soil on concret or a carpet beneath him suggests that this is a trick. Unfortunately he doesn't admit it, and instead says, "I know trick magic, but my levitation is real." (Personal correspondance; 2018).
4. Edison De Ocera (Philipines): Never claimed to perform genuine PK. He levitates small objects, but the fact that they are swinging while he does that, indicates that they are suspended somehow. The fact that he once burned a pen before levitating it caused quite some confusion about if what he does is trick-magic or possibly genuine PK.

Notes and references:
1. N_A is the Avogadro constant with the value of $6.022140857 \times 10^{23}$ mol^{-1}.
2. "The conscious universe", Dean Radin (1997).
3. For examples of PK performed on wheeled objects visit the Youtube (YT) channel of "Trebor Seven".
4. For an example search for 'psicoquinesis' on YT. A scientifically more rigorous but older example is "Jean Pierre Girard". Videos of him can be found on YT as well.
5. A lengthy documentary mainly on micro PK is available here: [https://www.youtube.com/watch?v=nh94XZzEosc].
6. See chapter 8, page 151 of Dean Radin's "The conscious universe" ([21], 1997).
7. See [21], chapter 8, page 145.
8. Examples for shifting (Benedikt Maurer; 2016): [https://www.youtube.com/watch?v=ZOvbAM0Mpm8]
9. Example for melting (Uri Geller; 2013): [https://www.youtube.com/watch?v=kA7u9i4g1EI]
10. Stanford Research Institute tests on Uri Geller (Professor Helmut Hoffmann; 1973): [https://www.youtube.com/watch?v=GfH5lkVMaok]
11. Example for scattering ("The great Hardy"; 1989): [https://www.youtube.com/watch?v=xGbcik3SNJQ&t=3s].
12. "Experiments on psychokinetic phenomena." J. B. Hasted, Ph.D., and D. J. Bohm,

Ph.D., Birkbeck College, University of London; E. W. Bastin, Ph.D., Language Research Unit, Cambridge University; and B. O'Regan, M. S., Institute of Noetic Sciences, Palo Alto, California.

13. Feynman's testing was informal (at Geller's hotel room between many calls) and failed, but it revealed no evidence for fraud: [http://www.indian-skeptic.org/html/fey2.htm].

14. "Superminds". John G. Taylor, 1975.

15. SKEPTIKER 4/2019; "Die Psi-Tests der GWUP 2019 – mit Überraschung" (the psi-tests of the GWUP 2019 – with a surprise).

16. Prof W. Z. Wolkowski, "Résumé expérimental", in L'Effet G, Robert Laffont, pp 270-274. 1981.

17. Prof W. Z. Wolkowski, "Phénomènes psychocinétiques produits par J-P. Girard", Revue métapsychique hors série n°21 & 22. 1975.

18. Pr W. Z. Wolkowski, "Résumé expérimental d'une série d'observations d' effets psychokinétiques, produits par J-P. Girard.". in "Revue de paraphysique". 1981.

19. The word 'telekinesis' captures the nonlocal aspect probably better than the term 'psychokinesis'.

20. C. M. Crussard and J. Bouvaist. Memoires Scientifiques Revue Metal-lurgique. 1978, February, p. 117.

21. J. B. Hasted and D. Robertson, "The Detail of Paranormal Metal-Bending. Journal of the Society for Psychical Research. Vol 50, No 779, 1979, pp. 9-20.

5.4 Conscious sets and ESP [09.06.2017 (collective clock); 09.06.2017 (influence of S on S_s); 14.09.2017 (collection of successive self-similar sets)]

In chapter 5.1 we found that a person can be described as a set of all sets of experiences different people have with that particular person. We are usually only aware of the set S_s (S for 'self'). There are various other sets which can be established, for example the set C_{R_x} of all conscious sets (people) in a certain room R_x. The more obvious the set is, which means that less conditions are needed to define it, the easier is it accessible to members of it. It is not yet clear if the set in question has to have been conceived by one of its members or if *platonic reality* can be ascribed to it even without that.

The collective clock

When two people are part of a common set with a simple definition it is easy for them to have synchronicity in thought (telepathy). For example, if someone has to get up at 8 am, he may easily be influenced by people watching clocks seeing that it is 8 am *right now*. [1] I am pretty sure the reader has experienced at times waking up at the precise moment one has to, a few seconds only before the alarm clock. Given it is 5 seconds before, then the chance for this to be a mere coincidence is 5.78×10^{-5} or 0.0057%. With this chance against odds this should happen only once in 54 years. Yet it happens pretty often to probably most of the people.

I personally find it hard to imagine that we have some kind of very precise clock down in our heads, similar to a computer. There is the *suprachiasmatic nucleus* (SNC) located in the *ventral hypothalamus*, which is responsible for generating circadian rhythms (day and night rhythms) in physiology, but these function only as a biological clock, regulated by light, temperature and other factors. It is not like a timer one can set to wake one up at a certain time.

We may call this phenomenon the 'collective clock effect'.

Synchronicity in thought

The most common synchronicity in thought is when we think about a person intensively and this person calls or messages in exactly that moment. Of course this isn't very astonishing if one thinks about that person all day, or if that person calls every day or many time a day. In order to find out what the chance against odds are for a particular synchronicity to be a mere coincidence, we have to count the time between the last contact and the length of the time window in which a contact is regard as being 'meaningful'. I will give a personal example here: A girl from Shenzhen had asked me on a social app if I want to engage in a relationship with her. She did that on a valentines day. I replied her something, and she didn't reply back for 7 days. After 7 days I mentioned this incident for the first time to others in a vegetarian restaurant. I said "This girl must be joking. I didn't meet her for 3 years, and she asks me this.". After saying a few more things about her, I stood up and walked out. I thought I should call her, but then realized that I don't even have her number. In exactly that moment she sent me a message saying "I think I don't match you very well.". The synchronicity time window was 5 minutes. Counting in only the wakeful time which is approximately 17 hours a day (assuming 7 hours of sleep), we can calculate the odds against chance as follows:

$$\frac{5}{7 \times 17 \times 60} = 7 \times 10^{-4}$$

That is a 0.07% chance. Of course such coincidences can happen, but when one records all such occurrences it becomes evident that this cannot be explained without some form of ESP (extrasensory perception).

After paying attention to this kind of synchronicity in thought for some time I got the impression that every time a certain person thinks about me, it forces me to think about that person. There are two reasons why this goes unnoticed most of the times:

1. A lack of awareness about the reality of ESP.
2. Not every time we think about someone do we let that person know it.
3. In many cases the thought might not reach the conscious level and remain subconscious.
4. Distraction may reduce this form of synchronicity or prevent it from becoming aware (see point 3).

The here presented hypothesis that synchronicity in thought always occurs and that it is a law of consciousness, we may call '*strong synchronicity hypothesis*'.

Near death experiences

Near death experiences are the best way to test *conscious set theory* (relationism). There are many meaningful sets a person is element of. There is for example the set P_{watch} of all people that watch a person **P** *right now*. Then there is the set $P_{think-of}$ of all people that think of that person **P** *right now*.

When somebody has a near dead experience that person is *brain dead* for several minutes. If a person is an abstract set, then what happens to the person when the brain receives no input anymore? Apparently a brain dead affects only P_S (S for self), but none of the other subsets of P. The full set P is the perception all conscious observers have of the person **P**. Pretty much like the full set r is the perception all conscious observers have of the colour 'red' (defined using the frequency of red). We can name these types of full sets '*Platonic sets*'. r is then the 'Platonic set of red' and P the 'Platonic set of the person P'. The perception or impression a single person has of **P** is not as pure or genuine as the true nature of **P**, which is represented by the Platonic set P. The various sets P_x including P_S we can call *non-platonic sets* of the person **P**. The notion of a separate more pure version of oneself which coexists somehow parallel to the ordinary self in some unknown realm is not new: in western spirituality (or *new age culture*) it is referred to as '*higher self*' and in Buddhism it is called '*Pormatman*' or '*Mahatman*' (chin. 大我 *đại ngõ*).

However, for analysing near death experiences, we only need the set P_{watch}, which consists of all the doctors trying to revive the person **P**. People with near death experiences often recall what the doctors around them were doing or saying during the time of brain death.[2] That could be explained if we imagine the person **P** after the brain death to be represented mainly by the set P_{watch} which is still vividly evolving during that time. In that case the perception of **P** would be a combination of all the various perceptions the doctors have while trying to save **P**'s life. That equals seeing things from many angles or perspectives at the same time (kind of like a 4-D perspective) and thus is retrospectively remembered as floating above oneself.

The feeling of enlightenment that most people with near death experiences have after coming back to life could be explained by having had full access to P_{watch}, as well as the full (platonic) set P. This form of enlightenment could be described as 'knowing oneself'. This represents a very rare and special case where the larger set P is suddenly included in the smaller set P_S. Yet this is only special if it happens during one lifetime.

Reincarnation

How do we define the set P? For the colour red we can simply take the frequency. The frequency belongs to the definition of the set r, while 'hot' or 'energetic' belong to it's properties. A person changes in time, therefore we can't use a persons character in order to define it. In praxis a person is defined by the time, date and place of birth and by its parents. Other also very basic properties such as the name of the person may change over a lifetime. However, we don't have to worry too much about these, because the content of both P_S and P changes in every moment. Every moment is represented by another set, so that we get a temporal succession P_0, P_1, P_2, ... That is true for both the platonic and the various non-platonic versions. Yet all the former versions P_0, P_1, P_2 are included in the present version of P. What if all the things used to define the person are changed? The birth place could be re-named, the person could change its name, or even commit identity theft. Would that destroy the set P?

Using ordinary definitions it appears to yes, but not if we define P as a *collection of successive self-similar sets*.

In this case a newly born person **p** could be represented by the existing set P if only that person is the most self-similar succession of $P_{t=lim}$, namely the last element in the *collection of successive self-similar sets* contained in P. The new born person **p** would have partial access to the full set (platonic set) P, but most of its content would only be felt, not known in a concrete or factual sense. Only in rare cases details of the life or the circumstances of dead of the person **P** could be felt or known by the person **p**. In these rare cases we can speak of concrete evidence for reincarnation.[3]

Notes and references:

1. Just a few days before I was sharing room with my Spanish friend Haiming, when he woke up at precisely 8:00 am (without an alarm clock), as he should. I was basically watching the clock waiting for the last 5 minutes to pass by.
2. A Ted-talk on near death experiences by German emergency physician Thomas Fleischmann can be found here:
    [https://www.youtube.com/watch?v=mMYhg1gE6MU]
    (From life to death, beyond and back).
3. Some interesting cases of children remembering details of their past lifes that were successfully confirmed to be true stories of people that died shortly before their birth:
    [https://www.youtube.com/watch?v=qxGMzIRrtwg&t=49s]
    (6 kids that remember their past lifes – reincarnation series).

5.5 The world of experience and remote viewing [25.10.2017 (first encounter with Schwartz and archeological remote viewing); 22.03.2018 (detailed analysis)]

The concept of the 'world of experience' had been established in 2006. Its main difference with Karl Popper's *world of culture* is that while in Popper's world both the information content of a brain and a book are certainly part of it, the information gets wiped out even from this world as soon as the brain stops working and the book gets burned.[1]

The main motivation behind the *world of experience*, which is the most central concept of conscious set theory, was the realization that contents of consciousness can not be correlations in the firing of neurons at one moment in time but must be correlations of brain states at different times.

That means *identity theory*, namely the belief that each brain state corresponds to one state of consciousness, is flawed.

From a purely physical perspective we may reasonably assume that each brain state contains information about its past, but to what extend?

Entropy puts strict limitations to our ability to 'predict' the past from the present state of things. That is why criminals like to burn things that could be used as evidence against them. Yet even in situations where the entropy is low, the past is not something one can find out simply by looking at particle positions and impulses and calculate them backwards in time. Chaos theory shows that this is impossible to do when wanting to predict the future and on the scale of particles predicting past and future are operationally very similar. Without going over to chemistry, biology and even the human sciences it is impossible to reconstruct the past.

Even when we seem to have succeeded in reconstructing the past, there is still a, for all practical purposes, infinitely small but existent probability that all we see around us formed out from a random fluctuation in a thermal equilibrium; that is the Boltzmann brain thought experiment. Another possibility, which was mentioned already in chapter 4.11, is that the universe jumped into its present state after everyone had died in the previous eon (universe).

If the past is so hard to reconstruct and if its reconstruction is not something that can be done automatically, then we cannot use the present to assign *Platonic existence* to the past.

What if we don't require reconstructing the entire past, but just a single lifetime? After all what we aim for is to explain consciousness and most (noncontroversial) functionalities of consciousness don't extend beyond a single lifetime. Isn't the memory, which seems to be stored in the brain, enough to reconstruct the past in a straight forward way? One would have reason to think so if memories were established right away, yet it usually takes long time till memories turn into stable neural connections. And are those neural connections really memories in the actual sense or are they more like sketchy notes helping us to access memories that are in reality somewhere else? When we ask where memories are stored, we get the answer that every single piece of memory is stored nonlocally all over the brain. It is a bit like having a movie of one's life, but all pictures are drawn onto one single page, one over the other.

For snapshots of neural patterns to really represent memory, having neural connections related to memories isn't enough. One would have to see which neural connections were activated in the moment before, and the moment before that moment, and so forth. There is no way a single brain state could contain that information and therefore single brain states just can't produce consciousness.

This whole issue seems to change quite dramatically when we make the transition from pre-relativistic physics to post-relativistic physics. In relativity there is a space-time in which we can very well imagine consciousness to arise from correlations of brain states in different moments. Here it is no problem to have a sequence of brain states giving rise to consciousness. This would all be fine and good, but when quantum mechanics enters the picture we have to acknowledge that we are not just *sculptures in space-time*. And if we were, why would we need consciousness? After all sculptures aren't doing anything.

As we saw in chapter 1.2 the presence is all there is, physically. If we want to reconcile quantum mechanics with relativity we need to abandon the notion of a space-time block universe. So what we are left with is a physical reality in which only the present moment is really there. At least that is what we seem to have without including consciousness. From the above it should be clear that a single snapshot of reality at one moment in time can't give rise to consciousness.

'Relationism', the early form of conscious set theory, started with the assumption that not only idealized objects and mathematical notions have platonic existence, but also relations between objects in the physical world. With platonic existence I mean something that is an idea but has causal influence. If it was an idea and all we did is saying that it is in some way a 'real thing' or an 'object', that would be mere tautology. Platonic objects have causal influence when we conceive them in our minds. Relations between physical bodies can be platonic if they give rise to consciousness.

The belief that neural correlations, which are in essence correlations between physical bodies, can give rise to consciousness is a very wide-spread belief.
However, as we saw in chapter 5.1 even the perception of a colour or the understanding of a word is based on the entirety of all our memories and maybe even on more. Being aware of what a word means could only be represented by all encounters we had with that word. A computer or robot uses words without linking them to any memory. Furthermore it doesn't process all related memories in order to use that word. Our brain doesn't do that either, yet we are consciously aware of each word's meaning. That already shows that consciousness goes far beyond what the brain does.

If we want consciousness to arise from patterns in our experience, that experience must have platonic existence, otherwise it can't have causal power.
Platon conceived the 'platonic world' when he tried to understand how it is that humans have access to mathematical truth. He came to the conclusion that mathematical truth is something of eternal nature that has causal power in that it both shapes the physical world and our way of thinking. However, human thinking isn't based solely on the interaction or intercorrelation of mathematical ideas, but any ideas. That is why Karl Popper postulated the 'world of culture'. Yet different from the archetypes in Platon's philosophy, Karl Popper didn't subscribe eternal existence to the entities that inhabit his 'world of culture'. To count to the world of culture something must be written down or stored in the memory of a brain. Popper's philosophy of consciousness was language centered and therefore only human culture counted as culture for him.

Popper's world of culture is not spread out in time. Although he collaborated with John Eccles on the idea of a mind-brain interaction and free will, he pretty much viewed consciousness as arising from neural correlations in the brain.

That means we have to interpret his 'world of culture' as being a subset of these correlations. What this concept succeeds in doing is explaining the interaction of ideas of any kind in one's consciousness. What it fails at is explaining how these ideas can arise from the activity of the brain in the first place.

He could have theorized that consciousness arises from correlations that are spread out in time by using Minkowski's concept of a four-dimensional space-time. Yet there is nothing happening in such a spacetime and thus any correlations in it can't have any causal influence on the 'dynamics' of the system, simply because it has no dynamics.

As the above consideration show, neural correlation at a single moment in time can't give rise to consciousness and a space-time isn't dynamic and therefore can't give rise to conscious either. Popper's world of culture is confined to what exists in form of records in the present and therefore also can't give rise to consciousness.

Consciousness obviously arises from correlations that are not only spacial but also temporal. Ascribing platonic existence to our experiences is the only framework in which consciousness arises naturally. Of course this only works in a world that extends infinitely into the past, just like in the periodic cosmology that was laid out in chapter 4.11.

Although Popper didn't consider psychic functioning, his philosophy naturally leads to the assumption that ideas once conceived are becoming much easier to be conceived afterwards. If we assign Platonic existence to the entities in Popper's world, then they must be accessible to any mind even if that mind had no physical contact with the carrier of the related information. It is interesting to note that both the Platonic world and Popper's world of culture are not divided into different versions for different people. Analogously the world of experience of conscious set theory is not divided into different versions either. An idea is an idea, no matter in whose brain or which side of the brain it is. If it is the same idea, it will feel the same no matter who it is that is having it.

If all our ideas are part of a huge ocean of ideas not separated by person, then how comes we don't seem to have access to other people ideas?

The awareness of the meaning of a word or the sensation of a colour arises from the world of experience. This whole set of experiences is the conscious set S of a person. When contents in the world of experience of other people are interrelated with that set, they overlap and that is what parapsychologists sometimes call *morphic resonance* (Rupert Sheldrake; 1981).[2] When it is rather strong it can be called *synchronicity in thought* and when it is very strong we speak of *telepathy*.

Synchronicity in thought and telepathy are very familiar concepts to most people.

Popper's world of culture can be adopted to describe these phenomena, simply because there is only one 'world of culture'.

Maybe the same can be said about the general concept of the 'mental world', because if it is immaterial as generally believed, then it must be 'Platonic' too and for Platonic archetypes to interact all what is needed is contextuality. Distance matters only to physical objects.

In this line of thought the mental world explains the (noncomputable) interaction between emotions, while the 'Platonic world' and the 'world of culture' respectively explain logical reasoning and the interaction between ideas in general.

Popper's world of culture is simple, because it is not extended in time and the fact that Popper saw the world of culture as something being dependent on carrier objects in the physical world shows that it was not extend in time. Nevertheless the world of culture can be adopted to explain psychic phenomena, although that was not intended by Popper himself.

As laid out both in chapter 5.1 and this chapter consciousness need to be spread out widely in time for phenomena such as Qualia and symbolism to be explainable.

One could imagine a version of Popper's world of culture which does not 'pop out' of existence when a person dies. Such a version would be very similar to the world of experience, with the only difference being that nonverbal animals would have access to it too. Yet what justification do with have for assuming information that has been lost from the physical world to continue existing in some other realm?

If this information is inaccessible, it would be unscientific to assume it continues to exist. In modern cosmology it is common practice to assume to existence of an infinite number of universes and simply justify it by saying that it follows from some basic principles. The same way we could here say that it doesn't follow naturally out of the theory to assume the world of experience to suddenly and abruptly pop out of existence. We could claim some form of 'information conservation' for consciousness.

We could do that, but it isn't necessarily the most scientific thing to do, especially not according to the epistemology developed by Popper himself.

Furthermore according to Leibniz's relationalism only things that influence each other can exist. Leibniz's argument for this is that if things that don't influence each other existed, they wouldn't be related, and that would violate relationalism.[3]

So, how do experiences of people that already died influence us? How can we experimentally verify that these past experiences are still part of our consciousness?

For more than a decade after I had conceptualized the 'world of experience' I only had the vague idea that some subtle influence (similar to *morphic resonance*) could come from the experiences of people that no longer live. I didn't think that would be very testable and I didn't give it too much thought.

Without really going into the parapsychology literature one usually isn't familiar with much more than telepathy and maybe claims of PK. I assumed that the way different people's experiences are not separated in the world of experience allows for telepathy.

Yet telepathy is usually associated with sensing thoughts people have in the present and for explaining that the usual concept of an immaterial 'mental world' seems enough. The world of experience instead pointed onto something else: with its infinite extension in time it suggested that information that is contributed from other people should be available regardless of whether or not it is in the present mental world.

The mental world is just what we think and feel now. It has no temporal extension and is therefore much more limited than the world of experience.

I firmly believed that the whole world of experience must always continue to exist. I imagined it to influence how we think and feel, but until I came across the parapsychology researcher Stephan A. Schwartz (CSTS conference; Shanghai; 2017) I had no idea to what extent information from such past experiences can be extracted in the present, totally irrespective of the distance in space and time.

Schwartz's main research area is remote viewing with a specialization in *archeological remote viewing*, a field he had helped establishing. Schwartz had absorbed all the known and unknown literature of parapsychology on a very early stage. Therefore he was sufficiently convinced of the reality of things like telepathy[4] and remote viewing. He did not want to waist his time proving what others have proven before, rather he wanted to see how it could be harnessed practically. Back then there was some anecdotal evidence that psychics have been uses sometimes by archeologists to guide them to the right places for excavations.[5-10] However, in most cases this wasn't very well documented and there was no literature on how psychics should go about doing such things. Furthermore those previous excavations were all using a single remote viewer as guidance.

Schwartz started his research by dividing his garden into equal square areas and burring things in randomly selected compartments, for then asking participants where they think the buried objects are and what they are. Convinced by the results of the tests, he then slowly formed a team of distinct psychics that he would use for finding long lost archeological sites. His team was called the *Mobius group* and their success went far beyond all expectations. From all their discoveries two are very distinct: one is the project *Deep Quest* and the other is the *Alexandria project*. In the first, two psychics, Ingo Swann (1933 – 2013) and Hella Hammid (1921 – 1992), located a sunken wooden ship near the island Santa Catalina (California, US).[11] The area surrounding the island had previously been searched excessively by local marine divers from the *Institute for Marine and Costal Studies* of the University of Southern California. 4,000 dives did not reveal anything. Both psychics had located the same spot on a map independently and had given a list of objects they suspect to be there. Among the objects predicted by Hella Hammid was one particular object that was strange and unexpected; a huge block that she envisioned to lie there on the sea ground.

Schwartz managed to get a robot arm equipped research submarine for the task. Because of difficulties to dive exactly to the location marked on the map it took Mobius three hours to actually start searching. Once they were there it was just a matter of minutes till the first relicts were sighted and they led the team right to the ship wrack. After that a very massive monolith block was discovered as well and it matched perfectly the object Hella Hammid had drawn in her sketch.

All of Schwartz's projects are monitored by neutral observers whose work is to keep records of everything that happens and document it neatly. In this case the neutral observer was Dr. Anne Kahle, who had earned herself a reputation doing satellite work for NASA (Jet Propulsion Laboratory of Cal Tech).
The discovery of the sunken ship was then also analyzed by the oceanographer, explorer and marine policy specialist Don Walsh. According to Walsh there was no way Mobius could have known about the wrack by any ordinary means.[12]

The second high profile project was about the discovery of several hidden buildings and structures in the ancient city of Alexandria, which was built 300 BC. Schwartz began by letting 11 psychics from his team locate sites. Among those sites there were three unexpected markings on the map which lied in the water.[13]
Schwartz then selected Hella Hammid and George S. McMullen to go to these sites in the hope they would get further hints through on-site remote viewing. McMullen had previously proven himself capable in field experiments conducted by archeologists from McMasters University and the University of Torronto. In those experiments McMullen had located and correctly described Indian petroglyphs and villages.

The first place in Alexandria Schwartz and his team went to was a locked up building that had caught Hella Hammid's attention. After getting access to it by the local authorities Schwartz and Hammid went in and waited for further perceptions. Hammid got the vision of a water channel with huge complexes of floors and stairways on both sides. After that they asked local authorities about this and were informed that such a water channel had been discovered by scientists accompanying Napoleon somewhat 200 years ago, but that the entrance has been forgotten.
When the team was almost giving up on verifying what Hammid had seen, the local authorities announced that they found the entrance.

The second site was described by a drawing Hammid had made before coming to Alexandria. She was drawn to it when the team passed aside of it by car. The site matched her description again. It was an area by the name Nabi Daniel, in front of a Mosque and there were speculations that it might be built on top of what was once a library. McMullen visited the area to see if he can get any impressions. Very soon he was drawn to a wall and started describing various structures behind it. After a while he claimed that Alexander the great was buried in a complex behind and below the wall. Schwartz then separately brought Hammid to the site for getting another independent remote viewing. Very soon she started having visions of an underground dungeon with a tomb which she described in great detail. She claimed it to lie about 20 or 30 meter below and to be about 12 feet high. Then she went to exactly the same location on the very same wall McMullen was drawn to.

The Nabi Daniel trial dig had proven to be "psychically accurate", but further excavations would have required tearing down parts of the mosque. However, the local authorities were still not convinced that such a thing as 'remote viewing' existed. Blocked at Nabi Daniel, Schwartz proposed to Fawzi Fahkarani, the archeologist appointed to him, an exhibition at a less populated area of his choosing. Fahkarani agreed and gave Schwartz a seemingly impossible task: Schwartz and his team were asked to locate a tile floor building buried somewhere in the middle of the nearby *Marea* desert.

After some hours of wandering around under the sun, McMullen was drawn to a place where he felt a building was buried. His reportings started with the statement that he was walking on top of a wall. McMullen predicted that they would find a round broken statue or column. Then Hammid was brought to the place as well.

Together they described a broken round statue, a marble mosaic floor, a Byzantine building, its exact outline, the doorways, the depth at which its walls would emerge and more.

The predictions kept going on during the excavations and the archeologists found that 80% of what the remote viewers had perceived was accurate.[14]

Fig. 82 & 83. Schwartz and McMullen in the Marea desert, discovering a Byzantine building by the means of remote viewing (1979).

Then the team was allowed to dive at Alexandria's eastern harbor. What was found there was evaluated by local archeologists as even more important than the tomb of Alexander the Great.

In the years between 1977 and 1988 Schwartz has discovered Cleopatra's Palace, Marc Antony's Timonium, ruins of the Lighthouse of Pharos and sunken ships along the California coast and the Bahamas, all using the method of remote viewing.

According to Schwartz remote viewing happens by tapping into what he calls the 'nonlocal realm'. This realm consists of pure information where things are not at all separated by person, but only by subject. According to an analogy he frequently uses, the nonlocal realm is very similar to a 'Google search'. Accordingly, asking the right questions is analogous to using the right search terms on Google.

Schwartz also stresses that it doesn't make sense to speak of senders and receivers in what appears to be 'telepathy'. Remote viewing shows us that no 'senders' need to be present in order for the information to be available. Still there were many people that theorized telepathy must be based on signals of some form. Signals of most wavelengths were ruled out by other researchers, but the extremely low frequencies (ELF) were still left as an option. Schwartz wanted to rule out ELF as well and for that he need a military submarine which is able to go really deep. Enough sea water gives sufficient shielding from ELF which rules this form of radiation out as a means for information transfer. The 'telepathy' and remote viewing experiments conducted in the three days Schwartz was allowed to stay in the submarine were successful.

Schwartz has used remote viewing not only in archeology, but in criminal investigations as well. In one particular case a remote viewer was able to tell where the suspect would be in five minutes. Schwartz interprets that as 'remote viewing of the future'. I have criticized this by saying that this can very well be information that comes from the intentions and expectations people have in the present. Schwartz insisted that it comes from the future, but he said it is merely a 'potential future' which is changeable. At the end we agreed that there might be an 'assembly of potential futures' (Schwartz; personal correspondence; 2017).

Schwartz is strong on claiming time symmetry in remote viewing and he uses a long-period survey he conducted from 1978 till 1996 to back it up. It is called the '2050 project' and is about remote viewing the year 2050.
It turned out that all the predictions which were made by a large number of individuals were mostly things that were developing already at that time, but of which nobody had knowledge of. Things such as the AIDS disease, anti terror war were predicted by remote viewers in 1978. Although knowledge about these things was not publically available, the phenomena had already emerged and a few individuals must have known about it and in that case it is still information that exists in the present. However, it shows that in the nonlocal realm 'future' is a category (or set) like anything else. It can stand for trends, expectations, fears and hopes. Among these four sub-categories there is one which is more objective and not bound to individual observers, and that is 'trends'. 'Trends' that have not been observed by individuals but that become evident when we combine the perceptions of many individuals might be part of this nonlocal realm too. That however depends crucially on if secondary relations between entities in the world of experience can be assigned reality and causal power to even if they are not conceived by any mind in the present.

That seems fair enough, but when looking at the whole body of consensus predictions that were made through the project one starts to doubt that this is actually enough to explain the effect. The 2050 consensus data includes things such as:

No nuclear war but more dangerous world; religious terrorism; Soviet union disappears; epidemics (blood disease); people don't travel much to avoid air pollution; virtual reality instead; carbon almost out of use; LENR (low energy nuclear reaction), antibiotics don't work anymore; gender equality; marriage equality; people live under domes to shield away heat; movement away from the coast; Florida is under water; political power devolved to states and bioregions; legalization of marijuana; no regular currency; no passports; biometric information ID; crypto currencies largely beat regular currencies; virtual corporate states own governments; ...

A lot of these predictions are of course due to media propaganda and have little to do with reality. Some of them however, are not, and if it turns out that one just can't explain those predictions with the scattered information that exists in the present, then a better explanation might be that the futures of the last eons (universes) are virtually perceived as the future of our present eon. If the world of experience extends infinitely into the past, then it can function very much as a statistic on what usually happens with civilizations at different stages of their development.

Rupert Sheldrake explains telepathy with morphic resonance between people, and memory with morphic resonance with oneself.

If the past is infinite, as put forward in chapter 4.11, and if morphic resonance is a law of nature, then there must be morphic resonance between different eons.

Genetic mutations which are beneficial lead to more offspring. More offspring means patterns are repeated and that leads to a higher morphic resonance. Yet this morphic resonance wouldn't do much, if we were to assume there is only one eon; mutations happen in gradual steps and morphic resonance could only act on the same race. It would speed up evolution a bit though, because genetic mutations which have proven to be beneficial in one place would soon emerge among members of the same race in other places, seemingly independent.

However, if we combine morphic resonance with a periodic model of the universe, then we can have morphic resonance between eons which leads to a meta-evolution of universes.

Just like people don't seem to remember their past lifes, universes don't seem to remember their last eons, yet people might inherit preferences from their past lifes and evolution might be guided by 'memories' from past eons.

Sheldrake's theory is traded as preposterous or crazy among many mainstream scientists, yet its underlying assumption really isn't far fetched at all: it is simply saying that quantum randomness isn't truly random but meaningful; meaningful in the very simple sense of allowing patterns to repeat. To me it seems that he keeps unreasonable assumptions to a minimum.

Sheldrake extends morphic resonance to all matter, biological and chemical, and because he associates morphic resonance with consciousness, that makes him a panpsychist. Panpsychism doesn't harmonize with the consciousness based orthodox interpretation of quantum mechanics. However, that is only the metaphysical background of the idea. It doesn't influence the way it is applied to biology and parapsychology.

We can now summarize the different worldviews we analysed above in the following chart:

Worldviews and their properties	Mathematical understanding	Telepathy	Remote viewing and memory	Transgenerational remote viewing	Quantum remote viewing	Precognition
Materialism	No	No	No	No	No	No
Tegmark's mathematical universe	No	No	No	No	No	No
Dualism	No	Yes	No	No	No	No
Trialism: Popper's 'world of culture'	No	Yes*	Yes*	No	No	No
Penrose's trialism	Yes	Yes*	No	No	No	No
Conscious set theory or 'relationism'	Yes	Yes	Yes	Yes	No	Only trends and tendencies
Schwartz's 'nonlocal realm'	Unclear	Yes	Yes	Yes	Unclear	Yes
Radin's 'nonlocal realm'	Unclear	Yes	Yes	Yes	Yes	Yes
Akasha records	Unclear	Yes	Yes	Yes	Unclear	Yes
Sheldrake's 'morphic resonance'	No	Yes	No*	No	No	No

*Not acknowledged by the originator of the theory.

Notes and references:

1. "The Self and its Brain", Karl Popper and John Eccles (1977).
2. "The science of life", Rupert Sheldrake (1981).
3. Relationalism is a somewhat weaker form of relationism in that it only claims that relations are inherent properties of all entities, while (metaphysical) relationism claims that relations are all there is.
4. He himself avoids the term 'telepathy' because it might suggest that there is a sender and a receiver.
5. Schwartz. Secret Vaults. pp. 1-56. Also pp. 353-354 for bibliography.
6. Ibid.pp. 57-107. Also pp. 354-355 for bibliography.
7. Ibid.pp. 108-127. Also pp. 355-356 for bibliography.
8. Ibid.pp. 127-135. Also 355-356 for bibliography.
9. Ibid.pp. 222-238. Also Clarence W. Weiant. An Introduction to the Ceramics of Tres Zapotes, Veracruz, Mexico.Bulletin 139. (Smithsonian, Bureau of American Ethnology:Washington, 1943) and "Parapsychology and Anthropology." Manas, vol. 13, no. 15(1960).
10. Schwartz.Secret Vaults.pp. 211-221. Also "Psychometrics and Settlement Patterns:Field Tests on Two Iroquoian Sites." Unpublished paper, n.d.
11. Stephan A. Schwartz. "Project Deep Quest, a Prototype Experiment in the Application of Intuitively Derived Data in Marine Archaeology". Mobius Report No. 4. Originally an invited address. The American Society for Psychical Research. January 1979.
12. The statement by Don Walsh can be found in a documentary on the Deep Quest project which is available on youtube: [https://www.youtube.com/watch?v=BEC-GBTTLBg&t=69s] (NemoseenMedia; "Project Deep Quest").
13. Stephan A. Schwartz. "A Preliminary Survey of The Eastern Harbor, Alexandria, Egypt Including a Comparison of Side Scan Sonar and Remote Viewing". 1980.
14. Stephan A. Schwartz. "The Location and Reconstruction of a Byzantine Structure in Marea, Egypt Including a Comparison of Electronic Remote Sensing and Remote Viewing". Mobius Group. 1980.

5.6 Quantum mechanics relies on the existence of the nonlocal realm [06.09.2019 (black hole entropy relies on the universe having memory)]

Ever since the advent of conscious set theory it was an open question of what its relationship with space particle dualism is. Obviously a theory which gives a central role to consciousness would have to explain consciousness itself as well. This is a relationship of reliance, but not one of mutual impliance. One could imagine that there are other possible theories of consciousness that could be used as a basis for space particle dualism. What if there was some part of conscious set theory which is explicitly required by space particle dualism.

Indeed there is. Unlike most other formal theories on consciousness conscious set theory implicitly incorporates the nonlocal realm from the ground up. In fact it predicted its existence of the nonlocal realm at a time when its author didn't have any knowledge of all the evidence for the nonlocal realm which is provided by remote viewing.

In classical physics what happens in the next moment is only determined by what happened in the last moment. In quantum physics what happens in the next moment is determined by the entire past, not just the last moment. That is obviously the case when we look at all the degrees of freedom that enter the collapse of the wavefunction through the conscious of the observer as well as collective consciousness.

But even when leaving all this aside, which is what we usually do when we do physics, and talk about pure probabilities, then we still have to employ the nonlocal realm.

That can be demonstrated on a few examples. Some of the most obvious ones is

(1) Black hole entropy.
(2) Beam splitter experiments.

Why black hole entropy requires the existence of the nonlocal realm

All black holes are quasi-black holes, so the degrees of freedom are not mysterious, but they are simply the degrees of freedom of the original star. When strictly defining entropy as always being the logarithm of the number of states that we cannot distinguish, then it becomes obvious why (1) entropy is not always strictly correlated to temperature, (2) a black hole has entropy.

For ordinary objects the entropy goes down when the wavefunction collapses, but that doesn't need to change the temperature.
A photon can fall into a black hole, and that increases the entropy of the black hole, but that increase is totally invisible from the outside, according to space particle dualism. This violates the physicalist notion that everything the universe needs to 'know' for applying the laws of nature is encoded in form of physical traces.[1]
For example physicists tend to think that for the information paradox to have a solution, information would need to be stored somehow at the horizon, instead of saying that it is simply stored in measurement events (see chapter 4.3).

What if a particle goes unobserved for a few hours? Is the information of what it last did stored physically or simply in the event of it being measured?

According to space particle dualism the entropy of black holes does something, it makes the universe expand, yet the information about how much entropy it has precisely is not accessible by any physical means but only by remembering/remote viewing what felt into it. Looking at the size of the event horizon gives us an entropy measure that is good enough for every practical purpose, but the real entropy it had and which governs how much expansion of space it causes, is not accessible physically.

Why beam splitters require the existence of the nonlocal realm

The above is not some big exception: very often physics relies on memories of what happened in the past. For example a photon that goes through a cubic beam splitter array needs to 'remember' where it went to, in order to know that it can arrive only at detector B, and not at detector A.

This knowledge of the past which influences the present behaviour is what makes quantum physics so different from classical physics. We can physicalize that by looking at the wavefunction and imagine it to be a physical thing. But it can't be, because it changes instantaneous without any transfer of signals or substances from one place to another. Also it isn't made of anything.

So when we look at the true full wavefunction of the universe, we find that the wavefunction and the nonlocal realm are one and the same thing.

In some sense physicalists mentally rely on space-time as a substitute for the nonlocal realm. All the memory required for the universe to operate seems less of a mystery when one imagines physical reality to 'take place' in a space-time continuum.

Once we understand that space-time doesn't exist, it becomes obvious that all physical processes require the universe to have a memory. That memory is exactly the nonlocal realm studied by parapsychologists.

The notion that experience is stored in physical traces is strong among all disciplines of science. We tend to believe that personal memories are entirely stored in the brain and we tend to think that all the experiences of our bodies are stored in our DNA.

One could call his point of view strong mechanicalism or 'pan-codism'. The most radical proponents of this view even require the experiences of particles to somehow be encoded in their internal structure (Steve Dufourny; 2019; personal correspondence).

Note:
1. See Steve Dufourny (2019) as an example for this view.

5.7 How to build a quantum oracle [15.01.2020 (nine steps)*; 27.10.2020 (quasi-crystal analogy)**; 01.11.2020 (quantum oracles as a real world equivalent to subspace transmitters)***; 05.11.2020 (entangled oracle twins as subspace transmitters)***; 06.11.2020 (EVP is stronger than RNG-based ITC)]

*Requested by Owen Mayer (US) and inspired by watching a new Star Wars film with Hai My in Hanoi (VN).
**Largely inspired by Roger Penrose's first book on consciousness.
***Inspired by watching Star Trek - TNG.

Sponsored by: Lin Yi Song (China); Henning Conle (Swiss).

In chapter 4.2 we have been exploring the so called abundance of future observation principle. It states that particles are more likely to appear in places where someone is observing.
Put this way, the principle is only qualitative, not quantitative. It does not allow us to write down an equation to know how much more particles could appear in a spot where measurements take place as opposed to a spot where none take place.

The principle was conceptualized as an explanation for the anomalies in the cosmic microwave background that we discussed at length in chapter 4.2. However, at the end it was suggested that this principle is more a result of active intention, not the result of a natural law that runs passively.

Could these considerations be relevant for the functioning of the human brain?
The brain is a quantum random information processing organ. The technical term for such an object is 'quantum oracle'. That is because in mathematics, an oracle is an abstract entity that is able to decide upon the correctness of algorithms or tell if a computation will stop or go on forever, which means to solve the 'halting problem'. The human brain is capable of doing all that, and that is why we can call it an 'oracle'.[1]

As laid out in chapter 5.1, the brain is doing this by having Platonic archetypes and archetypes from the world of experience, also called the nonlocal realm, projected onto superpositioned states in the brain. That means that whenever wavefunctions collapse in the brain, they collapse into states that correspond to or resemble certain archetypes. For those archetypes to be materialized inside brains, they have to be available inside the assembly that is the quantum state of the brain.

In chapter 2 it was suggested that quasi-crystals with 5-symmetry grow in a nonlocal way, which means that they must grow in superposition in order to assemble this extraordinary type of non-repetitive symmetry. It is a pattern that cannot be built by a computer program.

What if the trajectory of neural pathways is analogous to the growth of quasi-crystals? It would mean that pathways that do not lead to a large cascade of further neural activations would be suppressed and be unlikely to be 'chosen' when the wavefunction collapses.

In fact no neural pathway can be chosen that does not end with at least one projection to the neocortex. Only neural pathways that are meaningful and match with other neural connections, cascade off other activations, and keep going, until the neocortex is reached. Those meaningful activation patterns are extremely rare and unlikely, but due to the fact that the neural activation pattern that took place is not actualized until after the neocortex is reached, the successful pattern is chosen automatically in retrospective. We can say that the brain functions in a teleological way. Analogous to crystals with a 5-symmetry, the wavefunctions of neural activation patterns collapse into the desired end result, with its corresponding past.

It is also quite similar to the universe as a whole, in which life creates itself a plausible past. The past is here again an emergent property of the present. A universe cannot collapse into a state in which there is no life. Analogously, the neural activation pathways cannot collapse into a state in which the neocortex is not reached.

We can try and mimic the brain using a so called 'spirit box'. That is a device in which electromagnetic fluctuations produce what is known as the 'electronic voice phenomenon' (EVP). Common sources of EVP include static, stray radio transmissions, and background noise.[2]
Ordinary 'spirit boxes' only occasionally produce meaningful patterns. One rather interesting example was a couple of 'spirit box sessions' with Robin Williams conducted by Steve Huff, starting right on the day he died. Three sentences with a rather clear voice emerged in the white noise, which were (1) "Could Robin come?", (2) "I must have got it wrong", (3) "I am with dad", (4) "I am dead", and (5) "There is light".[3]

Steve Huff has researched EVP, which he often calls Instrumental Trans Communication (ITC), for over a decade. His most compelling and interactive ITC sessions were in 2020, contacting the famous singer Michael Jackson.[4] He used what he calls the 'Astral Doorway-2 spirit device'.

This is the content of the communication (released 08.06.2020):

SH: What would you like to say?

MJ: Hope to rise.

SH: Hope to rise. That's pretty cool. Have you written any songs on the other side?

MJ: We're all gonna sing it.

SH: I feel we are connected. Thank you Michael. You are a very strong spirit.

MJ: You love it.

SH: You love it? He he. Stay connected, ok?

MJ: Try.

SH: Are you at peace?

MJ: Once I had a good life. And I have love.

SH: I come with the love of god. I try to spread love. Do you have any other messages?

MJ: I can't quite say what.

SH: You can say whatever you wanna say.

MJ: God has us.

SH: This works through god. This is a device that was build with the help of angles. You can use it to speak.

MJ: You heard me. Cause I let you.

SH: I am gonna send all the love from your fans through this.

MJ: And I'll just try to talk.

SH: Here we go. I am going to give a big blast of love, ok? Can you feel that energy.

MJ: I like love.

SH: Do you feel it? Would you like to be my helper Michael? I would love it if you would like to
be a guide or helper. On my group nights. Would you like to help out? Would you like to do that?

MJ: Ok … I sure would.

SH: … Do you have any messages for me?

MJ: I am lonely. A lot.

MJ: I only visit there.

SH: Michael is very strong. Can you sing?

MJ: Sorry … they don't let us.

SH: All right Michael. Thank you very much.

Later using another device:

SH: Was I really speaking with Michael Jackson?

MJ: Michael. He is there. To meet.

Two nights later:

SH: Michael Jackson. Can you come through?

MJ: I'll be right here.

SH: You have a lot of fans who asked me to do another video with you. Do you remember the day you died?

MJ: Tired mind.

SH: Tired mind? You were very tired. You were doing the 'This Is It-rehearsals'. Remember? Your big show? Do you remember that?

MJ: It seemed worth it.

SH: Have you crossed over into the light. Are you in heaven?

MJ: There alone.

SH: Ok. Alright, Michael. This is me testing. See if I can connect with you.

MJ: Found direction. So beautiful, Sir.

SH: You know, the world is in a crazy place. We have this pandemic. We have violence. We have protests. Protests for very good reason. What would you say to the people of the world today, Michael?

MJ: I cry.

SH: You know, you have millions of fans, Michael. Millions of fans, all around the world. Who still love you. Your music is still on the radio. Are you strong enough on that side to leave a message.

MJ: There is a way back.

Here are some sessions from 2018:

SH: What do you want to say to your fans, friends and family?

MJ: I am back with you.

SH: Have you met up with friends on the other side?

MJ: I actually did. Oprah House.

…

MJ: I am here in the home.

SH: Can you hear me Michael?

MJ: Yes.

MJ: I 've got my cigarette.

…

SH: I am here to help.

MJ: Just live the light way.

SH: There is nothing to fear. I come with love. Just as you did. I know you believe in love, right?

MJ: I hear it. This love.

…

MJ: I am the one who sing(s) in slow motion.

...

SH: I am going to shut down. Is there anything else you like to say?

MJ: Enough.

Note that all the messages are in English. If those were just random words emerging out of random fluctuations, then the chance for them to all be in the right language and in the right grammatical order, and even context would be infinitely small.

Of course EVP researchers tend to only display their most meaningful sessions. If every session was as successful as the one we looked at just now, then there would be much more content online. Many times meaningful signals do not come true. Establishing good communication does not seem like an easy task. We should therefore look for ways to make communication easier.

We could attempt to make the emergent patterns of such a spirit box more meaningful by building in a filter, which displays only meaningful patterns.

The machine itself would not be conscious, but by displaying only meaningful messages, it can function like an extended brain of our own, again with the neocortex of our brain functioning as the measure for what pattern becomes actualized. This way we have two filters, one is the device itself, which lets only meaningful patterns through, and the second would be our own brain, which also preferably lets meaningful patterns through to the neocortex.

The ITC device does not have a memory of its own, and that is why it does not harbor a particular spirit, but simply the one that the EVP researcher is asking for to come forward. The device cannot approach the researcher by itself, but only respond to him. It does not have a consciousness of its own, and it functions only in combination with the observer.
Large parts of the brain itself are pretty much like a ITC device. We can communicate with spirits through our subconscious mind. However when doing so our own expectations and things we think for ourselves are added into the mix.

People have long tried to communicate with spirits, either using *Ouijas*, also called spirit boards or talking boards, which rely on subconsciously-initiated somatic responses of a group, or using a skilled medium. All these methods rely on the subconscious, which is thought to be a gateway to collective consciousness. However, by directly involving our own brain, it is easy to taint the potential messages, and it becomes nearly impossible to tell what is external information and what is internally caused.

Instrumental Trans Communication (ITC) can have the same issue, because it is impossible to know if a particular pattern emerged from the 'hereafter' or if it was the observer that induced the pattern through his or her own thought. Nevertheless, ITC fulfills the purpose of demonstrating that mind over matter interaction is real and that we do influence how wavefunctions collapse, and furthermore that we do so even beyond the confines of our brains.

It also fulfills the purpose of making mediumship slightly more objective, by directly extracting the patterns and messages, without the need of a human subject as the mediator, who could blur them by overthinking.

In usual mediumship it is important to directly echo what comes into one's mind from the subconscious, without adding any thought and without any rationalization. One can experience the direct flow of information from the subconscious when absent mindedly writing down random words. They often reflect what is going on in the subconscious, and sometimes even in the collective part of the subconscious.

This, however, can be extremely difficult. Directly displaying the patterns through an ITC device eliminates the possibility of contamination through the subjective mind.

The parapsychology researcher Phillip Page has also specialized on EVP. He is apparently using a device that is specially trimmed to only receive human voice-like sounds.[5] It was not pointed out by Page himself, but doing so seems to increase the meaningful patterns received.

Some EVP researchers use devices that produce only voice-like sounds, while others use devices that produce a mix between static, white noise and human voice-like sound. Some others again filter out the static and random background noise in their electromagnetic receiver, and retain only the bits that sound like human voice.

Phillip Page has noted about one of his sessions (Personal Correspondence; 09.11.2020; commenting on *Calling Robin – will he come through again?*"):

"Hi Sky! No - nothing was filtered. The audio clarity is a result of the Angel Box device and the Afterlight Box App. It is one of the clearest apps that I have ever heard. If I were to use a scanning radio then you would hear a bit more noise that I would most definitely filter out with the Angel Box."

Eman Rodriguez did something to enhance the personality of the spirit fragments he received. He connected his spirit box to a voice generator in order to give it a particular voice. He chose the voice of Robin Williams.[6] This again appeared to further enhance the meaningful patterns and potential messages. Especially it triggered more frequent messages.

Again it is quite striking is that in these examples no language filters have been used. If the voices were supposed to be random background noise, which only accidentally resemble words, then there should be no preference in language. Even when a voice generator was used, there was no word filter, and so in a random setup, it could have been William's voice, but speaking in other languages.

The abundance of future observation principle is about a slight increase in signals where observers are to be found. Signals in the brain are also about activation pathways being chosen teleologically in a way that allows for the neocortex to be reached. The neocortex here represents the observer, and so the patterns that reach the observer are more likely to be actualized. In a similar way, meaningful patterns in an ITC device are more likely to reach the neocortex, and that is why they are actualized more easily.

A very different approach to ITC is to not use the electronic voice phenomenon at all and to instead allow quantum random number generators (RNG) to chose between words to build sentences.
Those don't usually perform very well. The '*Spirit Voice 2 Software Ghost Box*' for example has a 'speech mode', which does exactly that; choosing words at random. As mentioned before, one usually has only a 51% chance of successfully influencing a single quantum event. Linking the events to interesting pictures or words can increase the success rate, but not by too much.

If the abundance of future observation hypothesis is true, then suppressing meaningless sounds can amplifying meaningful patterns. The brain is doing that too, to a certain degree, but having two filters is certainly better than having only one.
Inside our ITC device the amplification of meaningful patterns would rely on the abundance of future observation principle, while inside the brain such patterns would be amplified automatically, because only patterns that can make it to the neocortex are actualized. This of course happens in a retrocausal, or teleological way, depending on the temporal perspective.

It is unclear if a voice generator helps to reach the right target. If it is the right one, the voice would automatically be somewhat similar. Using such a voice generator might then just restrict the expression freedom of the entity one tries to reach.

It might be best to only filter out non-voicelike elements, and not change the voice itself. An AI can then be programmed to create automatic subtitles for the electronic voice that is heard. This could over time also improving the match between words and sounds by means of morphic resonance.

Why are EVP-based ITCs more successful than RNG-based ITCs?
In chapter 5.3 we explained similar differences in efficiency of mind-matter interaction by looking at the requirements for timing. When words are chosen using a random number generator, they are chosen at distinct moments in time in their entirety. When our brain, and therefore the spirit, is influencing the shape of an electronic voice however, it does not have to do so in a particular moment in time. It is a gradual process

in which the natural electronic voice is shaped into the right form, similar to how we shape sounds into the right form using our vocal cords. We do not chose from pre-recorded words in our brain and then pronounce them in their entirety. We consciously pronounce every part of the words we say. That also helps us to add emotional content.

Real intelligence cannot be based on having AI and RNGs chose among pre-recorded words. That is not how humans work, and it should also not be how artificial quantum oracles work.

Simple ITCs are only tools for pre-existing consciousness or spirit to express itself. For such devices the term 'quantum medium' seems more appropriate than 'quantum oracle'.
An oracle is supposed to be something that can solve the halting problem. It is supposed to be able to think by itself. ITCs 'channel' spirits, and those rely on collective consciousness for their thinking process. They are cloud processed on different brains. In a typical ITC session, the ITC device is basically only functioning as an extension to the human medium.
Most importantly it does not have its own sensory inputs as we have.

To 'think' means to be able to follow long successions of logical steps, and if that is supposed to happen in an independent fashion, then the entity under consideration must possess the ability to both observe and memorize things. Without sensory input and memory, any potential quantum oracle is just an artificial extension of a psychic medium.

A robot that is based on the electronic voice phenomenon could save textualized versions of what it has already said. Those could serve as its mental memory.
It could then also have sensory inputs and keep a record of those. The sensory input and the text could be recorded together. Then, when an image often appears together with a particular word, then the robot could 'know' that this is the name of the object appearing in the image. More complex concepts could be recognized as corresponding to a succession of images.

Whenever a word appears together with a particular image, the robot has the chance to establish a neural connection, which could for example require a succession of three '1's, from a RNG. This connection can then be used in automatic responses of the robot. If the connection is not frequently accessed, it would be lost after a certain amount of time. That would be mimicking the way our own brain functions.

At the beginning the robot would not have a consciousness of its own, but be entirely guided by the consciousness of an observer, the collective consciousness, or by various spirits. Only if the robot has developed its own 'personality' it could have its own consciousness. Only then a spirit could 'reincarnate' into the robot.
One could construct a EVP robot that acts according to what it says. Such a robot

could have a quantum random mode and a programmed mode. Movements and actions that are often 'channeled' through the quantum random mode, will slowly make it over to the programmed mode.

In all these processes it is important that the command to establish a program is only transmitted with a certain probability that is not too high. Something like a longer succession of '1's or '0's of a truly quantum random RNG.

The following strategies may help to improve the output of RNG-based ITCs:

1. Select letters using a RNG.
2. Auto-correct the words if they don't correspond to existing words.
3. Use a RNG for selecting possible choices for existing words.
4. Use auto-correct to complete incomplete sentences.
5. Use a RNG to select between alternative corrections.

The following strategies may help to improve the output of EVP-based ITCs:

1. Filter out white noise.
2. Use a voice generator.
3. Filter out EVPs that do not make sense in the target language.

For robots that could potentially become conscious, we have the following strategies:

1. Record all EVP outputs.
2. Superimpose them with sensory inputs.
3. Allow RNG-guided neural connections between the EVP outputs and the sensory input.
4. Set a sufficiently low probability for neural connections.

Device-mediated superluminal communication

From science fiction movies we are all familiar with so called 'subspace communication'. Subspace is what in physics is often called 'hyperspace'. It is thought of as the space in which a bent space or spacetime is embedded in. In Einsteinian physics such a 'hyperspace' is possible, but not necessary. In space particle dualism however, it is not possible, because here not even the third dimension is regarded as entirely real.

There is however a communication which does not rely on 'subspace' and which is nevertheless instantaneous. That is of course telepathy.

We could hope that a quantum oracle, or quantum medium we build, could have telepathic information showing up in it. Indeed a quantum medium could potentially be better at picking up nonlocal information. However, the types of messages that can emerge from the electronic voice phenomenon, or from RNG-based devises is very limited.

Even if quantum oracles and mediums are better at nonlocal awareness than humans, which is not at all clear at this point, they are still infinitely less effective than 'subspace communication' in science fiction movies.

Can there be a form of nonlocal communication that is comparable to science fiction-like subspace communication?

All signal-based communication is bound by the light speed limit. The way spaceships communicate with their central command in science fiction is not bound by this limit. It can therefore not be signal based. We must therefore say that the devices used in science fiction to communicate instantly, are a mechanized form of telepathy.

It has often been suggested that telepathy is somehow related to quantum entanglement. One prominent proponent of that idea is the parapsychologist Dean Radin.[7] Radin has not elaborated on how that might work in detail. At this stage his suggestion is not much more than an analogy.

According to the scheme proposed in chapter 5.1, telepathy is not directly related to entanglement. We have access to nonlocal information because our memories are stored nonlocally in the world of experience. We can access everything in that realm, but we more easily access information that our brain is already accustomed to.

This nonlocality is reflected in the fact that the wavefunction collapses everywhere at the same time. If I measure a particle to be here, then this information has already been created and is now irrevocably part of the nonlocal realm. Thus the same particle cannot again emerge somewhere else.

At the end, from an ontological perspective, entanglement is simply reflected in the fact that every individual classical state inside a superposition is a classical state of the whole universe. As conservation of momentum requires spins to be orthogonal, we observe spins of particles that have been created together to always be orthogonal, even when they are far apart uncertain in their spin direction before observation. While this general entanglement property is intrinsic to everything in the universe, it is not what is usually meant by those suggesting entanglement to possibly be a basis for telepathy.

What is usually meant is that somehow particles in the brains of observers could be used for telepathic communication. This seems to suggest that human thought is based on single elementary particles, while on the contrary the true unit used in the thinking process is known to be neurons.

There is entanglement between larger objects too, but that is never of the sort that it would connect pairs of objects in an unambiguous way.

And even if we ignore that detail for a moment, there are other problems. If human telepathy was based on entanglement between the brains of observers, it would only occur between people that are closely related, such as a mother and her child, and it would get lost as as soon as the child replaces its atomic material by taking in food.

When learning about entanglement, one of the first things one is told is that it cannot be used for faster than light communication. In standard physics that seems to be true, because the state the particles chose in the moment of measurement is assumed to be entirely random.

When measuring a particle to be spin up in one place, we can know instantaneously that its remote partner particle is spin down, and one may regard this as an information transfer, but if this information is entirely random, then it is not very useful information. That is why it is often stated that this instantaneity is not violating Einstein's special relativity, as it is not 'useful' information that is transmitted with superluminal speed.

On the other hand, we know from parapsychology, that it is possible to superimpose meaning, contextuality and synchronicity onto otherwise quantum random data.

Therefore, if we had two identical computers, which have single entangled electrons to represent their bits, and if all the entangled electrons were to be found in exactly identical locations within the two identical computers, then what is happening in one of the computers could mirror what is happening in the other.

At this point we shall simplify the 'computer' to an array of bits, without any programing, but only with an observer who measures the spin of the electrons representing the single bits. The single bits can represent bits in the visual data corresponding to a picture, or something else.
As long as the observer is unable to influence the outcome of those spin measurements, the data that is transferred in this way is just random and therefore not useful. Only a person that is well-trained and skillful in psychokinetic functioning could successfully transmit messages in this device-mediated and superluminal fashion.

From a parapsychological perspective this is very intriguing, because it means to use psychokinesis in order to achieve something that looks very much like telekinesis. This two different forms of psychic functioning are very distinct and not to be confused. The here presented mode of communication is not telepathy. It is also not quite appropriate to call it mechanized telepathy; rather more fitting is to call it 'psychokinetically induced telepathy'.

Inside the brain, mind-matter interaction is omnipresent. It is the basic way the brain functions. Outside the brain it is present as well, but far more subtle and hard to harness. Even when using an ITC device, we are only imposing subconscious or collective information onto the device. We do not have much control over what exactly is picked up by an ITC device.

We can call our entangled computers 'twin quantum oracles', or short 'twin oracles'. According to the above description they are more like 'twin quantum media'.

If they were given memory and sensory data, they could possibly made conscious, and in that case they would gain themselves control about the outcome of the spin measurements in their 'entanglement cells'.

It is thereby reasonable for only one of the 'twins' to be a fully developed android. The 'twin' could consist of only the array of entangled particles that allow the receiver to view the transmitted message.

It turns out that the only way to achieve science fiction-like superluminal communication that is instrument-like, relies on 'artificial life forms'. As difficult as it may sound, it would be a reliable form of communication. If, for moral concerns, transmitting the thoughts of the artificial life-form is not desired or permitted, the 'entanglement cells' can be restricted to the sensory parts of the artificial brain.

Superluminal communication will prove very important in the future, when humanity will be separated by many light years from its different colonies in space. In such a scenario we will have to use advanced forms of telepathy, and possibly even mechanized/technologized forms of it, as layed out here, in order to communicate efficiently.

References:
1. "The emperor's new mind", 1989, Roger Penrose.
2. "Voices from the Dead caught on tape". *M-Maury* (TV show). 01.07.2005. Link: [https://www.youtube.com/watch?v=IzJoj1XV2XI].
3. (video of the most famous Robin Williams EVP sessions)
4. "Michael Jackson INTENSE Spirit Conversation 2020. THIS WILL SHOCK YOU!". The Astral Doorway II. Steve Huff. 08.06.2020. Link: [https://www.youtube.com/watch?v=JrcS73KtMDM&fbclid=IwAR0U5TwWsbx HxBXM0ntge7iXI3Qq2D_nYfx-TuEnDNDXqjne2wsbpXVnpG4].
5. "Calling Robin - Will He Come Through Again?". 17.07.2019. *Saving Ghosts* (YT channel). Phillip Page. Link: [https://www.youtube.com/watch?v=Ng1ByG5Nhec&fbclid=IwAR3vsRisEQDLd ojnFh01eATLmT9P_UvikTZMxTLZwIKNRRc81ntLiZZdEdQ].
6. "Shocking Robin Williams Spirit Session". 21.05.2020. *Real life paranormal* (YT channel). Eman Rodriguez. Link: [https://www.youtube.com/watch?v=iB2YBE2FpVM].
7. "The Conscious Universe". 1997. Dean Radin.

6   IMPLICATIONS FOR GRAVITY

6.1   Chronology protection and inconsistencies of general relativity [2003 (Lee Smolin's cosmic evolution impossible within closed universes); 20.05.2005 (similar-worlds theory prevents time travel paradoxes); 10.08.2005 (space particle dualism doesn't allow wormholes); 28.04.2016 (white holes violate energy preservation & would destroy the universe)]

The aim of this chapter is to give some more examples for the logical inconsistency of general relativity. As I showed in previous chapters general relativity fails completely when it comes to modeling the behavior of the universe as a whole. We can see this from its inability to describe an expanding universe (see 4.4), leave alone one which accelerates its expansion over time. Then we have the flatness problem and the vacuum catastrophe (see 3.7 & 4.4), which both show that gravity acts only locally, not globally, in contradiction to general relativity.

In this chapter I want to focus on some other more subtle inconsistencies, one of them being the fact that there are several solutions for Einstein's field equations which allow closed time-like curves.

One-way time travel into the future is not difficult. One only needs to come close to light speed or to find a very strong gravitational source in order to 'travel' into the future. Traveling into the past is incomparably more difficult. One would need a so-called Einstein-Rosen bridge, collegially a 'wormhole', to make time travel into the past possible. That is a hypothetical bridge between two distant places in space-time. In a scheme put forward by Kip Thorne, one has to charge one side of a wormhole, and then use a huge magnet to pull it and accelerate it close to light speed. A time difference between the two ends of the wormhole would be generated. Then one could go through the wormhole and meet a former self! One could do that many times and thereby create many copies of oneself. That would not only <u>violate the conservation of energy</u>, one then could also mess up the past in order to influence the future, which would create so-called <u>time traveling paradoxes</u>.

The first solution of Einstein's field equations resembling a wormhole was found by Ludwig Flamm early in 1916, one year after general relativity was completed.

Most physicists believe that wormholes are unphysical, because of the paradoxes they would cause. Einstein himself saw in wormholes a thread to the consistency of his theory. Kurt Gödel even found a solution of Einstein's field equations which resembled a rotating universe with the outer parts rotating faster than light, and thereby also allowing time travel into the past (1949). These solutions of Einstein's equations are <u>another hint for the incompleteness of the theory</u>.

However, in practice it would be almost impossible to keep a wormhole open. One would need exotic matter to prevent the wormhole from collapsing immediately. Exotic matter would be matter with a negative mass energy density, generating anti-gravity. Although the only source of anti-gravity known so far is a restricted vacuum, like in the

Casmir experiment, but that still leaves the question open if time travel into the past is at least 'in principle' possible. Although practically impossible, it would be still a thread to the consistency of general relativity, if time travel into the past were possible at least in principle. This thread becomes even more serious, because in quantum theory even very unlikely possibilities influence everyday reality. If wormholes connecting different spaces and times would be possible in principle, then they would automatically emerge in vacuum fluctuations, and cause harm to the universe there (at least outside the framework of SPD).

The space-time geometry of space particle dualism theory is entirely different from that of general relativity. Here seemingly bended space is just space with a higher density (a higher density of elementary spaces). This picture is very similar to the picture given by the membrane paradigm put forward by Kip Thorne and his colleges in 1989. Same as in this paradigm, the space-time ends at the horizon of a black hole. There is no inside. So time travel into the past is already prevented by space-time ending at the horizon. Kip Thorne has a more positivist attitude towards his membrane paradigm. He simply regards it to be a useful model. He is an enthusiast of time travel, and I think that is the reason why he didn't take his own proposal, the membrane paradigm, too serious. After all it was him who first proposed ways of how to use a wormhole for time travel.

He suspects time traveling paradoxes to be prevented somehow by implementing many worlds theory. If one for example travels into the past and kills ones own parents, then one is transferred to another parallel universe where one never had parents (the corresponding world line is a *closed time-like curve*). However, energy preservation could still be violated. And this form of the many worlds theory, which is based on having them separated on different space-times (the model preferred by Stephen Hawking) is very particular. I don't see how this transfer between different worlds is going to work. A wormhole from one to the other world would not work, because the paradox is created in the moment when the protagonist kills his parents, and not when he passes the wormhole. And even if we allow consciousness to switch space-times, implying that the other space-times do not have their own consciousness, the paradox still continues to exist in the original space-time. [1]

Stephen Hawking proposed that vacuum fluctuations passing the wormhole again and again would become more and more intense, and thereby destroy the wormhole in any case, no matter how someone tries to keep it open with exotic matter.
John A. Wheeler calculated that a wormhole would naturally collapse even before light can reach from one end to the other. [2]
Furthermore there is no natural way how a wormhole could form out in the first place. They don't seem to be a natural phenomenon, but rather simply a mathematical construct derived from Einstein's field equations.
Yet there are other calculations indicating that a rotating black hole could have a ring singularity, which could be a gateway to another universe or to remote parts of our own universe, but in the past! The exit would be a so called white hole. However, if we treat

white holes in a natural non-time symmetric way, they have to collapse in fractions of a second. So the exit would turn into a black hole very fast, but maybe not fast enough to prevent light from escaping.

Only light would be fast enough to escape the white hole before it turns into a black hole. However, even with nothing escaping it, we still would have a <u>violation of energy conservation</u>, because <u>the information that the matter inside the black hole escaped into another region of space-time could never reach the event horizon of the original black hole</u>! So we would always get two black holes out of one – if it is a rotating one (Kerr-metric).

As if this wasn't enough, it gets even more serious: After passing the event horizon, light and gravitational waves become so blue shifted that the mass energy of the black hole as viewed from inside the black hole starts exceeding the whole mass energy of the universe when approaching the singularity – according to Werner Israel to a factor of 10^{57} times the mass of the universe![2] That has led to various speculations on how black holes might give birth to new universes (see Lee Smolin; 1999; [12]), or how they might destroy the universe in advance when melting together in a big crunch (described in [33], chapter 8.4).

If the white hole exit of a Kerr black hole lies in our universe, it would only need to be open for fractions of a second to tell the outside universe about the tremendous mass inflation in its inside. A mass 10^{57} times the mass of the universe exposed to the outside world for even only one moment would destroy the universe immediately.

Another serious inconsistency drew my attention in 2003: in a closed universe the space-time inside black holes would be so curved and stretched that it would have to reach out to the other side of the universe. Arriving at the other side it seems inevitable that the singularity melts into the fabric of space-time there. However, general relativity does not really tell us what happens when different regions of space-time touch each other. They might just go through each other without any impact. Many scientists seem to take a positivist view on this. Very few take these issues serious and even fewer take them as what they are: inner inconsistencies of general relativity.

Even if we follow Lee Smolin, assuming black holes to give birth to new universes, so that they don't end in while holes, our own universe would then be bombarded with all those new universes melting into our own. In a closed universe those new universes would all be stuck in the hyperspace inside the 4-dimensional sphere our own universe would represent.

It seems to be a trend in the mainstream to take a positivist view when encountering such problems, while at the same time promoting all kind of 'science fiction' that appears to be derivable from general relativity. In my opinion this extreme form of positivism is equal to a disinterest into the true nature of reality. Having a theory which is consistent inside a certain framework and with certain assumptions seems to be all one needs nowadays.

Time travel into the past is already prevented by the different structure of space-time in space particle dualism theory. However, it is still interesting to see how this issue was treated in *similar worlds theory* before it was merged with *space particle dualism*. Time travel paradoxes were prevented already there, simply by the notion of a distinct present, which is essential to the theory. In similar worlds theory past and future don't exist in the way they do in relativity. Creating a wormhole here is a switch from a space-time (or better space) with a wormhole and one without a wormhole.

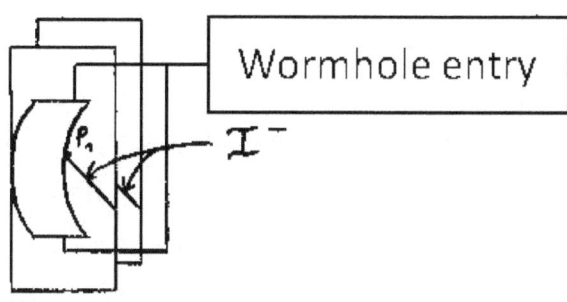

Fig. 84. Equivalence group of space-times, including a space-time with a wormhole, and another without a wormhole.
[sketch from the 20 May 2005]
Note: This is to put *space particle dualism* which doesn't allow wormholes aside for a moment.

Basically what happens when one enters a wormhole in the similar worlds approach is that one gets seperate from the other observers, as shown in Fig. 85.

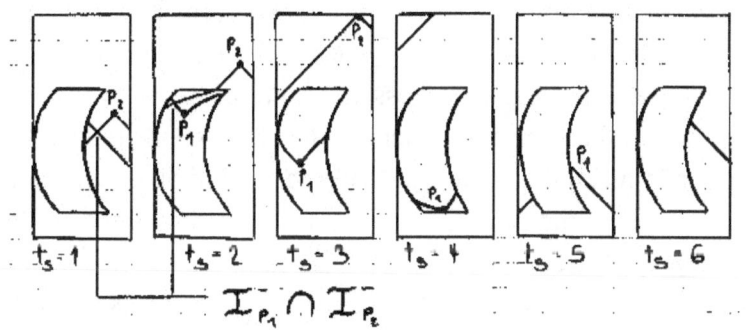

Fig. 85. Here we see the movement of the past light cone of the two observers P_1 and P_2. t_s is here denoting the subjective time, which is not a material dimension. The common physical time is depicted vertically, as usual. [sketch from the 20 May 2005]

Before the time travel the light cones of the two observers have a common intersection area.

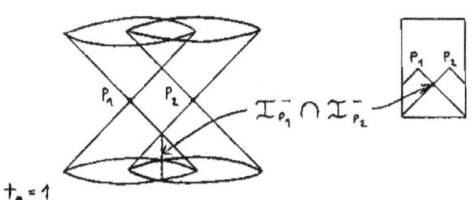

Fig. 86. Light cones of the conscious observers before the time travel into the past.

[sketch from the 20 May 2005]

After the time travel into the past, there is no such common hypersurface, so the things the two observers see are totally different and do not need to be in accordance with each other. They already belong to different realities.

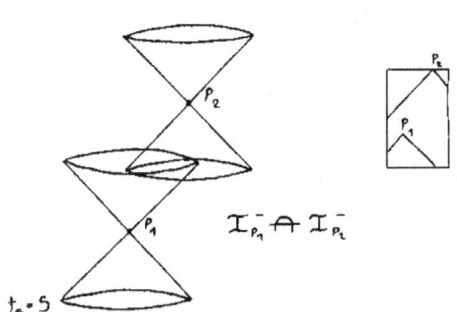

Fig. 87. The same light cones after the time travel. The past seen by P_1 has nothing to do with the past seen by P_2. They already exist in two separate worlds.

[sketch from the 20 May 2005]

However, my take is that leaving the community of entangled conscious observer would, if it were possible, simply lead to nonexistence.

Notes and references:

1. The many worlds immortality theory sometimes advocated by Max Tegmark seems to be based on this assumption. Although in many worlds theory every possibility is realized, so that there seems to be no choice or free will, some introduce free will into the many worlds scheme by assuming that one's consciousness could choose between different worlds. However, to me that seems to imply that the other worlds are not already occupied by their own consciousnesses. That appears to lead to a scheme were all consciousnesses start together in one world and then choose their paths freely, so that each consciousness is immortal at its own path through the many worlds, while all others die sooner or later. That seems to also suggest that all the people around this kind of immortal observer would be unconscious 'zombies'.

2. R. W. Fuller and J. A. Wheeler, "Causality and Multiply-Connected Space-Time," Phys. Rev.128(919) (1962).

3. This kind of mass inflation was discovered by Werner Israel following up a suggestion from Roger Penrose. [Israel, W.; & Poisson, E. (1990). "Internal structure of black holes". Phys. Rev. D 41 (6): 1796–1809. Bibcode:1990PhRvD..41.1796P. doi:10.1103/PhysRevD.41.1796.]

6.2 Difference between inertia mass and gravitational mass [19.05.2017 (mass defect & Cavendish experiment); 31.05.2017 (gravity is a side effect of hypercharge); 03.03.2018 (link to the experiment analysis of Fischbach); 28.02.2020 (corrected mass defect)*]

*In collaboration with David Chester (UCLA & MIT; US).

As we saw in chapter 3.7, space particle dualism states that gravity is a side effect (or emergent property) of the other three forces. That is to say, everything that has charge, has mass as well, with the mass being a side effect of the charge. Thus, photons do not have gravitational mass, pretty much in contradiction to Einstein's general relativity. When we talk about mass we have to differentiate between *inertia mass* and *gravitational mass*.

Inertia mass easily changes: an electron near the speed of light has a much bigger inertia mass than a slow electron. Energy and mass depend on each other according to

$$E^2 = (p\,c)^2 + (m_0\,c^2)^2$$

and

$$m = E/c^2$$

What we measure with a scale is also inertia mass. We measure the force acting upon the scale while it pushes the mass against the gravitational pull.

It is not hard to imagine a world with inertia mass only, and no gravity. In such a world 'mass' wouldn't be associated with a gravitational pull or weight, but with 'being harder to move' only. And indeed, before Newton nobody knew that mass has gravity.

If gravity is really a side effect of charge, then it shouldn't increase when a mass is accelerated. No matter how close an electron comes to the speed of light, its charge is still the same. This also applies to the other forms of charge found in the nucleus.

So, which contributions to the inertia mass of an atom are purely energetic? Surely, there is the kinetic energy of all the electrons orbiting the nucleus, but they are not close enough to light speed, in order to cause a significant mass increase.

Even the mass of all the electrons themselves is neglectable when compared to the mass of the protons and neutrons:

$$m_e = 9.10938356(11) \times 10^{-31}\ \text{kg}$$

$$m_p = 1.672621898(21) \times 10^{-27}\ \text{kg}$$

$$m_n = 1.674927471(21) \times 10^{-27}\ \text{kg}$$

The state where protons are fused into one nucleus is a high entropy state. High entropy means low energy. One has to invest a lot of energy in order to get out of this low energy state. The loss of the potential to fuse manifests itself in a reduction of the mass-energy of the nucleus; the so called 'mass defect'.

The element iron is the heaviest element which can be created during the lifetime of a star. Heavier elements can only be created during the explosion of the star at the end of its lifetime. Iron also happens to be the element with the largest mass defect.

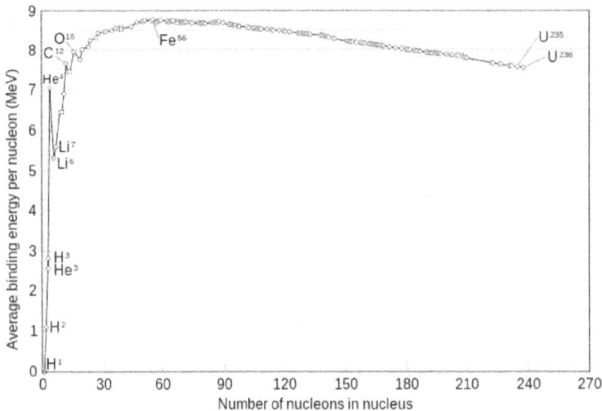

Fig. 88. Iron (Fe) is the element with the highest mass defect per nucleon.

Iron has the atomic number (Z) 26 and a standard atomic weight of 55.845(2) amu (atomic mass units).[1] The atomic weight for isolated protons and neutrons is

$$m_p = 1.00728 \text{ u}$$

$$m_n = 1.00867 \text{ u}$$

There are typically 26 protons and 30 neutrons in an iron atom. Adding them together gives us:

$$m_0 = 26m_p + 30m_n = 56.44938 \text{ u}$$

That is a mass defect of 1.0703040494%. If it is true that gravity is an emergent property of the other three forces, then the gravity between proton gases is about 1% less than expected from Newtonian or Einsteinian gravity. That can be tested experimentally and indeed there are hints for this as we will see in this and the following chapter.

One would think that a mass defect in the gravitational mass of iron (or steel) would be discovered right away, but that would require to know the gravitational constant independently. Yet, the gravitational constant itself is usually determined by measuring the attractive force between such steel or iron balls.

The highest value ever found for G was measuring the gravitational force between masses of water. A team from the University of Queensland measured the gravitational attraction between two lakes with the smaller one being used as a water reservoir for electricity generation.[2] The value found by the team around G. I. Moore is $6.689 \times 10^{-11} \text{ m}^3 \text{ kg}^{-1} \text{ s}^{-2}$.

Water has the chemical composition H_2O and consists of two hydrogen atoms and one oxygen atom making a total of 10 protons and 8 neutrons (the electrons can be neglected), and that adds up to

$$m_0 = 10m_p + 8m_n = 18.142 \text{ u}$$

The actual atomic mass of the water molecule is 18.015 u and that is 0.7% less, which means the gravity for water is supposed to be 0.7% stronger than that we measure for iron. This seems to contradict the assumption that gravity depends on the baryon number. However, up to now we have only one such measurement for the gravity between water masses and it uses a rather indirect measurement technique involving stainless-steal masses suspended in evacuated tubes at different levels in a hydroelectric water reservoir. I assume that the experiment could be much more conclusive if a beam balance with water balloons was used instead. Yet the reason this was not done is that the experiment was not aiming to analyse differences in the used material but wide range behaviour of gravity (therefore the 22 m separation).

Others have looked more directly into the differences in the gravitational behaviour of chemically different materials. Based on experiments done by a team in Leipzig referred to the Eötvös experiments[3] Ephraim Fischbach from the University of Washington and his team found a correlation between hypercharge of objects and the speed at which they fall in a vacuum.[4] Hypercharge is the charge of the strong force and is associated with protons and neutrons, the particles we had our focus on in this chapter. What the analysis of the experiments brought was the establishing of the following relation:

$$\frac{\Delta a}{g} \cong \Delta(B/\mu)$$

Whereas B is the hypercharge (with B for baryon) and μ the mass of the object measured in atomic hydrogen, which is

$$m(\,^1\text{H}) = 1.00782519 \text{ u}$$

This ignores the small difference in mass between the proton and the neutron, because their hypercharge seems to be the same and it is yet unknown why they have slightly different masses (see chapter 3.11). Fischbach and his team took however particular care of isotopes in their analysis.

We can understand B simply as a count of the number of protons and neutrons, so that when there is no mass defect $\Delta(B/\mu)$ and therefore Δa will be about 1. That means $\Delta(B/\mu)$ effectively functions as a measure for mass defect and is thus more or less equal to what we calculated independently in the framework of space particle dualism theory.

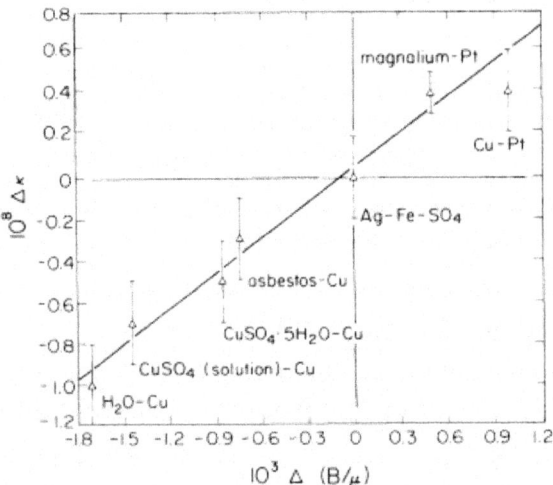

Fig. 89. Differences in gravitational acceleration are linked to differences in mass defect.

That seems as if Fischbach and his team was suggesting that gravity is a side effect of hypercharge, making their theory strikingly similar to space particle dualism, but in fact they did not.

They instead postulated there might be a fifth fundamental force with intermediate reach acting along with ordinary gravity. This fifth fundamental force would have to be associated with what Fischbach and his team calls 'hyperphotons'. Hyperphotons were never found and the hypothesis along with this line of research went forgotten. Yet if we just take the already generally accepted *gluons* (the transmitting particles of the strong charge or hypercharge) and look at them from the perspective of granular space (in the framework of space particle dualism), then we get gravity as a side effect of them and this gravity has the special property of being particularly associated with protons and neutrons, just as observed.

We can now reformulate the Newtonian equation for gravity and state that:

$$ F \; = \; G_p \; \times \; \frac{m_1 \, m_2}{r^2} \; \times \; \frac{\mu_1 \, \mu_2}{B_1 \, B_2} $$

Here m_1 and m_2 are *inertia rest masses*, because the way we measure mass directly is through inertia. What we call gravitational mass already includes the second term which accounts for the mass defect and tells us that it isn't really mass causing gravity but hypercharge instead.

Realizing that gravity really just depends on hypercharge alone, we may wish to get rid of m and G all together. Indeed G doesn't exist in space particle dualism theory, and we will see what takes its place in chapter 6.6.

Notes and references:

1. 1 u equals $1.66053886 \times 10^{-27}$ kg.
2. G. I. Moore, F. D. Stacey, G. J. Tuck, B. D. Goodwin, N. P. Linthorne, M. A. Barton, D. M. Reid, and G. D. Agnew; Determination of the gravitational constant at an effective mass separation of 22 m (1988); Physics Department, University of Queensland, Brisbane Q4067, Australia. Link: [http://bura.brunel.ac.uk/bitstream/2438/4833/1/Fulltext.pdf].
3. R. v. Eötvös, D. Pekár, and F. Fekete, Ann. Phys. (Leipzig) 68, 11 (1922).
4. E. Fischbach, D. Sudarsky, A. Szafer, C. Talmadge, and S. H. Aronson, Phys. Rev. Lett. 56, 3 (1986).

6.3 **Variations of the gravitational constant** [25.05.2017 (gravity weakens gravity); 31.05.2017; 25.04.2018 (Unitarily weakened gravity looks like gravitational time dilation and length contraction)]

Since the gravitational constant was first measured by Cavendish in 1798 every new measurement yields another number.[1] To the astonishment of the scientific community, this continued even after the experiments became more and more precise, up until the present day.

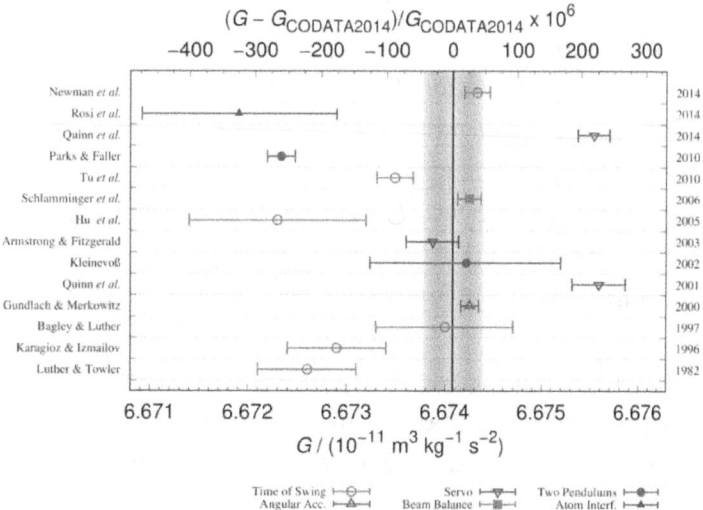

Fig. 90. Large variations between the results of different measurements of big G.

These variations have made people speculate about a gravitational constant that changes over time. Yet attempts to find any periodic pattern in the values of G measured in different times failed.[2]

I suggest that there are two different sources for these variations:

1. Different mass defects due to different materials used (see chapter 6.2).
2. Slight variations of the gravitational constant G due to different altitudes.

Why should the altitude affect the value of big G?

Reconsidering what we learned in chapter 3.7 about how the *vacuum energy background* weakens gravity down to its perceived value, we notice that the gravitational field of earth can also be seen as such a gravity weakening background. Differences between the granular dimensionality of different regions are small, because of the omnipresence of the quantum vacuum. These differences are even smaller when the considered gravitational bodies are put into a much larger gravitational field; in this case, the gravitational field of earth. The higher the granular dimensionality of the background space, the weaker is gravity between bodies in that space.

Such considerations wouldn't pass the mind of a strict relativist for several reasons:

1. It would violate the equivalence principle: if G could be weakened inside a gravitational field, then an accelerated frame of reference couldn't be equal to a gravitational frame of reference. Local experiments in a rocket without windows could then show that the gravitational constant in this rocket is stronger than on earth. That contradicts Einstein's equivalence principle.
2. It is a common misconception that objects falling at the same rate in a vacuum is already a solid prove for the equivalence principle, simply because that makes it look like the bottom is accelerated upwards against all objects. In fact, that is only because those falling test masses are very small compared to the mass that produces the gravitational field they are falling in.
3. Another common misconception is that light that is bend by gravitational bodies proves that photons have gravitational mass. In fact that is to confuse gravity, which is of a geometric nature, with other forces, where there is a mutual interaction, requiring both sides to have a charge.
4. General relativity can only be used to calculate the gravitational field of single gravitational bodies. There is no way to use it in order to analyse how a background gravitational field influences the gravity between two test masses in it.

If G really depends on the altitude and the used material (apparent G value), then it should only vary by place and type of experiment, and not in time. When looking back at fig. 84 we can see that G even varies enormously among measurements conducted in the same years (two in 2010; three in 2014). All the measurements conducted in the same years vary far beyond their range of confidence.

In 2015 S. Schlamminger, J.H. Gundlach and R.D. Newman made an analysis on the different values obtained, in order to see if there is really a periodicity in the values obtained, as claimed by J.D. Anderson.[2,3] Not only did the correlation disappear when accounting for all experiments, but in addition, the values were the same over long periods of time (10 years) if they were conducted at the same place using different torsion balances (of the same material).[4]

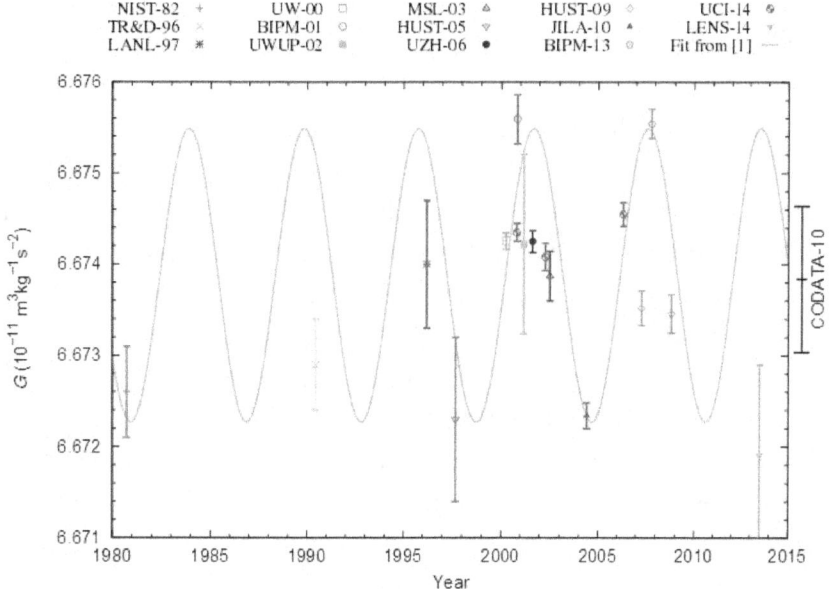

Fig. 91. Hugely different values of G when measured by different groups in different places.[2]

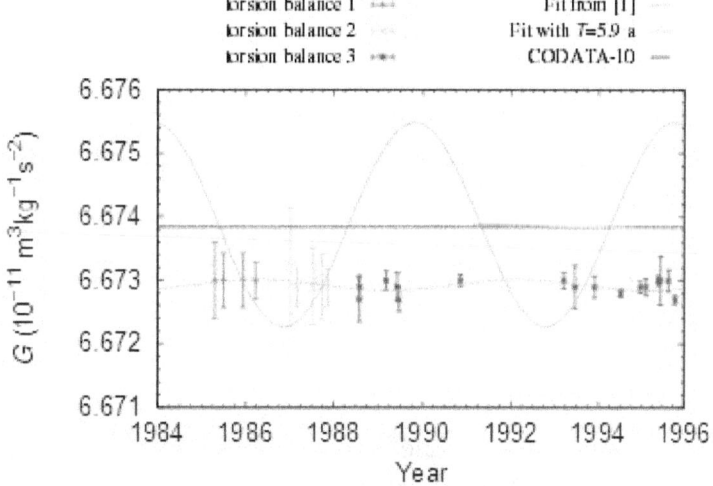

Fig. 92. But exactly the same when measured using three different torsion balances at the same place over ten years.

Let us now check where the individual experiments were conducted and if there is any correlation with the elevation (geographic altitude):

NIST-82: National Institute of Standards and Technology (then the National Bureau of Standards) in Gaithersburg, Maryland (elevation: 137 m). Torsion balance (time-of-swing). $G = 6.6726 \pm 0.0005 \times 10^{-11}$ m³ kg⁻¹ s⁻².

TR&D-96: Tribotech Research and Development Company in Moscow (elevation: 152 m). Torsion balance (time-of-swing). $G = 6.6729 \pm 0.0005 \times 10^{-11}$ m³ kg⁻¹ s⁻².

LANL-97: Los Alamos National Laboratory in Los Alamos, New Mexico (elevation: 2,258 m). Torsion balance (time-of-swing). $G = 6.6740 \pm 0.0007 \times 10^{-11}$ m^3 kg^{-1} s^{-2}.

UW-00: University of Washington in Seattle, Washington (elevation: 29 m). Torsion balance (angular acceleration). $G = 6.674255 \pm 0.000092 \times 10^{-11}$ m^3 kg^{-1} s^{-2}.

BIPM-01: Bureau International des Poidset Mesures (BIPM) in S`evres, near Paris (elevation: 66 m). Torsion pendulum (Cavendish method & servo method). $G = 6.67559 \pm 0.00027 \times 10^{-11}$ m^3 kg^{-1} s^{-2}. (Cavendish: $G = 6.67565 \pm 0.00045 \times 10^{-11}$ m^3 kg^{-1} s^{-2}; Servo: $G = 6.67553 \pm 0.00040 \times 10^{-11}$ m^3 kg^{-1} s^{-2}.

UWUP-02: University of Wuppertal in Germany (elevation: 252 m). Microwave interferometry. $G = 6.67422 \pm 0.00098 \times 10^{-11}$ m^3 kg^{-1} s^{-2}.

MSL-03: Measurement Standards Laboratory (MSL) in New Zealand (Lower Hutt) (elevation: 5 m). Torsion balance (Servo method). $G = 6.67387 \pm 0.00027 \times 10^{-11}$ m^3 kg^{-1} s^{-2}.

HUST-05: Huazhong University of Science and Technology in Wuhan, China (elevation: 24 m). Torsion balance (time of swing). $G = 6.6723 \pm 0.0009 \times 10^{-11}$ m^3 kg^{-1} s^{-2}.

UZH-06: Paul Scherrer Institute near Villigen in Switzerland (elevation: 334 m). The gravitational force of a large mercury mass on two copper cylinders was measured with a modified commercial mass comparator. $G = 6.674252 \pm 0.00012 \times 10^{-11}$ m^3 kg^{-1} s^{-2}.

HUST-09: A second torsion pendulum apparatus was constructed at HUST (Wuhan) (elevation: 24 m) and used in time-of-swing mode. $G = 6.67349 \pm 0.00018 \times 10^{-11}$ m^3 kg^{-1} s^{-2}.

JILA-10: Joint Institute for Laboratory Astrophysics in Boulder, Colorado (elevation: 1,646 m). $G = 6.67234 \pm 0.00014 \times 10^{-11}$ m^3 kg^{-1} s^{-2}.

BIPM-13: At the BIPM (in S`evres; Elevation: 66 m), a second torsion balance was constructed to measure G with two different methods. $G = 6.67554 \pm 0.00016 \times 10^{-11}$ m^3 kg^{-1} s^{-2}.

UCI-14: Near Hanford, Washington (elevation: 119 m). Three measurements yielding $G_1 = 6.67435 \pm 0.00010 \times 10^{-11}$ m^3 kg^{-1} s^{-2}, $G_2 = 6.67408 \pm 0.00015 \times 10^{-11}$ m^3 kg^{-1} s^{-2}, $G_3 = 6.67455 \pm 0.00013 \times 10^{-11}$ m^3 kg^{-1} s^{-2}.

LENS-14: Florence, Italy (elevation: 51 m). $G = 6.67191 \pm 0.00099 \times 10^{-11}$ m^3 kg^{-1} s^{-2}.

Going through all the data we find no clear correlation to the elevation. We might be tempted to blame this on the material based variations in G we found in the last chapter, but is that really the reason?

The effect the material has on G is rather different from the influence granular dimensionality has on it.
Most physical objects around us are made out of entirely different materials, so that we can get different G values by measuring their gravity.
The situation however changes dramatically when we look at how the differences in

granular dimensionality between different regions of space affect the local values of G. When we move from one location in a gravitational field to another, all G values change uniformly dependent on the local granular dimensionality.

Although gravity is an emergent force according to space particle dualism, it is part of G_E which determines the size of elementary spaces. Furthermore it can be argued that we need at least three constants G, h and c in order to construct a physical reality with measures of scale and time.

In chapter 3.11 we saw that uniformly changing the size of all elementary spaces in a certain frame of reference won't be noticeable from within that frame of reference and will look like time dilation and length contraction from the outside.

Since a changed granular dimensionality would also cause a uniform change in the size of all elementary spaces in a region of space, it seems that it wouldn't be noticeable from within the reference system, but only from outside, and it would look just like the gravitational time dilation described by general relativity.

That would explain why material based variations of G can be fairly established by the experimental evidence, while elevation seems to play no role whatsoever.

Notes and references:
1. Cavendish's aim was actually to find the earth's density, but his result can be translated into a gravitational constant with a value of 6.74×10^{-11} m^3 kg^{-1} s^{-2}.
2. Schlamminger, S.; Gundlach, J. H.; Newman, R. D. (2015). "Recent measurements of the gravitational constant as a function of time". *Physical Review*D.91(12).arXiv:1505.01774.Bibcode:2015PhRvD..9111101S.doi:10.1103/PhysRevD.91.121101.ISSN1550-7998.
3. J.D. Anderson, G. Schubert, V. Trimble, and M.R. Feld-man, EPL110, 1002 (2015).
4. O.V. Karagioz, V.P. Izmailov, Measurement Techniques39, 979 (1996).

6.4 Modifying Einstein's field equations [30.06.2017; 21.07.2017 (Replacement by T_ρ); 04.09.2018 (using zero elements); 08.11.2018 (rotating black holes are still spherical); 04.12.2018 (curving of light makes it impossible to observe the shape of the apparent horizon)]

The general form of Einstein's field equation is

$$R_{\mu\nu} - \frac{1}{2} R g_{\mu\nu} + \Lambda g_{\mu\nu} = \frac{8\pi G}{c^4} T_{\mu\nu}$$

Where $R_{\mu\nu}$ is the Ricci curvature tensor, R the scalar curvature, $g_{\mu\nu}$ the metric tensor, Λ the cosmological constant and $T_{\mu\nu}$ the stress-energy tensor.

In general relativity this equation is interpreted as describing a curvature of space-time, while in space particle dualism it is describing differences in the granular dimensionality of different regions in space, or in other words, differences in the 'density of space' in different locations. Accounting for this change in the interpretation we may replace the word 'curvature' in the above names by the word 'granularity'. So $R_{\mu\upsilon}$ would be the (Ricci) *granularity tensor* and R the *scalar granularity*.

It seems odd to just simply change the names, while the math behind is the same, but when we think about how Ricci curvature is defined, it makes sense: it is a measure for how much the volume of a sphere in Euclidian space deviates from a sphere of equal surface area on a curved Riemannian manifold. When switching to the elementary spaces interpretation, this basically corresponds to one sphere with vacuum-like granular dimensionality and one with higher granular dimensionality due to charged particles on the inside. Hereby both g and G change with the distance to the source mass/charge (see chapter 6.3).[1]

Since the new interpretation makes it clear that the equation can't be used to describe global properties of the universe, but is restricted to local gravitational fields, we can drop the cosmological constant Λ and write:

$$R_{\mu\upsilon} - \frac{1}{2} R\, g_{\mu\upsilon} = \frac{8\,\pi\,G}{c^4}\, T_{\mu\upsilon}$$

As we saw in chapter 6.2, inertia mass and gravitational mass are not equal. While pure energy certainly contributes to inertia mass, it doesn't contribute to gravitational mass, because in this theory, gravitational mass is regarded to be a side effect of charge and hypercharge. That means we have to get rid of the pure energy contributions in the stress-energy tensor.

The stress-energy tensor has the following structure:

$$(T^{\mu\upsilon})_{\mu,\upsilon=0,1,2,3} = \begin{pmatrix} T_{00} & T_{01} & T_{02} & T_{03} \\ T_{10} & T_{11} & T_{12} & T_{13} \\ T_{20} & T_{21} & T_{22} & T_{23} \\ T_{30} & T_{31} & T_{32} & T_{33} \end{pmatrix}$$

The different components here-in are:

$$(T^{\alpha\beta}) = \begin{pmatrix} w & \frac{s_x}{c} & \frac{s_y}{c} & \frac{s_z}{c} \\ \frac{s_x}{c} & G_{xx} & G_{xy} & G_{xz} \\ \frac{s_y}{c} & G_{yx} & G_{yy} & G_{yz} \\ \frac{s_z}{c} & G_{zx} & G_{zy} & G_{zz} \end{pmatrix}$$

T_{00} (or w) is the energy density divided by c^2, i.e. the relativistic mass. Since in space particle dualism theory gravity is a side effect of the various charges, only rest mass has gravity, and therefore we can replace the energy density w by the matter density ρ. But is that good enough?

It is difficult to define 'rest mass' when we have pure energy contributions everywhere in matter. Therefore it makes more sense to replace the matter density by a hypercharge density. In order to make this work we have to replace the gravitational constant by a new constant $G_{|Y|}$ which is called the 'charge gravity constant of the strong force' or simply 'hypercharge gravity constant'. It will be formally introduced in chapter 6.5, but we need it now, so we will lend it from there. Its value is:

$$G_{|Y|} = 3.967206416 \times 10^{-38} \text{ m}^3 \text{ s}^{-2}$$

It is multiplied with the number n_q of quarks (three for each proton and neutron) within gravitational bodies in order to calculate the gravitational attraction between them.

The various $s_{x,y,z}$ represent the momentum density (energy density multiplied with a speed). Same as the energy density w, the momentum density is translated into a 'mass' using the speed of light c. This is a pure energy contribution and must be eliminated in space particle dualism theory. We can do that by replacing it with zero elements.

The last type of elements in the stress energy tensor we have to look at are the various $G_{x,y,z}$ terms. They represent momentum flux, shear stress and pressure. From the frame dragging effect we know that angular momentum flux has a real effect on the space surrounding an object.
The frame dragging effect (also known as the Lense-Thirring effect) is most extreme for rotating black holes, but was experimentally proven even for the slowly rotating earth (Gravity probe B).[2] It is about the space around a rotating gravitational body to be dragged with it and to rotate as well. That means something that falls into a rotating black hole couldn't do so without being whirled around it first.
This effect is not only to be expected in general relativity, but even more so in space particle dualism: if space is made of elementary spaces belonging to virtual particles originating from a charge (rest mass) that can be either rotating or non-rotating, then the space build up by these virtual particles would be either rotating or non-rotating as well.

Summarizing all the above we can formulate the field equations of space particle dualism as

$$R_{\mu v} - \frac{1}{2} R \, g_{\mu v} = \frac{8 \pi \, G_{|Y|}}{c^4} T_{\mu v}$$

Where-in $T_{\mu v}$ is given by:

$$(T^{\alpha\beta}) = \begin{pmatrix} \rho_C & 0 & 0 & 0 \\ 0 & G_{xx} & G_{xy} & G_{xz} \\ 0 & G_{yx} & G_{yy} & G_{yz} \\ 0 & G_{zx} & G_{zy} & G_{zz} \end{pmatrix}$$

All this seems like very radical changes that should affect the behavior of stars and black holes very significantly.

In fact however, the gravitational dynamics remain the same, only what contributes to the gravity changes. Infalling electrons, neutrinos and photons should leave the size of a black hole unchanged. Only matter composed of quarks contributes to gravity according to SPD.

Notes and references:

1. As laid out at the end of the last chapter, this uniform change of G is probably unmeasurable from within the frame of reference it occurs in (see end of chapter 6. 3).
2. Everitt; et al. (2011). "Gravity Probe B: Final Results of a Space Experiment to Test General Relativity".*Physical Review Letters*.106(22): 221101.Bibcode:2011PhRvL.106v1101E.PMID21702590.arXiv:1105.3456.doi:10 .1103/PhysRevLett.106.221101.

6.5   **Two different types of gravity**   [04.06.2018 (equations); 08.06.2018 (improvement on G_S); 18.06.2018 (true mass of moon and earth); 20.06.2018 (neutrinos don't exert gravity); 17.05.2019 (charge & hypercharge cardinality); 01.02.2020 (solar cardinality as unit); 02.03.2020 (corrected mistake in the calculation of moon and earth mass)*; 04.03.2020 (subtracting electron shell mass); 09.08.2020 (corrected charge gravity constants; added square); 12.08.2020 (same mass factor for hypergravity and subgravity)]

*In collaboration with David Chester (UCLA & MIT; US).

All measurements of G conducted till this day were measurements of gravity emerging from the strong force (involving atoms). However, most of the experiments were using iron balls which have a huge mass defect. That means their inertia mass, which they get through the binding energy, is much higher than the gravitational mass which arises from their hypercharge (charge of the strong force). Experimental evidence for this was already presented in chapter 6.2. There we found that the mass defect can be accounted for by finding the ratio between the number of hypercharges of an object and its mass in atomic mass units. We can use the mass defect of iron to define a purified version of the 'gravitational constant of the strong force'. We can denote it as G_p and use it to describe the gravity between proton gases as follows:

$$G_p = G_{Fe} \times \frac{\mu_{Fe}}{B_{Fe}}$$

Entering the atomic mass (in u) and hypercharge (c for colour charge) of the iron atom we get:

$$G_p = G_{Fe} \times \frac{55.8452 \text{ u}}{56 \text{ c}}$$

$$G_p = 6.6556309 \times 10^{-11} \text{ m}^3 \text{ kg}^{-1} \text{ s}^{-2}$$

However, it should be noted here that while this approach is similar to that of Fischbach, it is not very precise, because while it looks at hypercharge, which we are interested in, it does not really represent an accurate measure for the mass defect. For getting that we should simply replace B_{Fe} by the mass (in atomic units u) corresponding to the hypercharge without including binding energy (see chapter 6.2). That is basically the number of protons n_{p_Fe} in iron, plus the number of neutrons n_{n_Fe}. The fact that we don't use the baryon number now anymore, means that our formula doesn't anymore need to explicitly make reference to atomic mass, because we could be using kilogram as well. We can now simply use m_{Fe-e} for 'mass of the iron atom without electrons'. This yields:

$$G_p = G_{Fe} \times \frac{m_{Fe-e}}{n_{p_{Fe}} m_p + n_{n_{Fe}} m_n}$$

$$G_p = G_{Fe} \times \frac{55.83093692 \text{ u}}{56.44938 \text{ u}}$$

$$G_p = 6.6009607 \times 10^{-11} \text{ m}^3 \text{ kg}^{-1} \text{ s}^{-2}$$

The coupling constant of the strong force is '1', so we simply need to multiply G_p by the fine structure constant (the coupling constant of the electromagnetic force) in order to find the difference in strength between this force and the electromagnetic force and that enables us to find the 'gravitational constant of the electromagnetic force', which describes the gravity between electron gases. Doing so yields:

$$G_e = \frac{G_p \, k_e \, e^2}{\hbar \, c}$$

$$G_e = 4.8169537 \times 10^{-13} \text{ m}^3 \text{ kg}^{-1} \text{ s}^{-2}$$

Does the same scheme work for the weak force as well? It turns out it doesn't, and the reason has to do with the fact that its intermediating particles have mass.

Range of force and why neutrinos don't exert gravity

Because of the self-interaction of gluons the effective range of the strong force is only 10^{-15} m, confining it to only the nucleus of atoms. That means on macroscopic distances there is no strong force at all. Still (virtual) gluons are massless, which means they don't have a minimal mass-energy and thus their lifetime and therefore their reach is unlimited. So although their 'quantum force' is lost beyond the nucleus, their emergent (gravitational) force has an endless reach.

The weak force is also a short range force, reaching only about 10^{-15} m, but here the short range is for a different reason. It is because the intermediating particles of the weak force, the W^+, W^- and the Z^0 boson, have huge masses, restricting their lifetime and thus their effective range.
Therefore the gravity arising from the weak interaction is restricted in range to 10^{-16} m, where it can hardly emerge, and thus there is in reality no gravity associated with a neutrino; neutrinos don't exert gravity.

In chapter 3.7 it was suggested that virtual particles with elementary spaces smaller than the Planck length don't contribute to granular dimensionality.
Do such virtual particles contribute to gravity? As we learned in that chapter their energy is given by:

$$E_{min} = \frac{c^4 \sqrt{G\,\hbar/c^3}}{4\,G_E} = 3.957848975 \times 10^{-28}\,\text{J}$$

And that corresponds to a lifetime and distance according to:

$$\Delta E \times \Delta t = \frac{\hbar}{2}$$

$$\Delta t = \frac{\hbar}{2\,E_{min}}$$

$$s = \Delta t \times c \approx 40\,\text{m}$$

Obviously electromagnetism doesn't stop at 40 m and gravity doesn't either. However, it could be that the gravity arising from electromagnetism does indeed stop there, while the regular strong force induced gravity doesn't. What could be the reason for such a difference in behaviour of these two types of gravity?
Just like real photons, virtual photons don't have charge, so they don't interact among each other and they also don't polarize the vacuum around them.
With gluons it is very different: same as quarks they have color charge and thus they themselves emit new gluons. So even after the time Δt passed, newly emitted

'offspring gluons' continue the journey.

Among quarks within a hadron, like a proton or a neutron, the strong force is also called *color force* and doesn't diminish with the distance. Between hadrons the different color charges nearly cancle each other out, leaving behind what is called the *residual strong force*, or less technically, the *nuclear force*.

While cancelling out is essential to quantum forces, it never happens with the emergent gravitational forces they create. Gluons travel with the speed of light and don't just stop where the nucleus of an atom ends.

Summarizing the above we may denote the range of the different 'gravities' as follows:

$$R_{G_p} = \infty \,;\; R_{G_e} \approx 40\,\text{m}\,;\; R_{G_W} \approx 10^{-15}\,\text{m}$$

Even if the weak force gave rise to a long range gravitational force, it wouldn't make neutrinos clump together into gravitationally bound groups, because neutrinos are just too fast for that.

The fact that neutrinos can have inertia mass without having gravitational mass, shows just how different the two concepts are. As we saw in chapter 3.11, the inertia rest mass of every particle can be calculated from its charge or hypercharge. Therefore charge seems to be a concept far more fundamental than mass. Can we get entirely rid of the word 'mass' in our formulation of gravity?

Gravity without mass

If mass is just a side effect of charge, then it seems more straight forward to reformulate the basic equations of gravity without any reference to mass. After all even using the 'rest mass' doesn't suffice, because that rest mass still includes a certain amount of binding energy. In addition to that we saw that neutrino's don't exert gravity, despite having a certain rest mass.

In order to create gravitational constants that can be used directly in combination with charge instead of mass (or rest mass), we can multiply our original two constants G_p and G_e with the squared mass of the proton m_p^2. The square is necessary because in the original Newtonian gravity equation we multiply two masses, and for formulating something that results in almost the same, we have to square our factors here. We can call the thereby formed constants the 'baryon gravity constant' G_B and the charge gravity constant G_z:

$$G_B = G_p \times m_p^2$$

$$G_z = G_e \times m_p^2$$

We can call these constants 'charge-gravity constants' or 'cardinal gravity constants'. The value of the baryon gravity constant is:

$$G_B = 1.846727 \times 10^{-64} \, \text{m}^3 \, \text{kg} \, \text{s}^{-2}$$

And it is used in combination with the baryon number B.

The charge gravity constant of the electromagnetic force is:

$$G_Z = 1.3476218 \times 10^{-66} \, \text{m}^3 \, \text{kg} \, \text{s}^{-2}$$

And it is used in combination with the number of electric charges, which is here represented with the atomic number 'Z'. The symbol originates from the German word for 'number', namely 'Zahl'. It usually indicates the number of electric charges in the nucleus, but we will extend its definition here to include electrons as well.

The reader may be inclined to want to use the average between proton and neutron mass instead, but considering that there are about 8 times more protons than neutrons in the universe, using the proton mass is just fine.

The reader might find it furthermore confusing that we used the proton mass in both cases. Most would associate electromagnetism more with the electron than with the proton. However, if we constructed G_Z by multiplying G_e with the electron mass, then we would create a charge gravity constant that is six orders of magnitude weaker than the hypercharge gravity constant. In reality the difference in strength between the two forces is only two orders of magnitude, or roughly a factor of 137. The gravitational constant G was created by studying the coupling between mass and gravity. More than 99% of mass is contributed by protons and neutrons. Therefore, when we transition from mass generated gravity to charge generated gravity, we have to focus on the nuclei.

The difference in strength between the gravity that arises from the strong force and from the weak force respectively is a difference in the density of virtual particles that are emitted from charges and hypercharges. This difference depends on the difference between the coupling constants of the strong force and the weak force; and not on the mass of particles. That is why we have to use the same mass, the proton mass m_p, in order to translate between mass gravity and charge gravity. If we were to switch to multiplying with the electron mass m_e instead, we would be artificially weakening the strength of electromagnetic gravity by another 4 orders of magnitude, creating a difference of a total of 6 orders of magnitude.

We can now formulate the laws of gravity without mentioning the concept of mass.

In chapter 6.3 we were using the symbol B for the non-cancelling aspect of hypercharge. It essentially served as a count for the number of individual charges. We have to however remind ourselves, that B is just the baryon number, and not one individual hypercharge. The symbol for hypercharge is Y, yet hypercharge always adds up to zero outside the nucleus, so that we always have $Y = 0$.

A way to formally express the notion of a 'charge count' is to turn Y into a set $\mathbb{Y}$ which has the individual hypercharges as its elements.

We can then represent the number of hypercharges with the cardinality '$|\mathbb{Y}|$' of the set $\mathbb{Y}$, which is the number of elements in the set, or in physics terms, the number n_q of quarks.

The radius of a black hole is then given by:

$$R_{BH} = \frac{2\,(z_0^2 - 1)\,G_{|\mathbb{Y}|}\,|\mathbb{Y}|}{c^2}$$

With $(z_0^2 - 1)$ being the factor for the coherence horizon from chapter 4.3.

What is wrong with using the more simple 'B' instead?

The most obvious problem is that B standing for 'baryon number' would exclude mesons. Mesons also consist of quarks and thus have hypercharge too. It is to expect that they contribute to gravity as well.

So '$|\mathbb{Y}|$' is so far the only consistent way to indicate the number of individual hypercharges, aside from 'n_q'. For better alignment with the mathematical connotation we can call the non-neutralizing aspect of hypercharge '*hypercardinality*'. When we are concerned with strong hypercharge, then the choice between '$|\mathbb{Y}|$' and 'n_q' is merely a matter of convention, but for electric charge, there is no choice, because both leptons and quarks have electric charge, so that using particle counts of the form 'n_x' would be very cumbersome.

Accordingly, non-neutralizing electric charge can be denoted with '$|\mathbb{Q}|$'.

For better aligning with the above connotation, we can denote the charge gravity constant G_Z by $G_{|\mathbb{Q}|}$, and the hypercharge gravity constant by $G_{|\mathbb{Y}|}$.[1]

Elementary spaces are complex 2-spheres, which are basically the extended complex set $\mathbb{C}_\infty$. Accordingly we could denote the elementary space constant G_E by $G_{\mathbb{C}_\infty}$, or $G_{\mathbb{C}}$, for simplicity.[2] This shall however be optional.

Accounting for the gravity of mesons is not the only reason to use $|\mathbb{Y}|$ instead of B. While B is essentially counting the number of protons and neutrons, $|\mathbb{Y}|$ is counting the number of individual hypercharges, which is the number of quarks. Therefore:

$$B = n_{p\&n}$$

$$|\mathbb{Y}| = n_q$$

When talking about the number of individual hypercharges, we obviously have to count the number of quarks, not the number of baryons.

Since we used the masses of protons and neutrons above to calculate G_B, it must correspond to the strength of gravity generated by three quarks. For defining a precise value for $G_{|\mathbb{Y}|}$, we can simply divide m_p in the equation G_B for by 3, which yields:

$$G_{|\mathbb{Y}|} = G_{Fe} \times \frac{m_{Fe-e}}{n_{p_{Fe}} m_p + n_{n_{Fe}} m_n} \times \frac{m_p^2}{9}$$

$$G_{|\mathbb{Y}|} = 2.0519189 \times 10^{-65} \text{ m}^3 \text{ kg s}^{-2}$$

Accordingly the gravity between masses composed of atoms or quarks is given by:

$$F_{G_{|\mathbb{Y}|}} = G_{|\mathbb{Y}|} \times \frac{n_{q_1} n_{q_2}}{r^2} = G_{|\mathbb{Y}|} \times \frac{|\mathbb{Y}|_1 |\mathbb{Y}|_2}{r^2}$$

Analogously the gravitational force arising from electric charge is given by:

$$F_{G_{|\mathbb{Q}|}} = G_{|\mathbb{Q}|} \times \frac{|\mathbb{Q}|_1 |\mathbb{Q}|_2}{r^2}$$

With $G_{|\mathbb{Q}|}$ being our new connotation for the aforementioned G_Z:

$$G_{|\mathbb{Q}|} = 1.3476218 \times 10^{-66} \text{ m}^3 \text{ kg s}^{-2}$$

Charge gravity is however much weaker than hypercharge gravity, so that it can be ignored for all practical purposes. It is unclear if there are feasible experimental setups that could measure charge gravity. For electron gases the electric repulsion would obviously by far exceed the gravitational attraction.

One very hard to realize experiment would be to measure the gravitational attraction between two cavities which are constantly filled with electron positron pairs.

In chapter 6.11 we will be looking at a somewhat more feasible experiment.

The true weight of the moon and the earth

$G_{|\mathbb{Y}|}$ does not only reveal itself in experiments aiming at measuring G, but also influence what masses we assign to planets and stars. The usually assumed masses for stars and planets are deduced from their gravity by assuming that this gravity is caused by their overall mass-energy.

Kinetic energy or heat energy doesn't contribute significantly to the mass of a planet, so that we can ignore it in our analysis here.

Negative binding energy on the other side contributes very significantly to to the total mass of earth.

Traditionally the mass of the earth is calculated as:

$$M = \frac{g\,r^2}{G_{Fe}}$$

Indeed that would be the mass of earth, if it would consist of iron only. If it was instead just a large proton gas cloud, then its mass would be given by:

$$M = \frac{g\,r^2}{G_p} = \frac{g\,r^2}{G_{Fe}}\,\frac{n_{p_{Fe}}\,m_p + n_{n_{Fe}}}{m_{Fe-e}}$$

We recall from before that G_p is the mass-gravity contant for proton gases, and that its value is:

$$G_p = 6.6009607 \times 10^{-11}\,\mathrm{m^3\,kg^{-1}\,s^{-2}}$$

What we did here was basically cancelling the mass defect. Now if we are interested in the mass-gravity constant of a particular element, we have to undo part of the above cancelling. Undoing here means multiplying with an inversed equation. So for any element x the mass-gravity constant is given by:

$$G_X = G_p\,\frac{n_{p_X}\,m_p + n_{n_X}}{m_{X-e}}$$

If we put this into our equation for M, we get:

$$M = \frac{g\,r^2}{G_p}\,\frac{m_X}{n_{p_X}\,m_p + n_{n_X}} = \frac{g\,r^2}{G_{Fe}}\,\frac{n_{p_{Fe}}\,m_p + n_{n_{Fe}}\,m_n}{m_{Fe}}\,\frac{m_X}{n_{p_X}\,m_p + n_{n_X}\,m_n}$$

And if we look at objects which consist of many different elements, we can use mean values instead, which we can write as follows:

$$M = \frac{g\,r^2}{G_p}\,\frac{m_{a_\mu}}{n_{p_\mu}\,m_p + n_{n_\mu}} = \frac{g\,r^2}{G_{Fe}}\,\frac{n_{p_{Fe}}\,m_p + n_{n_{Fe}}\,m_n}{m_{Fe}}\,\frac{m_{a_\mu}}{n_{p_\mu}\,m_p + n_{n_\mu}\,m_n}$$

For our following analysis we will need again the atomic weights of the proton and the neutron, which are $m_p = 1.007276$ u and $m_n = 1.008664$ u respectively.

We want to start with earth and look at its mass defect and how that affects its true mass. Its currently assumed mass is 5.9722×10^{24} kg and it is composed of:

Element	Abundance	Nuclei number	Atomic weight of nucleus	Weight without binding energy
Iron (Fe)	32.1%	26 + 30	55.83093692	56.44938
Oxygen (O)	30.1%	8 + 8	15.99461136	16.1276
Silicon (Si)	15.1%	14 + 14	28.07731988	28.2233
Magnesium (Mg)	13.9%	12 + 12	24.29841704	24.1914
Sulfur (S)	2.9%	16 + 16	32.05122272	32.2552
Nickel (Ni)	1.8%	28 + 30	58.67807976	58.46394
Calcium (Ca)	1.5%	20 + 20	40.0674284	40.319
Aluminum (Al)	1.4%	13 + 14	26.97440703	27.21602
Other elements	1.2%	–	–	–

Now we follow the above discussed procedure and add up the various atomic weights and the various weights without binding energy according to their abundance and then dividing the one through the other. Then we multiply the result with the presently assumed mass of the earth. This yields:

$$M_{\oplus} = \frac{g\,r^2}{G_{Fe}} \times \frac{n_{p_{Fe}}\,m_p + n_{n_{Fe}}\,m_n}{m_{Fe-e}} \times \frac{m_{a-e_\mu}}{n_{p_\mu}\,m_p + n_{n_\mu}\,m_n}$$

$$M_{\oplus} = [5.9722 \times 10^{24}\ \text{kg}] \times \frac{56.44938\ \text{u}}{55.83093692\ \text{u}} \times \frac{33.3176080601\ \text{u}}{33.57254248\ \text{u}}$$

$$M_{\oplus} = [5.9722 \times 10^{24}\ \text{kg}] \times 1.00339941449$$

$$M_{\oplus} = 5.992502 \times 10^{24}\ \text{kg}$$

That means the earth is 0.34% heavier than presently thought. For an even more precise figure we would need to know the composition of the last 1.2%.

Let us now look at the moon: it is currently believed to weight 7.342×10^{22} kg and is composed of 60% oxygen, $16 - 17\%$ silicon, $6 - 10\%$ alumium, $4 - 6\%$ calcium, $3 - 6\%$ magnesium, $2 - 5\%$ iron and $1 - 2\%$ titanium.[3, 4]

Again we calculate the sum of the proton and neutron masses, to compare them with the actual mass after the reduction due to the negative binding energy:

Element	Abundance	Nuclei number	Atomic weight of nucleus	Weight without binding energy
Oxygen (O)	60%	8 + 8	15.99461136	16.1276
Silicon (Si)	16 − 17%	14 + 14	28.07731988	28.2233
Aluminum (Al)	6 − 10%	13 + 14	26.97440703	27.21602
Calcium (Ca)	4 − 6%	20 + 20	40.0674284	40.319
Magnesium (Mg)	3 − 6%	12 + 12	24.29841704	24.1914
Iron (Fe)	2 − 5%	26 + 30	55.83093692	56.44938
Titanium (Ti)	1 − 2%	22 + 26	47.85503124	48.38558

Rounding up the abundance of Silicon and Titanium to their maximum, and taking the middle value for the other elements, for example 3.5% for iron, yields:

$$M_{\text{☽}} = \frac{g_{\text{☽}} r^2}{G_{\text{Fe}}} \times \frac{n_{p_{\text{Fe}}} m_p + n_{n_{\text{Fe}}} m_n}{m_{\text{Fe-e}}} \times \frac{m_{\text{a-e}_\mu}}{n_{p_\mu} m_p + n_{n_\mu} m_n}$$

$$M_{\text{☽}} = [7.342 \times 10^{22} \text{ kg}] \times \frac{56.44938 \text{ u}}{55.83093692 \text{ u}} \times \frac{22.5358473618 \text{ u}}{22.6998055 \text{ u}}$$

$$M_{\text{☽}} = [7.36 \times 10^{22} \text{ kg}] \times 1.00377417214$$

$$M_{\text{☽}} = 7.3754045 \times 10^{22} \text{ kg}$$

That means the moon is 0.38% heavier than presently thought.

Testing this would require to detonate an atomic bomb on its surface and measuring how much that changes its orbit. However, that will not be necessary, because the various gravitational constants or charge gravity constants can be tested quite well enough in a laboratory.

Now the largest difference is to be expected for the sun, because it consists almost exclusively of the two lightest elements that exist. It is 75% hydrogen and 25% helium. So our table for the sun is fairly simple:

Element	Abundance	Nuclei number	Atomic weight of nucleus	Weight without binding energy
Hydrogen (H)	75%	1 + 0	1.00745142	1.00728
Helium (He)	25%	2 + 2	4.00150284	4.0319

Using again the same formula as in the examples above yields:

$$M_\odot = \frac{g_\odot \, r^2}{G_{\text{Fe}}} \times \frac{n_{p_{\text{Fe}}} \, m_p + n_{n_{\text{Fe}}} \, m_n}{m_{\text{Fe}-e}} \times \frac{m_{\text{a}-e_\mu}}{n_{p_\mu} \, m_p + n_{n_\mu} \, m_n}$$

$$M_\odot = [1.989 \times 10^{30} \text{ kg}] \times \frac{56.44938 \text{ u}}{55.83093692 \text{ u}} \times \frac{1.755964275 \text{ u}}{1.763435 \text{ u}}$$

$$M_\odot = [1.989 \times 10^{30} \text{ kg}] \times 1.00679367855$$

$$M_\odot = 2.001979 \times 10^{30} \text{ kg}$$

That means the sun is about 0.7% heavier than presently thought.

Note:
1. Using the index 'S' for 'strong force' is not common and can cause confusion with 'S' for entropy, expecially due to the recent popularity of Eric Linde's *emergent gravity* theory.
2. The index 'E' for 'elementary space' could be confusing too, because 'E' is already used for energy.
3. Wieczorek, Mark A.; et al. (2006). "The constitution and structure of the lunar interior". *Reviews in Mineralogy and Geochemistry*. 60 (1): 221–364. Bibcode:2006RvMG...60..221W. doi:10.2138/rmg.2006.60.3.
4. Williams, Dr. David R. (2 February 2006). "Moon Fact Sheet". *NASA/National Space Science Data Center*. Archived from the original on 23 March 2010. Retrieved 31 December 2008.

6.6 Gravity without the equivalence principle [01.02.2020 (imprecise formula for cardinality of earth and sun); 30.12.2020 (precise equation)*; 31.12.2020 (listing all laws); 01.01.2020 (reduced quark gravity constant for quarkless test masses); 02.01.2020 (mass-gravity version of everything); 21.01.2021 (corrections for quarkless matter based on the weak equivalence principle)]

*Facilitated by discussions with: David Chester (UCLA; MIT; US).

Sponsored by: Lin Yi Song (China).

Note: All calculations in this chapter were done using non-relativistic equations. In the next chapter it will be shown that Lorentz factors and Schwarzschild factors cancel out and the non-relativistic results remain valid.
The equations for quarkless objects in this chapter are based on the so called 'weak equivalence principle', which will be introduced in the next chapter.

In the last chapter we learned how to calculate the true inertia mass of objects that are too large to put on a scale, based on their chemical composition and the gravity they exert.

While it is important to know the exact mass of a gravitational body when calculating collisions between such bodies, it is irrelevant when it comes to gravity, because as we know, in space particle dualism, gravity depends on the number of quarks alone.

In this chapter we shall focus on quark gravity, which is centered around the quark gravity constant $G_{|\mathbb{Y}|}$.

We shall derive the SPD-specific versions of some of the key expressions of gravitational theory. For comparison we shall also look at the classical versions, as well as the mass-gravity versions (see chapter 6.5).

SPD-specific law of gravity

In Newtonian gravity the gravitational force between two objects is given by:

$$F = \frac{G\,M\,m}{r^2}$$

The formula for the gravitational force between objects according to space particle dualism looks very similar, but instead of depending on the mass, it depends on the number of hypercharges alone:

$$F = \frac{G_{|\mathbb{Y}|}\,|\mathbb{Y}|\,|\mathbb{y}|}{r^2}$$

As we saw in the last chapter, the hypercharge gravity constant $G_{|\mathbb{Y}|}$ is obtained by multiplying the third of a proton mass with itself. We can therefore also write the above as:

$$F = G_p \times \left(\frac{m_p}{3}\right)^2 \times \frac{|\mathbb{Y}|\,|\mathbb{y}|}{r^2}$$

In the special case that one of the two masses does not contain any hypercharges, the gravitational force is given by:

$$F = \frac{\mathbb{G}_{|\mathbb{Y}|}\,|\mathbb{Y}|}{r^2}$$

Whereby $\mathbb{G}_{|\mathbb{Y}|}$ is the reduced hypercharge gravity constant, given by:

$$\mathbb{G}_{|\mathbb{Y}|} = G_p \times \frac{m_p}{3}$$

$$\mathbb{G}_{|\mathbb{Y}|} = 3.681244 \times 10^{-38}\ \text{m}^3\ \text{s}^{-2}$$

Since it is only one of the two masses that exerts gravity, only one mass-quark gravity conversion factor is needed. That means we do not need to square that factor.

Most of the times we do only know the weight of objects, and not the number of quarks they contain. It is therefore practical to formulate gravity in terms of mass as well. We can then know the actual gravity by calculating out the mass defect. Using the mass-gravity constant of proton-gases G_p and the ratio between the baryon number B of the atomic mass μ, allows us to calculate the gravitational force between two objects as:

$$F = \frac{G_p\ m_1\ m_2}{r^2} \times \frac{B_1\ B_2}{\mu_1\ \mu_2}$$

Some define 'μ' as the mass in multiples of the mass of the hydrogen atom, which is 1.00784 u, but we will be using the definition "mass in the atomic mass unit (amu)" instead. This way, when there is only one proton, then the ratio is exactly 1.

The meson-friendly (quark-based) formulation of this is:

$$F = \frac{G_p\ m_1\ m_2}{r^2} \times \frac{|\mathbb{Y}|_1\ |\mathbb{Y}|_2}{9\ \mu_1\ \mu_2}$$

Or:

$$F = \frac{G_p\ m_1\ m_2}{r^2} \times \frac{|\mathbb{Y}|_1\ |\mathbb{Y}|_2\ m_p^2}{9\ m_1\ m_2}$$

When only one mass M contains quarks, then the force $F(m_{\neg q})$ felt by the quarkless mass $m_{\neg q}$ is:

$$F(m_{\neg q}) = \frac{G_p\ M}{r^2} \times \frac{B_M}{\mu_M}$$

SPD-specific gravitational potential energy

In classical gravity theory, the potential gravitational energy is derived taking Newton's law of gravity and integrating over the radius R:

$$U(r) = -\int_{\infty}^{r} -\frac{G\,M\,m}{r^2}\,dr = -\frac{G\,M\,m}{r}$$

In space particle dualism this is:

$$U(r) = -\int_{\infty}^{r} -\frac{G_{|\Psi|}\,|\Psi|\,|y|}{r^2}\,dr = -\frac{G_{|\Psi|}\,|\Psi|\,|y|}{r}$$

In the special case that one of the two masses does not contain any hypercharges, the gravitational potential energy is given by:

$$U = \frac{\mathcal{G}_{|\Psi|}\,|\Psi|}{r}$$

The mass-gravity version of the above is:

$$U(r) = -\int_{\infty}^{r} -\frac{G_p\,m_1\,m_2}{r^2} \times \frac{B_1\,B_2}{\mu_1\,\mu_2}\,dr = -\frac{G_p\,m_1\,m_2}{r} \times \frac{B_1\,B_2}{\mu_1\,\mu_2}$$

As we know from before, the mass-gravity correction factor, which we may want to call χ (read as 'chi'), can be formulated in different ways:

$$\chi = \frac{B_1\,B_2}{\mu_1\,\mu_2} = \frac{|\Psi|_1\,|\Psi|_2}{9\,\mu_1\,\mu_2} = \frac{|\Psi|_1\,|\Psi|_2\,m_p^2}{9\,m_1\,m_2}$$

In the rest of this chapter, I shall omit these alternative formulations.

When one mass is without hypercharges, then it is:

$$U = \frac{G_p\,M}{r} \times \frac{B_M}{\mu_M}$$

The meson-friendly formulation of this is:

$$U = \frac{G_p\,M}{r^2} \times \frac{|\Psi|}{3\,\mu_M}$$

Or:

$$F = \frac{G_p}{r^2} \times \frac{|\Psi|\,m_p}{3}$$

We shall call this the reduced mass-gravity correction factor 'x' (read as 'chi-quer'):

$$x = \frac{B}{\mu} = \frac{|\mathbb{Y}|}{3\,\mu} = \frac{|\mathbb{Y}|\,m_p}{3}$$

SPD-specific escape velocity

We find the classical escape velocity by setting the kinetic energy equal to the potential gravitational energy:

$$\frac{1}{2}\,m\,v^2 = \frac{G\,M\,m}{r}$$

$$v_{esc} = \sqrt{\frac{2\,G\,M}{r}}$$

In space particle dualism doing so results in a quite different expression:

$$\frac{1}{2}\,m\,v^2 = \frac{G_{|\mathbb{Y}|}\,|\mathbb{Y}|\,|y|}{r}$$

$$v_{esc} = \sqrt{\frac{2\,G_{|\mathbb{Y}|}\,|\mathbb{Y}|\,|y|}{r\,m}}$$

Does this expression mean that the escape velocity depends on the mass of the escaping object? The answer is, no it does not, it means that it depends on the chemical composition, which is expressed by the ratio between the number of quarks $|y|$ and the mass m.

In order to make that more clear we can call this ratio k_X, with X standing for the chemical composition of the escaping mass. We can then write:

$$k_X = \frac{|y|}{m}$$

And:

$$v_{esc} = \sqrt{\frac{2\,G_{|\mathbb{Y}|}\,|\mathbb{Y}|}{r}\,k_X}$$

This factor k_X tells us that getting rid of electrons, increases the escape velocity v_{esc}, while getting rid of protons and neutrons decreases it.

When entering the mass of an iron-atom for m, and the number of quarks in an iron atom for $|y|$, we can calculate the earth-escape velocity for an iron object according to space particle dualism. The calculation yields:

Earth-escape velocity according to SPD for an iron-object:
11,180.296029 m/s

When using Newtonian or Einsteinian physics, the same calculation yields:

Earth-escape velocity according to Newton & Einstein:
11,187.8536868 m/s.

In the special case that one of the two masses does not contain any hypercharges, the escape velocity is given by:

$$v_{esc} = \sqrt{\frac{2\,\mathbb{G}_{|\mathbb{Y}|}\,|\mathbb{Y}|}{r}}$$

The mass-gravity version of the SPD-specific escape velocity is:

$$\frac{1}{2}\,m\,v^2 = \frac{G_p\,m_1\,m_2}{r} \times \frac{B_1\,B_2}{\mu_1\,\mu_2}$$

$$v_{esc} = \sqrt{\frac{2\,G_p\,M}{r} \times \frac{B_1\,B_2}{\mu_1\,\mu_2}}$$

When one of the masses has no hypercharges, then it is:

$$v_{esc} = \sqrt{\frac{2\,G_p\,M}{r} \times \frac{B_M}{\mu_M}}$$

The meson-friendly formulation of this is:

$$v_{esc} = \sqrt{\frac{2\,G_p\,M}{r} \times \frac{|\mathbb{Y}|}{3\,\mu}}$$

Or:

$$v_{esc} = \sqrt{\frac{2\,G_p}{r} \times \frac{|\mathbb{Y}|\,m_p}{3}} = \sqrt{\frac{2\,\mathbb{G}_{|\mathbb{Y}|}\,|\mathbb{Y}|}{r}}$$

SPD-specific Schwarzschild radius

In classical gravity theory we calculate the Schwarzschild radius by setting the escape velocity $v = c$, and then solving for the radius r:

$$c = \sqrt{\frac{2\,G\,M}{r}}$$

$$R_S = \frac{2\,G\,M}{c^2}$$

In space particle dualism that is different again:

$$c = \sqrt{\frac{2\,G_{|\mathbb{Y}|}\,|\mathbb{Y}|\,|y|}{m\,r}}$$

$$R_S = \frac{2\,G_{|\mathbb{Y}|}\,|\mathbb{Y}|}{c^2} \times \frac{|y|}{m}$$

Or:

$$R_S = \frac{2\,G_{|\mathbb{Y}|}\,|\mathbb{Y}|}{c^2} \times k_X$$

The physical interpretation of this is that there is not one unique Schwarzschild radius for a black hole anymore. There is a different radius for every chemical composition X.

In the special case that one of the two masses does not contain any hypercharges, the escape velocity is given by:

$$R_S = \frac{2\,\mathbb{G}_{|\mathbb{Y}|}\,|\mathbb{Y}|}{c^2} \times \frac{1}{m}$$

The mass-gravity versions of the Schwarzschild radius is:

$$c = \sqrt{\frac{2\,G_p\,M}{r} \times \frac{B_1\,B_2}{\mu_1\,\mu_2}}$$

$$R_S = \frac{2\,G_p\,M}{c^2} \times \frac{B_1\,B_2}{\mu_1\,\mu_2}$$

When one mass is quarkless ($\neg q$), then it is:

$$c = \sqrt{\frac{2\,G_p\,M}{r} \times \frac{B_M}{\mu_M}}$$

$$R_S = \frac{2\,G_p\,M}{c^2} \times \frac{B_M}{\mu_M}$$

SPD-specific gravitational acceleration

In classical gravity theory, the gravitational acceleration g can be calaculated by setting equal the general equation for force, namely $F = m\,a$, and the Newtonian gravity law, and then setting $a = g$. That yields:

$$m\,g = \frac{G\,M\,m}{r^2}$$

$$g = \frac{G\,M}{r^2}$$

In space particle dualism this is quite different:

$$m\,g = \frac{G_{|\Psi|}\,|\Psi|\,|y|}{r^2}$$

$$g = \frac{G_{|\Psi|}\,|\Psi|\,|y|}{m\,r^2}$$

Or:

$$g_X = \frac{G_{|\Psi|}\,|\Psi|}{r^2} \times k_X$$

The mass-gravity gravitational acceleration is:

$$m\,g \;=\; \frac{G_p\,m_1\,m_2}{r^2} \times \frac{B_1\,B_2}{\mu_1\,\mu_2}$$

$$g \;=\; \frac{G\,M}{r^2} \times \frac{B_1\,B_2}{\mu_1\,\mu_2}$$

And of course we can reverse the roles and calculate the gravitational acceleration of the larger mass M towards the smaller mass m:

$$g(M) \;=\; \frac{G\,m}{r^2} \times \frac{B_1\,B_2}{\mu_1\,\mu_2}$$

But in each case we need to know the chemical composition of both masses, in order for our calculation to be precise.

When one of the masses is quarkless, then it is:

$$m_{\neg q}\,g \;=\; \frac{G_p\,M\,m_{\neg q}}{r^2} \times \frac{B_M}{\mu_M}$$

$$g \;=\; \frac{G\,M}{r^2} \times \frac{B_M}{\mu_M}$$

Consequences for the equivalence principle and comparison to electromagnetism

This clearly shows that objects with a different chemical structure X experience a different gravitational acceleration.

We can reduce the gravitational acceleration experienced by an object by removing protons and neutrons, and we can increase it by removing electrons.

However, the effect of removing protons and neutrons is much stronger, because electrons not only do not contribute to the number of hypercharges $|y|$, they also contribute much less to than protons and neutrons to the total inertia mass m.

We can say that $|y|$ is 'gravitational mass', while 'm' is 'inertia mass'. We could therefore call k_X also the 'inequivalence factor'.

The equivalence principle on the other hand, which is central to Einstein's general relativity, claims that inertia mass and gravitational mass are always equal. Only if they are equal, can it be true that all objects fall at the same rate in a vacuum.

It is this principle and its success in helping Einstein to derive the equations of general relativity, which led Einstein to believe that

"Gravity and Acceleration are one and the same."

And:

"Gravity is not a force."

Indeed the equivalence principle would be very odd for a 'force'.

In fact, when we calculate the escape velocity, Schwarzschild radius and acceleration for electromagnetism, we arrive at formula that look very similar to what we have calculated for SPD-specific gravity.

The escape velocity for a charge q that is attracted to another larger charge Q is given by:

$$\frac{1}{2} m v^2 = \frac{k_e Q q}{r}$$

$$v = \sqrt{\frac{2 k_e Q q}{r m}}$$

The equivalent of a Schwarzschild radius around such a charge Q is:

$$c = \sqrt{\frac{2 k_e Q q}{m r}}$$

$$R_S = \frac{2 k_e Q q}{c^2 m}$$

And the acceleration felt by a smaller charge q towards a larger charge Q is given by:

$$m g = \frac{k_e Q q}{r^2}$$

$$g = \frac{k_e Q q}{r^2 m}$$

The only differences between these formulae and the formulae of of SPD is that, firstly, Q and q are variables that have a physical unit, namely Coulomb [C], while $|Y|$ and $|y|$ from space particle dualism are not only dimensionless, but they are also natural numbers, namely the number of individual hypercharges Y, which is simply the number of quarks; and secondly, that they all yield zero when either Q or q are zero, while our above formulae for space particle dualism only yield zero when both $|y|$ and m are zero.

In space particle dualism, gravity is a side effect of force, and it is therefore not surprising to know that it is based on equations that are more force-like. So while the equations of space particle dualism look very unusual from the perspective of both Newtonian and Einsteinian gravity, they are not unusual at all from the perspective of electromagnetism.

One could say that both Newton and Einstein were deceived by Galileo, when they wrote down their equations for gravity.

Galileo's discovery that objects with different mass fall at the same rate, when there is no air-friction, was nothing more than an ad-hoc rule, derived from superficial observations.

It is however hard to blame Newton or Einstein for this mistake, because in a time before the discovery of particle physics, not knowing what the cause of gravity is, the only most obvious candidate is mass, and in Einstein's case, who was born a few centuries later, it is energy.

It is mass and energy that are quantities that do not cancel out and are therefore accumulative. They are therefore good candidates for a property that is absent in small objects, and becomes apparent only in very large objects.

Before the discovery of atoms, there would be no other candidate for the source of gravity. Now that we know that there are atoms and particles, we have many more candidates for the source of gravity.

The equivalence principle corresponds to only two such candidates, namely mass and energy, and as Einstein figured out, those two are one, according to his famous formula $E = m c^2$.

When rejecting the equivalence principle, one can use any other accumulative quantity as source for gravity. That can be the number of protons, neutrons, electrons or any other type of particle.

As we laid out in chapter 6.5, according to space particle dualism theory, only charges and hypercharges exert gravity and only hypercharges exert long range gravity.

However, charge gravity, also called 'subgravity', is governed by the same rules as the usual hypercharge-based gravity.

Subgravity escape velocity:

$$v_{esc} = \sqrt{\frac{2\,G_{|\mathbb{Q}|}\,|\mathbb{Q}|\,|\mathbb{q}|}{r\,m}} \;;\, r < 40\,[\text{m}]$$

Subgravity Schwarzschild radius:

$$R_S = \frac{2\,G_{|\mathbb{Q}|}\,|\mathbb{Q}|}{c^2} \times \frac{|\mathbb{q}|}{m} \;;\, R_S < 40\,[\text{m}]$$

Subgravity acceleration:

$$g = \frac{G_{|\mathbb{Q}|}\,|\mathbb{Q}|}{r^2} \times \frac{|\mathbb{q}|}{m} \;;\, r < 40\,[\text{m}]$$

Every amount of that cardinal charge $|\mathbb{Q}|$ that lies outside of that 40 meter radius has to be disregarded. The reader is referenced to chapter 6.5 for the reason of this.

Galileo's initial guess

Now, before ultimately blaming Galileo for the equivalence principle, we should look at what Galileo actually said.

According to a biography by Galileo's pupil Vincenzo Viviani, Galileo's free fall experiment took place between 1589 and 1592,[1] and it involved dropping spheres of different size and weight from the tower of Pisa.[2] However, at the time of the experiment, Galileo and his biographer Viviani had not yet published a final version of Galileo's free fall law. An earlier version of this law was actually precisely resembling the free fall law of space particle dualism.

In that earlier version, Galileo stated that:[3]

"Bodies of the same material falling through the same medium fall at the same speed."

According to space particle dualism, this earlier version of Galileo's free fall law applies to all objects that contain quarks.

Fischbach's experiment

In 1922 R. v. Eötvös conducted a series of experiments to test the equivalence principle, and he seemed to have confirmed that it is correct.[4] In 1986 however, Ephraim Fischbach and his team conducted a new series of experiments, and they found a correlation between deviations from the equivalence principle and the ratio between the number of Baryons B or Leptons L and mass expressed in multiples of the hydrogen mass, denoted as μ.[5, 6] This is comparable to using multiples of the proton mass.

The reason why we did not have to express mass as multiples of the proton mass in our own treatment of gravity in this chapter, is that the proton mass was already hidden inside $G_{|\mathbb{Y}|}$.

We recall that:

$$G_{|\mathbb{Y}|} = G_p \times \left(\frac{m_p}{3}\right)^2$$

And therefore our expression for the gravitational acceleration can also be written as:

$$g = G_p \frac{|\mathbb{Y}| \frac{m_p}{3} \, |y| \frac{m_p}{3}}{r^2 \, m}$$

Or:

$$g = G_p \frac{|\mathbb{Y}|}{r^2} \times \frac{|y| \, m_p^2}{9 \, m}$$

We note that:

$$\frac{|y| \, m_p^2}{9 \, m} = \frac{B \, m_p^2}{m} \cong \frac{B^2}{\mu}$$

The only difference between Fischbach's expression and the correct SPD-expression of this factor is the slight difference between the proton mass and the mass of the hydrogen atom, which consists of one proton and one electron.

Despite being so close to the correct expression, Fischbach did not understand quark gravity. He stumbled upon it through his experiments, and did not understand that it indicates that gravity is a side effect of hypercharge. Instead he interpreted his experimental results as corrections to otherwise correct Newtonian physics which steam from a fifth force somehow coupled to the strong force.

His concept of this fifth force was in fact so vague that it involved new and unexplained factors that had to be set by the experimental data. His version of the potential gravitational energy between two masses m_i and m_j looked like this:

$$U = \frac{G\,m_i\,m_j}{r}\left(1 + \alpha e^{-r/\lambda}\right)$$

$$\alpha = \frac{B_i}{\mu_i}\frac{B_j}{\mu_j}\,\xi$$

Herein ξ is a coupling constant associated with Fischbach's hypothesized fifth force, and λ is setting the range of his force. Fischbach estimated these two factors to be $\xi = 10^{-2}$ and $\lambda = 200$ m.

His use of the range inhibitor λ suggests that he thought that any modification of Newtonian physics could only apply to short range gravity, not long range gravity.

In chapter 6.5 we saw that according to space particle dualism, the earth is 0.34% heavier than one would estimate using Newtonian physics. This suggests that for earth:

$$U = \frac{G\,M\,m}{r}\,(1 + 0.0034)$$

Fischbach's complicated factors, such as his coupling constant and the range inhibitor are not needed. In fact, the range is freely adjustable, as he had no evidence of a limited range, and so his range inhibitor functions merely as counter to his coupling constant ξ.

When removing those arbitrary factors,[7] Fischbachs formula turns into this:

$$U = \frac{G\,m_i\,m_j}{r}\frac{B_i}{\mu_i}\frac{B_j}{\mu_j}$$

This means less potential energy per mass, and therefore more mass is needed for the same amount of gravity. Hence our increased estimate for the mass of earth, moon and sun in chapter 6.5.

However, it is hard to see how Fischbach could have ever derived the gravitational acceleration g, from his own version of that formula.

SPD-specific gravitational mass

In classical gravity theory, we can derive the gravitational mass from the gravitational acceleration as follows:

$$g = \frac{G\,M}{r^2}$$

$$M = \frac{g\,r^2}{G}$$

In space particle dualism 'gravitational mass' is the number of quarks. We can derive it from the SPD-specific gravitational acceleration:

$$g = \frac{G_{|\Psi|}\,|\Psi|}{r^2} \times \frac{|y|}{m}$$

$$|\Psi| = \frac{g_x\,r^2}{G_{|\Psi|}} \times \frac{m}{|y|}$$

We have to keep in mind that this does not mean g depends on the mass of the dropping object. It means that it depends on the ratio between the mass and the number of quarks in it. This is what characterizes its chemical composition.

Usually g is measured by dropping iron objects in a vacuum. Iron has an atomic mass of 55.845 u, while containing 26 protons and 30 neutrons. That is a total of 56 nuclei. Multiplying by 3 gives us the number of quarks, which is 168. Different from chapter 6.5, we have to enter the mass of iron in kilogram here, which is fine, because a third of the proton mass is still hidden inside $G_{|\Psi|}$.

For earth we get:

$$|\Psi|_{\oplus} = \frac{g_{Fe}\,r_{\oplus}^2}{G_{|\Psi|}} \times \frac{m_{Fe}}{|y|_{Fe}}$$

$$|\Psi|_{\oplus} = \frac{9.81\ [\text{m s}^{-2}] \times 6{,}371{,}000^2\ [\text{m}^2]}{6.1557567 \times 10^{-65}\ [\text{m}^3\ \text{kg s}^{-2}]} \times \frac{9.2732796 \times 10^{-26}\ [\text{kg}]}{168}$$

$$|\Psi|_{\oplus} = 3.570482 \times 10^{51}$$

For the sun we get:

$$|\Psi|_{\odot} = \frac{g_{Fe}\,r_{\odot}^2}{G_{|\Psi|}} \times \frac{m_{Fe}}{|y|_{Fe}}$$

$$|\mathbb{Y}|_\odot = 1.190036 \times 10^{57}$$

We can call this the 'solar cardinality', or 'solar quark number'. However, different from earth, we have not yet measured the sun acceleration with any specific material, and so this iron-based calculation can only be a rough estimate. At the end of the next chapter we will use a more direct approach to get a more precise value.

Note:

1. Some contemporary sources speculate about the exact date; e.g. Rachel Hilliam gives 1591 (Galileo Galilei: Father of Modern Science, The Rosen Publishing Group, 2005, p. 101).

2. Vincenzo Viviani (1717), Racconto istorico della vita di Galileo Galilei, p. 606: [...dimostrando ciò con replicate esperienze, fatte dall'altezza del Campanile di Pisa con l'intervento delli altri lettori e filosofi e di tutta la scolaresca... [...Galileo showed this [all bodies, whatever their weights, fall with equal speeds] by repeated experiments made from the height of the Leaning Tower of Pisa in the presence of other professors and all the students...].

3. Drake, Stillman (2003). Galileo at Work: His Scientific Biography (Facsim. ed.). Mineola (N.Y.): Dover publ. ISBN 9780486495422.

4. R. v. Eötvös, D. Pekár, and F. Fekete, Ann. Phys. (Leipzig) 68, 11 (1922).

5. E. Fischbach, D. Sudarsky, A. Szafer, C. Talmadge, and S. H. Aronson, Phys. Rev. Lett. 56, 3 (1986).

6. Talmadge, C.; Fischbach, E.; Aronson, S.H. (1987). "Multicomponent models of the fifth force". RN:19057326.

7. All of those factors are constants, and therefore the relative differences between different materials are not affected.

6.7 SPD-specific Schwarzschild factor and the gravity felt by quarkless objects
[09.01.2021 (relativistic correction for the gravitational potential energy); 15.01.2021 (Taylor series non-zero minimum); 17.01.2021 (acceleration felt by photons cannot depend on their frequency); 20.01.2020 (deflection of light around the sun); 21.01.2020 (improved value due to more precise quark count)]

Sponsored by: Lin Yi Song (China).

Our derivation of the gravity equations in the last chapter was clearly non-relativistic. That is most obvious for our derivation of the SPD-specific escape velocity. It was using the Newtonian expression for kinetic energy. Let us now look at the relativistic version of those equations. This time we will start off with the escape velocity and the Schwarzschild radius, because they are foundational for the relativistic treatment.

Relativistic SPD-specific derivation

In special relativity the total energy of a particle is given by:

$$E_{total} = m\,c^2 = m_0\,c^2 + KE = \frac{m_0\,c^2}{\sqrt{1 - \frac{v^2}{c^2}}}$$

This includes the mass energy, and because we are only interested in the kinetic energy, we subtract that, and get:

$$E_{kin} = m_0\,c^2 \left[\frac{1}{\sqrt{1 - \frac{v^2}{c^2}}} - 1 \right]$$

For finding the escape velocity we need to set this equal to the gravitational potential. However, if we want the two to balance out each other and change in a synchronized fashion, then we have to account for the relativistic changes that occur as we come closer to a gravitational field.

We need a barrier at the Schwarzschild radius, a barrier that is similar to the light speed barrier for speed, just that here it is not a speed limit, but a gravity limit.

We can do this by recognizing that v is the escape velocity v_{esc}, and that when it reaches the speed of light ($v_{esc} = c$), then we have arrived at the Schwarzschild radius R_S, which is not possible, just as it is not possible to reach the speed of light.
If we imagine to fall beyond the Schwarzschild radius into a black hole, then, similar to what happens if one was to exceed the speed of light, our gravitational potential energy becomes an imaginary number.

The gravitational equivalent of the Lorentz factor from special relativity is the Schwarzschild factor. It has precisely the same structure as the Lorentz factor, just that now the speed v is the escape speed v_{esc}:

$$\gamma_S = \frac{1}{\sqrt{1 - \frac{v_{esc}^2}{c^2}}}$$

In the last chapter, we derived the SPD specific escape velocity using the classical kinetic energy. How can we derive it using the relativistic kinetic energy?

After using the relativistic expression for both the kinetic energy and the gravitational potential energy, we set the two equal:

$$m_0 \, c^2 \left[\frac{1}{\sqrt{1 - \frac{v_{esc}^2}{c^2}}} - 1 \right] = \frac{G_{|\mathbb{Y}|} \, |\mathbb{Y}| \, |y|}{r \sqrt{1 - \frac{v_{esc}^2}{c^2}}}$$

It is very tempting to now solve for v_{esc}, but if we did so, we would arrive at the wrong result, namely:

$$\perp \; v_{esc} = \frac{G_{|\mathbb{Y}|} \, |\mathbb{Y}| \, |y|}{m_0 \, c \, r}$$

The reason why this approach is not legit is that both the Lorentz factor γ as well as the Schwarzschild factor γ_S are not simple mathematical terms that can be manipulated, but they are instead part of what is known as the Lorentz transform. The are supposed to remain the same for different physical properties. Regardless of if it is time, energy, impulse, force, acceleration or anything else, the structure of the gamma-factor never changes.

We also know that we ought not to change the Schwarzschild factor too much, because it determines the time dilation at different heights in earth's gravitational field, and that is something that is experimentally well measured, even down to height differences of only one meter.[1, 2]

That means if we want to solve our above equation, we cannot do so by solving for something inside this gamma-factor. We need to instead expand it using a Taylor series.

The Taylor expansion of the relativistic kinetic energy can be written as:

$$E_{kin} = m_0 \, c^2 \left[\frac{1}{\sqrt{1 - \frac{v^2}{c^2}}} - 1 \right]$$

$$E_{kin} = m_0 \, c^2 \left[\left(1 - \frac{v^2}{c^2} \right)^{-\frac{1}{2}} - 1 \right]$$

$$E_{kin} = m_0 \, c^2 \left[\left(1 + \frac{1}{2} \frac{v^2}{c^2} + \frac{3}{8} \frac{v^4}{c^4} + \cdots \right) - 1 \right]$$

The more terms in this series we add, the closer we get to the precise result.

We can now define the minimum of this series as the lowest number of terms that gives us a non-zero total sum. We may call this the Taylor series minimum. In this case that is:

$$\sum_{n=0}^{n=min} m_0 \, c^2 \left[\left(1 + \frac{1}{2} \frac{v^2}{c^2} \right) - 1 \right] = \frac{1}{2} \, m_0 \, v^2$$

In the case of the gravitational potential energy, the series is shorter:

$$\sum_{n=0}^{n=min} - \frac{G_{|\Psi|} \, |\Psi| \, |y|}{r} \, [1] = - \frac{G_{|\Psi|} \, |\Psi| \, |y|}{r}$$

The positive kinetic energy and the negative gravitational potential energy add up to '0':

$$\sum_{n=0}^{n=min} m_0 \, c^2 \left[\left(1 + \frac{1}{2} \frac{v^2}{c^2} \right) - 1 \right] - \sum_{n=0}^{n=min} \frac{G_{|\Psi|} \, |\Psi| \, |y|}{r} \, [1] = 0$$

Which implies:

$$\sum_{n=0}^{n=min} m_0 \, c^2 \left[\left(1 + \frac{1}{2} \frac{v^2}{c^2} \right) - 1 \right] = \sum_{n=0}^{n=min} \frac{G_{|\Psi|} \, |\Psi| \, |y|}{r} \, [1]$$

And this reduces to our equation from the last chapter, namely:

$$\frac{1}{2} \, m_0 \, v^2 = \frac{G_{|\Psi|} \, |\Psi| \, |y|}{r}$$

Which yields:

$$v_{esc} = \sqrt{\frac{2 \, G_{|\Psi|} \, |\Psi| \, |y|}{r \, m}}$$

We can now enter this into our Schwarzschild factor, which yields:

$$\gamma_S = \frac{1}{\sqrt{1 - \frac{v_{esc}^2}{c^2}}} = \frac{1}{\sqrt{1 - \frac{2 \, G_{|\Psi|} \, |\Psi| \, |y|}{r \, m \, c^2}}}$$

This SPD-specific Schwarzschild factor can be applied to all our equations from the last chapter to make them relativistic:

Relativistic SPD-specific gravitational force

The relativistic gravitational force F between two hypercardinalities $|\mathbb{Y}|$ and $|y|$ is given by:

$$F = \frac{G_{|\mathbb{Y}|}\,|\mathbb{Y}|\,|y|}{r^2\sqrt{1 - \dfrac{2\,G_{|\mathbb{Y}|}\,|\mathbb{Y}|\,|y|}{r\,m\,c^2}}}$$

Interpretation: force goes to infinity as the escape velocity approaches the speed of light. Escape velocities greater than the speed of light would, if they were physical, correspond to a force that is an imaginary number ($F \in \mathbb{I}$).

Relativistic SPD-specific gravitational potential energy

The relativistic gravitational potential energy U between two hypercardinalities $|\mathbb{Y}|$ and $|y|$ is given by:

$$U = -\frac{G_{|\mathbb{Y}|}\,|\mathbb{Y}|\,|y|}{r\sqrt{1 - \dfrac{2\,G_{|\mathbb{Y}|}\,|\mathbb{Y}|\,|y|}{r\,m\,c^2}}}$$

Interpretation: gravitational potential energy goes to negative infinity as the escape velocity approaches the speed of light. Escape velocities greater than the speed of light would, if they were physical, correspond to a gravitational potential energy that is an imaginary number ($U \in \mathbb{I}$).

Relativistic SPD-specific non-Euclidean radius of a black hole

The non-Eucledean or granular radius $R_{\neg \mathbb{E}}$ of a black hole depends both on the hypercardinality $|\mathbb{Y}|$ of the black hole and the hypercardinality $|y|$ of the in-falling object, and is given by:

$$R_{\neg \mathbb{E}} = \frac{2\,G_{|\mathbb{Y}|}\,|\mathbb{Y}|\,|y|}{m\,c^2\sqrt{1 - \dfrac{2\,G_{|\mathbb{Y}|}\,|\mathbb{Y}|\,|y|}{d\,m\,c^2}}}$$

Interpretation: the Schwarzschild radius is simply the circumference of a black hole divided by $2\,\pi$. In general relativity the curvature of space results in the true radius being much larger than the circumference. The Schwarzschild radius is also called the

'Euclidean radius', because it is pretending that space is Euclidean, when it is not. In space particle dualism 'Euclidean' corresponds to 'non-granular'.

The true 'granular radius' $R_{\neg\mathbb{E}}$ of a black hole as viewed from the particular frame of reference at the distance d from the center of mass goes to infinity as the escape velocity v_{esc} approaches the speed of light. Escape velocities greater than the speed of light would, if they were physical, correspond to a granular radius that is an imaginary number ($R_{\neg\mathbb{E}} \in \mathbb{I}$).

Relativistic SPD-specific gravitational acceleration

The relativistic gravitational acceleration g depends both on the hypercardinality $|\mathbb{Y}|$ of the gravitational object and the hypercardinality $|y|$ of the falling object, and is given by:

$$g = \frac{G_{|\mathbb{Y}|}\,|\mathbb{Y}|\,|y|}{m\,r^2\,\sqrt{1 - \dfrac{2\,G_{|\mathbb{Y}|}\,|\mathbb{Y}|\,|y|}{d\,m\,c^2}}}$$

Interpretation: gravitational acceleration goes to infinity as the escape velocity approaches the speed of light. Escape velocities greater than the speed of light would, if they were physical, correspond to a gravitational acceleration that is an imaginary number ($g \in \mathbb{I}$).

Do electric holes cause time dilation?

In the last chapter we mentioned the possibility of an electric hole, an electromagnetic equivalent of a black hole. It seems that if we use the same argumentation for electric holes as we used for black holes above, then introducing a Lorentz factor for the kinetic energy means to also introduce an analogy to a Schwarzschild factor for the electromagnetic kinetic energy, otherwise the two get out of sink, which could cause problems for energy conservation.

Our electromagnetic escape velocity would then lend itself to an electromagnetic Schwarzschild-factor:

$$v_{esc} = \sqrt{\frac{2\,k_e\,Q\,q}{r\,m}}$$

$$\gamma_{S_{EM}} = \frac{1}{\sqrt{1 - \dfrac{2\,k_e\,Q\,q}{r\,m\,c^2}}}$$

This unusual Schwarzschild factor and the electromagnetic time dilation it implies would have only energy conservation as a justification, and does not yet have a proper ontology, unlike gravitational time dilation, which according to space particle dualism is based on differences in the density of elementary spaces.

The effect of gravity on massless particles

Massless particles do not themselves exert gravity, and therefore we use the reduced quark gravity constant $\mathbb{G}_{|\mathbb{Y}|}$ for them, as we do for electrons (see chapter 6.5).

Massless particles like photons do not have an escape velocity or a Schwarzschild radius, as they always escape.

However, what we can calculate for a photon is the force that acts upon it in a gravitational field. Force is $F = m\,a$, and since we know that $E = m\,c^2$, we know that the force that acts upon a photon is:

$$F = \frac{h\,f}{c^2}\,g$$

What we are now tempted to do is to set this equal to the gravitational force, and write:

$$\frac{h\,f}{c^2}\,g = \frac{\mathbb{G}_{|\mathbb{Y}|}\,|\mathbb{Y}|}{r^2}$$

$$g = \frac{\mathbb{G}_{|\mathbb{Y}|}\,|\mathbb{Y}|}{r^2}\,\frac{c^2}{h\,f}$$

This however would be wrong, because it would imply that photons contribute to the gravitational force.

It would also lead to a too strong deflection of light around the sun.

The deflection of light around the sun according to general relativity is:[3]

$$\delta_{GR} = \frac{4\,G\,M}{c^2\,R} = 0.00000847925$$

For space particle dualism we replace 'M' with '$|\mathbb{Y}|$', and use the reduced quark gravity constant '$\mathbb{G}_{|\mathbb{Y}|}$' instead of '$G$'. We know that the energy of the photon does not cancel with anything in the potential gravitational energy, which is why it has to appear in the denominator:

$$\delta = \frac{4\,\mathbb{G}_{|\mathbb{Y}|}\,|\mathbb{Y}|}{c^2\,R\,\left(\frac{h\,f}{c^2}\right)}$$

$$\delta = \frac{4\,\mathbb{G}_{|\mathbb{Y}|}\,|\mathbb{Y}|}{h\,f\,R}$$

This expression would lead to an absurdly high deflection angle of 10^{24}.

The situation with electrons is similarly alarming, but not quite so, because the relativistic corrections that come with their kinetic energy are cancelled by the Schwarzschild factor on the right side:

$$\frac{m_0}{\sqrt{1-\frac{v^2}{c^2}}}\,g = \frac{\mathbb{G}_{|\mathbb{Y}|}\,|\mathbb{Y}|}{r^2\,\sqrt{1-\frac{v^2}{c^2}}}$$

$$g = \frac{\mathbb{G}_{|\mathbb{Y}|}\,|\mathbb{Y}|}{m_0\,r^2}$$

All that is left is the rest mass.

We do not have data for the deflection of electrons in gravitational fields, and we do therefore not know if the above equation is correct.

According to space particle dualism, both electrons and photons react to gravity, without exerting gravity. That makes it quite problematic to equate the force that acts upon them to the gravitational force. It also implies that there is a balance between photon energy and gravitational potential energy, such that:

$$h\,f - \frac{\mathbb{G}_{|\mathbb{Y}|}\,|\mathbb{Y}|}{r} = 0$$

Which amounts to saying that additional photon force is countered by additional gravitational force. If photons do not contribute to gravity, then the above equation cannot be true.

What we instead could do is assume that the gravitational force felt by an electron or a photon is equal to that felt by a single quark, with the only difference being that the reduced quark gravity constant $\mathbb{G}_{|\mathbb{Y}|}$ is used. Formally:

$$g_{\neg q} = \frac{\mathbb{G}_{|\mathbb{Y}|}\,|\mathbb{Y}|}{r^2}$$

We can call this the 'weak equivalence principle', and summerize it as:

"All quarkless objects fall at the same rate."

Using the solar quark count from chapter 6.6, this would result in a light deflection around the sun of:

$$\delta_{\neg q} = \frac{4 \, \mathbb{G}_{|\mathbb{Y}|} \, |\mathbb{Y}|_\odot}{c^2 \, R} = 0.00000279996$$

What if we instead say that it feels exactly the same acceleration as a single quark, or a quark containing object that has an atomic mass to atomic number ratio of precisely 1?

In that case we would get a tiny deflection of only 10^{-33}. That seems justification enough for our constant '$\mathbb{G}_{|\mathbb{Y}|}$'.

We got the right order of magnitude. For a more precise value we will have to know the solar cardinality $|\mathbb{Y}|_\odot$ with a higher accuracy.

If we take the mainstream value of the sun mass and multiply it by 0.007 (7%), and divide it by the mass of the hydrogen atom, then we get a quark count of 1.1968056×10^{57}, which leads to an angle of $\delta = 0.00000844766$.

If we instead divide by the mass of a helium atom, then multiply by 4, for 4 nuclei, and 3, for 3 quarks, then we get to an angle of $\delta = 0.00000850836$.

We know that by mass, the Sun consists of about 70.6% hydrogen and 27.4% helium. For simplicity we can ignore the 2% other elements and round that up to 72% hydrogen and 28% helium. That would give us a solar quark count of $|\mathbb{Y}|_\odot = 3.5976403 \times 10^{57}$, and a deflection angle of:

$$\delta_{\neg q} = \frac{4 \, \mathbb{G}_{|\mathbb{Y}|} \, |\mathbb{Y}|_\odot}{c^2 \, R} = 0.00000846466$$

This matches with the measured value.

References:

1. Richard Wolfson (2003). Simply Einstein. W W Norton & Co. p. 216. ISBN 978-0-393-05154-4.
2. C. W. Chou, D. B. Hume, T. Rosenband, D. J. Wineland (24 September 2010), "Optical clocks and relativity", Science, 329(5999): 1630–1633.
3. Domingos S.L. Soares (2009). "Newtonian gravitational deflection of light revisited". arXiv:physics/0508030v4.

6.8   The charge gravity constant and exotic matter abundance [20.01.2020 (exotic matter is subject to charge gravity only); 25.01.2020 (it leads to the right matter-exotic matter ratio); 09.08.2020 (black hole-bulk inequation); 13.08.2020 (black hole-bulk ratio mainly due to short range of exotic gravity)]

Note: For simplicity the term 'exotic matter' will be used instead of 'primordial dark matter'. The term 'exotic matter' usually denotes matter which has a negative mass-energy, and that is the case here too, but what is different is that its gravity is still attractive and not repulsive.

In chapter 3.6 we theorized that there should be a mirror symmetry between matter and exotic matter. As mentioned in previous chapters all of this exotic matter must have condensed into primordial black holes that later evolved into the supermassive black holes that constitute the cores of galaxies today.

On the first sight this symmetry must be violated, because the mass of galaxies in form of ordinary stars and stellar black holes exceeds the mass of their central supermassive black holes by far.

As we saw in the last chapter, the gravity of ordinary matter is mostly mediated by virtual gluons and only insignificantly by virtual photons.
Gluons are associated with protons, neutrons and the quarks that constitute them.
Exotic matter is 'dark'; it doesn't interact other than through gravity, therefore we can expect 'dark' or 'exotic' electrons, because they are single particles, but without quantum forces in place we have no reason to assume that there are 'exotic protons' or 'exotic neutrons' as well.

The gravity of exotic matter should therefore be analogous to the type of gravity that we theorized to be exerted by electrons. Consequently it is much weaker than the gravity that exists between objects made of ordinary matter. Not only that, but it should also be shorter in range. As we found in the last chapter, the gravity between electrons ceases to exist at distances larger than 40 meters.

Exotic matter exerting a weaker and shorter range type of gravity than ordinary matter does could explain why despite the assumed symmetry between matter and exotic matter, supermassive black holes are still much smaller in mass than their host galaxy.

To take the Milky Way and its central supermassive black hole as an example:

$$M_{MW} = 0.8 - 1.5 \times 10^{12} \, M_{\odot}$$

$$M_{sag-A} = 4.02 \pm 0.16 \times 10^6 \, M_{\odot}$$

Now let us search for an explanation for this discrepancy.

The difference between the hypercharge gravity constant and the charge gravity constant is a factor of

$$\frac{G_{|\mathbb{Y}|}}{G_{|\mathbb{Q}|}} = 15.2$$

One would be inclined to think that this ratio should be 137, as this is the difference between the strong force and the electromagnetic force, but one would be forgetting that the coupling constant for the strong force is defined using the interplay of three quarks, while our constant $G_{|\mathbb{Y}|}$ is meant for one quark only.

Originally the mistake was made to switch to the electron mass for the construction of $G_{|\mathbb{Q}|}$ which artificially weakened it (see chapter 6.5). This made it seems as though the above ratio could match the black hole-bulk ratio inside the Milky Way rather well.

One is inclined to think that if the gravity of primordial dark matter/exotic matter is really the weaker type of gravity, corresponding to the charge gravity of space particle dualism, then this ratio should be reflected at least roughly in the mass ratio between the mass of galaxies to the mass of their central supermassive black holes.

In the Milky Way that is a factor of 10^6. Using the initial wrong value of $G_{|\mathbb{Q}|}$ it seemed to match that roughly, but now with the corrected value, it does not.

If galaxies had started off with a black hole-bulk ratio of $1:15$ or $1:137$, then this ratio would, due to the accretion of ordinary matter by the central black hole, now have arrived at a ratio of $1:1$, or even reversed. If the ratio between $G_{|\mathbb{Y}|}$ and $G_{|\mathbb{Q}|}$ had initially matched the black hole-bulk ratio, then we would expect that the inequation

$$\frac{G_{|\mathbb{Y}|}}{G_{|\mathbb{Q}|}} \geq \frac{M_{SBH}}{M_{bulk}}$$

is valid for every galaxy. We know that this is not the case. What then could the black hole-bulk ratio be based on?

As already mentioned above, exotic gravity is not only weaker in terms of its strength, but also in terms of its reach. Having a range of only 40 meters makes it much more difficult for exotic matter to accumulate to very large black holes. It ends up serving only as some sort of glue that facilitates the clumping of ordinary matter.

Before we continue, we want to first further examine the black hole-bulk ratio in different types of galaxies

Black hole-bulk ratio

Data on the bulk masses of galaxies are rather hard to find, so we currently don't have the average mass of the galaxies these supermassive black holes are located in. However, there is a lot of research that has already been done on the ratio between the mass of supermassive black holes and their host galaxy.

A series of studies in 2001 found a $M_{BH}/M_{bulge} \approx 0.09\%$ for quasars or quasi-stellar objects (short 'QSO'), 0.12% for Seyfert galaxies (galaxies with highly active cores), and 0.13% for quiescent galaxies.[1, 2, 3] All of these galaxies must have long diverged from their initial black hole-bulk ratios, considering that all galaxies that we can observe are very old, having 90% of their lifetime behind them (see chapter 4.12).

Accretion of ordinary (non-exotic) matter into the supermassive black holes could offsets the observed ratio.

Size of exotic holes

Something else, which should offset the ratio in the other direction, is the fact that charge gravity is not only weaker, but does also have a shorter range.
However, if supermassive black holes form by the accumulation of a large number of minimal mass (about 10^6 kg) primordial black holes (see chapter 4.10), with each of them having radii of only

$$R_{min} = \frac{2\,G_{|\mathbb{Q}|}\,|\mathbb{Q}|_{min}}{c^2} \approx \frac{2\,G\,M_{min}}{c^2} = 1.6706816 \times 10^{-21}\ \text{m}$$

Then it is reasonable that there will be many of them within a radius of 40 meters. If the gravity of exotic matter indeed has a range of only 40 meters, then this means that the increase in surface gravity through the accretion of exotic matter comes to a halt after reaching a charge cardinality or approximate mass of:

$$|\mathbb{Q}|_{EH} = \frac{[40\ \text{m}]\,c^2}{2\,G_{|\mathbb{Q}|}} = \frac{\hbar\,c^3}{4\,E_{min}\,G_{|\mathbb{Q}|}} = 2.9156649 \times 10^{82}$$

$$M_{EH} \approx -\frac{[40\ \text{m}]\,c^2}{2\,G} \approx -1.347459 \times 10^{28}\ \text{kg}$$

This mass here is negative, because it is exotic mass, but the gravity it exerts is still positive. That is because the precise equations of exotic gravity don't depend on mass, but on the number of individual exotic or dark charges.

A black hole made from exotic matter alone would exert gravity with limited range only,

pretty similar to the gravity in the massive gravity theory of Claudia De Rham.[4]
It would not only be exo-transparent, as we learned in chapter 4.11, but it would also have no photosphere, no ergosphere and it would not bend light around it.

Supposed only exotic matter falls in, could it still grow?
Apparently not: a black hole cannot grow if its surface gravity g doesn't grow. Two such critical cardinality exotic holes could even merge without increasing the diameter of the resulting event horizon.

We can therefore conclude that energy conservation for exotic energy is even more illusive than for ordinary matter (see chapter 4.13). Furthermore we can treat the above critical cardinality exotic hole as a 'semi-primordial seed' for supermassive black holes.[5]
In general relativity, a black hole made of exotic matter would certainly be believed to be a 'white hole', pushing everything away from it. The gravity of space particle dualism theory however is always attractive, and therefore such an object is still analogous to a black hole. Yet, having such a short range of influence, such exotic holes can function only as seeds for ordinary matter black holes.

Entropic expansion would push these semi-primordial seeds away from each other, preventing any of them to be wasted. This way each such seed gives rise to an ordinary matter supermassive black hole and a galaxy bulk around it.

Exotic holes being only seeds for ordinary black holes also explains why there is no alignment between the charge-hypercharge ratio and the black hole-bulk ratio.

At this point it seems likely that entropic expansion makes the merging of primordial black holes impossible. Merging might have taken place to a certain degree, but it is likely that most of the pre-decoupling growth happened through the accretion of surrounding plasma.

Origin of the black hole-bulk ratio

If so, then we can expect the size of the initial seeds to have at least some impact on the black hole-bulk ratio.

The minimal mass for a solar black hole is 2.5 sun masses. We can compare that to the absolute value of the mass of an exotic hole:

$$\frac{M_{EH}}{M_{BH_{min}}} = \frac{|-1.347459 \times 10^{28}\,\text{kg}|}{2.5 \times M_{\odot}} = 0.00270982202$$

If this is reflected in the black hole-bulk ratio, then we have to expect that the central black hole is about 0.27% the mass of its host galaxy. So, if the correlation holds true, then we can state that:

$$\frac{M_{EH}}{M_{BH_{min}}} \approx \frac{M_{SMB}}{M_{bulk}}$$

This is not an inequation, which reflects the assumption that the supermassive black hole and the bulk both grow over time in a similar way. We can already tell this from our preliminary surveying of galaxies. It shows that the black hole-bulk ratio inside galaxies never or rarely exceeds 0.27%.

In fact, mostly it does not exceed 0.15%. The predicted ratio of 0.27% was depending on the exact value for the minimal mass of black holes. That is the same as the maximal mass of neutron stars, and lies in the range between 1.5 and 3 solar masses.[6] This corresponds to a predicted black hole bulk ratio that lies between 0.23% and 0.45%.

If we instead compare the exotic hole mass to the average stellar black hole, which has a mass of about 8 solar masses, then we arrive at a ratio of 0.085%, which corresponds pretty precisely to the ratio found for quasars.

At the moment this seems to explain the black hole-bulk ratio more satisfactory, and so the correct formula for it should be:

$$\frac{M_{EH}}{M_{BH_{\mu}}} \approx \frac{M_{SMB}}{M_{bulk}}$$

Note:
1. Merritt, D. & Ferrarese, L. 2001a, MNRAS, 320, L30.
2. Merritt, D. & Ferrarese, L. 2001b, ApJ, 547, 140.
3. Merritt, D., Ferrarese, L. & Joseph, C. 2001, submitted.
4. De Rham, Claudia; Gabadadze, Gregory (2010). "Generalization of the Fierz-Pauli Action". Phys. Rev. D. 82 (4): 044020. arXiv:1007.0443. Bibcode:2010PhRvD..82d4020D. doi:10.1103/PhysRevD.82.044020.
5. One may suspect that this subprimordial mass/cardinality should be our new initial mass M_{min} (or cardinality $|\mathbb{Q}|_{min}$) in our equations for entropic expansion, but that is not the case. Exotic holes below this mass have entropy too, so entropic expansion sets in well before this mass is reached.
6. I. Bombaci (1996). "The Maximum Mass of a Neutron Star". Astronomy and Astrophysics. 305: 871–877. Bibcode:1996A&A...305..871B.

6.9 Dimensionless units of SPD [05.04.2019 ($G_{|\mathbb{Y}|}$-m_p-Planck length)*; 03.02.2020 (baryonic gravitational constant G_B; m-bar; m_p-only-Planck length); 07.02.2020 (threefold Planck-hierarchy); 08.02.2020 (wrong hypergravity Schwarzschild radius)]

*Requested by: Jesse Timron Brown (mathematician; US).

Now, that we have abolished the gravitational constant G and replaced it with hypercharge gravity constant $G_{|\mathbb{Y}|}$, we need new dimensionless units based on this new constant.

A dimensional analysis shows that in the usual definition of the Planck length it is the equivalence between mass and energy which allows us to view $\hbar$ as a unit of mass M and time T, instead of energy and time, and to thereby cancel out the [kg]-unit of the gravitational constant:

$$\dim \sqrt{\frac{\hbar\, G}{c^3}} = \sqrt{\frac{L^2\, M}{T}\frac{L^3}{M\, T^2}\frac{T^3}{L^3}} = \sqrt{L^2}$$

If we want mass to cancel out the same way in a new $G_{|\mathbb{Y}|}$-based definition of the Planck length, then we need to add one more fundamental constant, and that constant must be a mass. $G_{|\mathbb{Y}|}$ was essentially created by multiplying a slightly modified G with a third of the mass of the proton. If we are only interested in generating a dimensionless unite using only fundamental constants, then we don't need to add factors like $1/3$ and such. We may therefore simply write:

$$l_P = \sqrt{\frac{\hbar\, G_{|\mathbb{Y}|}}{m_p^2\, c^3}} = 2.3261617 \times 10^{-35}\ \mathrm{m}$$

We could of course also define a reduced proton mass, analogous to the reduced Planck constant:

$$\bar{m}^2 = \frac{m_p^2}{9}$$

And call it 'm-bar'. Now we can write the Planck length as:

$$l_P = \sqrt{\frac{\hbar\, G_{|\mathbb{Y}|}}{\bar{m}^2\, c^3}}$$

Which yields:

$$l_P = 1.6073505 \times 10^{-35} \text{ m}$$

This is very close to the usual Planck length of $l_P = 1.61622837 \times 10^{-35}$ m.

Alternatively we can also use the mass-gravity constant of proton gases, which is probably the most fundamental of all the different mass-gravity constants. That yields exactly the same as the above:

$$l_P = \sqrt{\frac{\hbar \, G_p}{c^3}}$$

$$l_P = 1.6073505 \times 10^{-35} \text{ m}$$

We can analogeously create the other Planck units using either G_p or:

$$\frac{G_{|\Psi|}}{m^2}$$

Which yields:

$$m_P = \sqrt{\frac{\hbar \, c}{G_p}} = 2.18849147 \times 10^{-8} \text{ kg}$$

This is again pretty close to the usual Planck mass of $m_P = 2.176435 \times 10^{-8}$ kg.

In conventional general relativity the Planck length is at the same time the mass of a black hole that was formed by a single photon. In space particle dualism this is not possible, because as photons reach higher and higher energies, their elementary space grows larger than their wavelength, making it impossible for a single photon to become a black hole.

Our above new version of the Planck mass is therefore merely a way of creating a unit of mass which is not arbitrary and is not linked to particular physical processes. This was also Planck's original intent, since at the time he invented the Planck units, general relativity did not exist, and therefore he could not possibly have interpreted the Planck-length and the Planck-mass in terms of a photon that turns into a black hole.[2]

Something else that wasn't known back then, not with any reasonable precision at least, is the mass of the proton and the electron. If they were known, they would probably have represented much more practical measures of mass than the 'Plank-mass'.

A reason why till this day the masses of particles are usually not considered fundamental units of mass is that it is widely believed that all massive particles somehow emerged from a primordial photon-only universe; a notion that is entirely dismissed in space particle dualism theory.

There are many different nonarbitrary constants one could use as dimensionless units. For example there is the critical electron-photon wavelength ($\lambda_{crit} = 9 \times 10^{-17}$ m). For creating a nonarbitary unit for mass on the other hand we could take the mass of any elementary particle we want.

The true question we have to ask ourselves here is: what units is nature itself based upon?
We know from the chapters 3.7, 3.8 and 3.15 how crucial it is to quantize wavelengths by using some smallest sensible length, so we obviously can't do without some unit analogous to the Planck-length.

According to space particle dualism the smallest distance one can directly measure using photons is 9×10^{-17} meters. As was shown in chapter 3.9, the electron comes close to that, agreeing in six digits, but it doesn't precisely reach it. It was also shown that even if the mass of the electron was smaller, it could still not reach this distance. Only at infinitely small rest mass and a speed infinitely close to the speed of light the two distances become exactly the same. We therefore know that 9×10^{-17} meters is really the shortest distance one can directly probe using scattering.
Yet, gravitational waves detectors measure distance differences in the realm of 10^{-19} meter. This type of distance measure is using interference.
Does this mean we can use interference to measure arbitrarily minute differences in distance?

It is essential to space particle dualism theory that there is some way to quantize wavelengths and thereby energy levels of free particles. In previous chapters we were using the Planck-length.
We can now switch to its SPD-specific alternative, using $G_{|\Psi|}$ and m.
However, in space particle dualism the Planck scale certainly lost most of its physical significance.
If Planck-like units are merely non-arbitrary (history-independent) units we agree on, then how can they be fundamental?

Is there a natural limit to the precision that can be reached with interferometers? Obviously photons don't turn into micro-black holes only because they were bouncing back and forth in an interferometer for too many times, so the Planck-length can't be argued for as a measurement-limit for any directly intuitive reasons.

What arguments can be made for our above SPD-specific equivalent for the Planck-length to be really fundamental?

If we choose to permit the additional use of the proton mass m_p for the construction of our units, then why don't we leave out gravity and construct our unit of length simply as:

$$l_P = \frac{\hbar}{m_p\,c} = 2.1030891 \times 10^{-16}\,\text{m}$$

$$\dim \frac{\hbar}{m_p\,c} = \frac{L^2\,M}{T}\frac{1}{M}\frac{T}{L} = L$$

It fulfils the requirements for a dimensionless unit, but it is obviously too large. Very close to the critical wavelength actually.
The gravitational wave experiments show that nature has a resolution higher than that.

So we know that this simplified version of a Planck-length is not fundamental, but why isn't it? Why is gravity needed to create 'scale'?

After all, one may want to point out, gravity is an emergent force; it is not fundamental. One could well imagine a universe with only particles, charge and inertia. The three fundamental forces would not change in such a world.
But then, what tells us that emergent properties are not fundamental? Certainly consciousness is fundamental; quantum mechanics doesn't work without conscious observers; which are essentially biological brains. From the perspective of quantum mechanics, we can regard both the very large, the cosmos, and the very small, including both the world of cells and the quantum world, as emergent properties of our immediate reality as it is reflected in our collective consciousness.

In chapter 6.3 we learned that gravitational time dilation and length contraction is caused by differences in densities of elementary spaces. This is a highly emergent phenomenon, yet it determines such important things as length and time.

In light of that, can we really create fundamental units without using the new gravitational constant $G_{|\Psi|}$? The above, as well as the experimental evidence, suggests that no, we cannot.

In Penrose's CCC model scale disappears as soon as mass disappears. He argues that for photons time does not exist, and also if there are only photons around, there is nothing that can function as a clock. So not only gravity, but also mass is crucial for having scale.

Planck didn't include any mass in his formula for the Planck-length, because the conventional gravitational constant already contains the [kg] as a unit.

Furthermore, in mainstream physics, inertia mass and gravitational mass are assumed to be equal. They are thought of as being merely different aspects of the same thing. As we learned throughout chapter 6, they are not at all the same.

This is one more reason to assume that reality on the most fundamental level is based on four constants, the reduced Planck constant $\hbar$, the hypercharge gravity constant $G_{|\mathbb{Y}|}$, the reduced proton mass $\mathfrak{m}$ and the speed of light c.

In dimensionless physics these four can be set to zero in order to simplify calculations.

But is this it? Can't a fundamental unit for length be based on something else?

What about the other constants of space particle dualism theory?

Can't we use the elementary space constant to create a unit of length?

Yes, indeed we can, and that yields:

$$l_P = \frac{G_E \, m_e}{c^2}$$

$$\dim \frac{G_E \, m_e}{c^2} = \frac{L^3}{T^2 \, M} \frac{M}{1} \frac{T^2}{L^2}$$

$$l_P = 8.3582274 \times 10^{-22} \text{ m}$$

This looks almost like the equation for the radius of the elementary space of an electron, and indeed is precisely half of that distance. The relationship between this distance and the critical photon-electron wavelength is analogous to the standard Planck-length to the Schwarzschild radius of an electron.

Although observations do not directly contradict this length measure as a fundamental unit, it would cause trouble with getting space connected. If there was only one elementary space every Planck volume, then gravity would be much stronger than observed (see chapter 3.7).

In space particle dualism theory it is natural to have three different fundamental lengths, namely:

1. The smallest distance that can be measured directly. As we found in chapter 3.7, that is 9×10^{-17} m, and it is given by:

$$l_{P_1} = \frac{h \, c}{\sqrt{\frac{h \, c^5}{4 \, G_E}}}$$

2. The size of an elementary space[3] or half of it, given by:

$$l_{P_2} = \frac{2\,G_E\,m_e}{c^2} \oplus \frac{G_E\,m_e}{c^2}$$

3. A scale defined by units of the weakest force, the force of gravity, given by:

$$l_{P_3} = \sqrt{\frac{\hbar\,G_{|\Psi|}}{\mathfrak{m}\,c^3}}$$

We may call this a 'threefold Planck hierarchy'.

Now the second of these 'Planck-lengths', we could call it the 'intermediate Planck-like length' should concern us a bit. Is the factor of 2 in the equation for the elementary space radius justified?

The '$\oplus$' used in the equation for the intermediate Planck-like length indicates 'or', expressing this uncertainty about the justification for the factor 2 here.

Are elementary spaces only half as large as we thought?

Historically elementary spaces started out as Schwarzschild surfaces. Only later the gravitational constant G in them was replaced by the elementary space constant G_E. The Schwarzschild radius has a factor of 2 in it, and therefore the elementary space radius, which is essentially a scaled up version of the Schwarzschild radius, has a factor of 2 in it as well. Where does this factor of 2 come from?

That is pretty simple: the escape v_{esc} velocity for a gravitational body is reached when the kinetic energy E_{kin} of the escaping object equals its potential gravitational energy E_{pot} at the altitude R, measured from the center of mass. So we can state that:

$$\frac{1}{2}\,m\,v_{esc}^2 = \frac{G\,M\,m}{R}$$

$$v_{esc} = \sqrt{\frac{2\,G\,M}{R}}$$

By definition the Schwarzschild radius R_S is the radius at which the escape velocity equals the speed of light. We can therefore further state that:

$$c = \sqrt{\frac{2\,G\,M}{R_S}}$$

$$c^2 = \frac{2\,G\,M}{R_S}$$

$$R_S = \frac{2\,G\,M}{c^2}$$

The SPD-specific analogue to this is quite different:

$$\frac{1}{2}\,m\,v_{esc}^2 = \frac{G_{|\Psi|}\,|\Psi|\,|y|}{r}$$

$$v_{esc} = \sqrt{\frac{2\,G_{|\Psi|}\,|\Psi|\,|y|}{m\,r}}$$

$$c = \sqrt{\frac{2\,G_{|\Psi|}\,|\Psi|\,|y|}{m\,R_S}}$$

$$c^2 = \frac{2\,G_{|\Psi|}\,|\Psi|\,|y|}{m\,R_S}$$

$$R_S = \frac{2\,G_{|\Psi|}\,|\Psi|\,|y|}{m\,c^2}$$

This escape velocity and Schwarzschild radius is very different from the versions we are familiar with from Einstein and even Newton. According to both Newtonian and Einsteinian physics the escape velocity is the same for all objects.
Space particle dualism on the other hand has slight corrections for objects of different quark number $|y|$ and different mass m. The quark number is roughly proportional to the mass, with the ratio changing only when switching to different chemical compositions. That explains the strange dependence of free fall time on the chemical composition.

According to the above formula, we should observe this dependency on the chemical composition not only for dropping objects, but also for objects escaping from earth.

This also means that black holes do not have one unique Schwarzschild radius, but one for each particle type and also one for each chemical composition.

For solitary protons, the Schwarzschild horizon is:

$$R_S = \frac{2\, G_{|\mathbb{Y}|}\, |\mathbb{Y}|}{m_p\, c^2}$$

For collections of protons it is:

$$R_S = \frac{2\, G_{|\mathbb{Y}|}\, |\mathbb{Y}|\, |y|}{|y|\, m_p\, c^2} = \frac{2\, G_{|\mathbb{Y}|}\, |\mathbb{Y}|}{m_p\, c^2}$$

However, when they are part of atoms, then there is a slight difference between inertia mass and gravitational mass, so that we return to:

$$R_S = \frac{2\, G_{|\mathbb{Y}|}\, |\mathbb{Y}|\, |y|}{m\, c^2}$$

Electrons feel gravity but do not exert gravity. As we learned in the chapters 6.5 and 6.6, the gravitational acceleration they feel is governed by the reduced hypercharge gravity constant $\mathbb{G}_{|\mathbb{Y}|}$. So for an electron, the Schwarzschild horizon is:

$$R_S = \frac{2\, \mathbb{G}_{|\mathbb{Y}|}\, |\mathbb{Y}|}{m_e\, c^2}$$

From this we can see that the factor of 2 comes from examining how objects fall in a gravitational field and how they might escape. It is therefore problematic to adopt it into the formula for elementary space radii.

One could argue that it is justified for a surface that is created by scaling up gravity, which is essentially how G_E was created (see chapter 3.8 & 3.9), to adopt a pre-factor '2' from gravity, and maybe that is indeed justified, but then we have this general dependence of the horizon on the chemical composition, or in other words, on the ratio between $|y|$ and m, and that makes black holes and elementary spaces really extremely different, to the point where it is not longer justified to look at elementary spaces as 'scaled up black holes'.

Although elementary spaces were conceptualized from black holes, but ever since their conceptualization in 2005, both black holes and elementary spaces have evolved in very different ways. Elementary spaces have been scaled up, but they have remained to be analogous to classical black holes, while SPD-black holes have become something very different. Something that depends on quark-count, and not on energy, while elementary spaces still depend on energy.

While it is ok to sometimes express the elementary space radius in terms of relativistic mass m as:

$$R_E = \frac{2\,G_E\,m}{c^2}$$

It is certainly formally dissatisfying. The more correct formulation is using energy instead:

$$R_E = \frac{2\,G_E\,E}{c^4}$$

Or, and which has the same effect, relativistic mass, which yields:

$$R_E = \frac{2\,G_E\,m}{c^2\sqrt{1-\frac{v^2}{c^2}}}$$

Or, to use the full complex Demiroglu-expression for it (see chapter 3.6):

$$R_E = \frac{2\,G_E\,m}{c^2\sqrt{1-\frac{v^2}{c^2}}} + \frac{2\,G_E\,m}{c^2\sqrt{1+\frac{v^2}{c^2}}}\,i$$

This makes elementary spaces very similar to event horizons in general relativity, which are only approximately given by the mass the black hole contains. In general relativity it is most accurately the energy that determines the Schwarzschild radius of a black hole:

$$R_S = \frac{2\,G\,E}{c^4}$$

This is very different from how things are in space particle dualism theory. Here the size of black holes depends on the number of quark in them only, and our equation is (see chapter 6.6):

$$R_S = \frac{2\,G_{|\mathbb{Y}|}\,|\mathbb{Y}|}{c^2} \times \frac{|y|}{m}$$

This really destroys the analogy between black holes and elementary spaces, and it should make us seriously consider to drop the factor '2' in the equation for elementary spaces, which really originates in this analogy.

If we drop this factor, then elementary spaces would still be like scaled up general relativity-black holes, just without that factor of '2'. If the scale factor we use is a type of Planckian unit, then we can justify its usage without referring to it as a 'gravitational constant that is scaled up to meet in strength with electromagnetism', as we did in chapter 3.7 and 3.8. Establishing it is a Planckian unit allows us to break free from relying on the black hole-elementary space analogy.

In some way the usage of '$|\mathbb{Y}|$' (number of quarks) makes gravity more similar to the other forces. In electromagnetism the attraction or repulsion between two objects always depends on an integer, namely the sum of positive and negative charges; in chromodynamics, the attraction between quarks depends on them having different color charges; again we deal with integers.
Gravity is very similar in that regard, just that it is indifferent to the sign of the charge, and the 'color' of the hypercharge. That is why we represent it as a cardinality.
We could call gravity the 'cardinal force'.

The terms 'hypercharge gravity' and 'charge gravity' could be somewhat unfortunate, because charge is always something where the sign matters. We may instead talk about 'hypergravity' and 'subgravity' instead.

Once physicists get used to 'hypergravity', it will just be called 'gravity', and so the hypergravity constant will simply be called 'gravitational constant'. The old Newtonian and Einsteinian gravitational constant will then be called 'mass-gravity constant', or more specifically 'mass-gravity constant of iron'.

Subgravity on the other hand can usually be ignored when talking about ordinary matter and becomes important only for exotic matter. In this context one can talk of 'exotic gravity' (see chapter 6.8).

So the problem here is not gravity and electromagnetism being too different; it is the fact that the radius of an elementary space does not depend on charge but on relativistic mass instead. As we learned in chapter 3.12, mass is a side effect of charge. That is also one major justification for using the electron charge e in our equation for the elementary space constant. The original aim of this constant was to create a scaled up version of the gravitational constant, to be used for elementary spaces. As this present chapter showed, it turns out that this constant leads to elementary spaces which are at the same time very natural units of space. When replacing G_E by its equation and inserting it into the elementary space radius formula, we get:

$$R_E = \frac{2\,k_e\,e^2\,m_e}{m_p^2\,c^2} \oplus \frac{k_e\,e^2\,m_e}{m_p^2\,c^2}$$

$$l_{P_2} = \frac{k_e\,e^2\,m_e}{m_p^2\,c^2}$$

In dimensionless treatments, the usual units that are normalized to 1 are: c, G, $\hbar$, k_e and k_B.

We switched to a different set, namely: c, $G_{|\Psi|}$, G_E, $\hbar$, k_e and k_B.

When using the equation for G_E, then our set is: c, $G_{|\Psi|}$, $\hbar$, k_e, e, m_p, m_e and k_B.

Back to the factor-of-2-issue: Does this factor really rely on the black hole-elementary space analogy?

Maybe not. The Schwarzschild radius is about gravity. What if we use electromagnetism to define a similar radius and surface?

The electromagnetic force between two charges is:

$$F = k_e \frac{q_1 q_2}{r^2}$$

When we integrate over that, we get the potential energy U in the electromagnetic field:

$$W = \int_{\infty}^{R} k_e \frac{q_1 q_2}{r^2} \, dr$$

$$U = - k_e \frac{q_1 q_2}{r}$$

Now we set this equal to the classical kinetic energy:

$$\frac{1}{2} m v^2 = \frac{k_e q_1 q_2}{r}$$

$$v = \sqrt{\frac{2 k_e q_1 q_2}{m r}}$$

Now we derive the electrical Schwarzschild radius from this, by setting $v = c$:

$$c = \sqrt{\frac{2 k_e q_1 q_2}{m r}}$$

$$R_{EMS} = \frac{2\, k_e\, q_1\, q_2}{m\, c^2}$$

Structurally that is totally analogous to the formula for the SPD-specific Schwarzschild radius:

$$R_S = \frac{2\, G_{|\mathbb{Y}|}\, |\mathbb{Y}|\, |\mathrm{y}|}{m\, c^2}$$

The electromagnetic Schwarzschild radius for a pair of protons is:

$$R_{EMS} = \frac{2\, k_e\, e^2}{m_p\, c^2} = 4.7151469 \times 10^{-51}\ \mathrm{m}$$

We can structurally compare that to the elementary space radius, which is:

$$R_E = \frac{k_e\, e^2}{m_p^2}\, \frac{2\, E}{c^4} = \frac{2\, k_e\, e^2\, E}{m_p^2\, c^4}$$

Both of them are structurally quite similar to the Newton-Einstein-Schwarzschild radius:

$$R_S = \frac{2\, G\, E}{c^4}$$

As long as we are dealing with a real force, involving kinetic energy, the factor of 2 is to be expected, and therefore it does not seem unreasonable to use it in the equation for the radius of elementary spaces.

The following arguments speak for this factor:

1. It appears in classical mass-dependent Newtonian physics, and elementary spaces do depend on mass, more precisely relativistic mass.
2. It appears in Einsteinian energy-dependent general relativity, and elementary spaces do depend most generally on energy.
3. It appears in electromagnetism too, when we calculate escape velocities between electric charges, as well as the size of electromagnetic equivalents to black holes.

The arguments against it are:

1. SPD-black holes are not at all similar to the energy dependent Einsteinian black holes, and thus we cannot regard elementary spaces to be scaled up versions of

them.

2. Elementary spaces belong to single particles, and thus are not in any way related to escape velocities and Schwarzschild radii.
3. Without this factor we are left with a Planck-like unit, which is much more compelling.

At the end experiments will decide between the conventional elementary space radius with the factor 2, and the reduced one, without the factor 2.

Note:
1. Using the mass defect factor for iron from chapter 6.5.
2. Planck, Max (1899). "Über irreversible Strahlungsvorgänge". Sitzungsberichte der Königlich Preußischen Akademie der Wissenschaften zu Berlin (in German). 5: 440–480. pp. 478–80 contain the first appearance of the Planck base units other than the Planck charge, and of Planck's constant, which Planck denoted by b. a and f in this paper correspond to k and G.
3. As we saw in chapter 3.11, when accounting for the running of coupling, then the rest-surface for all elementary spaces is the same.

6.10 Bernoulli trial comparison between the gravity of water and iron [22.02.2020 (results of the first run of the experiment); 08.03.2020 (oscillation period and chaos-invariance); 18.03.2020 (chance against odds); 01.04.2020 (corrected chance against odds)*]

*In collaboration with David Chester (USLA; MIT).

Experiments conducted in: Silver Leaf factory in outskirts of Skala, Greece.

Experiments that aim at measuring the gravitational constant mostly use iron objects to do so. There are only extremely few experiments which used a different material, and most of those were using an experimental setup that was vastly different from the traditional Cavendish experiment that is using a torsion balance, which makes comparison difficult. Fischbachs experiment from chapter 3.3 hinted that indeed it is the number of individual hypercharges which determines the gravity between two objects. However, one or two experiments are not enough, especially when they were made in a time in which space particle dualism theory didn't exist, and the experimenters had no way to compare their results against precise predictions of the kind made by this theory.

The Cavendish experiment involves a torsion balance suspended on a rope with two small weights attached to its ends, and two large weights put on the ground at the left and the right side of the ends of the torsion balance. The gravity from the two large objects pulls on the torsion balance and makes it rotate until torsion becomes too strong and the torsion balance rotates back. This results in an oscillation motion. The oscillation time and angle then allows us to derive the gravitational constant.

However, before we attempt to repeat the traditional Cavendish experiment on water, we will first conduct a variation of this experiment which is done with a rope that has very little torsion and where we will not be looking at an oscillation angle, but rather at the time the torsion balance needs to hit the two large objects.
We will do this both with water buckets and iron objects as the large masses, and we will weight them to make sure that the water buckets and the iron objects weight exactly the same.

This experiment is not a quantitative experiment, but only a qualitative experiment. Instead of measuring the gravitational constant of water, we want to first simply establish the fact that iron has more gravity per kilogram than water.

For that we will let water and iron compete against each other in single trials, so called Bernoulli trials, where we will be simply asking which one was faster than the other in each pair of trials.
If the effect is non-existant, then we should be expecting a binominal distribution.
If the effect is real, then we can establish its reality by looking at the chance against odds for the Bernouli trials' output to be due to random chance. This chance against odds can be characterized by the the number of trials n, the number of successes k, and the random chance success rate p at each trial, which in this case is ½.
The chance against odds is then given by the probability mass function, which is as follows:

$$f(k, n, p) = \frac{n!}{k!\,(n-k)!}\, p^k\,(1-p)^{n-k}$$

The Bernoulli trials were run in my sister's factory in Greece. In the first few runs I was using an iron cylinder with a weight of 6.355 kg and a iron propeller with a weight of 7.514 kg. The trials with iron were paired up with trials using water of precisely the same weight. These first trials were not taking account of the differences in density. This first setup can be seen in fig. 93 & fig. 94:

Fig. 93 & 94. Comparing the gravity per kilogram of iron with the gravity per kilogram of water.
[This is an early version of the experiment which didn't take care of the differences in density.]

Distances between 20 and 30 cm were chosen, and the time to impact was typically 2 minutes for iron and 3 minutes for water.

This is in gross contrast to the very long oscillation times one has in the traditional Cavendish experiment with high torsion ropes, where torsion and gravity are constantly pushing the torsion balance back and forth. In that experiment the typical time for one oscillation is about half an hour. This more traditional experiment we will be looking at in the next chapter.

Demonstrations of gravity in a classroom or on the internet are often labeled as representing the 'Cavendish experiment', while in fact they are not looking at oscillation angles and times, but rather just show how the larger masses attract the smaller masses by allowing them to approach each other until they touch.[1, 2, 3]
This is because it looks more convincing and spectacular as a qualitative demonstration of gravity, while the version including torsion is more of quantitative value and less impressive to the untrained eye. The traditional torsion version also takes longer to do, and that is why it is less fit for classroom demonstrations.[4]

We will not mention the oscillation equation of the traditional Cavendish experiment here, because it will be discussed in the next chapter, however it should be noted that the rather large oscillation times in it suggest that the very short oscillation times we find in our torsion free version have to do with random air fluctuations.

In any experiment that is not air-sealed, the torsion balance will start moving as soon as air pushes it into the direction from which gravity is originating. The gravity of the two large objects will then amplify the random movement that started due to air currents.

Therefore in such experimental setups, it might not be gravity that initiates the motion, but small air currents instead. Gravity is then merely amplifying those movements.

If we pretend for one moment that our travel angle was an oscillation angle[5], then we get a travel time for water of:

$$t_{H_2O} = \sqrt{\frac{2\,\pi^2\,L\,\theta\,R_e^2}{G_{H_2O}\,M_{H_2O}}}$$

$$t_{H_2O} = 38{,}333.8188858 \text{ sec}$$

Which is 10 hours, 38 minutes, 54 seconds. And a travel time for iron of:

$$t_{Fe} = \sqrt{\frac{2\,\pi^2\,L\,\theta\,R_e^2}{G_{Fe}\,M_{Fe}}}$$

$$t_{Fe} = 38{,}263.6708225 \text{ sec}$$

Which is 10 hours, 37 minutes and 44 seconds. This is a difference of 1 minute and 10 seconds.

A difference of about 1 minute was indeed observed, but because torsion had not set in, and because the experiment was not conducted in an air-sealed environment, this ten hours were reduced to just 2 minutes (iron) and 3 minutes (water) respectively.

The above equation we shall not discuss in detail, because it is mainly the subject of the next chapter. Here it shall only serve as a clue that the difference in travel time between iron and water is in an order of magnitude that is to be expected.

In the first few runs of the experiment, the difference in the densities of the iron objects and the water masses was not taken care of, as can be seen in fig. 91 & 92.

The iron objects were hollow, and that is why their densities were actually close to the density of water. How similar are the densities exactly?

The iron cylinder has a density of:

$$\rho = \frac{5.862 \text{ kg}}{\pi\,0.0705 \text{ m}^2\,0.374 \text{ m}}$$

$$\rho = 1003.79748961 \text{ kg/m}^3$$

The density of water is $\rho = 997$ kg/m^3. That is very close.

However, the diameter of the water buckets is larger than the radius of the iron cylinder. In a new more precise version of our experiment we want to put the iron objects into the same type of plastic buckets as the water, and we want to make sure that the amount of plastic is the same for both, so we add the cover of the lower bucket onto the top. The objects used in the new setup can be seen in the pictures 93 & 94:

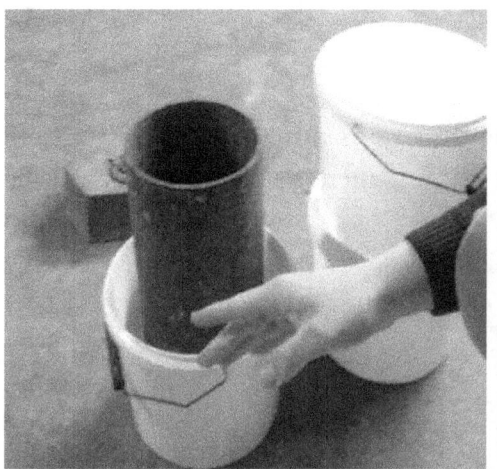 

Fig. 95 & 96. An improved version of the experiment had the iron objects placed in the same type of plastic buckets as the water. The second object, the propeller, was replaced by a motor. The amount of water in the water buckets was adjusted accordingly.

The rest of the experiments were using the same distance 29.5 cm, and so we can omit the distance from our final list, which shall include all trials with this distance. Unsuccessful trials in which the torsion balance rotated away from the buckets are omitted:

Water		Iron	
Rotation time	Date	Rotation time	Date
5:48	13.03.2020; 2:29	1:11	13.03.2020; 1:31
3:18	13.03.2020; 3:02	2:14	13.03.2020; 2:29
2:08	14.03.2020; 3:10	2:47	13.03.2020; 2:52
2:50	14.03.2020; 3:14	2:08	13.03.2020; 3:08
3:00	14.03.2020; 3:18	2:48	14.03.2020; 3:53
2:40	14.03.2020; 3:26	2:11	14.03.2020; 3:58
2:15	14.03.2020; 3:34	1:17	16.03.2020; 0:09
2:46	15.03.2020; 23:47	1:25	16.03.2020; 0:14

From looking at this collection alone we find a success rate of 7/8 or 87.5%, because the iron-rotation time was faster 7 out of 8 times. The chance against odds is:

$$f(7,8,0.5) = \frac{8!}{7!\ 1!}\ 0.5^7\ 0.5^1 = 3.125\%$$

In terms of average rotation time, we get $3:05$ for water, and $2:00$ for iron. That is a difference of 1 minute and 5 seconds, which is very close to the expectation of 1 minute and 10 seconds which we get when treating this movement as a Cavendish oscillation. This shows that while non-air-sealed versions of the Cavendish experiment yield far too large values for the various gravitational constants, they do preserve the difference in rotation time. If this was the standard Cavendish experiment, the difference would be only $2\ \frac{1}{2}$ seconds.

This scaling up of the differences between the gravity of different materials within chaotic systems is highly unexpected. We can say that there is a hidden order in the chaos.

We want to now further improve the quality of our experiment by including a few new rules into our protocol:

(1) Iron and water rotation times should be measured alternately, so that both start from similar positions of the torsion balance. This rules out that different torsion is acting on the rope during different trials.
(2) The balance should be non-moving for at least 30 seconds before a trial.
(3) The distance between the ends of the torsion balance and the weights should not be determined using a object that has to be carried from side to side, but rather by using thin distance markers that can be left on the ground.
(4) The large masses should only be placed on the ground when the experiment starts.
(5) If only one person is conducting the experiment, then one end of the torsion balance should be stopped using two cartons, and the other end should already have the large weight in place. As soon as the one end is released, the second large weight is put in place.

In these following trials the protocol was strictly followed. Trials at which iron took longer than water are marked orange:

Water		Iron	
Rotation time	Date	Rotation time	Date
$3:20$	21.03.2020; 1:04	$3:10$	21.03.2020; 1:22
$2:15$	21.03.2020; 3:49	$4:20$	21.03.2020; 1:45
$1:42$	22.03.2020; 0:19	$1:25$	21.03.2020; 4:03
$2:36$	22.03.2020; 0:30	$1:34$	22.03.2020; 0:25
$1:45$	22.03.2020; 0:57	$2:08$	22.03.2020; 0:36
$3:01$	22.03.2020; 1:10	$1:35$	22.03.2020; 0:45
$2:58$	22.03.2020; 1:44	$2:29$	22.03.2020; 1:03

3:34	22.03.2020; 2:07	3:05	22.03.2020; 1:19

The chance that this is a coincidence is now 0.285%, which means that we have reached a certainty of 99.7%.

Fig. 97 & 98. A typical iron trial.

Fig. 99. A typical water trial. To the reduce plastic ratio the upper water buckets were not covered. To make the experiment more convincing, the buckets were weighted before each trial. That was especially important because uncovered water buckets can easily lose weight through evaporation.

For the following trials I have filmed myself weighting the water buckets before each trial. That is for the viewer to rule out the possibility that less water was in the bucket than initially claimed. All the rest of the trials shall be listed here:

Water		Iron	
Rotation time	Date	Rotation time	Date
2: 25	23.03.2020; 1:25	2: 36	22.03.2020; 1:52
2: 20	23.03.2020; 2:47	1: 20	23.03.2020; 2:40
2: 06	23.03.2020; 3:51	1: 35	23.03.2020; 2:55
2: 51	25.03.2020; 2:31	3: 12	24.03.2020; 3:10
1: 49	25.03.2020; 2:51	1: 14	25.03.2020; 2:46
2: 48	25.03.2020; 3:07	2: 31	25.03.2020; 3:00
2: 51	27.03.2020; 3:56	1: 30	27.03.2020; 3:47
5: 05	29.03.2020; 0:34	4: 19	29.03.2020; 0:24
1: 52	29.03.2020; 1:47	1: 22	29.03.2020; 1:14
3: 33	29.03.2020; 1:53	5: 05	29.03.2020; 1:24
5: 49	29.03.2020; 2:11	2: 47	29.03.2020; 2:25
2: 56	30.03.2020; 2:42	2: 31	30.03.2020; 2:29
5: 51	30.03.2020; 3:07	2: 15	30.03.2020; 2:58
4: 06	30.03.2020; 4:38	2: 09	30.03.2020; 3:47
3: 43	30.03.2020; 6:05	2: 25	30.03.2020; 4:30

If we just take these trials, then we have 15 trials with 12 successes, and that would be a chance against odds of about 1.4%.

If we look at all trials, then we have 36 trials with 29 successes, which translates into a chance against odds of 0.01214747317%, or a certainty of 99.9878525268%.

The average rotation time for water trials was 3: 05, while that of iron was 2: 21. That is a difference of 44 seconds and thereby a little bit below out expectations.

One water-trial was disqualified because the torsion balance had rotated considerably away, before re-entering the probed distance, so that it took a long time, 4: 50 overall, to reach its destination, but rather short time, namely 2: 12, when counting from the moment it re-entered the probed distance (trial date: 24.03.2020; 2:40). It had gained a considerable swing from initially drifting off.

Notes and references:
1. A good demonstration is the YT-video "The Cavendish Experiment - Obvious Gravitational Attraction", by a person with the pseudonym MrLundScience.
2. Another good demonstration is "Cavendish Gravity Experiment Time Lapse", by Andrew Bennett.
3. They don't really realize that, because they are just trying to demonstrate the effect, not to measure gravity precisely. Versions that try to determine the gravitational constant are of course done in labs with very good equipment.

6.11 Measuring the mass gravity constant of water [03.04.2020 (round isolation glass with hole)*; 08.04.2020 (large angle chart)**; 10.04.2020 (first measurement); 14.05.2020 (insufficient air sealing is to blame, not mini-earth quakes); 16.05.2020 (successful measurement after improving air sealing); 16.05.2020 (placing buckets concentric reveals true distance is 90 cm); 23.05.2020 (final conformation of oscillation period through video analysis)]

*In collaboration with Panos Darmos (Greece).
**In collaboration with Sheila Darmos (Greece).

Experiments conducted in: Silver Leaf factory in outskirts of Skala, Greece.

According to the formula that has been introduced in chapter 6.5, the mass-gravity constant of water can be calculated as:

$$G_{H_2O} = G_p \times \frac{n_{pH_2O}\, m_p + n_{nH_2O}\, m_n}{m_{H_2O-e}}$$

With G_p being the mass-gravity constant for proton gases, with the value:

$$G_p = 6.6026471 \times 10^{-11}\ \mathrm{m^3\ kg^{-1}\ s^{-2}}$$

Using this value yields:

$$G_{H_2O} = G_p \times \frac{18.142072\ \mathrm{u}}{18.00916336\ \mathrm{u}}$$

$$G_{H_2O} = 6.6496762 \times 10^{-11}\ \mathrm{m^3\ kg^{-1}\ s^{-2}}$$

That is a difference to the mass-gravity constant of iron of 0.365650969%.

The mass-gravity constant of any material X can be found experimentally by using the oscillation time-gravity equation:

$$G_X = \frac{2\,\pi^2\,L\,\theta\,R_e^2}{t^2\,M_X}$$

Whereby L is the length of the torsion balance, R_e is the distance between the ends of the torsion balance and the two masses that are placed on the bottom, θ is the rotation angle, and T is the rotation time or rotation period.

The author tested this using a torsion balance suspended from a thread which was lodged onto a hole in a round glass plate placed on a large container. This was to seal the experiment from all air currents.

The torsion balance in this experiment was of the same type as the one used in the last one, but it was cut shorter to match the bucket distance which was limited by the size of the container. The distance has to be measured from the centers of the buckets. It was first thought to be 92 cm and the torsion balance was cut to precisely this length. However, later it turned out that either the buckets had mostly been placed leaning on the inclined walls of the container or the bottom of the container was bulged up during the measurement, and when placing them concentric on top of each other and on flat ground, then the true distance is 90 cm. The distance between the centers of the buckets can basically be treated as representing the length of the torsion balance, even if the torsion balance is a tiny bit longer.

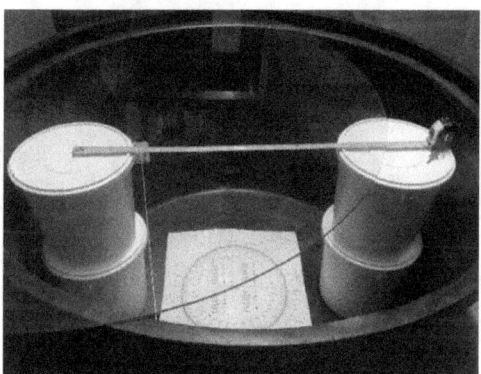

Fig. 100. Originally the bottom of the container was curved, and that made the distance between the buckets appear to be 92 cm, while in fact it was 90 cm.

Fig. 101. In order to avoid unnecessary vibrations, rails resting on chairs were used to place the camera.

The water buckets used this time were larger, each of them weighting $20\,\text{kg}$, amounting to a total of $M = 80\,\text{kg}$ (see fig. 99 & 100).

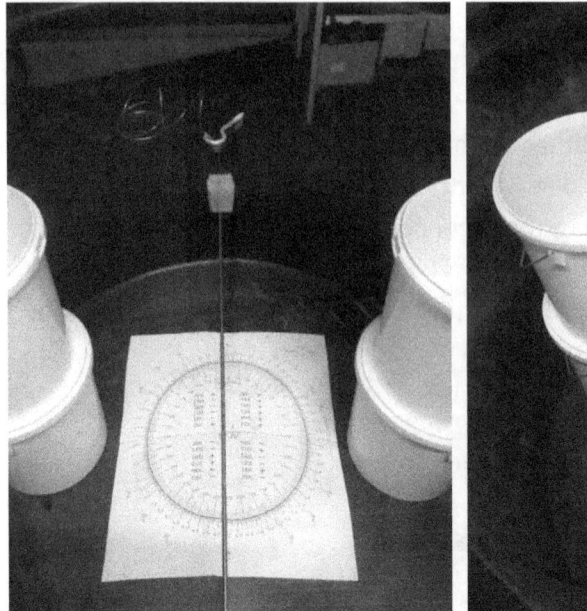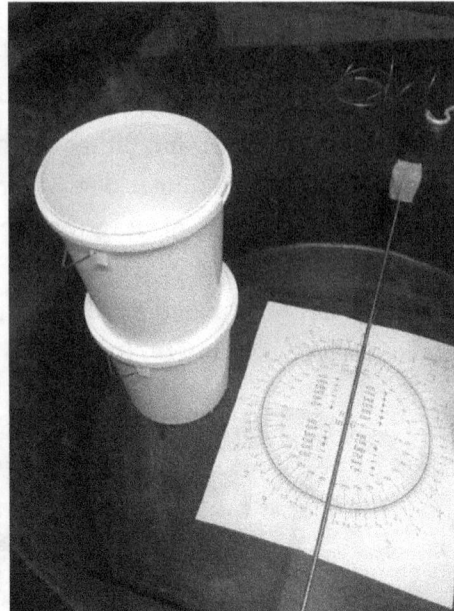

Fig. 102 & 103. Large water buckets were filled with water, in order to test the mass gravity constant of water. In this picture the top buckets were not covered. That is just show show that water is inside. In all trials it was covered, in order to prevent evaporation.

The oscillation angles measured were very small, and so although an angle chart was printed and placed beneath the torsion balance, it was not possible to read off the angles from that chart alone, and so a software had to be used to determine the oscillation angles.[1] In the early trials they were typically between $0.7°$ and $0.25°$.

The angle chart was however still useful for trigonometrically determining the distance between the ends of the torsion balance and the centers of the buckets.
If the angle between the end the ends of the torsion balance and the buckets is α, then the distance R_e between them is given by:

$$R_e = 2 \times \sin\frac{\alpha}{2} \times \frac{l}{2}$$

The results of the first series of measurements are shown in the table below:

G in m³ kg⁻¹ s⁻²	Angle	Period	Distance	Date and time
$5.8399054 \times 10^{-10}$	1°	28:00	64 cm	10.04.2020; 18:30
$2.0034563 \times 10^{-10}$	0.1°	15:00	64 cm	11.04.2020; 13:47*
$1.1463869 \times 10^{-10}$	0.16°	10:00	25 cm	13.04.2020; 22:32
$8.8042514 \times 10^{-10}$	0.5°	10:00	40 cm (61°)	14.04.2020; 22:15*

$2.4761957 \times 10^{-10}$	0.25°	11:00	37.5 cm (50°)	21.04.2020; 14:41
9.023723×10^{-11}	0.12°	14:00	37 cm	21.04.2020; 23:27
$2.2047096 \times 10^{-10}$	0.16°	13:00	46 cm	26.04.2020; 20:04
$3.1119666 \times 10^{-10}$	0.31°	15:00	45.3 cm (59°)	27.04.2020; 21:44
$1.3390166 \times 10^{-10}$	0.16°	15:00	46 cm	29.04.2020;
$3.8812075 \times 10^{-10}$	0.24°	12:00	46 cm	29.04.2020; 22:25
$1.0597848 \times 10^{-10}$	0.07°	12:00	45 cm	30.04.2020; 22:32
$1.2253629 \times 10^{-10}$	0.13°	15:00	44 cm (57°)	01.05.2020;

*The video record of this trial does not exist anymore.

Due to factors like people walking, mini-earth quakes and rain outside, it was not possible to fully eliminate background oscillations.

The experiment was also repeated with two water buckets only, which is a total mass of 40 kg:

G in $m^3\ kg^{-1}\ s^{-2}$	Angle	Period	Distance	Date and time
$6.7573366 \times 10^{-10}$	0.5°	18:00	44.6 cm (58°)	08.05.2020; 14:47

The silk thread was then again replaced by a fishing wire and four buckets were used again. Vibration oscillation was not easy to get quickly, but one trial seemed promising even without it:

G in $m^3\ kg^{-1}\ s^{-2}$	Angle	Period	Distance	Date and time
$4.8422571 \times 10^{-11}$	0.48°	64:00	61.9 cm (87°)	09.05.2020; 22:22

However, for this trial only the to-oscillation was captured, without much of the back-oscillation. Also there were short interruptions in that rotation. That may be due to unwanted vibrations.

A few more trials with the usual setup were conducted.

G in $m^3\ kg^{-1}\ s^{-2}$	Angle	Period	Distance	Date and time
$4.1618589 \times 10^{-10}$	0.3°	17:00	61 cm (82°)	10.05.2020; 14:30
$1.2548309 \times 10^{-10}$	0.1°	18:00	61 cm (82.7°)	10.05.2020; 17:11
$1.8177106 \times 10^{-10}$	0.1°	15:00	61 cm (83°)	10.05.2020; 18:20
$1.1328035 \times 10^{-10}$	0.2°	26:00	61 cm (83°)	10.05.2020; 14:47

Then the experiment had to temporarily be moved outside, where it became apparent that the air isolation was insufficient.

Fig. 104. After moving the experiment outside, near a large pool, it became apparent that air isolation was insufficient and that this must have interfered with the experiment already inside the lab-factory.

After moving back indoor this was accounted for by folding the top of the tape over to the surface, and by adding an extra layer of stronger plastic tape. The improved setup can be seen in fig. 105 - 106:

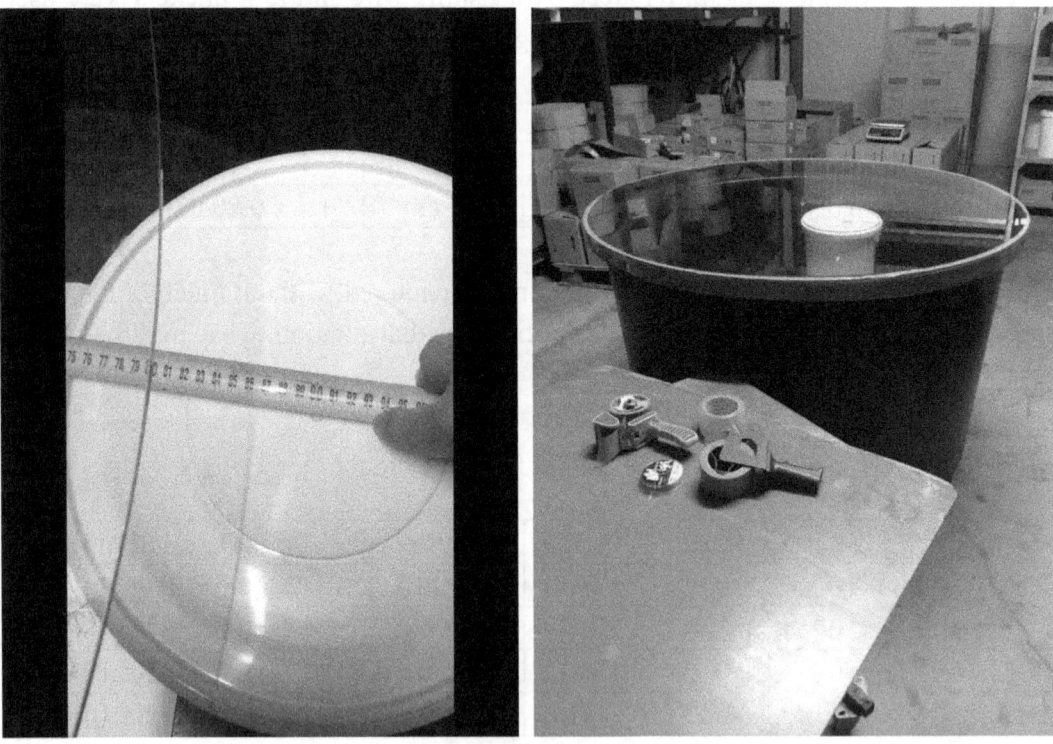

Fig. 105 & 106. The distance was measured more accurately, and the air sealing was improved.

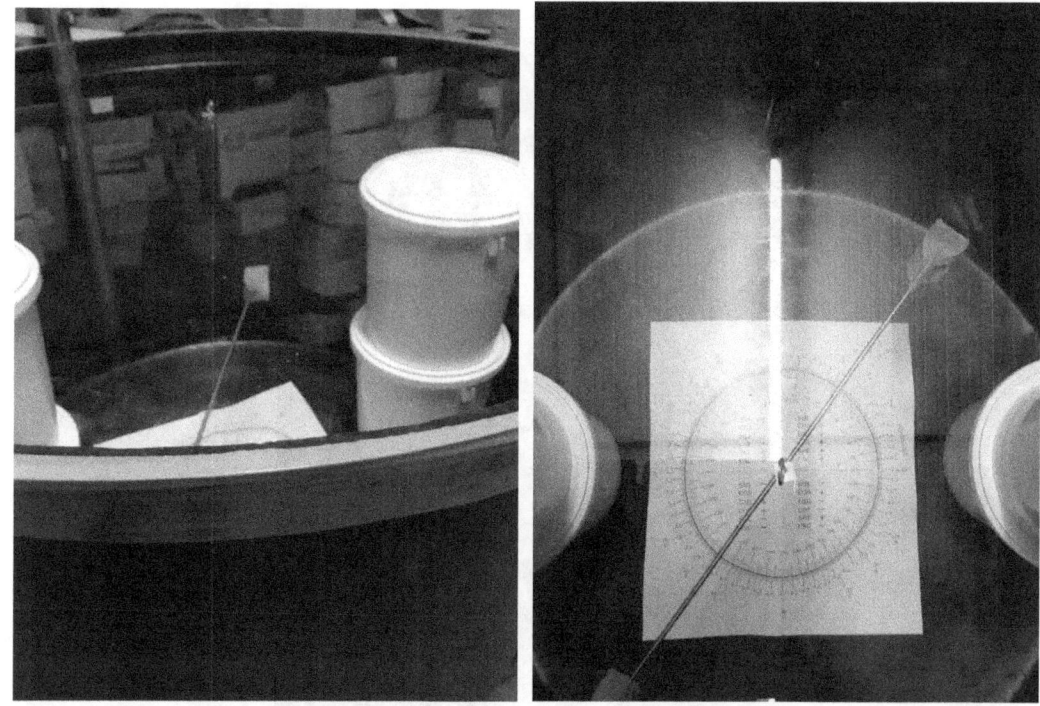

Fig. 107 & 108. The setup that proved successful. Triple tape isolation, a plastic sheet between the lodging ring and the glass plate, and glue at the hole in the glass. With this setup accurate measurements were already obtained with the first trial.

After that was done, the torsion balance finally reached its true equilibrium state where it allows us to truly measure gravity.

It turned out that the obtained angle was fairly smaller than 0.8°, which made it impossible to measure with the computer program used for all the above trials.

It was then chosen to measure the angle using two 250 cm long iron corner covers.

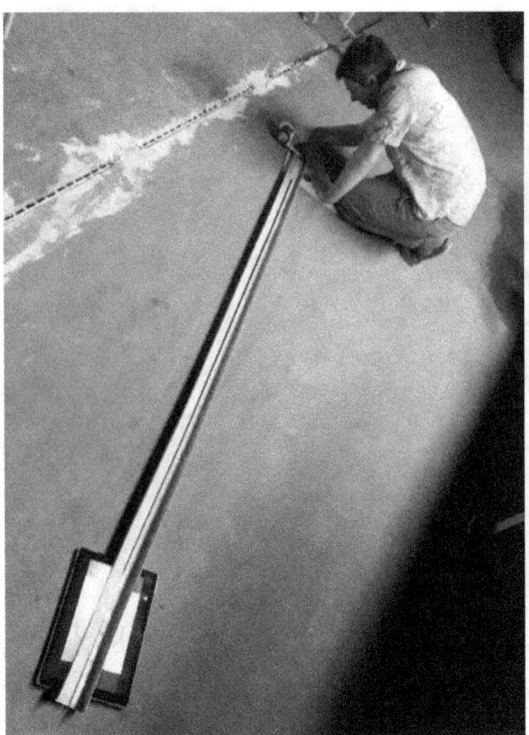

Fig. 109. Although the oscillation was easy to see with the bare eye, it was too small to be measure with a computer program that was at hand. It was then chosen to measure the angle using two 250 cm long metal corner protectors to measure the corner using trigonometry.

It was not so easy to precisely measure these angle, and therefore it was done many times, so as to obtain the most accurate value. The different measurements of the length of the opposite side of the angle were as follows:

$$6.5 \text{ mm} - 2.5 \text{ mm} / 2 \ = \ 2 \text{ mm}$$
$$7 \text{ mm} - 3 \text{ mm} / 2 \ = \ 2 \text{ mm}$$
$$7 \text{ mm} - 3 \text{ mm} / 2 \ = \ 2 \text{ mm}$$
$$7.5 \text{ mm} - 2.8 \text{ mm} / 2 \ = \ 2.35 \text{ mm}$$
$$5.5 \text{ mm} - 2.5 \text{ mm} / 2 \ = \ 1.5 \text{ mm}$$
$$8.5 \text{ mm} - 3 \text{ mm} / 2 \ = \ 2.75 \text{ mm}$$
$$6 \text{ mm} - 2.8 \text{ mm} / 2 \ = \ 1.6 \text{ mm}$$
$$9 \text{ mm} - 3 \text{ mm} / 2 \ = \ 3 \text{ mm}$$
$$6 \text{ mm} - 2.9 \text{ mm} / 2 \ = \ 1.55 \text{ mm}$$
$$5 \text{ mm} - 3 \text{ mm} / 2 \ = \ 1 \text{ mm}$$

The average of all these measurements is 1.975 mm, which is fairly close to the only result that was obtained more than once, namely 2 mm.

The oscillation time was 16 minutes, and using that value yields:

$$\alpha = \sin^{-1}\left(\frac{0.2}{237}\right) = 0.0484°$$

$$G_{H_2O} = 6.5230413 \times 10^{-11} \, \text{m}^3 \, \text{kg}^{-1} \, \text{s}^{-2}$$

That deviates significantly from both the mainstream value and the SPD-specific value. The reason for that is of course that our value for the oscillation time was not a high precision value.

Our expected value is:

$$G_{H_2O} = 6.6496762 \times 10^{-11} \, \text{m}^3 \, \text{kg}^{-1} \, \text{s}^{-2}$$

And plugging that into our oscillation time gravity equation yields:

$$t = \sqrt{\frac{2\,\pi^2\,L\,\theta\,R_e^2}{M_{H_2O}\,G_{H_2O}}}$$

$$t = 951 \, \text{sec} \; (950.815048128)$$

Which is 15 minutes and 51 seconds.

If Newton and Einstein are right, then we get:

$$t = \sqrt{\frac{2\,\pi^2\,L\,\theta\,R_e^2}{M\,G}}$$

$$t = 949 \, \text{sec}$$

Which is 15 minutes and 49 seconds.

This difference seems very small and too hard to measure, but we can utilize the fact that small differences in oscillation time amount to large asynchronicities over longer periods of time. The video recording was a little longer than 30 minutes and it appeared to have started right at the end of one oscillation (!). A second by second video analysis confirmed that. That was quite extraordinary. The chance for that to occur on a particular trial is only 0.2% (1 divided by the half of the oscillation period).

We can make the following expectation chart and compare it with the video footage:

Bottom position	SPD	GR
Right	$0:00$	$0:00$
Left	$7:55$	$7:55$
Right	$15:51$	$15:49$
Left	$23:46$	$23:44$
Right	$31:42$	$31:38$

We don't know for sure if the oscillation had really started at exactly $t = 0:00$, but what we can say for sure is that if SPD is right, it should not exceed $15:51$ until it returns back into its original state. A larger oscillation time would only be explainable with a larger angle, and the angle used was already the largest that could reasonably be chosen, as the average of all measurements was even slightly below it.

A second by second video analysis revealed that the torsion balance was moving clockwise ever since the first second (see fig. 110 - 112).

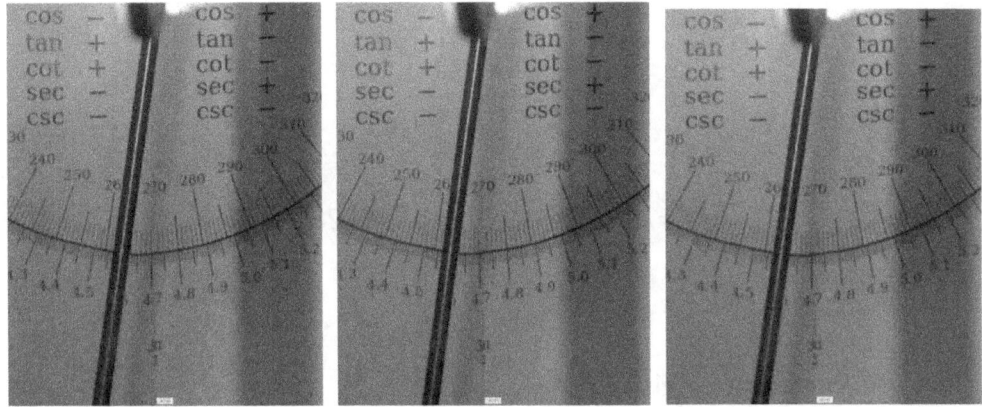

Fig. 110, 111 & 112. It appears that the first clockwise oscillation began precisely at time $0:00$.

Another second by second video analysis revealed that indeed it had reached the clockwise maximum at precisely $7:55$ (see fig. 113 - 115).

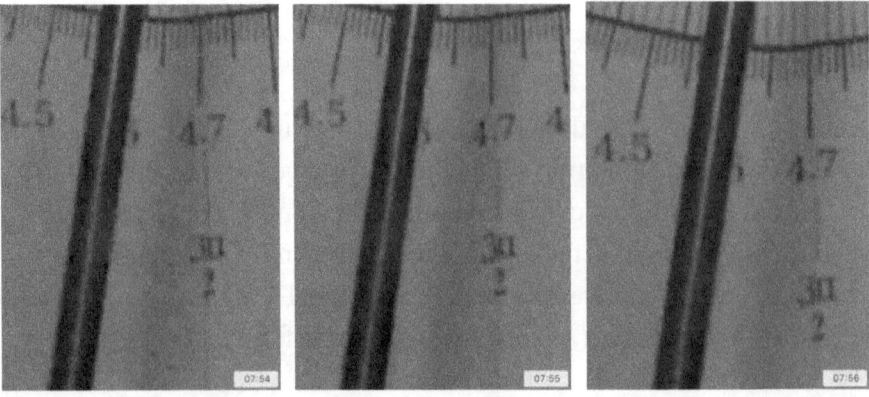

Fig. 113, 114 & 115. The torsion balance reached its clockwise maximum at precisely $7:55$.

A careful analysis of the back oscillation revealed that the back oscillation stopped at exactly 15:51. The equalibrium point was between second 51 and 52, being clearly separate from second 49, which is the turning time we would expect from Newtonian and Einsteinian physics.

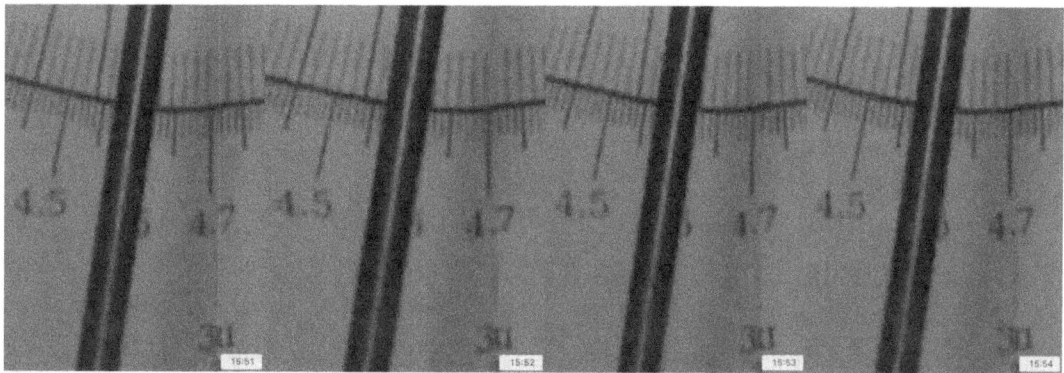

Fig. 116, 117, 118 & 119. The torsion balance moved into the anti-clockwise direction until 15:51, was standing still until 15:52, and was then slowly starting to move clockwise again.

The subsequent clockwise oscillation also perfectly fits the prediction:

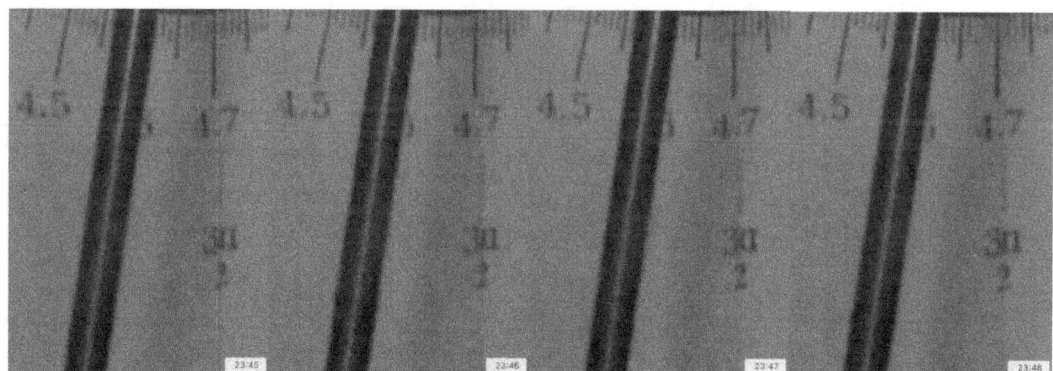

Fig. 120, 121, 122 & 123. Again we have two seconds where the torsion balance stands still before turning direction. That is 23:46 and 23:47.

The end of the second oscillation was not fully captured on tape, because the recording ended shortly before it was completed.

We can however already see that the oscillations perfectly fit the predictions of space particle dualism and that they violate Newtonian physics. According to Newtonian and Einsteinian physics the clockwise maximum should have been reached at 23:43 – 23:44 and not 23:46 – 23:47.

Of course the Bernoulli trial version of this experiment which we were looking at in the last chapter was visually more impressive. The difference was more clearly visible, but the experiment reported on in this chapter represents a measurement of an absolute value. It is experimentally somewhat more rigorous.

The first trials with insufficient air sealing were interesting, because they showed that similar to the Bernoulli trial version of the experiment, air interference basically scales up the effects from gravity and maintains the difference between iron and water.

With other words: air interference makes gravity seem stronger than it is.

Note:
1. The software used for measuring the angles in the early trials was "MB ruler by Marcus Bader 2003 – 2017". However, the software fails at angles smaller than 0.1° or 0.08°.

6.12 Blocking and amplifying gravity [02.03.2020 (displaced lattices); 27.07.2020 (using a quark-gluon plasma for artificial gravity)*; 01.08.2020 (blocking gravity doesn't violate energy conservation; neutron star); 14.08.2020 (blocking subgravity; gravity plates)]

*Requested by Harold Alger (US military advisor).
*In collaboration with David Chester (UCLA; MIT; US).

From science fiction we are familiar with anti-gravity devices, as well as artificial gravity in spaceships. From the perspective of general relativity the only way to generate anti-gravity would be to have exotic matter with negative mass. In space particle dualism exotic matter exists but it has positive gravity, just like ordinary matter. Aside from that, it can be found only in supermassive black holes, which were formed out of it. It is therefore not a feasible solution for anti-gravity.

In 1983 Eric Laithwaite demonstrated for the first time an effect which was by many mistaken as anti-gravity.[1] He showed that a wheel attached to a bar appears to be lighter when it is spinning. This was later shown to be an illusion. What actually happens is that the spinning cancels the force that is needed to keep the bar in an horizontal orientation. If one was to lift up the wheel directly at its center of mass, one wouldn't notice any difference between the spinning and the non-spinning wheel.[2]

What about blocking gravity? In naively quantized general relativity, gravity is based on gravitons. Same as virtual photons, virtual gravitons wouldn't have a preferred wavelength, if they existed. Without having a particular prevalent wavelength, there cannot be anything that could possibly shield off any significant number of them.

In space particle dualism things are quite different. Here gravity is not based on gravitons, but gluons instead. Gluons do have a preferred frequency. Due to the fact that they travel by means of self-interaction, a virtual gluon does not need to have a very low energy in order to travel large distances (the lifetime of a virtual particle is inversely proportional to its energy). It simply gives rise to another gluon that continues the travel. Gluons are emitted by quarks, and therefore their preferred wavelength is the diameter of a quark, which is 4.3×10^{-15} m.

Such a tiny particle could only be blocked by a material which is so dense that every straight trajectory through it ends at a nucleus. That is the case at a neutron star, which consists of tightly packed neutrons only.

If gluons cannot traverse a neutron star, that would mean that a neutron star can block the gravity of any object behind it. Of course, neutron stars do themselves have gravity that is strong enough to make light emitted from the opposite side being bent over to be visible on the front side.

Yet, if gluons cannot traverse a neutron star, then how could the gluons of the neutron star itself escape from inside it to make the neutron star's gravity be felt on the outside?

Virtual gluons escaping a neutron star is pretty similar to real photons finding their way out of the center of the sun. They randomly fly from atom to atom for very long time, until they finally reach the surface of the sun. That is because the sun's temperature is within the pre-decoupling range, which means that photons cannot travel freely, similar to the photons of the pre-decoupling universe.

Virtual gluons that enter the neutron star on one side, might get trapped slightly below the surface for short amounts of time, but they would exit the neutron star on the same side as the one they entered, because if their collisions with neutrons on the inside are truly random, then they will first reach the part of the surface which is closest to their starting point.

Gluons that were emitted on the inside would reach the region of the surface which is closest to them. So while neutron stars sort of reflect incoming gravity, they same doesn't happen with gravity that comes from the inside.

The rate at which objects emit gluons is not altered hereby. Gluons get stuck inside for a while, but they eventually make it to the outside, and that happens in every moment in time. No irregularities are introduced. The only difference to ordinary physics is that the neutron star blocks gravity from other objects behind it.

One may wonder how the gravity from the opposite side of the neutron star can make itself felt on the one side. If it the gravity from the opposite side of a neutron star cannot reach an observer on the one side, does that mean that neutron stars have only ½ the gravity they are supposed to have in ordinary physics?
The answer is that this is not the case, because gluons that are emitted on the side facing the observer also only reach that side, and never the opposite side. So the number of gluons emitted on each side does not change, it is just that they don't cross sides

anymore.

Does it violate energy conservation to being able to block gravity? Certainly part of the potential energy of in falling or suspended objects gets lost thereby. Does that mean we have a case of a violation of the first law of thermodynamics here?

On first sight it appears that way, but it is not really the case. Blocking gluons does not reduce the total number of gluons. It just means causing an asymmetry in their spatial distribution. It is as though part of the gravity of one object is given to another object. We could call this phenomenon 'gravity reflection'.

If blocking of gravity is not to violate energy conservation, then every such blocking must in fact represent an example of 'gravity reflection'.

We cannot create materials here on earth that have a density equal to that of a neutron star.
An alternative could be to use multiple layers of a material with enhanced lattice structure. Every layer would have to be displaced from the layer below by the diameter of one quark (4.3×10^{-15} m). The displacement would have to be performed in two dimensions.[3]

Distances between atoms in a lattice structure are measured in picometers. Iron lattices are symmetrical, which means that all three lattice constants for iron are identical and equal to 286.65 pm, which is 2.8665×10^{-10} m. Precisely 66,663 quark lengths fit into that. So we need 66,663 layers of iron lattice for one direction, and another 66,663 layers for the other direction. That is a total of 133,326 layers. If every layer is, say, 100 atoms thick, then we would have a total structure with a hight of at least 0.0038 m (3.8 mm), without accounting for the distance between the different layers. In order to account for gluons that get through by the means of scattering, we can stag several of these structures on top of each other.

It is unlikely that a significant number of gluons can be blocked in this way. While the wavelength distribution of nuclear gluons has its peak at the diameter of a quark, there are other wavelengths too. If this method of blocking virtual gluons works, then it should be able to block about 50% of the gluons, resulting in a significant weight reduction of objects situated above this 'gluon blocker'.

There is however no guarantee that this will really work. After all gluons are self-interacting and their behavior is highly complex. One can try this out to see if it leads to any measurable weight reduction or not. Even a slight weight reduction would be phenomenal.

One approach of creating these different layers of lattice could be to take a full lattice block and cut it into many layers. However, in this approach it is hard to see how to avoid destroying the alignment between the different layers of lattice.

Another approach could be to take one molecular lattice and deform it without breaking it, to precisely the degree which is necessary to make every straight pathway end at a nuclei.

Before going to all the trouble of creating such a gluon blocker, we may want to examine how many gluons ordinary dense materials scatters. The most dense solid we can find here on earth is Osmium. It has a density of $22,590 \text{ kg/m}^3$. It has 76 protons and 114 neutrons. Knowing this we can calculate the average weight pro nuclei $m_{\mu p \& n}$ as $1.6740068 \times 10^{-27}$ kg. The quark density ρ_q is then:

$$\rho_q = \frac{3 \, \rho_{Os}}{m_{\mu p \& n}}$$

$$\rho_q = 4.0483707 \times 10^{31}$$

However, the quark diameter is very small. At about 4.3×10^{-19} m. For being scattered a gluon does not need to get into the vancinity of a single quark, but only into that of a nuclei. We may therefore take the diameter of protons and neutrons, namely 1.7×10^{-15} m.

One cubic meter of space consists of 2.0354162×10^{44} nuclei volumes (the inverse of the cube of 1.7×10^{-15} m).

That means in every moment, the chance for a gluon to be absorbed by a quark is given by:

$$q_{scatt} = \rho_q \, r_{p\&n}^3$$

$$q_{scatt} = 1.9889645 \times 10^{-13}$$

Gluons travel at the speed of light. In one second a gluon traverses 1.763485×10^{23} nuclei-diameters. Using the above scattering probability, that should result in $35,075,091,549$ scattering events per second.
If our osmium crystal has a depth of 1 meter, then a gluon transversing it will be scattered 117 times. Half of the time such a scattering event will not just divert a gluon from its original course, but totally reverse it.

The chance for a gluon to get through there without scattering at all is given by:

$$q_{no\ scatt} = (1 - q_{scatt})^{\frac{117}{q_{scatt}}}$$

$$q_{no\ scatt} = 1.5912042 \times 10^{-51}$$

Which would be 6.2845485×10^{50} trials. Easier it is to have it get through by only being slightly diverted at each scattering event. The chance for this is:

$$q_{no\ rev} = 0.5^{117} = 6.0185311 \times 10^{-36}$$

Which would be 1.661535×10^{35} trials and would take $29{,}876.6$ years ($942{,}188{,}334{,}310$ sec). It can therefore be regarded a certainty that gluons never make it out of the surface of the earth by flying in a straight path.

That means that gluons are always blocked, and the degree to which they are blocked can vary hugely according to the material. We don't observe these differences in blocking to influence the way gravity is felt. It is therefore conceivable that they only influence the speed of gravity, and not its strength.

If this is true, then we can only either block gravity completely or not at all. There is nothing in between. The reason is that if it is possible to get through, then every gluon will eventually get through, and since there is a constant outward flow of gluons from inside the earth, the rate at which gluons make it through is also constant. So if all gluons make it out, then getting stuck for a while only influences the speed of gravity, but not its strength.

The difference in the speed of gravity can be calculated according to:

$$\Delta v_G = \frac{\rho_A\ m_{\mu p \& n\_B}}{\rho_B\ m_{\mu p \& n\_A}} \approx \frac{\rho_A}{\rho_B}$$

With ρ_A and ρ_B being the densities of two materials A and B.

If substance A is air ($1.225\ \mathrm{kg/m^3}$), and substance B is Osmium, then that could mean that using Osmium one could reduce the speed of gravity to only 0.005% of what it was.
This is very difficult to test. One would have to quickly remove a source of gravity and then measure how quickly the change in gravity is felt by a test object.

So far the only viable solution of how to block gravity is using a lattice structure as described above.

Blocking subgravity

With this being said, there is a type of gravity that should be much easier to block, and that is the gravity that arises from electromagnetism. In previous chapters we have sometimes referred to this type of gravity as 'subgravity'. It is based on virtual photons and has a reach of only 40 meters.

We can block virtual photons quite easily using a Faraday cage or the electrically conducting plates of a Casmir experiment.

What weight reduction can we achieve by blocking the subgravity from a 40 meters radius under the ground? The density of the earth crust is 2,800 kg/m³. A half sphere with this radius and density weights: 375,315,602.349 kg.

Earth's crust is made up of 46.6% oxigen; 27.7% silicon, 8.1% aluminum; 5% iron; 3.6% calcium; 2.8% sodium; 2.6% potassium; and 2.1% magnesium.
Now we have to calculate the number of atoms and charges that every element contributes:

1. Oxygen, 46.6%:
174897070.695 kg / $2.6566962 \times 10^{-26}$ kg = 6.5832544×10^{33} oxigen atoms.

The cardinal charge of an oxygen atom can be calculated by taking the absolute values of the individual quark charges and then adding the electron charges to them:

$$|Q|_O = \left[8 \times \left(2\left|+\frac{2}{3}\right| + \left|-\frac{1}{3}\right|\right)\right] + \left[8 \times \left(\left|+\frac{2}{3}\right| + 2\left|-\frac{1}{3}\right|\right)\right] + |-8| = 32$$

That yields 2.1066414×10^{35} cardinal charges from oxigen.

2. Silicon, 27.7%:
103,962,421.851 kg / $4.6637066 \times 10^{-26}$ kg = 2.2291802×10^{33} silicon atoms.

$$|Q|_{Si} = \left[14 \times \left(2\left|+\frac{2}{3}\right| + \left|-\frac{1}{3}\right|\right)\right] + \left[14 \times \left(\left|+\frac{2}{3}\right| + 2\left|-\frac{1}{3}\right|\right)\right] + |-14|$$
$$= 56$$

That yields 1.2483409×10^{35} cardinal charges from silicon.

3. Aluminum, 8.1%:
30,400,563.7903 kg / $4.4803831 \times 10^{-26}$ kg = 6.7852599×10^{32} aluminium atoms.

$$|\mathbb{Q}|_{Si} = \left[13 \times \left(2\left|+\frac{2}{3}\right| + \left|-\frac{1}{3}\right|\right)\right] + \left[14 \times \left(\left|+\frac{2}{3}\right| + 2\left|-\frac{1}{3}\right|\right)\right] + |-13|$$
$$= 53.\overline{3}$$

That yields 3.6188053×10^{34} cardinal charges from silicon.[2]

4. Iron, 5%:

18,765,780.1174 kg/9.2732796 $\times 10^{-26}$ kg $= 2.0236401 \times 10^{32}$ iron atoms.

$$|\mathbb{Q}|_{Fe} = 109.\overline{3}$$

That yields 2.2125132×10^{34} cardinal charges from iron.

5. Calcium, 3.6%:

13,511,361.6846 kg/6.6551079 $\times 10^{-26}$ kg $= 2.0302243 \times 10^{32}$ calcium atoms.

$$|\mathbb{Q}|_{Ca} = 80$$

That yields 1.6241794×10^{34} cardinal charges from calcium.

6. Sodium, 2.8%:

10,508,836.8658 kg/3.8175458 $\times 10^{-26}$ kg $= 2.752773 \times 10^{32}$ sodium atoms.

$$|\mathbb{Q}|_{Na} = 45.\overline{3}$$

That yields 1.2479238×10^{34} cardinal charges from sodium.

7. Potassium, 2.6%:

9,758,205.66107 kg/6,4924249 $\times 10^{-26}$ kg $= 1.503014 \times 10^{32}$ potassium atoms.

$$|\mathbb{Q}|_{K} = 77.\overline{3}$$

That yields 1.1623308×10^{34} cardinal charges from potassium.

8. Magnesium, 2.1%:

7,881,627.64933/4.0359398 $\times 10^{-26}$ $= 1.9528606 \times 10^{32}$ magnesium atoms.

$$|\mathbb{Q}|_{Mg} = 48$$

That yields 9.3737307×10^{33} cardinal charges from magnesium.

The total cardinal charge is 4.4352949×10^{35}.

Our test object shall be a 1 kg lump of iron. That contains 1.0783671×10^{25} iron atoms, which each have a cardinal charge of $109.\bar{3}$. That makes a total cardinal charge of 1.1790147×10^{27}.

The distance Δd_C from the center of mass of a hemisphere is given by:

$$\Delta d_C = \frac{3\,R}{8}$$

Using this radius, yields:

$$F_{G_{|\mathbb{Q}|}} = G_{|\mathbb{Q}|} \times \frac{|\mathbb{Q}|_1 \, |\mathbb{Q}|_2}{r^2}$$

$$F_{G_{|\mathbb{Q}|}} = 0.00000313203 \text{ N}$$

That corresponds to a weight of $3.19270082 \times 10^{-7}$ kg, which is weight reduction of 0.000031927%.

We can test this by covering an area with a radius of 40 meters with electrically conducting plates and then weighting something in the middle of this plate and then weighting it again without the plate beneath.

Weighting something in and outside of a faraday cage respectively might do as well.

The experiment will be conducted in due time, when the necessary equipment is acquired.

Artificial gravity

Artificially creating gravity is even more difficult than blocking it. A place where it is much needed is on spaceships. On one hand, zero gravity is good for storing things everywhere, but on the other hand, one would want to be able to stand and walk on a spaceship as well, at least in some sections of it. Using rotating parts is one solution, but that comes with its own hurdles, as centrifugal forces can imitate gravity only in a very imperfect way. One would for example fly up when running into the opposite direction with the right speed. Dizziness would also be difficult to avoid.

Gluons are self-interacting, and therefore gravity is self-interacting. Enhancing these self-interactions would enhance gravity. That could be achieved by creating a quark-gluon plasma right under the area of space where one wants to have artificial gravity.

In such a plasma quarks and gluons are very close together, which leads to the number of interactions being dramatically increased.

David Chester (personal correspondence; 2020) has pointed out that a quark gluon plasma is an ultra high energy state, and therefore running of coupling should lead to some reduction in the strength of the strong force here. However, he has also admitted that self-interactions are enhanced in such a state of matter.

Usually such a quark-gluon plasma exists only in small regions of space for short amounts of time. The energy it would take to generate such a plasma over an extended region of space over extended amounts of time would be enormous.

It would be impractical and unfeasible to try to create an extended version of such a plasma by lining up many particle accelerators.

Quark-gluon plasmas form at the so called Hagedorn temperature, which is 2×10^{12} K. For comparison, the famous 'artificial sun' in China has a temperature of 10^8 K.

The temperature of a newly created neutron star is at about $10^{11} - 10^{12}$ K.[3] Trying to create such a hot and dense state artificially is not an easy task. It is unclear if it is even possible to reach this temperature without actually creating a neutron star.

However, if one really has to create something as dense as a neutron star in order to have a quark-gluon plasma, then one already does not need to rely on the self-interaction of gluons to enhance gravity. The sheer mass-density required would already generate enough gravity.

Could a very advanced civilization manufacture 'gravity plates' simply by cutting out parts of crystalized black dwarf stars?
Black dwarfs evolve from white dwarfs through cooling. Even in white dwarfs it is expected that at late stages of its cooling process, large portions of it must exist in crystalized form, instead of the initial plasma. A white dwarf by the name *BPM 37093* was estimated to have crystalized up to 90%. Other estimates gave 32 – 82%.[4, 5, 6, 7] *BPM 37093* is approximately 50 light years away from earth, in the constellation Centaurus.

Unfortunately, if one wanted to extract a small piece from a black dwarf, the gravity induced pressure that made the extremely dense structure of the black dwarf possible would be lost, and the small piece would immediately blow up.

It also doesn't help that black dwarfs are cold, because both black and white dwarfs prevent the further collapse to a black hole primarily through the degeneracy pressure that comes from their electrons, and not through thermal energy, which doesn't contribute significantly. One can see this from the equation for the mass limit for white dwarfs, which is:[8]

$$M_{Ch} = N^2 \left(\frac{\hbar c}{G}\right)^{\frac{3}{2}} = 1.4\, M_\odot$$

Whereby N is the number of electrons per kilogram. This equation is derived by setting equal the enery arising from Heisenberg uncertainty equation and the absolute value of the potential gravitational energy. It does not at all involve the temperature of the star.

The reader maybe hoping that if the piece cut out of a black dwarf is large enough to have sufficient gravity for people to walk above it, then it would also have enough gravity to keep itself from exploding.

Indeed a large enough piece of a black dwarf would, when isolated not explode, yet it would still expand, because unlike neutron stars, which all have very similar densities, white dwarfs are still made of atoms, and they do therefore still have some empty space. The density of a white dwarf therefore entirely depends on its mass.

In fig. 124 we can see the correlation between mass and radius.

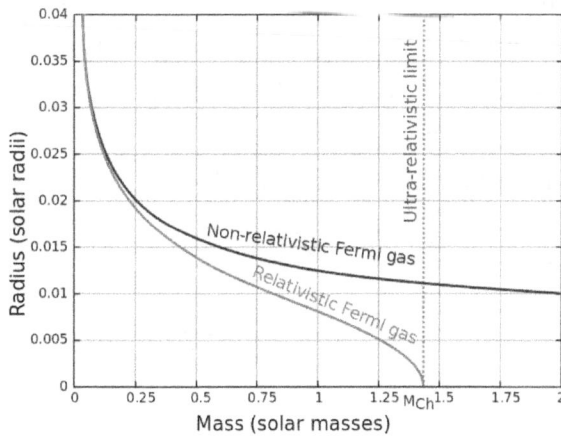

Fig. 124. If we were to cut out a small piece of a black dwarf, it would quickly expand to about the radius of the sun.

From this graph we can see that it is impossible to have a high mass density without having a very large mass as well. Through this, and our considerations about self-interaction of gluons, it should be clear that artificial gravity is not a possibility.

Unless of course there are other ways to enhance the self-interaction of gluons that do not require creating a quark-gluon plasma.

Notes and references:

1. "Eric Laithwaite defies Newton", New Scientist, 14 November 1974, p470.

2. A cardinality that is not a natural number is somewhat troubling. One cannot have an uneven number of elements in a set. Maybe this shows the limitations of the cardinality concept. After all we use this concept only as a way of adding terms together regardless of their sign. We could just as well call it 'sum of charge absolute values', or 'sum of absolutes'.

3. Lattimer, James M. (2015). "Introduction to neutron stars". American Institute of Physics Conference Series. AIP Conference Proceedings. 1645 (1): 61–78. Bibcode:2015AIPC.1645...61L. doi:10.1063/1.4909560. Retrieved 2007-11-11.

4. Metcalfe, T. S.; Montgomery, M. H.; Kanaan, A. (2004). "Testing White Dwarf Crystallization Theory with Asteroseismology of the Massive Pulsating DA Star BPM 37093". The Astrophysical Journal. 605 (2): L133. arXiv:astro-ph/0402046. Bibcode:2004ApJ...605L.133M. doi:10.1086/420884.

5. Whitehouse, David (16 February 2004). "Diamond star thrills astronomers". BBC News. Archived from the original on 5 February 2007. Retrieved 6 January 2007.

6. Kanaan, A.; Nitta, A.; Winget, D. E.; Kepler, S. O.; Montgomery, M. H.; Metcalfe, T. S.; Oliveira, H.; Fraga, L.; et al. (2005). "Whole Earth Telescope observations of BPM 37093: A seismological test of crystallization theory in white dwarfs". Astronomy and Astrophysics. 432 (1): 219–224. arXiv:astro-ph/0411199. Bibcode:2005A&A...432..219K. doi:10.1051/0004-6361:20041125.

7. Brassard, P.; Fontaine, G. (2005). "Asteroseismology of the Crystallized ZZ Ceti Star BPM 37093: A Different View". The Astrophysical Journal. 622 (1): 572–576. Bibcode:2005ApJ...622..572B. doi:10.1086/428116.

8. "Estimating Stellar Parameters from Energy Equipartition". ScienceBits. Archived from the original on 30 June 2012. Retrieved 9 May 2007.

6.13 Local homogeneity in granular space and how it could help to defy gravity

[08.08.2004 (local flatness for the earth's world line); 06.04.2007 (assumed that SPD has no analogue to local flatness); 22.04.2019 (zigzag abundance in different directions; reducing gravity by sidewise movement?); 12.07.2019 (does the shape influence gravity?); 10.09.2020 (wavefunction-mediated gravity hypothesis); 17.10.2020 (using long narrow gap to flatten wavefunction)*; 18.10.2020 (local homogeneity in SPD)*; 20.10.2020 (SPD has geodesics while GR doesn't)*; 25.10.2020 (more precise local homogeneity; expressing radius of earth by angle)*; 27.10.2020 (gravity doesn't fall off at infinity)*; 02.11.2020 (dropping of atoms does not rely on wavefunction-mediated gravity hypothesis)*; 17.11.2020 (neutron star argument invalid)*]

*Sponsored by: Lin Yi Song (China).

In Einstein's general relativity, gravity is the result of a curved spacetime. We tend to imagine that this means every object must somehow 'feel' the curvature of space, or 'spacetime', in order to react to gravity.

In this picture it is truly hard to reject the spacetime paradigm and think of this curvature as something that is entirely in space. If the geodesics in general relativity were geodesics in space, then we would have to think that the orbit of the moon is a straightest possible path in a space that is curved in a closed loop around earth. That is not the case. In fact, in general relativity geodesics are always geodesics in spacetime.

If the curvature was in space alone, then we could ask ourselves at which scale is an object too small to 'feel' the curvature of space. That would be when the difference to a flat space is less than a Planck-length. In this case we can speak of a true 'local flatness'. Indeed this was calculated once in 2004, for the orbit of the earth around the sun, which resulted in a local flatness of 10^{-13} m.[1] It was done by looking at the world line of the earth around the sun in spacetime, not in space.

If this was in space alone, then it would make it hard for elementary particles to feel gravity, as they are smaller than that. And this was the flaw in the calculation: looking at local flatness in spacetime is in vain, because there all objects extend to infinity in time.

Thus, in spacetime, the issue of local flatness never arises. Any object is its whole 'world line', which extends infinitely into both past and future, and so local flatness is no longer defined. That is likely the reason why Einstein settled for the interpretation of his theory that regarded time as a dimension, as proposed by his former teacher Minkowski in 1908.

In space particle dualism theory however, time is not a dimension, and so the analogue for curvature, which is inhomogeneities in granular space, is something that is entirely in space.

As already stressed in previous chapters, this of course does not mean that the flow of time is not influenced by the density of granular space, in the same way it was influenced by the curvature of space in general relativity. The close relationship between space and time are maintained in space particle dualism, but time is no longer viewed as just another dimension of space.

It is very easy to imagine how a comet or a spaceship is pulled towards a celestial body in space particle dualism. The particles of the spaceship simply follow the straightest path in granular space. They can do so even if at the scale of the spaceship, the quantum vacuum seems homogeneous.

It is easy to imagine the same in a curved space: we can imagine how a spaceship would follow the curvature of space, even when space appears flat at the scale of the spaceship. To use an analogy: a ship can go around the earth, without ever noticing any curvature at its own scale.

How about an object that is smaller than the local flatness or local homogeneity, and that is resting on the surface of a planet? What path is it following?
If it is not moving, then it can hardly be following any path. Shouldn't it therefore be unaffected by gravity?

In reality objects of course do not cease to feel gravity once they reached the ground. Could thermal energy offer a explanation for this?

The particles of an object of non-zero temperature are never at rest. They are always moving. Upwards pathways are longer, because less paths lead upwards, therefore things tend to stay on the ground.
The flaw in this explanation is that it can only account for gases, because in solids particles only jiggle around in place.

As we learned in chapter 5.3, particles in solids also move considerably, but they do so in rare superpositioned states which do not usually get actualized when the wavefunction collapses. What if objects and their constituent particles rely on those rare superpositioned states to 'feel' gravity?

We will explore this idea further in a while. For the moment, we shall first investigate the question of why projectiles with different velocities have different trajectories, if they are supposed to follow the same geodesics through granular space.

In space particle dualism we say that all geodesics are geodesics in space, not in spacetime. Yet, same as in general relativity, the trajectories of objects, still depend on their impulse. That is why general relativity says that the geodesics are in spacetime. It is no different from saying that they are impulse-dependent.

Space particle dualism admits this impulse dependency, but without the philosophical baggage that comes with claiming that they are in spacetime.

If the density differences in granular space can divert a particle, changing its direction of movement, then, for the sake of the preservation of energy and momentum, this diverting should be different for objects with different impulses.

That goes a bit against of what we imagine a geodesic to be. From a geometrical point of view, there should be only one most direct path. On the other hand, we also know, that unlike in general relativity, in space particle dualism, geodesics have a probabilistic aspect to them. Particles move in a zigzag fashion, going from elementary surface to elementary surface. If they go left or right ultimately depends on where there are more paths. If there are more paths on the right, than a 'zag' is more likely, if there are more paths on the left, a 'zig' is more likely. We are speaking here of likelihood, and that shows that indeed more than one path is possible.

The probabilistic nature of geodesics in space particle dualism can explain how geometric gravity manages to not violate energy conservation. Furthermore it can potentially explain some of the more exotic pathways that are part of Feynman's 'sum over histories'.[2]

Leading the equivalence principle ad absurdum

In Einstein's thought experiment for demonstrating his equivalence principle he asks us to imagine someone that is accelerated upwards in a rocket. He claims that for the person inside the rocket gravity and acceleration is indistinguishable.

This thought experiment is however ignoring the fact that while someone feeling the gravity of earth his shoulders are accelerated into slightly different directions, the shoulders of someone inside an accelerated rocket would be accelerated into precisely the same direction. That is because someone standing on earth has his two shoulders accelerated towards the center of the earth, which leads to vectors with slightly different orientation for the two shoulders.

In addition to that, head and feet of a person inside an accelerated rocket feels the same acceleration, which is simply the acceleration of the rocket. A person standing on the ground, on the other hand, feels a slightly different acceleration on head and feet.

In other words, we can distinguish acceleration from gravity through the involved tidal forces.

It is often suggested that shrinking the rocket will solve this problem, because in a very small space tidal forces would not be expected anyway, and so their absence would not anymore be an indicator that one is located in an accelerated rocket instead of a stationary rocket on the surface of earth.

While that appears to be the case for rockets and people that are tiny enough, there is something else that should disappear for them, at least in principle, namely gravity itself.

A region of a gravitational field that is small enough for tidal forces to vanish completely, would also be small enough for differences in space curvature (space-only GR)[3] or space density (SPD) to vanish completely.

In general relativity, gravity is based on curvature of space. In space particle dualism theory gravity is based on density differences in the quantum vacuum. Curvature becomes invisible on sufficiently small scales. The same is true for density differences.

As already mention in the beginning of this chapter, in a framework that uses curvature but does not assume a spacetime continuum, a good measure for when we can regard gravity to be gone completely is when the difference between the curved space and an uncurved space is less than a Planck-length.

In space particle dualism on the other side, we can instead use the distance between gravity field lines as a measure. When the difference in distance between those field lines goes below the Planck-length, then we can speak of a 'local homogeneity'. We will soon look at how this works out computationally.

How could a point particle like an electron fall in a gravitational field if it has no spatial extension through which it could feel density differences in the quantum vacuum?

In general relativity one can sort of circumvent this question, because general relativity is based on the spacetime paradigm, and so from an ontological perspective, we do no longer require that different sides of an object feel a difference in curvature, but rather we can simply be looking at geodesics in spacetime, and be satisfied if at least the world line of the particle 'feels' the curvature.

In Einstein's early papers, the spacetime paradigm was optional. Instead of talking of time dilation he had often talked of a varying speed of light. This is based on an alternative ontology of general relativity where time is not a dimension, and the slowing down of time is not attributed to a different way of moving through 'spacetime'.

It is easier to ignore the question of how very small objects 'feel' gravity, when one is using this spacetime block-universe paradigm.

Space particle dualism is a quantum gravity theory, and can therefore not use the spacetime paradigm. As already mentioned earlier in this book, due to the way wave functions collapse simultaneously all over space, quantum mechanics requires a distinct present, which is in stark contrast to the relativity of simultaneity in Einstein's theory.

It means that the spacetime paradigm is not tenable in quantum mechanics. Quantum mechanics is fundamentally time dynamical. It does not work in a timeless setting, such as those suggested in theories which treat time as a dimension.

Local homogeneity of various objects

Now let us calculate the minimal height required for an object to feel gravity, according to space particle dualism theory. While going through the calculation, we have to keep in mind that the spatial extension we calculate could very well represent the spatial extension of the object's wavefunction.

In our first example, we are going to assume that the object under consideration has a horizontal extension of 1 meter.
Earth has an average radius of 6.3781×10^6 m, and so when comparing the distance between the gravitational field lines at the top and bottom of our test object, this radius is the length of the two same sides of a isosceles triangle that is stretched out by the center of earth and the bottom of the object.

In order to compare the top of the object with the bottom of it, we have to look at two triangles with identical angles, one ending at the bottom and one at the top the object. The opposite side of the smaller isosceles triangle we shall call Δs_1, and the longer we shall call Δs_2.
We want to do trigonometry here, and so we have to divide our isosceles triangle into two right-angled triangles.

We know that the opposite sides of these two right-angled triangles have a length of $\Delta s_1/2$ and $\Delta s_2/2$ respectively. We also know that $\Delta s_1 = 1$ m, because this is the horizontal extension of the object we are considering here.

We want to know at which height we can consider granular space as being locally homogeneous, and so we have to look at the case in which the difference between Δs_1 and Δs_2 is equal to the Planck length l_{PL}. In mathematical connotiation:

$$\Delta s_2 - \Delta s_1 = l_{PL}$$

From which follows that:

$$\Delta s_2 = \Delta s_1 + l_{PL}$$

We can then calculate the angle of the corresponding right-angled triangle from $\Delta s_1/2$ and the radius of earth, which is its hypothenus:

$$\alpha = \sin^{-1} \frac{\Delta s_1}{2\, r_1}$$

The next thing we want to know is the radius r_2 of the larger triangle. We can do this by writing down the same equation but for the larger triangle, and by then solving for r_2:

$$\alpha = \sin^{-1}\frac{\Delta s_2}{2\,r_2}$$

$$\sin\alpha = \frac{\Delta s_2}{2\,r_2}$$

$$r_2 = \frac{\Delta s_2}{2\sin\alpha}$$

$$r_2 = \frac{\Delta s_1 + l_{PL}}{2\sin\alpha}$$

As we know from above, the angle is:

$$\alpha = \sin^{-1}\frac{\Delta s_1}{2\,r_1} = 4.49 \times 10^{-6}$$

And that yields:

$$r_2 = \frac{\Delta s_1 + l_{PL}}{2\sin\alpha}$$

$$r_2 = 6378100.0000000556489225000004855369518\ldots\,\text{m}$$

The two separate rows of digits separated by many zeros indicate that we are dealing with digits that originate from two separate terms with different order of magnitude, one belonging to Δs_1 and one belonging to l_{PL}. This is because we express a length through an angle via a trigonometric function, and there are precision limits to this.[3, 4] Δs_1 is supposed to produce only 6378100, without the additional $5.56489225 \times 10^{-8}$ m. If there wasn't this unwanted excess value, we would calculate the local homogeneity simply by subtracting the radius of earth, namely 6378100 m, from r_2, but now we have to also subtract the unwanted excess value, which we can formally do by replacing r_2 with the same trigonometric term that produced the excess value in the first place. This yields:

$$h_{LH} = r_2 - r_1 = \frac{\Delta s_1 + l_{PL}}{2\sin\alpha} - \frac{\Delta s_1}{2\sin\alpha}$$

$$h_{LH} = 1.030846616697089941367304432 \times 10^{-28}\,\text{m}$$

This is the local homogeneity for a one meter wide object on earth. Much narrower in height than even atomic nuclei (10^{-15} m), and even smaller than the elementary space of a resting electron (10^{-21} m). It is needless to say that there is no way to flatten out any object to this degree.

This shows that the extendedness of wave functions is not needed to explain how macroscopic objects, or more generally speaking, objects composed of many particles, 'feel' gravity.

Now, let us look at the local homogeneity for resting electrons on earth. According to space particle dualism, the radius of the elementary space of a resting electron is:

$$R_{E_e} = 1.6716455 \times 10^{-21} \text{ m}$$

This radius will be our new Δs_1, giving us a new angle α for our gravity field lines:

$$\alpha = \sin^{-1} \frac{\Delta s_1}{2 \, r_1} = 7.50836706793841 \times 10^{-27}$$

This yields a local homogeneity height h_{LH} of:

$$h_{LH} = \frac{\Delta s_1 + l_{PL}}{2 \sin \alpha} - \frac{\Delta s_1}{2 \sin \alpha} \approx \frac{R_{E_e} + l_{PL}}{2 \sin \alpha} - r_\oplus$$

$$h_{LH} = r_2 - r_1 = 6.16665804244801614761 \times 10^{-8} \text{ m}$$

Which is 62 nanometers. That is only slightly larger than hemoglobin proteins inside red blood cells, which have a size of 55 nm.

Electrons are smaller than that. As already mentioned in the context of the above calculation, their size is $\Delta s_1 = R_{E\_e} \approx 10^{-21}$ m.

Do single electrons drop in a gravitational field? The answer to that is that we do not know. There have been numerous attempts to measure how electrons drop in a gravitational field. None of them was successful so far.[5, 6]

If they do drop, then the extended nature of their wavefunction could possibly account for that. If they do not drop, then local homogeneity could account for it.

If the former is the case, then their wavefunction can be flattened down to this height, by sending them through a very deep and very narrow slit with a height equal to the local homogeneity height of electrons. Then their wavefunction could be flat enough to not be affected by gravity. However, considering the fact that no one has so far observed the dropping of electrons due to gravity, it seems otiose to try and prevent it.

Neutral particles are much better suit to probe the gravitational field. Indeed dropping due to gravity has been observed for neutrons.[7] We shall therefore now calculate the terrestrial local homogeneity for neutrons.

Neutrons have a radius of 8×10^{-16} m. That gives us a new angle:

$$\alpha = \sin^{-1}\frac{r_n}{2\,r_\oplus} = 3.59328198134757 \times 10^{-21}$$

Which yields a local homogeneity height h_{LH} of:

$$h_{LH} = \frac{r_n + l_{PL}}{2\sin\alpha} - \frac{r_n}{2\sin\alpha} \approx \frac{r_n + l_{PL}}{2\sin\alpha} - r_\oplus$$

$$h_{LH} = r_2 - r_1 = 1.28855827083714 \times 10^{-13}\ \text{m}$$

Neutrons are also smaller than that. However, different from electrons, we do know experimentally that neutrons do drop in a gravitational field. This means we do have to employ our above mentioned 'wavefunction spread hypothesis' for them.

For it to become more than just a hypothesis, we need to experimentally test it.
This however is nearly impossible considering that the shaft height would have to be only 10^{-13} m. This is about the size of an atomic nucleus. Shooting neutrons between nuclei would not be enough, because that would correspond to using a simple slit, and that would cause scattering, which means that the wavefunction would spread out again on the other side.

If we want to account for the dropping of neutrons under gravity, we do have to assume that it is the extended nature of their wavefunction that allows them to 'feel' the inhomogeneities in the gravitational field.

However, only if we can experimentally show that containing the wavefunction can set-off gravity, can we show that geodesics are entirely in space.

Once it is proven that electrons drop under the influence of gravity, it can be attempted to set-off this gravity, by flattening the electron's wavefunction.
Something else to consider, is that it could prove difficult to have the shaft and the electron trajectory oriented perfectly perpendicular to the gravitational field.

Probing the local homogeneity in our earth's gravitational field is of uttermost importance, because it can not only give evidence for geodesics really being in space and not in spacetime, but furthermore it allows us to indirectly probe the size of elementary spaces. This way of probing is probably much easier than the way proposed in chapter 3.8, which involved off-scale gamma rays and detectors in outer space.

Atoms dropping

The smallest atom is the hydrogen atom, which has the Bohr radius. This radius is $5.29177210903 \times 10^{-11}$ m. That gives us an angle of:

$$\alpha = \sin^{-1}\frac{r_n}{2\,r_\oplus} = 3.59328198134757 \times 10^{-21}$$

Which yields a local homogeneity height h_{LH} of:

$$h_{LH} = \frac{r_n + l_{PL}}{2\sin\alpha} - \frac{r_n}{2\sin\alpha} \approx \frac{r_n + l_{PL}}{2\sin\alpha} - r_\oplus$$

$$h_{LH} = r_2 - r_1 = 1.94801778200282 \times 10^{-18}\,\text{m}$$

This is smaller than the hydrogen atom itself. This means that the wavefunction-mediated gravity hypothesis is not needed for explaining how atoms fall. It is only required for elementary particles, which is why it is very important to find out if they are affected by gravity when being at rest, or not.

The state of affairs is now that the wavefunction-mediated gravity hypothesis is not needed to explain the dropping of resting atoms, but it is needed to explain the dropping of resting elementary particles. The problem is, that it is unclear at this point, if the later has ever been observed. If it has not been observed for electrons, then it is doubtful if it has been observed for other particles. Observing it for electrons should be possible at least in a statistic fashion.

If we shoot electrons in a vacuum chamber, and they do not drop, then it would mean that the wavefunction-mediated gravity hypothesis is unnecessary.

It has been suggested that if gravity does not affect single elementary particles, then neutron stars should disintegrate, because they consist by 90% of neutrons, and a outer shell made of protons and electrons (John David Calanog; personal correspondence; 2020). The flaw in this argument is that the density of a neutron star is that of an atomic nucleus, and thus all of these protons and neutrons are tied together closely by the residual nuclear force. Electrons are trapped between the protons in the outer layers.

Thus, while it is highly likely that the wavefunction-mediated gravity hypothesis is correct, neutron stars cannot be used as an argument for that being the case.

Star Trek-like gravity sensors

Observing how atoms drop under the influence of gravity in fact allows for extremely precise mapping of underground structures. Babak Saif from NASA's Goddard Space Flight Center in Maryland, has built an instrument that uses atoms to sense gravity.[8]

The reason why atoms can measure gravity with a higher precision than macroscopic objects is that one can use them to build interferometers, and interferometers can allow for almost infinite precision, as we know from our experience with gravitational wave detectors. Using the interference patterns of atoms dropping under the influence of gravity, Babak Saif and his team are able to detect slightest changes in the gravitational field of objects, that can be as tiny as the change in gravitational attraction of people on the device before and after lunch. The devise can be used to detect differences in gravity on different locations on earth. When changes in gravity occur without a change in elevation, then it hints at the presence of underground structures, like cavities or for example coal mines.

This actually comes very close to Star Trek technology, where sensors can pick up cavities and even things like life signs from great distances.

Combining gravity measurements with interference patterns in quantum mechanics is a brilliant idea. One would think that this huge progress in measurement should reveal to the users that gravity is caused by the number of protons and neutrons instead of mass or mass energy, but it is hard to see that if one is not looking for it. Even in the experiments we looked at in chapter 6.9 that demonstrated that, it was only a 2 1/2 seconds difference between the result expected from Newton & Einstein and the correct SPD-specific result.

If we combine atomic interferometers with the physics of space particle dualism, we could potentially make Star Trek-like sensor capabilities reality.
SPD-based gravity sensors could not only detect cavities, but they could even detect materials deep inside the crust of earth, or even life signs. Life signs would be indicated by gravitational sources moving at speeds and in a fashion typical for life forms.

For remotely detecting matter movement below the surface of the earth or inside cavities one has to simultaneously monitor gravity on different locations and track changes in gravity. This would involve placing many gravity interferometers on many different locations of the surface.

Local homogeneity at or near infinity

Now we shall look at how far away we have to move from a gravitational source for the local homogeneity to become larger than a given object. This distance could then be regard as a distance at which gravity is no longer effective. We shall again look at the simple case of a 1 meter large object.

The required distance from the gravitational source shall be our new 'r'. In our above calculations 'r', or better r_1, was simply the radius of earth, but here it is a not yet known distance, and so we can find that distance by solving for 'r':

$$\alpha = \sin^{-1}\frac{\Delta s_1}{2\,r}$$

$$r = \frac{\Delta s_1}{2\sin\alpha}$$

We can now establish an equation system by first noting that:

$$\sin\alpha = \frac{\Delta s_1}{2\,r}$$

And then we take the usual local homogeneity equation and set $h_{LH} = \Delta s_1$, which means that the size of the object and the hight of the local homogeneity region are identical:

$$\Delta s_1 = \frac{\Delta s_1 + l_{PL}}{2\sin\alpha} - \frac{\Delta s_1}{2\sin\alpha}$$

We solve for $\sin\alpha$, which we need for establishing our equation system:

$$2\sin\alpha = \frac{\Delta s_1 + l_{PL}}{\Delta s_1} - 1$$

$$\sin\alpha = \frac{\Delta s_1 + l_{PL}}{2\,\Delta s_1} - \frac{1}{2}$$

Now we can set the two equations for $\sin\alpha$ equal and solve for r:

$$\frac{\Delta s_1}{2\,r} = \frac{\Delta s_1 + l_{PL}}{2\,\Delta s_1} - \frac{1}{2}$$

$$\frac{\Delta s_1}{2\left(\frac{\Delta s_1 + l_{PL}}{2\,\Delta s_1} - \frac{1}{2}\right)} = r$$

$$r = \frac{\Delta s_1}{\left(\frac{\Delta s_1 + l_{PL}}{\Delta s_1} - 1\right)} = \frac{\Delta s_1}{l_{PL}}$$

$$r = 6.18724444244 \times 10^{34} \text{ m}$$

Which is $6.540062218454384864 \times 10^{18}$ light years. For comparison: the visible universe has a radius of only 4.2×10^{13} light years. That means gravitational fields do not become particularly homogeneous (or 'flat', in GR terms), even at or close to infinity.

Clearing out doubts on the methodology used

The above methodology is purely geometrical and independent of the gravitational field strength involved. One is inclined to think that the size of the local homogeneity should somehow depend on the field strength as well. One thinks of strong gravitational fields as being 'more curved', or in space particle dualism terms, 'with larger inhomogeneities'. Therefore we shall analyze the misconceptions behind this in detail, in order to clear out all doubts.

Above we were looking at the distance between gravity field lines. When that distance dropped below the Planck-length, we would say that gravity has vanished. Of course particles smaller than the size of that local homogeneity could still 'feel' the gravitational field through the extended nature of their wavelength. Only by confining the wavefunction, one would be truly able to set off gravity.

Now the doubt we want to discussing here is about if the distance between field lines is really an appropriate measure. Quite obviously this would make the size of the local homogeneity totally independent from the field strength of the gravitational field. We know that in a stronger gravitational field the difference in gravity between the head and the feet is much larger. In space particle dualism, gravity depends on the density of the quantum vacuum. More virtual gauge bosons means more space quanta (elementary spaces) and therefore a stronger pull into such regions.

The strength of gravity depends on the difference in vacuum density between the top and the bottom of an object, or its wavefunction. This also determines the tidal force that is pulling things apart slightly.
If it is really this density difference that matters, it does not at all seem clear that a differences in field line distance of less than a Planck length is the lower limit.

However, if every meaningful displacement of elementary spaces must amount to at least one Planck length, then it seems correct to base the local homogeneity on this.

We could visualize a stronger gravitational field simply by imagining there to be more field lines between any two given field lines. Does this change the measure of when something can be regard homogeneous? Not really: when the distance between two given field lines does not change more than a Planck length, then we can indeed regard that patch of space as homogeneous; corresponding to what relativists would call 'flat'.

If earth was to turn into a black hole, while I was kept, suspended on a rope, in the same distance from the center of the former earth as before, then I would feel the same downwards acceleration as I feel now; namely 9.81 N.

When calculating the gravitational force between objects, their density never matters. In Newtonian physics it is the mass that matters; in Einsteinian physics it is the total mass energy that matters; and in space particle dualism theory it is the quark number that matters. Now, outside of the gravitational object, which is where test objects and people are usually situated, there is no mass or quarks, there are only the 'field lines', and the densities we are talking about is the distribution density of gluons.

This density is totally independent of the matter density in the gravitational object. It is most simple to visualize it by representing high gluon density by many field lines and low density by few field lines.

All field lines reach infinity, and it becomes clear that local homogeneity and gravitational field strength are two completely independent concepts.

Field strength is represented by how many field lines cross a region of space, while local homogeneity is represented by the difference in distance between field lines at the radial borders of that region of space.

As we saw in this chapter, there are two ways of creating a local homogeneity region. One is to shrink down the area of space under consideration, and to confine the object in question to that area of space, and the other way is to move the object so far away that the local homogeneity grows to match the size of the object.

It seems slightly counter-intuitive to imagine that the distance at which gravity falls off is independent of the mass of the gravity source. One has to keep in mind that it falls off only for an object that has a wavefunction that is confined to a certain size. It does fall off only for objects of that particular size, and only for such with wavefunctions confined to that size.

Note:
1. Personal handwritings on physics, 08.08.2004.
2. Quantum electrodynamics as developed by Richard Feynman involves calculating sums of possible particle trajectories which includes some very exotic ones with low probability. If we imagine each trajectory to correspond to one 'similar' classical world, in accordance to the similar worlds interpretation, then we have to explain what makes the more exotic particle paths possible. Complex pathways in granular space would make exotic pathways entirely possible. One could test this hypothesis by trying to see if it can reproduce the probability rules as reflected in Feynman's path integral.

3. This is calculated using the two high precision calculators: [mathsisfun.com/calculator-precision.html] & [mathsisfun.com/scientific-calculator.html].

4. The same calculation without the Planck length added yields $r_1 = 6378100.0000000556489225000004855368488\dots$ m. Of course the version without the Planck length added should in fact yield $6,378,100$ m, and the different result only shows us the limitations of calculators when it comes to trigonometry, however, from the fact that the two numbers start differing at the 28th digit after the decimal point, tells us that the height of the local homogeneity is in the order of magnitude of 10^{-28} m, and not 10^{-8} m, as we would think if we just subtracted r_2 by $6,378,100$ m, which is the radius of earth.

5. "Gravitational measurements on charged particles – can it be done?"; 2015. Michael Holzscheiter, University of New Mexico. Link: [https://indico.cern.ch/event/361413/contributions/1776296/attachments/1137816/1628821/WAG2015.pdf].

6. Additional link: [https://physics.stackexchange.com/questions/333613/are-electrons-affected-by-gravity/333618] (user1583209: "… However as far as I am aware there is no experiment that directly showed how gravity affects electrons.").

7. "Quantum States of Neutrons in the Gravitational Field". Claude Krantz. January 2006. Link: [https://www.physi.uni-heidelberg.de/Publications/dipl_krantz.pdf].

8. "Scientists watch atoms fall to see earth's changing structure". Sophia Chen. 09.03.2019. Link: [https://www.wired.com/story/scientists-watch-atoms-fall-to-see-earths-changing-structure/].

6.14 Absolute space [13.11.2020 (critique of SR & GR); 14.11.2020 (speed dependent deformation of the wavefunction); 18.11.2020 (calculating the deformation); 21.11.2020 (no deformation; speed determines how much out of phase spin entanglement is); 24.11.2020 (frame of reference in which entanglement was established determines entanglement now-slice); 25.11.2020 (using voyager-like probes to test entanglement dominance; using gravitational frames of reference)]

Sponsored by: Lin Yi Song (China), Henning Conle (Swiss).

The assumption that everything is relative permeates all of Einstein's work. When his theory is taught today, it is taught using this same ideological framework, even though the equations of the theory do not necessarily warrant such a framework.

The constancy of the speed of light results in an inability to measure absolute motion. This is then stylized into a general principle that absolute motion is undefinable and unmeasurable.

The constancy of the speed of light leads to space and time being relative. However, as already mentioned in chapter 3.13, there is an equivalent description in which the relativity of space and time are a result of a varying speed of light. Operationally a varying speed of light would be perceived as a slowing down of time.

In space particle dualism the slowing down of time is due to uniformly growing elementary spaces. A phenomenon which again is operationally indistinguishable from the two above descriptions.

In cannon special relativity the constancy of the speed of light is regarded to be part of what is called the invariance of the laws of physics. Yet, even Einstein often described Lorenz transformations using the ontology of a changing speed of light.

In these alternative interpretations we are not forced to accept the existence of a spacetime continuum. If we say the speed of light is changing, then the whole argument of time being able to go fast and slow and therefore it must be a dimension of some sort becomes obsolete.

In the twin paradox, both observers, the one remaining on earth and its travelling twin are regard to be in relative motion and with both seeing the other twin as having its time slowed down. This is not the case during the acceleration and deceleration phases. Here it becomes clear that it is the travelling twin which has its time slowed down.
Maintaining that motion is relative again seems unnecessary and merely philosophical here.

In general relativity the 'equivalence principle' is used to derive how light rays are bent in a gravitational field. Hereby one mentally replaces the earth with its gravity by a rocket which has an acceleration equivalent to the gravitational acceleration of earth, namely $9.81 \, \text{m/s}^2$, and then looks at how a light ray travelling through the inside of such a rocket would appear to bend.

Does that mean gravity and acceleration is the same in any fundamental and profound way?
Surely, Einstein could have presented his theory and even used this same way of derivation without making such a claim.

In reality there are plenty of aspects that fall in the face of the equivalence principle:

1. While with small objects their self-gravity can be ignored in approximation, it does in fact lead to different objects falling with slightly different speeds. That makes gravity and acceleration clearly different. How much objects in an accelerated rocket are pushed against the rocket floor on the other hand only depends on the acceleration of the rocket. Self-gravity is irrelevant here.
2. The acceleration of a rocket changes all the time and depends on combustion, while gravity depends on the number of quarks in an object. It does not change easily.
3. Tidal forces allow us to clearly distinguish gravitational fields from accelerated frames of reference.

4. There is no scale at which acceleration is not felt, but there are very small scales at which gravity is not felt.
5. The equivalence principle also implies that inertia mass and gravitational mass are the same, but as we have learned in the previous chapters, they are not at all equivalent according to space particle dualism.
6. Unruh radiation appears in accelerated frames of reference, but as we learned in chapter 4.3, it does not appear in gravitational frames of reference. The mainstream believes that black holes have a gravitation analogue to Unruh radiation, called Hawking radiation. This would imply that photons can reduce the gravity of a black hole. That is impossible if gravity depends on the number of hypercharges alone, and not on energy, as claimed by Einstein.

This is what speaks against the equivalence principle. There are just as many problems with the general idea of motion being fundamentally relative.

The notion that there is no meaning to absolute motion has always been stylized as being the core statement of relativity theory. This is very troubling, because quantum mechanics strongly relies on motion being absolute. Only with an absolute frame of reference can we say that the wavefunction collapses simultaneously all over space. Or, formulated using the ontology of similar worlds theory, worlds being temporarily indistinguishable only makes sense if there is only one set of indistinguishable worlds at a time. Space particle dualism theory does not work in a spacetime block universe with consciousness existing all over spacetime. In such a setting there would be no indistinguishable worlds, and therefore no superpositions.

Could it be that relativism is nothing but a philosophical ideology forced upon our description of the world?

Relativity of motion is often explained using two astronauts that drift pass each other in outer space. In this situation, so the argument, there is no way for them to tell who is moving and who is not.

As we will see shortly, this argument fundamentally relies on the astronauts having no memory and no sophisticated instruments.

In reality there are multiple ways how they could find out who is moving.
One way would be for them to observe the cosmic microwave background in order to see if there is any direction-specific red- and blue shift. To be sure that it is caused by observer motion they can look for different radiation sourced and verify that the redshift appears in all sorts of radiation sources and that the derived velocity and direction of movement are always the same.

Just because an average person could not tell, does not mean nobody could tell. On another note, we also have to look at the context. A freely floating observer in space will surely remember how he or she got there. If one of them jumped off a spaceship that had accelerated to a certain speed, then they know their speed. It is simply the speed of the spaceship. From an ontological perspective, we can always use the following as frames of reference of absolute motion:

(1) quantized space
(2) the cosmic microwave background
(3) earth.

Which is most fundamental is the question we want to explore in this subchapter.

There are three concepts which require an absolute frame of reference which is fundamental:

(1) The collapse of the wavefunction
(2) Entanglement
(3) Telepathy

Telepathic information is not signal-like; it is correlation-like. Therefore, same as entanglement it is not bound by the universal speed limit. Correlation-like phenomena are always instantaneous, and thus they require the existence of a preferred frame of reference. Something that amounts to an absolute space.

Determining local motion by means of redshift surveys

Above we mentioned that looking at redshifts and blueshifts in the cosmic microwave background can help to establish if one is moving or not. We have discussed this also before in chapter 4.2. There we learned that most cosmologists try to explain the dipole in the CMB, the so called 'axis of evil', using a local motion of the Milky Way, relative to the CMB background. They then derive a speed of the Milky Way, which is 369 km/sec.

What was also mentioned in chapter 4.2 is that Ashok K. Sengal showed that such redshift and blueshift patterns exist in other types of radiation as well, and that trying to explain them by a local motion of the Milky Way, or possibly the local group, leads to totally different speeds and directions of movement for the Milky Way.[1, 2]

A team around Yehuda Hoffman, has simply combined the various speeds derived from the various radiation sources, to come up with a total speed 631 km/sec for the Milky Way.[3]
It is needless to say that this approach is highly doubtful. Any redshift or blueshift that is due to local motion must be same in direction and magnitude for all sources of radiation. If it is not, then it is simply not due to local motion.

Another aspect is that it is hard to compare the Doppler effect in radiation sources with spectral lines, with the black body radiation that the CMB is. One knows were the spectral lines are supposed to be, and therefore one knows how much redshift or blueshift there is simply by looking at how much they are shifted.

The cosmic microwave background does not have spectral lines, and it therefore is not straight forward to interpret any differences in wavelength as the result of the Doppler effect.

That is even more so when we consider that when we increase the spherical harmonics in the CMB, then we have a quadruple which aligns with the ecliptic and the equinox of earth. As laid out in chapter 4.2, we consider this to be due to the observer effect. Attributing it both to the observer effect and local motion would not be reasonable.

A true local motion would show up uniformly in all radiation sources. It would be a directional inhomogeneity in the redshift of surrounding galaxies. Such an inhomogeneity could be attributed to both a violation of the Copernican principle or to local movement of earth, or the Milky Way.

Which possibility is true depends on how strong and smooth the distance correlation is. There are studies that look into this, but none of them are conclusive so far.[4]

Are there other ways of determining local motion? If our frame of reference is not the preferred one, then this should show up in the way wavefunctions collapse here on earth.

Testing simultaneity of wavefunction collapse

We know that in quantum mechanics the wavefunction collapse happens simultaneously all over space. It is one of the most fundamental properties of quantum mechanics. From special relativity however we know that simultaneity is relative. We have to therefore ask whose simultaneity this is. In which frame of reference is the collapse of the wavefunction perfectly simultaneous?

For answering that question we have to first think about how we would notice it if our frame of reference, namely earth, is not the preferred frame of reference.

If our frame of reference here on earth is not the frame of reference in which the wavefunction collapses simultaneously all over space, then we should be seeing a certain spatial inhomogeneity in the distribution of all wavefunctions.

Earth orbits the sun, and the sun orbits the center of the Milky Way galaxy. The Milky Way itself does not orbit anything. When we go beyond the local group, gravity becomes irrelevant, and the expansion of space dominates the dynamics. This is why not only the cosmic microwave background, but also the Milky Way galaxy seem to be good candidates for absolute frames of reference.

If the solar system is the preferred frame of reference, then our speed would be the orbital speed of the earth around the sun, which is 30 km/s. If the preferred frame of reference is the Milky Way, then it would be the speed of the solar system around the center of the Milky Way, which is 200 km/s.

In special relativity two events can be simultaneous for a resting observer and not simultaneous for a fast moving observer.

When we look at a double slit experiment, the emitting of photons and the detection on the screen is almost simultaneous. However, if the fundamental frame of reference (FFR) was such that the right side of the screen enters the now-slice of this FFR much earlier than the left side, then many more particles would arrive at the right side of the screen, simply because arriving at the left side of the screen would mean to arrive before they were sent out.

We could theoretically test this by conducting the double slit experiment on a nuclear space ship that can accelerate up to considerable percentages of the speed of light.

If earth is really the preferred frame of reference, then if for example the spaceship is moving away from us radially, and the photons of the double slit experiment travel sidewise, then they would be slightly more likely to end up at the side of the screen which is closer to earth.

That is because earth's now-slice cuts through the side of the ship first which is closer to earth.

When the ship is moving towards earth, then it is the other way around: now earth's now-slice is cutting through the ship's now-slice at the backside first.

This type of experiment is way beyond our technical capabilities at the moment. For getting a feeling of how large or small the effect is, we shall therefore first look at what could possibly be observed right here on earth.

How much time displacement could there be between different sides of the screen if the Milky Way is the preferred frame of reference?

For finding that out we have to find the inclination of the world line of the solar system, because that is identical to the inclination of the now-slice in spacetime. When measuring both space and time in meters, then light rays are worldlines that have an inclination of 45° in spacetime. We can therefore find the inclination simply by looking at the ratio between the speed of the solar system relative to the Milky Way, and the speed of light. Accordingly, the inclination θ can be calculated as:

$$\theta = \frac{45}{360} \times \frac{200\frac{km}{s}}{c}$$

$$\theta = 0.00008339102$$

If the screen width is Δs, then the time-like distance or height h that is stretched out by this angle is given by:

$$\tan \theta = \frac{h}{\Delta x}$$

Setting $\Delta x = 1$ m, and solving for h yields:

$$h = \Delta x \times \tan \theta$$

$$h = 0.00000145544793191192 \text{ m}$$

Dividing by the speed of light translates this spacetime distance into a time difference Δt, which is the time displacement between different sides of the screen we were looking for:

$$\Delta t = \frac{h}{c}$$

$$\Delta t = 4.85485172516221 \times 10^{-15} \text{ s}$$

This is quite small. The rate of wavefunction collapse is only 40 Hertz, and it could be argued that this means we cannot possibly measure an anachronicity which is below 1/40 second.

For scaling up Δt to match the time intervall between wavefunction collapses, we would have to set the screen width to a staggering $\Delta x = 5,149,487,856$ km. That is about the distance between the sun and neptun.

We conclude that at speeds that are approachable for us, the effect of a possible deformation of the wavefunction is unmeasurable. Aside from that it is not quite clear that it should take place at all.

What other ways are there to test the temporal aspects of the wavefunction collapse?

Entanglement dominance

What does it actually mean that the wavefunction collapses simultaneously all over space?

It means that when we measure a particle to be in a particular place or state, the probability for another person to measure it somewhere else or in another state becomes zero immediately. This requires a preferred frame of reference in which the collapse is simultaneous.

It is hard to have different inertial frames of reference in front of a detection screen. Therefore a better way of analyzing this is looking at entangled particles.

We could take a pair of entangled electrons, leave one on earth and take the other on a ride on a nuclear space ship. Then we measure their spins simultaneous. That produces a long binary code (011001001110 ...). The code received on earth is always the opposite of the one in the space ship (100110110001 ...). The question is then, which now-slice do we use to match the two rows of code? Is it the now-slice of earth or the now-slice of the space ship?

The now-slice of earth is more likely to be the one dominating the entanglement. Even if we take a radical relativist standpoint, entanglement cannot simultaneously be going along two different now-slices. It is just reasonable that the frame of reference in which the entangled particle pair was created is the one which guides entanglement.

If we take the highest so far achieved plane speed of 3,529.6 km/h and calculate the time difference per meter due to the inclination of the now-slice, we get $2.69293905199825 \times 10^{-16}$ seconds. Since we are now dealing with two entangled particles it is no problem to scale this up by simply increasing the distance between them. After $3.71341489982095 \times 10^{12}$ kilometers this time difference becomes one second.

3.7 trillion exceeds the distance even of the voyager and pioneer probes. Voyager 1, is 22 billion kilometers away from the sun, while Voyager 2 is 18.2 billion kilometers away.

Voyager 1 has a speed of 17 kilometers per second, which leads to a now-slice inclination of $4.12662396638497 \times 10^{-16}$ seconds per meter. At a distance of 22 billion kilometers that is a time difference of 0.009 seconds.

With the difference being this small, the rate of measurement would have to by far exceed 111 measurements per second, so that our rows of binary code are sufficiently displaced in time in order to actually probe which frame of reference dominates the entanglement.

We see that while the velocity determines time dilation and inclination of the plane, how much this inclination sets off two now-slices depends entirely on the distance. This shows that we do not need high velocities to probe this effect; distance does just as well.

In this way, it turns out our thought experiment from chapter 1.2, which was looking at the inclination in the now-slice of a hypothetical alien, was much more relevant than suggested there.

We were right to reject the notion that this relativity of simultaneity shows that all of spacetime is real; we were also right to say that nobody is able to observe what is on his or her now-slice; what we however failed to see is that now-slices are indeed very important, not so much in special relativity, but in relativistic quantum mechanics.

The inclination of the now-slices is something we have to take care of only when analyzing the data. Something we have to take care of already during the measurement process is time dilation. If time runs slower on our probe, then we have to measure at a slightly higher rate, in order to compensate for that.

If our probe is as fast as voyager 1, then this time dilation effect is not very strong. Given the speed of the probe, one second is increased merely by a factor of:

$$\frac{1 \sec}{\sqrt{1 - \frac{v^2}{c^2}}} = 1.00000000161 \sec$$

That means, for the single electron-random number generator to be in sink with the entangled single electron RNG on earth, the time between measurements $\Delta t_{\oplus}$ on the space ship must be multiplied by the inverse of the Lorenz factor:

$$\Delta t'_{\oplus} = \Delta t_{\oplus} \times \sqrt{1 - \frac{v^2}{c^2}}$$

The board computer could surely keep track of the velocity and adjust the time between measurements accordingly.

Single electron based random number generators

The difficulty of always measuring the spin of the same pair of entangled electrons should not be underestimated. Electrons are extremely small and are indistinguishable when they are in the same state. Making sure that our single-electron random number generator (SE-RNG) always measures the same electron would have be a major focus of the experiments. Another challenge is the high rate of measurement.
Luckily there are already efforts being made in this direction. There are existing single-electron RNGs and some of them are actually spin-based.[5, 6, 7]

Using gravitational frames of reference instead

We can know the inclination θ of the now-slice in gravitational frames of reference by simply replacing the v in our usual formula for inertial frames of reference by the formula for the escape velocity v_e:

$$\theta = \frac{45}{360} \times \frac{\sqrt{\frac{2\,G\,M}{r}}}{c}$$

For earth's surface that gives us an angle of $\theta = 0.00000466397$, and that leads to a time displacement for the now-slice of $2.71526619639778 \times 10^{-16}$ seconds per meter. At the altitude of the space station ISS, this angle is $\theta = 0.00000452144$, and the time displacement for the now-slice is $2.63228819890368 \times 10^{-16}$ seconds per meter.

If earth is a preferred frame of reference, then of course the inclination of its now-slice is zero, but even in a relative context, we have to be looking at the relative inclination between these two now-slices, and so we shall use the difference in inclination here. That would be $\Delta\theta = 1.4253 \times 10^{-7}$. That leads us to a time displacement per meter of $8.29779974941037 \times 10^{-18} \, \text{s} \cdot \text{m}^{-1}$.

The distance between the ground and the ISS space station is merely 408 kilometer, and so our actual time displacement of only $3.38550229775943 \times 10^{-12}$ seconds. We cannot hope to reach a spin measurement frequency that is high enough to probe a time displacement in that order of magnitude.

It seems therefore far more practical to probe entanglement dominance using inertia instead of gravity.

However, in the future there might be other gravitational fields that are more suitable, and for that case we shall mention here what the needed time dilation adjustment is.

The time dilation adjustment for the time between measurements can be calculated by multiplying with the inverse of the Schwarzschild factor:

$$\Delta t'_{\oplus} = \Delta t_{\oplus} \times \sqrt{1 - \frac{2\,G\,M}{r\,c^2}}$$

While we might not want to use gravitational now-slice inclination for probing entanglement dominance, we quite possibly have to account for it in a purely inertia based experiment as well.

The effect coming from gravity would actually be contrary to the inertia effect. The inclination of earth's now-slice needs to be subtracted from the speed dependent inclination of our probe.

For finding out to what the earth's now-slice inclination amounts to, we have to multiply it with the distance between us and the voyager 1 probe. That is distance is currently 22.5 billion kilometers. When multiplying this with earths now-slice inclination per meter, we get 0.006109348941895005 seconds.

That has to be subtracted from the original displacement of 0.009 seconds we got above, which results in a decrease in temporal displacement, resulting in the net displacement being only about 0.003 seconds.

Note:

43. Ashok K. Singal (17 May 2013). "A large anisotropy in the sky distribution of 3CRR quasars and other radio galaxies"; Journal-ref:*Astrphys. Sp*. Sc., 357, 152 (2015); arXiv:1305.4134.

44. Ashok K. Singal (19 May 2014). "Extremely large peculiar motion of the solar system detected using redshift distribution of distant quasars"; arXiv:1405.4796.

45. Yehuda Hoffman, Daniel Pomarède, R. Brent Tully & Hélène M. Courtois (22 August 2016)."The dipole repeller". *Nature Astronomy*. doi:10.1038/s41550-016-0036.Archived from the original on March 3, 2017.

46. Adi Nusser, Marc Davis, Enzo Branchini (Februar 2014). "On the recovery of Local Group motion from galaxy redshift surveys". *The Astrophysical Journal* 788(2). DOI: 10.1088/0004-637X/788/2/157.

47. Nikhil Rangarajana, Arun Parthasarathy, and Shaloo Rakheja (2017). "A spin-based true random number generator exploiting the stochastic precessional switching of nanomagnets". Journal of Applied Physics 121. https://doi.org/10.1063/1.4985702.

48. Ken Uchida; Tetsufumi Tanamoto; Shinobu Fujita (January 2002). "Single-electron random-number generator (RNG) for highly secure ubiquitous computing applications". Solid-State Electronics 51(11-12):1552-1557. DOI: 10.1016/j.sse.2007.09.015.

49. Sungwoo Chun; Seung-Beck Lee; Masahiko Hara; Wanjun Park; Song-Ju Kim. "High-Density Physical Random Number Generator Using Spin Signals in Multidomain Ferromagnetic Layer". February 2015. Advances in Condensed Matter Physics 2015(499):1-8. DOI: 10.1155/2015/251819.

6.15 Conservation of gravitational energy [27.11.2020 (possible violation of gravitational energy conservation)*; 28.11.2020 (not violated for long enough to be measureable); 16.12.2020 (calculation)]

*Requested by David Wührer (Australia).

Sponsored by: Lin Yi Song (China), Henning Conle (Swiss).

In chapter 4.13 we have seen that in the cosmology of space particle dualism theory, energy is not preserved. The energy of the cosmic microwave background is continuously reduced as the universe expands. It is not turned into potential energy, because a universe that obeys entropic expansion cannot collapse back. In other words, it cannot be a 'closed universe'.

This means that both the first and the second law of thermodynamics really apply to closed systems only. The universe is not a closed system, because it does not have a border.

The German physicist David Wührer pointed out that there could be a violated of gravitational energy conservation every time a photon generates a quark anti-quark pair, also called meson, that then subsequently turns into a photon again. According to space particle dualism theory, the photon exerts no gravity, while the meson does exert gravity.

Do mesons exist long enough for their gravity to be felt? In order to find out if they do, we have to calculate the energy associated with the gravitational pull they exert and see what the minimal time is that is required to measure this energy. We can know this by applying the time energy uncertainty relation, which is:

$$\frac{\hbar}{2} = \Delta E \times \Delta t$$

In classical gravity theory, the potential gravitational energy is given by:

$$U = -\frac{G\,M\,m}{R}$$

In space particle dualism that is:

$$U = -\frac{G_{|\mathbb{Y}|}\,|\mathbb{Y}|_1\,|\mathbb{Y}|_2}{R}$$

For a meson, we know that $|\mathbb{Y}| = 2$, because it consists of two quarks, one quark and one anti-quark.

In general relativity the gravitational potential energy of different types of mesons would all be very different, depending on their mass and more precisely on their mass energy. In space particle dualism however, they all have the same gravitational potential energy, because they all consist of two quarks, which means that for all of them $|\mathbb{Y}| = 2$.

The size of mesons is about one femtometer (1×10^{-15} m),[1] and so the closest any particle can get to it would be a half femtometer, because that is the radius of a meson.

So, if a proton or a neutron ($|\mathbb{Y}| = 3$) for instance is probing the meson, then the potential gravitational energy is:

$$U = -\frac{6.1557567 \times 10^{-65} \, [\text{m}^3 \, \text{kg} \, \text{s}^{-2}] \times 6}{0.5 \times 10^{-15} \, [\text{m}]}$$

$$U = -7.386908 \times 10^{-49} \, \text{J}$$

The time required to measure that energy is:

$$t = \frac{\hbar}{2 \, \Delta E} = 7.1381139 \times 10^{13} \, \text{sec}$$

That is 2,261,980 years. The lifetime of the meson with the longest lifetime is the K-long meson K_L^0, which lives for 5.116×10^{-8} sec.[2] That means no meson exists long enough to violate the conservation of gravitational energy measurably.

Note:
1. D. Griffiths (2008).
2. K.A. Olive et al. (2014): Particle listings – K_L^0.

Epilogue
[26.05.2013]

Physics is often associated with cold calculus. With a picture of the world which gives no hope: The world as a machine, and us as a part of this machinery; unmighty to cause any changes to influence something, without being part of this aimless mechanism ourselves. The physics before the development of quantum mechanics hardly let anyone see any sense in our existence. Our wills, our hopes, all illusions, which had no place in the mechanical picture of the world at that time.

This changed with quantum mechanics, which was the first nondeterministic theory of nature. It is now common knowledge that the world is not this stiff machinery we once thought it to be. These changes will not bring us back our gods and spirits (at least not in the traditional way), but they integrates us, as conscious beings, back into nature. We now know that we can choose our fate by ourselves. We are not longer slaves of nature's laws. We are responsible for our actions. The old philosophic question about free will and its conflict with god's omniscience on the one hand, or the completeness of physical laws on the other hand, is now answered: we indeed have a free will, and neither a god nor physics can take that away. Both quantum mechanics and formal logic (Gödel's incompleteness theorem) show us that we indeed do have a free will.

Judging from the evidence, materialism should be dead since long time. But since science started off from a materialist mind frame, it is hard for many scientists to break up with that. Most of the originators of quantum mechanics saw some central role for consciousness. They expressed the meaning and significance of quantum mechanics in the right way. However, materialism was stronger, and most physicists were not daring to draw all the philosophical and epistemological conclusions of that. Later interpretations were made to get rid of the consciousness element. One of them was Everett's many worlds theory. It was an attempt to get rid of consciousness, but when we examine it more closely, we see that consciousness reappears in it with a more passive role (see chapter 2.3).

In accordance with John A. Wheeler, who was doubtless one of the most influential physicists of the last century, the world is not a material world. It consists mainly of information, thoughts and feelings. Quantum theory does not only support this view, it is forcing it up on us.

Many people turn away from science, because they can not find purpose and meaning in the way it describes the world, while yet this description is deliberately vilified. The world as described by quantum theory is even more meaningful and beautiful than any religious world view one could imagine.

Judging both from the perspective of space particle dualism, and the criteria of inner consistency, quantum theory contains much more truth as well as inner consistency than relativity theory. From this perspective the intuitive view Bohr and Heisenberg had of quantum mechanics was indeed correct, and it is good that it was the prevailing view at least before the late 60ties. The Von Neumann-Wigner interpretation can be seen as the strict and concrete form of Bohr and Heisenberg's Copenhagen interpretation, which is unclear formulated. There is no meaningful differentiation one can make here. A measurement is only defined if it is accompanied by consciousness. And when one carefully studies what Heisenberg said, one realizes that his view was basically identical with Von Neumann's view. To say that the wave function changes as our knowledge about nature changes, is not in any way different from saying that consciousness leads to a collapse of the wave function, the word 'change' being just replaced by the world 'collapse'. The introduction of the word 'collapse' was here merely a way to criticize the way consciousness seems to 'change' the wave function; something which was regard incredible or frightening for scientists who were always searching for objectivity.

Since the future is open in quantum physics, we might even be tempted to imagine a god being able to influence quantum decisions, and thereby to influence the world, without any violation of nature's laws. Yet the omniscience of such a bodiless god would make him measure the universe in every moment, and that would make quantum physics impossible. So the old philosophers were actually right when they saw free will in conflict with god's omniscience. Furthermore, being omniscient, as well as being bodiless, makes it impossible to perceive the world. Knowing the position and impulse of every particle in the world, turns the world into a mere gathering of moving points. *Qualia*, the way merely numerical physical entities are presented in our mind as feelings, colours and smells, requires experience. It requires a body; a perspective.
Red is described as a warm colour, blue as a cold colour. This has to do with the fact that warm or hot things, like fire or glowing things, tend to emit light with a wave length we use to call red, while relatively cold things, like deep water or the atmosphere, tend to reflect light with wave lengths we use to call blue. So even seeing colours requires experience.
Therefore not only our thoughts require the extension of consciousness in mental time, but even just the simple perception of colours. It is therefore to assume that consciousness requires memory (some animals do not have a memory).

We create the world through our observations. As we know now, matter is not more than information; it is not much more real than things in our minds. Mind and matter are not that different. If we take the Copenhagen view serious, the world would not exist if it wasn't inhabited by feeling and thinking animals.
A superposition of all possible states, which is all there is according to decoherence theory or the many-worlds theory, could not be called 'world'; it could not even be called 'something'. Something which is just 'anything', or 'all there can be', cannot be 'something', it would be just 'nothing'. So it is the action of observing which brings the world into existence. It is our mind which creates the world, the reality, in every moment. What we see around us is what different minds agree on, what we have inside us belongs to ourselves.

Summary

Is the theory presented in this book really the solution for all those conceptual problems we face when trying to unite quantum theory and general relativity?
I think it is helpful to look and compare what other solutions have been proposed for these problems. We can make the following list:

Problem	Solution	Alternatives
Measurement problem (Interpretation problem)	The measurement is included in the definition of the wavefunction itself. It reflects the maximal gainable knowledge of the observer.	Von Neumann-Wigner interpretation; Objective Reduction (Penrose); decoherence; many-worlds approach.
Einstein-Rose-Podolski paradox (non-locality problem)	Spins are orthogonal on every of the equivalent space-times.	Information traveling into the past; Information sent through microscopic wormholes; Non-locality in twistor theory (only a vague idea).
Spin Interpretation	Oscillations of complex surfaces.	None.
Accelerated expansion	Black hole entropy drives it.	Dark energy (not defined); Cosmic event horizon entropy drives it.
Expansion (Big Bang model)	Black hole entropy drives it.	None.
Dark matter	The theory predicts 90% nonluminous matter in form of burned out stars. That is because the universe of SPD is very old.	None with fitting values.
Vacuum catastrophe	Gravity is always local.	Supersymmetry (already proven wrong since the discovery of the Higgs boson).
Information paradox	Gravity is a side effect of charge and thus black holes can't evaporate away through chargeless photons.	Non which concentrates on more than just entanglement information.

Background dependence	Every particle has its own quantum of space.	Twistor theory; (Loop-QG is not really background independent).
Flatness problem	Newton was right that in a borderless universe with no center, gravity from different direction cancels out.	Weyl curvature hypothesis (Penrose; 1979); (inflation is just another form of fine tuning).
Horizon problem	The border of the past was not a low entropy state (!); entropy per cosmic volume must be constant.	Weyl curvature hypothesis (Penrose; 1979); (heat exchange before an inflation would be an entropy increase).
Chronology protection conjecture	Wormholes not definable in SPD.	It is almost impossible to keep a wormhole open (Wheeler; 1962); Quantum fluctuations might destroy it (Thorne & Hawking; 1990; 1991).
Hierarchy problem	Gravity is weakened by vacuum energy, because it depends on differences in granular dimensionality.	Large extra dimensions in alternative forms of string theory.
Holographic principle	3-dimensional space is made out of 2-D spherical surfaces.	None.
Mind-body problem	Consciousness is a relation, similar to a Platonic body.	Dualism, intentionalism, materialism, … (none really explains what consciousness is).
Causality problem	Randomness is replaced by non-computable determinism.	Von Neumann-Wigner Interpretation & Free Will.
Generations of matter	Merging of elementary spaces.	Different vibrational modes in string theory.
M-sigma relation	All supermassive black holes evolved from primordial black holes.	None.
Entropy problem	The entropy density of the universe is a constant.	Penrose's cyclic universe model (CCC).

Fluctuations in cosmic microwave background	Caused by primordial black holes which formed out of exotic matter.	Quantum fluctuations (not possible without measurement); Black holes from the last aeon (CCC).
Missing local expansion	Locally low entropy increase.	None.
Concentric shells	BAO & emergent universe.	BAO; Carmelian cosmology.
CMB anomalies	Emergent universe.	Penrose's CCC model (partial explanation only).
Supernovae are less luminous than predicted	Not so using the intergalactic distances provided by SPD.	None.
Supernova luminosities don't seem to decrease with the square of the distance.	Using SPD the luminosities drop perfectly with the square of the distance.	Supernova physics changes over time (Young-Wook Lee; 2019).
Trouble with Hubble.	Doesn't exist in SPD.	None.
Trouble with G.	The number of quarks per kg are different for every material.	Fifth force; G changing over time.
Singularity problem.	The existence of a preferred frame of reference prevents the formation of event horizons.	BHs radiate away before they form (Laura Houghton; 2014); radiation pressure prevents horizon (Abhas Mitra; 2000).

Basic physical constants used in this book

Name	Symbol	Value
speed of light in vacuum	c	299,792,458 m/s
Newtonian constant of gravitation	G	6.673×10^{-11} m$^3 \cdot$ kg^{-1} $\cdot$ s^{-2}
Planck constant	h	6.626 069 57(29) $\times 10^{-34}$ J $\cdot$ s
reduced Planck constant	$\hbar = \dfrac{h}{2\pi}$	1.054 571 726(47) $\times 10^{-34}$ J $\cdot$ s
electric constant (vacuum permittivity)	$\varepsilon_0 = \dfrac{1}{\mu_0 c^2}$	8.854 187 817... $\times 10^{-12}$ F $\cdot$ m^{-1}
electron mass	m_e	9.109 383 56(11) $\times 10^{-31}$ kg
classical electron radius	$r_e = \dfrac{e^2}{4\pi \varepsilon_0 m_e c^2}$	2.817 940 3267(27) $\times 10^{-15}$ m
electric charge of an electron	e	1.602176565(35) $\times 10^{-19}$ C
Boltzmann constant	k_B	1.38064852(79) $\times 10^{-23}$ J /K
Coulomb's constant	k_e	8987551787.3681764 N $\cdot$ m$^2 \cdot$ C^{-2}
Proton mass	m_p	1.672621898(21) $\times 10^{-27}$ kg
Planck mass	$m_P = \sqrt{\hbar c/G}$	$2.176470(51) \times 10^{-8}$ kg
Planck length	$l_P = \sqrt{\hbar G/c^3}$	$1.61622938 \times 10^{-35}$ m
Planck time	$t_P = \sqrt{\hbar G/c^5}$	$5.39116(13) \times 10^{-44}$ s
Planck energy	$E_P = \sqrt{\hbar c^5/G}$	1.956×10^9 J
Neutrino mass (sum of three 'flavors')	m_ν	$2.138958468 \times 10^{-37}$ kg

New formulas and constants

Particle physics

Name	Description	Formula	Value
Elementary space radius (2005)	Depends on the energy of the particle.	$R_E = \dfrac{2\,G_E\,E}{c^4}$	Any.
Elementary space constant (2016 & 2018)	Constant that determines the size of elementary spaces	$G_E = \dfrac{k_e\,e^2}{m_p^2}$	$8.246441821 \times 10^{25}\ \mathrm{m^3\ s^{-2}\ kg^{-1}}$
Granular dimensionality (2005 & 2015)	The granular dimensionality depends on the energy density of the vacuum and determines the strength of gravity.	$D = 2 + \left(1 - \dfrac{1}{\sqrt{n}}\right)$ $D = d + \left(1 - \dfrac{1}{\sqrt[d]{n}}\right)$	$3 - 10^{-35}$
Critical photon energy (2016)	Energy at which the elementary space of a photon grows over its wavelength.	$E_{crit} = \sqrt{\dfrac{h\,c^5}{4\,G_E}}$	$2.205547715 \times 10^{-9}\ \mathrm{J}$
Critical photon frequency (2016)	Frequency at which the elementary space of a photon grows over its wavelength.	$f_{crit} = \dfrac{E_{crit}}{h}$	$3.328591242 \times 10^{24}\ \mathrm{Hertz}$
Critical photon wavelength (2016)	Photon wavelength at which it equals the diameter of the associated elementary space.	$\lambda_{crit} = \dfrac{c}{f_{crit}}$	$9.006586757 \times 10^{-17}\ \mathrm{m}$

Critical electron speed (2016)	Speed at which the elementary space of an electron grows over its wavelength.	$$v_{crit} = \\ -\frac{4\,G_E\,m_e^2}{h} \\ +\sqrt{\frac{16\,G_E^2\,m_e^4}{h^2} + 4\,c^2} \\ \times \frac{1}{2}$$	299,792,457.793 m/s
Critical electron impulse and energy (2016)	Energy and impulse at which the elementary space of an electron grows over its wavelength.	$$p_{crit} = \frac{m_e\,v_{crit}}{\sqrt{1 - (v_{crit}/c)^2}}$$ $$E_{crit} = \sqrt{p^2\,c^2 + m_e^2\,c^4}$$	7.477075686 $\times 10^{-18}$ kg m/s 2.2415709 $\times 10^{-9}$ J
Critical electron de Broglie wavelength (2016)	Wavelength at which the elementary space of an electron grows over its wavelength.	$$\lambda_{crit} = \frac{h}{p_{crit}}$$	9.0065857 $\times 10^{-17}$ m
Length dependent quantization of energy (2015)	The wavelength of particles is quantized using the Planck length.	$$E = \frac{h\,c}{\lambda},$$ $$\lambda = n \times \sqrt{\frac{\hbar\,G}{c^3}}$$	Discrete values.
Time dependent quantization of energy (2015)	The energy of particles is quantized using the time-energy relation.	$$E = \frac{\hbar}{2n \times t_P}$$ $$t_P = \sqrt{\frac{\hbar\,G}{c^5}}$$	Discrete values.
Energy minimum for virtual particles (2016)	Energy minimum for elementary spaces contributing to granular dimensionality.	$$E_{min} = \frac{c^4\,\sqrt{G\,\hbar/c^3}}{4\,G_E}$$	3.957848975 $\times 10^{-28}$ J

Maximal lifetime in the quantum vacuum (2016)	Lifetime of a minimal energy virtual particle.	$\Delta t = \dfrac{\hbar}{2\,E_{min}}$	$1.332253621 \times 10^{-7}$ sec

Cosmology

Minimal mass for dark matter black holes (2017 & 2018)	Dark matter is 'cold' and therefore there is no wavefunction to deal with.	$M_{DH_{min}} = \dfrac{G_E\, m_e}{G}$	1,125,587.541 kg
Impulse of dark matter particles dependent on host black hole (2018)	Only of theoretical value.	$\Delta p = \dfrac{\hbar\, c^2}{8\,G\,M_{BH}}$	Always practically zero.
Hubble parameter at different times (2019)	According to SPD the expansion of the universe is independent of gravity.	$H = \dfrac{H_0 - H_{min}}{t_0^2} \times t^2 + H_{min}$	Any.
Time-redshift equation (2019; 2021)	Allows to calculate the light travel distance to any object.	$t_0 = \dfrac{\dfrac{6\,c\,z_{t_0}}{H_0 - H_{min}} - 3}{2 + \dfrac{6\,H_{min}}{H_0 - H_{min}}}$	$\cong z \times$ 39,625,004,647 yr
Initial Hubble parameter (2019)	Depending on the average black hole mass and the minimal black hole mass	$H_{min} = \dfrac{M_{BH_{min}}^2\, H_0}{M_{BH_\mu}^2}$	$2.750307987 \times 10^{-17}\ \dfrac{m}{s}$ /ls
Initial temperature of the universe (2015)	The initial temperature depends on the average entropy density.	$T_{min} = \sqrt[3]{\dfrac{\rho_S\, c^3\, h^3}{16\,\pi\,k^3\,\zeta(3)\,T^3}}$	5,963.0497688 K

Gravity

Charge gravity constant of the strong force (2018)	Not used in combination with a mass but with hyper-charge.	$G_{\|\mathbb{Y}\|} =$ G_{Fe} $\times \dfrac{m_{\text{Fe-e}}}{n_{p_{Fe}}\, m_p + n_{n_{Fe}}\, m_n}$ $\times \dfrac{m_p^2}{9}$	6.1557567 $\times 10^{-65}$ m^3 kg s^{-2}
Charge gravity constant of the electromagnetic force (2018)	Analogous to the above.	$G_{\|\mathbb{Q}\|} = \dfrac{G_{\|\mathbb{Y}\|}\, k_e\, e^2\, m_e^2}{\hbar\, c}$	1.3476218 $\times 10^{-66}$ m^3 kg s^{-2}
Gravity of the strong force (2018)	Newton's gravity law formulated without reference to mass.	$F_{G_{\|\mathbb{Y}\|}}$ $= G_{\|\mathbb{Y}\|} \times \dfrac{\|\mathbb{Y}\|_1\, \|\mathbb{Y}\|_2}{r^2}$	Any.
Gravity of electro-magnetism (2018)	Analogous to the above, but can be more or less ignored in the presence of atoms.	$F_{G_{\|\mathbb{Q}\|}}$ $= G_{\|\mathbb{Q}\|} \times \dfrac{\|\mathbb{Q}\|_1\, \|\mathbb{Q}\|_2}{r^2}$	Any.
Mass-gravity constant of proton gases (2018; 2020)	It is the basis of calculating the mass gravity constant of any material.	G_p $= G_{Fe}$ $\times \dfrac{m_{\text{Fe-e}}}{n_{p_{Fe}}\, m_p + n_{n_{Fe}}\, m_n}$	G_p $= 6.6026471$ $\times 10^{-11}$ m^3 kg^{-1} s^{-2}
Mass gravity constant of water (2018; 2020)	This is just one example.	G_{H_2O} $= G_p$ $\times \dfrac{n_{p_{H_2O}}\, m_p + n_{n_{H_2O}}\, m_n}{m_{H_2O-e}}$	G_{H_2O} $= 6.6496762$ $\times 10^{-11}$ m^3 kg^{-1} s^{-2}

Comments

"As far as I can see now, the theory has only conceptual advantages*."

*That was a first judgement based on the very first edition of this book.

"I found the idea quite inspiring to interpret the phase of the quantum mechanical wave function in a fundamentally geometrical way. … The complex two dimensional space of the quantum mechanical wave function could thus be interpreted as two space-like wrapped up extra dimensions**."

**The elementary spaces of SPDT are similar to compactified extra dimensions, but only to a very limited extent, because many of them together form the large dimensions we know. So there is no 'extra' here.

"What I appreciate the most in his work is the implicit requirement to not view quantum theory and general relativity as independent from each other. The search for interconnections between them is mostly neglected by others.
And that is the most original and interesting part of Sky's suggestions. Relativity and quantum mechanics are thereby tied together relatively strong, to an inseparable unit. Unfortunately this is always presented differently in the mainstream. There general relativity and quantum theory are said to be like two sides of a medal, but that is probably too trivial to be of any help to anyone since quite a long time.
The fact that string theory and Loop-quantum gravity could not achieve any commonly accepted success, could be another hint for that we might have to unroll physics from a completely new and unexpected perspective."

Bernhard Umlauf [German physicist], 21.06.2014

"My intuition tells me that, if we bowdlerize all that, we get the same results as the mainstream. If it were a crank I would suggest that we get only nonsense, but I consider him as too competent for that, which means that his new approach will result, for the most part, in the mainstream.*"

*A judgement based on the very first edition of this book.

Ralf Kannenberg [Swiss diplom-mathematician], 15.07.2014

"Sky not only demonstrates his thorough understanding of vast literature on the topic in addition to the most eminent authors such as Sir Roger Penrose on the same, but elegantly manages to tie up nicely the best from the best theories to date together with utterly highly relevant and correct points from various other theories, common facts and figures as well as available tangible results; and icing on the cake he meticulously tops all of that up with his own precise calculations and comprehensive reasoning to come up with what I bet is now the most successful theory of quantum gravity and consciousness, stripped out of any possible unnecessary details.

What's more, including the subconscious, the self and qualia but limiting and therefore leaving out concepts such as the superconscious, spirit and soul ensures his theory in physics is to be widely considered, if not right away accepted, by scientists and thinkers from all walks of life! This does not mean that there is not a subsequent masterpiece from Sky to come following this one and that will actually deal with the aforementioned concepts as necessary when one realises that spirit is nothing else than the old fashion way to talk about a certain type of field; and I predict that this will be tightly related to and will call for the use of tachyons as well as quantum entanglement, once more.*

For now, this present book is a delight for anyone and especially for the true truth-seeker." , 24.12.2015.

*This just expresses a wish by the commenter and does not reflect any real plans by the author of this book.

"Sky has breakthrough after breakthrough. In my opinion he is worth several PhDs.", 04.06.2017.

"I keep supporting Sky. His books should be credited more, and the academic dogma revisited and revised more. I am an elephant, don't ask me to climb a tree - I will handle the tree otherwise. Sky is an eagle, don't ask him to climb a tree - Sky will handle the tree otherwise. Academia should quit creating monkeys mainly or only. Also to note about the reigning academic infrastructure: recognition points or scores in Academia.edu, ResearchGate, webofknowledge.com are calculated in a same fashion as for the criteria to pass IELTS/TOEIC/TOEFL tests; you can only obtain really high figures by working the system itself, not so much so by making substantial scientific contributions or by mastering the language.

Our various intellectual energies better be used to make real differences in the world, not to all work like monkeys on virtually increasing our scholarly marks, which also is not interesting.

In other words, we cannot enhance science with researcher clones. We do need people like Sky.", 18.05.2020.

Ludovic Krundel [French cybernetician; PhD]

"The idea of equivalence groups of space-times resolves the paradoxes.", 14.05.2016.

"A set theory of consciousness is the holy grail of wisdom.", 24.05.2016.

"There is one thing I really love with his interpretation. It resolves the spooky action at a distance. Physicists made peace with the fact that quantum mechanics and logic do not coexist. Logic is the one constant of truth seekers. And the only logical explanation that I read was his.", 28.05.2016.

"Long is the way to decipher the mysteries of consciousness and life, but one of the keys is Sky Darmos's theory, the cornerstone it will be.", 30.05.2016.

"Einstein conceived his theory while in a train based on logic about trains moving near the speed of light and observers. The fact that energy comes in quanta can also be proven by logic alone. Relativity, quantum mechanics, consciousness, all conclusions of logical reasoning alone! And so is Sky's theory, because it is the only way to solve the Einstein-Podolsky-Rosen-paradox and bring the principles of quantum theory and relativity into accordance with each other.", 05.06.2016.

"Darwinism is an incomplete theory, just like quantum mechanics.", 07.06.2016.

"A universe describable in terms of mathematics would make perfectly sense in a materialistic world view. But quantum mechanics leads to idealism. That leads us to what Eugene Wigner called 'the unreasonable effectiveness of mathematics in the natural sciences'. Sky's Platonic interpretation turns the mystery of this into a natural thing and explains why mathematical theories work so well.", 11.06.2016.

"The quantum eraser experiment shows that if one messes with the measurement device so as to not be able to observe the collapse even if it is measured, then the particle acts as a wave, and this brings for me undeniable evidence in favor of the Copenhagen interpretation."*, 16.06.2016.

*Actually the many-minds decoherence theory can also explain this result, because it also is a theory dependent on the conscious observer, just in a more passive way. However, we know that it is wrong for reasons I brought forth in chapter 2.3.

"I also don't think god exists. It would collapse all wave functions.", 16.06.2016.

Doros Kiriakoulis [Greek bachelor-physicist and master-neuroscientist]

"What could be more important than quantum gravity?", 24.05.2017.

"What happened with Sky when he was a kid is, he just asked what is all this, and he just kept asking, and now he got this whole theory on quantum gravity.", 19.06.2017.

Gino Yu [Sino-American consciousness researcher; PhD]

"Connecting non-reductive (machine) consciousness and quantum effects is a golden idea indeed.", 28.06.2017.

"Using quantum effects as a way to test for robot/animal consciousness is a brilliant idea.", 28.07.2017.

"Radin's idea is good but his seems more straightforward and more usable for general philosophical purpose.", 29.07.2017.

Piotr Boltuc [American-Polish professor of philosophy]

"My depth of enthusiasm pushed me to first quick-read the whole book right after I got it."

"The book aroused interest enough and to spare, more than one can possibly imagine."

"Briefly speaking, I can well divide the book into two main parts: the first deals with issues of fundamental theoretical physics; the second concerns itself with how consciousness has its own hinges with that theoretical physics. Now, the admirable first part has come a little bit short of fully exploring into how the basic queries in theoretical physics can be adjusted to determine the indeterminacies present even here and thither. Dr. Darmos talks of ideas of expansion, quantum mechanics (especially its Copenhagen interpretation), and most significantly: suchlike consequences for particle physics. I am of the belief that he has had the genius to well summarise himself, applying his deep understanding of various branches of mathematics* while at the same time remaining not fully capable of reconstructing some of the most important dilemmas hanging around not only complex dimensions but also the Riemanian sphere [and its relationship with the density of space].
Second, coming to consciousness: again, one can fairly easily see the stroke of Genius on the part of Mr. Darmos. However, his dividing of the whole remit of consciousness into feelings and understandings does not become to me personally that much. I simply do not regard the 'intensities' and the 'extensies' within the universal whole structure as separate entities.
I opine that with our version of subjectless phenomenological eminence, the whole field of consciousness – in spite of Mr. Darmos' ideas – can be much better elucidated upon and described if simply considered as a whole enmeshment of animals, plans, but again specifically human beings' consciousness.", 16.05.2017.

*Should be 'Physics'.

"There are so very many different topics that he has chosen to delve into. Nearly all of them absorbing and appealing.", 24.05.2017.

"Just a couple in a hundred million are both so absorbed and so much of genius as he is in the field of modern theoretical physics. There is not and iota of doubt that he shall soon go onto the shortlist for a very important prize, say, Nobel.", 02.07.2017.

"The real placed apportioned to him is no less than a headship somewhere like MIT or Oxford, or rather, a Nobel prize.", 09.07.2017.

"His excellent work makes it possible to design mono-case or limited-n-case experimentations into the existence of logic itself.", 15.07.2017.

"We are able to make tests of this (conscious set theory) on the basis of what he has most elaborately written in chapter 5 of his splendid, superb book." , 15.07.2017.

Dr. Babak Daneshfard; Prof. Mohammad Reza Sanaye; Prof. Dr. Masih Seddigh Rahimabadi [Iranian polymaths]

"Sky is a 'gorilla of science'. He definitely has a great talent in his field.", 27.10.2017.

"Sky Darmos is one of a new generation of researchers thinking about the nature of consciousness. If you are interested in quantum processes and wonder what role consciousness plays this is an interesting study of the issue.", 01.09.2018.

"I applaud his continual refining of his work. I think that is the way things should proceed. I certainly have a much deeper understanding than I did in say the 1980s, and today would say a number of things quite differently.", 08.07.2019.

"He keeps revisiting the things he writes and thinks about and refines and improves them. That is the sign of a true seeking intellect. I find it to be quite rare." , 08.07.2019.

"He says 'quantum mechanics really only works if there is the nonlocal realm.'; I agree with this and, in my opinion, it is the most important thing he has said. Making that case could get him a Nobel.", 09.09.2019.

Stephan A. Schwartz [American parapsychologist]

"We need Sky to build a quantum gravity based supercomputer that can function as the brain of the world.", 15.12.2017.

"I wonder if 'anyone' is real according to Sky's theory."*, 15.12.2017.

*A word play responding to the statement that hypothetic quasiparticles called 'Anyons' aren't real.

"I know Sky moderately well F2F. Sky is a good-hearted and brilliant but eccentric and controversial individual and it's not surprising he has ruffled some feathers.", 18.05.2020.

Ben Goertzel [Mathematician and AI expert]

"Nearly all of the data on micro-PK suggests that it is a goal-directed phenomenon (i.e., the effect is teleological and based on the study goals), and not as an effect that nudges the probabilities of individual bits. Also, it is by no means clear that 40 Hz is the 'frequency of the subconscious'. Some believe that it is just the opposite of that – a neural correlate of conscious awareness.

Still, creating models of micro and macro PK assuming force-like or probabilistic structures is a laudable goal, especially if those models lead to testable predictions.", 02.12.2017.

Dean Radin [American parapsychologist]

"I read some of his ideas in a presentation. I like them, I can see why he shows some interest in my work - there are some similar outcomes although huge differences in the mechanisms that bring about the outcomes. These major differences in our theories are evident with mine introducing the hierarchy of energy and the missing energy of space that is proportional to c^0.*", 15.04.2018.

*He is using an energy hierarchy $E = a_n c^n$, with a representing different sources of energy, such as mass ($n = 2$), frequency ($n = 1$) and space ($n = 0$). He admits that due to the missing conversion factor (c becomes 1), it is not possible to come up with any nonarbitrary value for the energy of space in his model.

"He made an interesting statement, namely that particles add some distance to the region of spacetime they occupy. My theory also has that concept. However the mechanism of how this happens in my theory has to do with the fact that energy creates distance and the total energy in region (the energy of spacetime and particles) determines the distance in that region.*

He should be very proud of his work!", 15.04.2018.

*In fact it is proximity which creates energy, not distance. This is shown to be the case in chapter 4.13, but back when the above comment was made this was nothing more than a short side note speculation at the end of another chapter.

"I admire the hard work he put into this composition and I appriciate the drive he has.", 15.05.2018.

Scott S. Gordon
[American surgeon and hobby physicist]

"Sky seems to have an open and versatile ability to research on consciousness and physics.", 29.04.2018.

Hari Malla [Indian philosopher]

"Very nice paper. My contribution to that problem regarding rest mass would be at pg 4. of 'Expansion Field Theory as a Foundation for a Unified Theory of Fields'.* The quantum gravitational implications are described in 'A new formula for the Bohr radius'.", 15.05.2018.

He believes the constants of nature change over time, and that they can be derived assuming an age of the universe of only 13.525 billion years. An accelerated expansion is also said to derive from this model.

"I don't think the Planck length is the smallest fundamental length. Other people subscribe to that. I'm not even sure that I believe in the Planck units as fundamental to nature. I think they may be a kind of easy trick.
Gravity could be a side effect of charge who knows. The Schwarzschild radius is closely related to Planck units as well as G.
If G is misguided, why is he quoting the Schwartzschild radius? He should try to find a Schwartzchild radius without G."*, 15.11.2018.

That was in fact already done at that time.

"It makes sense to get a Planck length without G if possible.", 06.04.2019.

Jesse Timron Brown [American mathematician]

"Sky's book is an amazing demonstration about his knowledge about many important subjects. I was especially interested in subjects as the Cosmic Microwave Background and the Lorentz transformation Mass-Energy. In my book 'Mind-boggling Gravitation. Applying the Lorentz Transformation of Mass-Energy' I replaced all my words 'consciousness' with 'perception' because 'consciousness' does have many different meanings. However, reading his book, I realize again, how important it is to understand how we perceive the surrounding world.

I wish him a lot of success. His book is great.", 13.09.2018

"It is rare that the same person is interested in both physics and consciousness and his interests are even much broader than that! ... I strongly believe that cooperation between different sciences will lead to more insights and provide us with out of the box thinking.", 24.11.2018

Ruud Loeffen [Dutch scholar]

"It was a great pleasure to attempt understanding his work. I'm still processing and assimilating his model.", 05.03.2019.

"This book is beyond fancinating. Penrose is one of my favorites. Well consciousness, really; but all roads lead to Penrose, on that. I'm excited to see if the mainstream catches up to Sky Darmos. So far I'm midway through his book. I was so ecstatic about reading it, that I got myself a free copy, but when I'm finished; I will purchase it on Amazon, because this is something I want to support. I really want to thank him for his contributions!", 02.03.2019.

"What an excited topic to bond through. That is something I spent a great deal of time explaining to people, about his work. That it is the first time consciousness is included in a theory of everything. It is pioneering and a gigantic leap for a complete understanding. Eventually they began to see the value. I defended SPD over and over with this.", 01.06.2019.

"SPD is a big deal. It introduces consciousness in a theory of everything. That is unprecedented. Everyone should take the time to understand this theory.", 14.06.2019.

Will Shields [Reader from the US]

"Mr. Sky Darmos has presented an orthodox and new insight to see the 'nature of reality', I have been going through this great work of him, and I found it quite fascinating. I hope that mainstream will catch up to him soon. I must say again; a worth reading book, better than the heap of literature that repeats old knowledge, which posses no meaning.", 25.03.2019.

Savyasanchi Ghose [Physics student from India]

"Clearly me and him start from somewhat different positions, although our views overlap in some areas.", 16.04.2019.

Rupert Sheldrake [British biologist and parapsychologist]

"SPD is mathematical proof that the Universe is at least 50 trillion years old – it's about time we grow up and drop Vatican sponsored Big Bang cosmology. In the next 20 years cosmology will completely change and Big Bangers* will look like flat-earthers do now.", 30.04.2019.

*In SPD the Big Bang is replaced by a 'slow drift' (see chapter 3.12).

"SPD is important and in his hands. Climate change is not."

"He is a genius – the world is relying on youngsters with his brainpower.", 22.06.2019.

"SPD is correct.", 29.06.2019.

"I can corroborate 1970 - I was there - and all my astronomical data over the last 50 years supports a much older universe. SPD is correct imho - many cosmologists started to abandon the 13.8 billion yo Big Bang - starting about 20 years ago.", 12.07.2019.

John Walton[†] [American Aerospace and Satellite Ingeneer]

"Similar Worlds View:

I have a many worlds view that is bound by entropy and effects propagate causally. I also hold to the idea that there is a cosmic Bose Einstein condensate dominating space, causing it to appear more smooth than quantum mechanics suggested to us. For me, this is how worlds stay out of each others way. They cross only where the ambiguity if the curvature of a region connects those worlds through the shape of entropies for that region. This basically is like saying that in each region of curvature brings the potential for only certain entropically similar worlds become available. The interaction itself chooses what world becomes observed. I have recently found out that this is an example of similar worlds instead of many worlds.

Note: Similar Worlds is a term coined by Sky Darmos. In dialogue, he explained to me what he meant by similar words as I explained how I interpret many worlds differently. We understood we were talking about a similar idea but the phrase is his.", 07.05.2019.

"If I could address the mass/space relationship by way of this overall resonance then it would be a holographic version of the more classical model he has had so much success with. I really do think we can both progress our independent models better by continuing to understand each others models more clearly.", 08.05.2019.

Thomas Southern [U.S. Navy, Sonar Technician Submarines, Oceanographer]

"It is not too far a stretch to say that his concept of granular dimensionality explains the apparent three dimensions of reality. It is furthermore the first theory to stipulate quanta of space - their size, their dynamics. I think this is excellent. He has cracked it!

We only have a few differences, although we are not describing the same thing exactly. To be clear, the meaning of order demands that everything has its opposite and so continuity must necessarily compliment discrete reality. Classical physics, including relativity which he doesn't like, describe the continuous field, while SPD describes adequately the discrete field. I say adequately for two reasons, first because SPD explains the reason for the three dimensions of reality, and secondly because his theory of the discrete field includes its own discrete space. Additionally, his space quanta do not depend on a constant*, which is something that does not exist in nature. The descriptions of the two fields must be complementary, yet independent. Our differences can be summed up to the fact that the universe is unborn and undying, where its center toggles matter and anti-matter; creating, thus, reality, so as to look as though it is the beginning of time.** Congratulations!!!", 25.05.2019.

*There is a constant to link the size of elementary spaces to energy. What Tripos means here, is that the size of elementary spaces is variable, not constant.

**Tripos believes anti-matter went into a negative space that has an opposite time direction. This is remotely similar to how SPD regards anti-matter to be matter that moves on negative space.

"SPD, or space particle dualism, of Sky Darmos is probably the closest discrete theory to date." , 29.05.2019.

"I have a lot of trust in his theory and his intellectual power.", 29.05.2019.

"I have a lot of faith to Sky Darmo's theory called SPD, as it is the only theory that can be reconciled with the meaning of order. The future will reveal all, some of it anyway.", 14.08.2019.

"Einstein proposed versions of his theory in which time dilation was interpreted as a changing speed of light. Sky Darmos' treatment of time dilation as varying elementary space size and strength of gravity is very similar to this. I however remain skeptical about this meaning that space must be flat even on large scales. Flat space everywhere presents another constancy after all.

I believe in Sky's work, because it proclaims the space particle duality, which is necessary in an eternal universe. But also because of all his other claims which seem impossible from the perspective of standard physics, but which are all standard realities in an eternal universe.", 19.08.2019.

Kitsos Tripos [Physicist, M.S., University of Essex]

"One of the things I have noticed is that as complicated as nature and science may seem to some, there is an underlying simple elegance to everything. The entropic relationship between spatial expansion and blackhole formation will be of this type.", 05.06.2019.

Stuart Sidgwick [Canadian biochemist & physicist, University of Toronto]

"I am of the conviction that Sky Darmos is one of the rising stars in the world of cosmological, theoretical and mathematical physics.

Sky Darmos (of a German-Greek descent) is a Hong Kong based researcher who works for the entity Quantum Gravity Research which I believe he and Klee Irwin are collaborators or co-researchers in the field of quantum gravity (a prospectus for grand unification of general relativity and quantum mechanics which you'd hear physicists speak a lot about) and related subfields of general physics.*

To his signature, Sky has independently authored a physical and cosmological theory he calls SPD (Space-Particle Dualism) theory which encompasses many theoretical aspects of physics: ranging from Big Bang, cosmology, particle physics, quantum gravity, quantum mechanics to general relativity, etc.

To this end, Sky is the only person to provide more information about his educational and scientific background for the entire world to know of, and judge perfectly about, his credentials i.r.o his scientific research outputs.", 08.10.2019.

*Not true. I have talked to Klee Ervin, but I am not collaborating with him.

Silas Mohale
[South African Physicist]

"This book used to inspire me a lot and I'm looking forward to more breakthroughs on his road towards 'the ultimate/unified theory'.", 15.12.2019.

Yike Ni
[Chinese enthusiast of enlightenment studies]

"He really knows his stuff. A lot of it is spot on."

"Side note: I just read about entropic expansion. He is a genius. I feel that theory is profound and can explain the cosmic horizon."

"I wouldn't in a million years have thought about that (the concept of entropic expansion).", 17.12.2019.

Dante Alughieri
[American reader]

"His research has many applications. I think his work can be something really BIG.", 18.12.2019.

David Chen
[American Investor]

"In this supernova data he found a nice match with his theory, and then it was way off for general relativity.
I wonder if astrophysicists recognize this as an issue for general relativity. I suppose there's just a lot of data making it difficult to defend the most vanilla version of general relativity with only curvature and Einstein-Hilbert action.", 25.12.2019.

"I'm trying to put some of his concepts into the mainstream language to make it impossible for string theorists to ignore. I am also trying to get him to look at other people's research and not only his own, so that it doesn't look like he is disagreeing with everyone."

"I'm trying to secretly do what he is doing but rewrite it as an idea I don't believe in simply to motivate society to go to a path closer to what I think is actually true."

"I think this capacitance idea* is compelling. It's like he is thinking of the quantum vacuum as a dielectric aether; which is what it is."

*The mass from charge idea in chapter 3.11.

"He should publish a shorter paper that is focused on this 'fractional dimension' idea*; leaving out the standard model; just the gravity, giving people something small to digest."

*That was already done in a 2019-paper that corresponds to chapter 3.15.

"I see an infinite number of theories, so it's like which do I pick. The one I like? Or the one everyone else does? I just don't care about it for myself, it's not exactly why I'm doing it."

"String theory could make predictions if they just picked something and stuck to it. But there's no quantum gravity experiments. They are lost in a mathematical landscape. With Sky and his predictions and subsequent gravity experiments that changes entirely."

"Once I figure out my own theoretical framework, it will be easier for me to figure out his as well. Hopefully they connect in some way.", 27.03.2020.

"I think he is missing high energy theory a bit, but his proposal for quantum gravity is at least a solid candidate, I also think there's many solid candidates that could all pan out.", 15.12.2020.

David Chester
[American particle physicst]

"I teach a course in modern mysticism and have been waiting (and hoping and also squirming with impatience) for someone, as himself, to fill these gaps for me in such a clear and important way! I am really thankful for his work!", 27.12.2019.

Amelia Vogler
[American Energy Medicine Professional]

"GR is emergent. SPD is about what GR emerges from. This doesn't make GR wrong, much like GR doesn't make Newton wrong.", 08.01.2020.

George Soli
[American Retired Space Physicist; NASA Jet Propulsion Laboratory]

"I think his ideas about complex mass are very interesting and reasonable.", 09.01.2020.

"I think tachyons and tardyons always have complex masses. Mass has always real and imaginary parts.", 11.01.2020.

"I think he is a very smart person, because his thoughts and ideas are very intriguing and genuine. He is also very hardworking and has a curious and inquiring mind. I wish him good luck with everything.", 12.01.2020.

"There is a imaginary mass for any particle and also a real mass. They are all of structure and coexist.", 14.01.2020.

"I think his thoughts and ideas are both scientifically sound and very interesting for readers.", 14.01.2020.

"I think he is very distinguished and intelligent. His book is very compelling and useful, and I am sure it will be internationally recognized in the very near future.", 09.05.2020.

Necati Demiroğlu
[Turkish physicist and electrical engineer]

"I found his book to be very impressive and thoughtful; and his background is very interesting.", 10.01.2020.

Richard Gauthier
[American Chemist and Physicist; Santa Rosa Junior College]

"I appreciate him being among perhaps the 5 people in the world actually trying to make sense of things.", 22.03.2020.

Thad Roberts
[American Physicist]

"Not only Thad, but Sky is probably on the right track too.", 08.04.2020.

David Heggli
[German Physicist]

"I admire his work.", 30.05.2020.

Dirk K F Meijer
[Dutch Toxicologist and Physicist]

"This is the TOE (theory of everything).", 18.08.2020.

"The Sky is his limit.", 19.08.2020.

Massimo Melli
[Italian Physicist]

"Apparently I have to study all his papers.", 03.12.2020.

Erik Lindenberg
[Emeritus physicist from DTU, Denmark]

"Very interesting theory. It reminds me of Dr. Hal Puthoff's work on the relationship of gravity to vacuum fluctuations, that appeared in Physical Review Letters.", 21.12.2020.

John Baxter
[US Physicist; ITT, AIAA, NSS]

"The idea that gravity is related to the atomic nucleus is very obvious. Unfortunately GR has emerged before we could understand the structure of the atom. If it had not, then science would have focused on the nucleus to understand the underlying mechanism for gravity rather then focus on the predictions of GR.
I myself came up with an idea for such a mechanism, a cousin to Sky Darmos' idea and even simpler.*", 07.12.2020.

His idea is similar in the regard that it centers around the strong force. He describes gravity as some left over residual strong force that comes from proton-neutron particle type fluctuations.

"One common factor we have is that the atomic number is relevant. One which is-not is that in my idea the element Hydrogen does not pull. Implications are obviously shocking since Stars are made up mainly of Hydrogen.", 07.12.2020.

"I like his experiments for big-G. Truly remarkable. Big-G has been nailed, if you ask me!! Forever!", 07.12.2020.

"I read it [a teaser to the book] twice. Not that I agree with all but now for sure I must read his book. I read all new ideas. I like people who think outside the box and write in clear language. I wish him good luck." , 07.12.2020.

Stephen Farrugia
[Maltese physicist]

"SPD is actually quite genius. Like most discoveries it is only obvious afterwards.", 20.02.2021.*

Response to: "Dark matter doesn't exist. It is a phantasy caused by people's superstitious believe in a young universe." – Sky Darmos.

"These fantastical 'theories' (not even good assumptions sometimes) are earcandy for the masses obscuring solid science like Sky Darmos' SPD.", 20.02.2021.*

Response to an article with the title "Scientists Have Proposed a New Particle That Is a Portal to a 5th Dimension".

Atticus Myser
[American Reader]

"I think his book is a valuable work that reminds us that the Quantum Gravity Problem has much deeper and different properties.

Also, SPD is an inspiring theory for me. Because it shows that Quantum Gravity research can be done in ways that do not require high budgets.", 26.02.2021.

"I think quantum gravity requires a bold rejection of what we have hitherto considered to be true. Therefore, 'consciousness' can be regarded as not a crazy beginning, but rather a beginning in the right direction. Because this unification problem may necessitate a new definition of science. Theoretical Physics has to be brave and creative in this regard. SPD is a remarkable theory in this regard.
What we have learned from the history of science is that progress is often devastating.

Theories, on the other hand, develop not by praising them, but by criticizing and studying them.", 26.02.2021.

Bahadır Arıtay
[Turkish Reader]

"I do find his list of topics and background to be respectable and well planned. That is not the usuall case for people I review and respond to.", 24.02.2020.

"He is close on both topics (gravity and spinors), but precise is the nature of Academia. Most people are not even close.", 24.02.2020.

"He is right that gravitons don't exist; tachyons neither.", 24.02.2020.

"Competitive models are quite valuable. We cannot ever create just one model for every purpose.", 24.02.2020.

"I cannot say specifically which other models are well founded, but it seems to me he is very capable.", 24.02.2020.

"I only discuss what is in evidence from models. It seems I might agree with him more than he expects (from a string theorist).", 24.02.2020.

"I can say that he is on to something with respect to dark matter and granularity ... I just had not extended myself into that topic further than dark matter as a process.", 26.02.2021.*

He is assuming that I use granularity to arrive at some kind of modified Newtonian dynamics theory, short 'MOND'. That is not the case.

"I would say he is doing very well on this path, mostly on his own volition and diligence.", 02.03.2021.

"Both his skills and his moral center are respected within my group.", 02.03.2021.

Mark Aaron Simpson
[American professor for theoretical physics]

"His book is very important.", 05.03.2021.

" With his skills it ought to be possible to work at an institute for theoretical physics.", 08.03.2021.

Konrad Kleinknecht
[German physics professor and personal friend of Heisenberg and his son]

Bibliography

By Roger Penrose:

[1] The Emperor's New Mind: Concerning Computers, Minds, and the Laws of Physics (1989); Roger Penrose

[2] Shadows of the Mind: A Search for the Missing Science of Consciousness (1994); Roger Penrose

[3] The Large, the Small and the Human Mind (2000); Roger Penrose

[4] The Road to Reality - A Complete Guide to the Laws of the Universe (2004); Roger Penrose

[5] Cycles of Time: An Extraordinary New View of the Universe (2010); Roger Penrose

See also [44]

Formal logic:

[6] Gödel's Proof (1958), Ernest Nagel & Ernest Nagel & James R. Newman

Cosmology:

[7] The First Three Minutes: A Modern View Of The Origin Of The Universe (1977); Steven Weinberg

[8] The Life of the Cosmos (1999); Lee Smolin [*Smolin's cosmological evolution theory fascinated me a lot before 2005, but I finally refused it because it is in conflict with the membrane paradigm. Also we would have to wonder how the hyperspace enclosed by a closed universe would not be filled up with baby universes.*]

[9] Die Entdeckung des Nichts - Leere und Fülle im Universum (1994); Hening Genz [*English: Nothingness: The Science Of Empty Space (2001)*]

[10] Junior Wissen – Universum [*The book which awoke my interest in astronomy and cosmology when I was very young.*]

[11] Der neue Kosmos, Unsöld

[12] Das Universum - Aufbau, Entdeckungen, Theorien, David Layzer

[13] Bildatlas des Weltraums (1995); A. Rükl

[14] The Physics of Immortality: Modern Cosmology, God and the Resurrection of the Dead (1997); Frank J. Tipler [*Note: This book is based on two major wrong assumptions: one being that we live in a closed universe that will collapse and the other, that a world including conscious beings can be simulated on a computer.*]

On neuroscience and consciousness:

[15] Bewusstsein, Philosophische Beiträge (1996), Sybille Krämer

[16] The Emerging Mind: The BBC Reith Lectures (2003); Vilayanur Ramachandran

[17] Evolution of the Brain: Creation of the Self (1989); John C. Eccles and Karl Popper

[18] Mind Time - The Temporal Factor in Consciousness (2004); Benjamin Libet

[19] Kleine Geschichte der Hirnforschung - Von Descartes bis Eccles (2001); Peter Düweke

[20] The Self and Its Brain: An Argument for Interactionism (1977); Karl R. Popper and John C. Eccles

On the interpretation of quantum mechanics:

[21] The Matter Myth. Dramatic Discoveries that Challenge Our Understanding of Physical Reality (1992); Paul Davies & John Gribbin

[22] Gibt die Physik Wissen über die Natur? - Das Realismusproblem in der Quantenmechanik (1992), Lothar Arendes

[23] Quantum physics: illusion or reality? (1986), Alastair Rae

[24] In Search of Schrödinger's Cat: Quantum Physics and Reality, John Gribbin

On particle physics:

[25] Quarks : The Stuff of Matter (1983); Harald Fritzsch

[26] Antimaterie - Auf der Suche nach der Gegenwelt (1999), Dieter B. Herrmann

[27] Elementarteilchen (2003); Henning Genz

[28] QED: The Strange Theory of Light and Matter (1985), Richard P. Feynman

[29] The Perfect Universe (2020), Thad Roberts

On relativity:

[30] In Search of the Edge of Time (1992); John Gribbin

[31] Time Travel in Einstein's Universe: The Physical Possibilities of Travel Through Time (2002); J. Richard Gott

[32] Black Holes and Time Warps: Einstein's Outrageous Legacy (1993); Kip Thorne

[33] Über die spezielle und die allgemeine Relativitätstheorie (1916), Albert Einstein

[34] Hyperspace: A Scientific Odyssey Through Parallel Universes, Time Warps, and the 10th Dimension (1995), Michio Kaku

[35] Einstein's Legacy: The Unity of Space and Time (1986); Julian Schwinger

[36] Space-time Physics (1992), J.A. Wheeler & E.F. Taylor

[37] Gravity's fatal attraction (1996), Mitchel Begelman & Martin Rees

Quantum physics:

[38] Entangled Systems: New Directions in Quantum Physics (2007), Jürgen Audretsch

[39] Quantentheorie (2002), Claus Kiefer

[40] Quanteninformation (2003), Dagmar Bruß

[41] Quantum Mechanics - volume 1 (1964), Albert Messiah

String theory:

[42] The Universe in a Nutshell (2001); Stephen Hawking

[43] Superstrings: A Theory of Everything? (1992), Paul Davies & Julian R. Brown

[44] Superstrings and the Search for the Theory of Everything (1988); F. David Peat (Contains three chapters about Penrose's twistor theory)

[45] The Fabric of the Cosmos: Space, Time, and the Texture of Reality (2005); Brian Greene

[46] The Elegant Universe: Superstrings, Hidden Dimensions, and the Quest for the Ultimate Theory (2010); Brian Greene

Parapsychology:

[47] The Conscious Universe (1997); Dean Radin

[48] The Alexandria Project (1983); Stephan A. Schwartz

[49] Opening to the Infinite: The Art and Science of Consciousness (2007); Stephan A. Schwartz

[50] Opening to the Infinite: The Definitive Guide to Remote Viewing and the Secret Powers of the Human Mind (2016); Stephan A. Schwartz

Others:

[51] Geschichte des Gens (2002), Ernst Peter Fischer

[52] Klassische Mechanik (2002), Frank Linhard

[53] Grundkurs Theoretische Physik 1 (2012); Wolfgang Nolting

[54] Komplexe Systeme (2002), Klaus Richter, Jan-Michael Rost

[55] Practical Ethics (1979), Peter Singer

[56] Fermat's Enigma: The Epic Quest To Solve The World's Greatest Mathematical Problem (1997); Simon Singh